T0229658

A Revised Handbook
to the
FLORA OF CEYLON

VOLUME VII

BASELLACEAE

BORAGINACEAE

FABACEAE

FLAGELLARIACEAE

HYDROPHYLLACEAE

JUNCACEAE

LEEACEAE

MALPIGHIACEAE

SALVADORACEAE

TAMARICACEAE

TILIACEAE

VAHLIACEAE

A Revised Handbook
to the
FLORA OF CEYLON

VOLUME VII

Sponsored jointly by the
University of Peradeniya,
Department of Agriculture, Peradeniya, Sri Lanka,
and the Smithsonian Institution,
Washington, D.C., U.S.A.

General Editor
M.D. DASSANAYAKE

Editorial Board
M.D. DASSANAYAKE and F.R. FOSBERG

1991
A.A. BALKEMA/ROTTERDAM

For the complete set of eight volumes, ISBN 90 6191 063 3
For volume I, ISBN 90 6191 064 1
For volume II, ISBN 90 6191 065 X
For volume III, ISBN 90 6191 066 8
For volume IV, ISBN 90 6191 067 6
For volume V, ISBN 90 6191 068 4
For volume VI, ISBN 90 6191 069 2
For volume VII, ISBN 90 6191 551 1
For volume VIII, ISBN 90 6191 552 X

Printed in India at Model Press Pvt. Ltd., New Delhi

FOREWORD

The Handbook to the Flora of Ceylon, by Henry Trimen, published in 1893–1900, was in its time, one of the most comprehensive and outstanding floras available for any comparable tropical area. In 1931 A.H.G. Alston added a volume of additions, updating, and corrections to the original five volumes. These six volumes for many years served their purpose very well.

However, the original Handbook was published in a very small edition, and the paper on which it was printed, as was usually the case at the time, was very poor and has deteriorated very badly. Hence the Handbook has for years been absolutely unobtainable, and there are very few copies available even in libraries in Ceylon. Furthermore, botanical science has made substantial progress since the Handbook appeared, and many of Trimen's taxonomic and nomenclatural conclusions are now outdated. Also, with more thorough botanical exploration, new plants have been found to be members of the Ceylon flora. Hence, a new edition of this magnificent work was long overdue.

For quite a number of years Professor B.A. Abeywickrama had in mind a revision of Trimen's Handbook. But heavier and heavier administrative duties consumed his time and there was little opportunity for work on the Ceylon Flora, though he did produce, in 1959, an updated Checklist of the Ceylon Flora, which has been most useful to botanists.

Fortuitously, in 1967, the Smithsonian Institution initiated a number of research projects in Ceylon, in cooperation with Ceylonese institutions and scientists. These included an investigation of several problems in plant ecology, with which Prof. Abeywickrama was associated as Co-Principal Investigator. These projects were financed by the Smithsonian using U.S. excess foreign currency under the provisions of Public Law 480.

While we were discussing the ecological investigations, Prof. Abeywickrama wondered if it might not be possible to initiate, using PL-480 support, a project for the revision of Trimen's Handbook to the Flora of Ceylon. I offered to work up a cooperative proposal and submit it to the Smithsonian Special Foreign Currency Program.

This was duly done, approved, and a year's tentative budget authorized. The project was started under the joint auspices of the Smithsonian Institution, the Ceylon Department of Agriculture and the University of Ceylon. I was appointed Principal Investigator. Co-Principal Investigators are Prof. B.A. Abeywickrama, Dr. J.W.L. Peiris, Mr. D.M.A. Jayaweera, Prof. M.D. Dassanayake, and Mr. K.L.D. Amaratunga. The plan was to enlist the cooperation of botanists, from wherever available, who were preferably experts in particular families represented in the Ceylon flora. These monographers would be given a period of field work in Ceylon, an opportunity to study the specimens in the Ceylon National Herbarium, in the Royal Botanic Gardens, Peradeniya, with expenses met by the Smithsonian Institution. In return, they would provide updated manuscripts of their families for the revised Handbook, which would then be published by the University of Ceylon, with Smithsonian financing.

This enterprise was initiated in February, 1968, and has been continued without interruption since that date. Quarters for the work, herbarium and library, and other facilities have been furnished by the Division of Systematic Botany of the Department of Agriculture, Peradeniya and by the Botany Department, Faculty of Science, University of Ceylon, Peradeniya. We have enjoyed the cooperation of the U.S. Embassy, Colombo, various Ceylon government departments and agencies, especially the Wildlife Department and the Forest Department, and of many plantations and individuals in all parts of Ceylon, too numerous to enumerate.

Special thanks must be offered to Professor Dieter Mueller-Dombois, of the Botany Department, University of Hawaii, who was for two years Principal Field Investigator for the plant ecology project, and who, on top of his duties in that capacity, supervised the activities of the flora project staff, facilitated the work of the visiting botanists, and acted as finance officer of the project. Without his help, the flora project could not have got started.

Special thanks are also offered to Dr. Marie-Hélène Sachet, Research Botanist, Smithsonian Institution, who, though in no official capacity in the Project, has carried much of the administrative burden, at a sacrifice of her own work. The members of the Flora Project staff at Peradeniya directed by Mr. F.H. Popham, Smithsonian Representative in Ceylon, also deserve great credit for their willing and enthusiastic assistance to the visitors, handling and processing of specimens, typing of labels and manuscripts, and keeping the Project's work going.

The materials on which the flora revisions are based are the visiting botanists' own collections, the herbarium at Peradeniya, personal collections of Mr. K.L.D. Amaratunga and the late Mr. Thomas B. Worthington of Kandy, and materials housed in various foreign herbaria, especially those of Kew, British Museum, and the Indian National Herbarium, Calcutta for the use of which we are grateful to those in charge. A large amount of valuable material was amassed as vouchers for the ecological observations mentioned above and has been utilised by the flora project botanists. Sets of the specimens collected under the auspices of the Smithsonian projects are deposited and permanently available in the U.S. National Herbarium and the Ceylon National Herbarium, Peradeniya. Partial sets are also being deposited in several other Ceylon institutions and in a number of herbaria with tropical interests in other parts of the world.

The resulting revised treatments of the families are to be published, as material accumulates, in volumes of convenient size, without regard to the order of families.

Those families, previously published in fascicles 1 and 2, are to be republished, in revised form, in the new format, as manuscripts are received from the authors. An editorial format has been suggested by us, but the content of each revision and the taxonomic conclusions are those of the various authors. An attempt has been made to have the nomenclature in accord with International Code of Botanical Nomenclature, but the application of the Code is, again, the final responsibility of the authors.

A comprehensive index to these volumes will be prepared and published as a separate volume.

It is hoped that, after the Handbook treatments are published, simplified versions can be prepared, suitable for lower school use and for the use of non-botanists.

We also have the hope that the new Handbook will stimulate active interest in the plant resources of Ceylon. Above all, it is hoped that this interest will bring about the establishment of more national parks and nature reserves in all parts of Ceylon, in order that the remarkable Ceylon flora may still have a suitable range of habitats in which to live. By this means, only, the species will be able to survive for the use and pleasure of many future generations of Ceylonese and of visitors from other parts of the world. Without a great increase in such reserves, at the present rate of deforestation and bringing land under agriculture, very many species will surely become extinct in the near future, as some probably already have.

F.R. FOSBERG,
Botanist, Emeritus,
Smithsonian Institution,
Washington, D.C., U.S.A.

CONTENTS

A Revised Handbook
to the
FLORA OF CEYLON

VOLUME VII

BASELLACEAE

(by Neil A. Harriman*)

Moq., Chenopod. Mon. Enum. x. 1840. Type genus: *Basella* L.

Perennial glabrous twining herbs, succulent, from fleshy underground parts (or perhaps sometimes behaving as annuals). Leaves alternate, simple, estipulate. Flowers sessile in axillary peduncled spikes, actinomorphic, perfect, incomplete, hypogynous, proximally with three tiny basal bracts, distally with two large bracteoles somewhat exceeding the 5-lobed perianth (= calyx), the perianth lobes imbricate, not fully opening at anthesis. Stamens 5, episepalous, opposite the calyx lobes. Ovary superior, styles 3, separate to base (not fused, as illustrated in Verdcourt, op. cit. infra), each terminated by a clavate stigma. Fruit dry, one-seeded, the bracteoles and perianth becoming fleshy, accrescent, the whole falling as a pseudo-berry.

The family comprises four genera and 25 species in the Paleo- and Neotropics; it is represented in Ceylon only by one species of *Basella*.

BASELLA

L., Sp. Pl. ed. 1, 272. 1753. Lectotype species: *Basella rubra* L.

Characters of the family. One species in Ceylon.

Basella alba L., Sp. Pl. ed. 1, 272. 1753.

Basella rubra L., Sp. Pl. ed. 1, 272. 1753; Trimen, Handb. Fl. Ceylon 3: 410. 1895.

Fleshy, glabrous, twining perennial herb. Leaves alternate, rhombic-ovoid, 2–5 cm long, 0.8–2.5 cm wide, rounded below and cuneate onto petioles 0.3–1 cm long. Flowers greenish-white, c. 3 mm long, in unbranched axillary peduncled spikes 5–8 cm long. Fruits small, one-seeded, enclosed in the enlarged, fleshy bracteoles and calyx, c. 0.5 cm in diameter.

Distr. Throughout the tropics of the Old and New World.
Ecol. Forests and shady places in the dry region, rather rare (Trimen).

*Biology Department, University of Wisconsin-Oshkosh, Oshkosh, Wisconsin 54901, USA.

Vern. Niviti, Gas-niviti, Pasalai.

Specimens Examined. ANURADHAPURA DISTRICT: Kotagala, *Jayasuriya et al. 626* (US). As a cultivated plant, and as an escape, it is doubtless much more widespread in Ceylon.

Taxonomic Note. The view is widely held that *B. alba* and *B. rubra* are conspecific; see, e.g.: Trimen, H. 1895. A Handbook to the Flora of Ceylon. Part 3: 409–410; van Steenis, C.G.G.J. 1957. Basellaceae in Flora Malesiana, series 1. 5(3): 300–304, which includes a very complete synonymy; Verdcourt, B. 1968. Basellaceae in Flora of Tropical East Africa. 4 pages; Walker, E.H. 1976. Flora of Okinawa and the southern Ryukyu Islands: Basellaceae, pp. 450–451; Burkill, L.H. 1935. A dictionary of the economic products of the Malay Peninsula, pp. 397–398. The contrary view is held by Bailey, L.H. 1949. Manual of cultivated plants: Basellaceae, pp. 367–368, in which the Ceylon plant will key to *B. alba*.

Nomenclatural Note. If it be agreed that there is but one Linnaean species, then Article 57.2 of the Code is very explicit on how the matter is to be handled: "The author who first unites taxa bearing names of equal priority must choose one of them. . . . The name he chooses is then treated as having priority." It appears that the union of the two names into one was first done by Graham, J. 1839. Cat. Pl. Bombay. 170, and he reduced *B. rubra* to the synonymy of *B. alba* (teste van Steenis, op. cit.); hence, the name adopted here. Further: lectotypifying a genus with a name that falls into synonymy is of course not forbidden by the Code, though it may seem contradictory at first glance.

BORAGINACEAE

(by Joan W. Nowicke* and James S. Miller**)

Type genus: *Borago* L., Sp. Pl. 137. 1753.

Herbs, shrubs, or trees, rarely lianas; frequently scabrous or strigose, the hairs simple, uniseriate, or rarely stellate, often barbed, sometimes with multicellular cystolith-like bases, or rarely glabrous. Leaves alternate, subopposite, fasciculate, more rarely opposite, simple, generally petiolate in woody taxa, obscurely petiolate or sessile in herbaceous taxa, the blade pinnately-nerved, entire or rarely serrate. Inflorescences cymes, generally helicoid or scorpioid, sometimes paniculate and open, raceme-like, spike-like, or glomerate, or rarely the flowers solitary; the bracts mostly absent. Flowers perfect or functionally unisexual, actinomorphic or rarely zygomorphic; bracteoles generally absent; calyx of five sepals, rarely four, free or connate basally, sometimes irregular; corolla of five petals, rarely four (-18), united, salverform, funnelform, or campanulate, the lobes distinct or obscure, the tubes sometimes with folds or faucal appendages in the throat; stamens five, rarely four (-7 or more), functional or not, epipetalous, alternate with the corolla lobes, sometimes sessile, the filaments simple or with dorsal appendages, glabrous or pubescent/fimbriolate basally, the anthers linear or deltoid, functional or not, ovary superior, syncarpous, 2-carpellate; bilocular and becoming falsely 4-locular, placentation axile, ovules 4, or fewer by abortion, anatropous; style one, gynobasic or terminal, simple, cleft or twice cleft; stigmas 1, 2, or 4. Fruit 4 nutlets, a 1–4 seeded nut, or a drupe; seeds without endosperm (except in Ehretioideae). $x = 4, 8, 10, 13$.

A widely distributed family with as many as 100 genera and at least 2,000 species. Nine genera are found in Ceylon but two are monotypic and two more are represented by only one species.

REFERENCES

Alston, A.H.G. 1931. Supplement to Handbook to the Flora of Ceylon by Trimen, H. 198–201. Dulau, London.

*Botany Department, Smithsonian Institution, Washington, D.C., U.S.A. (*Carmona, Coldenia, Cynoglossum, Heliotropium, Rotula, Tournefortia, Trichodesma*).

**Missouri Botanical Garden, St. Louis, MO., U.S.A. (*Cordia, Ehretia*).

Clarke, C.B. 1883. Boraginaceae. In: Hooker, J.D., Flora of British India 4: 134–179. Reeve & Co., London.

Hutchinson, J. 1918. *Cordia myxa* and allied species. Kew Bull. 1918: 217–222.

Johnston, I.M. 1951. Studies in the Boraginaceae XX. Representatives of three subfamilies in Eastern Asia. J. Arnold Arbor. 32: 1–26 and 32: 99–122.

Kazmi, S.M.A. 1970; 1971. A revision of the Boraginaceae of West Pakistan and Kashmir. J. Arnold Arbor. 51: 133–184; 51: 367–402; 52: 110–136; 52: 334–363; 52: 486–522; 52: 614–690.

Matthew, K.M. and Rani, N. 1983. Cordiaceae in: The Flora of the Tamilnadu Carnatic. The Rapinat Herbarium, St. Joseph's College, Tiruchirapalli 620002, India. Vol. 3, pt. 1, 996–1003.

Miller, J.S. 1988. A revised treatment of Boraginaceae for Panama. Ann. Missouri Bot. Gard. 75: 456–521.

Nowicke, J.W. 1969. Boraginaceae in: Flora of Panama. Ann. Missouri Bot. Gard. 56: 33–69.

Nowicke, J.W. and Ridgway, J.E. 1973. Pollen studies in the genus *Cordia* (Boraginaceae). Amer. J. Bot. 60(6): 584–591.

Nowicke, J.W. and Skvarla, J.J. 1974. A palynological investigation of the genus *Tournefortia* (Boraginaceae). Amer. J. Bot. 61(9): 1021–1036.

Roxburg, W. 1824. Boraginaceae. In: Fl. Indica 2: 1–14; 330–345. Mission Press, Serampore, India.

Saldanha, C. and Nicolson, D. 1976. The Flora of Hassan District, Karnataka, India. Boraginaceae 477–484 by: T.P. Ramamoorthy. Amerind Publ. Co., New Delhi, India.

Trimen, H. 1895. Handbook to the Flora of Ceylon. Boraginaceae 3: 192–203. Dulau, London.

KEY TO THE SUBFAMILIES REPRESENTED IN CEYLON

1 Styles four, stigmas four; cotyledons plicate; endosperm absent . . . .**Cordioideae** (*Cordia*)
1 Styles one or two, stigmas one or two; cotyledons not plicate
  2 Style arising from fruit, seated apically in pericarp, falling away with it; endosperm usually present
    3 Stigmas usually two, small, undifferentiated . . . .
    . . . .**Ehretioideae** (*Carmona, Coldenia, Ehretia, Rotula*)
    3 Stigmas one, differentiated into sterile cone and receptive, sharply delimited band at its base
    . . . . **Heliotropioideae** (*Heliotropium, Tournefortia*)
  2 Style gynobasic, arising independently from the floral receptacle or from a projection of the receptacle, and between the four nutlets . . . .**Boraginoideae** (*Cynoglossum, Trichodesma*)

KEY TO THE GENERA

1 Shrubs or trees (*Carmona, Cordia, Ehretia, Rotula, Tournefortia*)
  2 Leaves small, mostly < 4 cm long, attached in fascicles or crowded on short branches
    3 Style cleft below the middle; at least some leaves with retuse or 3-lobed apices . . . .
    . . . .**1. Carmona**

3 Style not cleft below middle; leaves with acute apices . . . . **2. Rotula**
2 Leaves larger, at least some > 5 cm long, attached alternately or sub-oppositely, but not crowded or in fascicles
  4 Style twice cleft with 4 stigmatic branches; fruit a drupe, unlobed . . . . **3. Cordia**
  4 Style once cleft with 2 stigmatic branches, or not cleft and with single conical stigma; fruit 2–4 lobed
    5 Style not cleft; plants with tubular corolla or with leaves densely silk-pubescent . . . .
    . . . . **4. Tournefortia**
    5 Style once cleft with two stigmatic branches; corolla campanulate; leaves more or less glabrous
    . . . . **5. Ehretia**
1 Herbs, sometimes wiry or slightly suffrutescent, but never trees or shrubs (*Coldenia, Cynoglossum, Heliotropium, Trichodesma*)
  6 Leaves markedly asymmetric, the margins coarsely dentate, the veins ending in the sinuses
  . . . . **6. Coldenia**
  6 Leaves + or − symmetrical, the margins entire, or if weakly dentate, then plants erect, veins not ending in sinuses
    7 Erect herbs; inflorescences raceme-like, open; at least some pedicels and mature sepals > 1.4 cm long . . . .**7. Trichodesma**
    7 Erect or prostrate herbs; inflorescences spike-like or raceme-like; pedicels and sepals < 1.0 cm long
      8 Erect herbs; inflorescence an elongate raceme, rarely a spike; flowers blue, remote; fruit of 4 nutlets with glochidiate spines . . . . **8. Cynoglossum**
      8 Erect or prostrate herbs; inflorescence a scorpioid spike; flowers blue or white, crowded; fruit of 2–4 nutlets without glochidiate spines . . . . **9. Heliotropium**

1. CARMONA

Cav., Icones 5: 22, t. 438. 1799. Type species: *Carmona heterophylla* Cav.

Shrubs or small trees. Leaves mostly fasciculate on short branches, the apices 2–7-toothed. Inflorescences mostly solitary and axillary. Flowers perfect, 5-merous; stamens with elongate filaments, the anthers exserted from the throat; ovary globose, style cleft below the middle, stigmas two. Fruit drupaceous, globose, 4-seeded, not separating at maturity.

  Carmona consists of a single species originally described as a species of *Cordia*. It has also been included in *Ehretia* but differs in habit, leaves, pollen and fruit structure (see Johnston, 1951).

Carmona retusa (Vahl) Masamune, Trans. Nat. Hist. Soc. Formosa 30: 61. 1940.

Cordia retusa Vahl, Symb. 2: 42. 1791.
Ehretia microphylla Lam., Tab. Enc. 1: 425. 1792; Alston in Trimen, Handb. Fl. Ceylon 6: 200. 1931.
Carmona microphylla (Lam.) G. Don, Gen. Hist. 4: 391. 1837.

  Shrubs or small trees, to 2 m, the twigs glabrous to sparsely strigillose. Leaves mostly fasciculate on short, spur-like branches, occasionally alternate, small, spathulate, to 4.0 cm long and 1.5 cm wide, retuse and/or with 2–7 teeth at apex, the bases attenuate, somewhat coriaceous, uniformly scabrate above, the trichomes with cystolith-like bases; subsessile. Inflorescences mostly solitary and axillary. Flowers small, subsessile to pedicels c. 2 mm long; calyx of 5 sepals, slightly connate basally, spatulate, 3–3.5 mm long, to 4 mm in fruit; corolla salverform, white, tube shallow, 5-lobed, the lobes c. 3 mm long; stamens exserted, the anthers c. 1.2–1.6 mm long, the filaments 1.5 mm long; ovary globose, the style slender and elongate, cleft below the middle (1.5 mm unbranched and 2.5 mm for branches), stigmas scarcely differentiated. Fruit drupaceous, globose, 3.5–4 mm in diam, 4-seeded, not separating at maturity.

Distr. All of southern Asia from India to Taiwan and the Philippine Islands, and eastwards to New Guinea and the Solomon Islands.

Ecol. Sandy soils and scrub forests.

Specimens Examined. VAVUNIYA DISTRICT: Mullaitivu-Puliyankulam Road, *D.B. Sumithraarachchi & D. Sumithraarachchi DBS 814* (US). ANURADHAPURA DISTRICT: Ritigala Strict Natural Reserve, ascent to Wannatikanda along eastern slope, alt. 1800 ft, *Jayasuriya 1059* (US); Ritigala Strict Natural Reserve, alt. 1300 ft, *Jayasuriya 1265* (US); Maradanmaduwa, Wilpattu National Park, alt. 0.5 m, *Ripley 37* (US); Ritigala (Una Kanda Hill), *Sohmer & Sumithraarachchi 10761* (MO, US); Wilpattu National Park, Timbiriwila, *Sohmer & Sumithraarachchi 10694* (MO, US); Ritigala, *Waas 317* (US); W. shore of Nuwara Wewa, behind the Nuwara Wewa Rest House, *Sohmer 8102* (US); Mihintale Sanctuary, 55.5 mile marker, *Sohmer 8115* (MO); Mihintale Sanctuary, 55 mile marker, *Sohmer 8137* (MO). POLONNARUWA DISTRICT: NE of Habarana, along road to Trincomalee, hills opposite mile marker 112, *Davidse 7509* (MO); Polonnaruwa Sacred Area, alt. 61 m, *Ripley 188* (US); Polonnaruwa Sacred Area behind archaeological labour gardens, alt. 61 m, *Ripley 204* (US); Polonnaruwa Sacred Area in roadside jungle east and north of Rankot, *Ripley 379* (US); Polonnaruwa Sacred Area, alt. 61 m, *Ripley 158* (US); Polonnaruwa Sacred Area, *Fosberg 51872* (US). MATALE DISTRICT: Area E. of Wewala Tank, 9.5 miles directly ENE of Dambulla, alt. 270 m, *Davidse & Sumithraarachchi 8120* (US). AMPARAI DISTRICT: E. side of Ulpasse Wewa, 2 mi. N. of Panama, *Fosberg & Sachet 52956* (US). HAMBANTOTA DISTRICT: Ruhuna National Park, plot R25, S. of road at mile marker 7, alt. 6–10 m, *Comanor 664* (US); Ruhuna National Park, Block 1, Andunoruwa Wewa, alt. 2–4 m, *Comanor 404* (US); Ruhuna National Park, Block 1, at Karaugaswala Junction, *Mueller-Dombois & Cooray 67121089* (US); Ruhuna National Park, Block 1, at Gonalabbe Lewaya, in grass cover plot R34, *Mueller-Dombois 68050319* (US); Ruhuna National Park, Patanagala Rocks, alt. 3–5 m, *Fosberg & Mueller-Dombois 50124* (US); Ruhuna National Park, Block 2 at Rakinawala, *Cooray 68102214R* (MO); Ruhuna National Park, Block 1, Rugamtota, *Cooray 69111406R* (US).

2. ROTULA

Lour., Fl. Cochinch. 121. 1790. Type species: *Rotula aquatica* Lour.

Subshrubs or trailing vines, submerged for part of year. Leaves crowded on short branches, the apices acute or slightly rounded, entire, veins obscure. Inflorescences cymes. Flowers perfect, 5-merous; the sepals unequal in basal width; corolla campanulate, the tube shorter than the lobes; stamens attached low in corolla tube, the anthers exserted; style simple, elongate, the stigma weakly bifid. Drupe 4-lobed, breaking into 4 one-seeded nutlets.

Rotula comprises two species, *R. aquatica* in SE. Asia, and *R. lycioides* (Mart.) I.M. Johnston in eastern Brazil and tropical West Africa.

Rotula aquatica Lour., Fl. Cochinch. 121. 1790; Alston in Trimen, Handb. Fl. Ceylon 6: 200. 1931.

Rhabdia aquatica (Lour.) Kuntze, Rev. Gen. 2: 439. 1891.

Shrubs or trailing vines, submerged for part of year, stems sparsely branched and flexuous, mostly glabrous. Leaves alternate, crowded on short branches, spatulate to oblong, small, to 2 cm long and 0.5 cm wide, rounded or slightly apiculate, entire, the bases cuneate, somewhat thickened, veins obscure, scabrous on both surfaces or glabrous above; petioles to 3 mm long or sessile. Inflorescences cymes, terminal or axillary, 1–4-flowered. Flowers subsessile, bracts two; calyx of 5 sepals, elongate-deltoid, to 4 mm long, unequal in basal width, strigose; corolla campanulate, pink, the lobes oblong, 3–4 mm long; stamens slightly exserted, the filaments 1.5 mm long, attached long in corolla tube, anthers 1.2 mm long; ovary globose-oblong, 4-loculed, style elongate, to 3.5 mm long, stigma weakly capitate and very slightly bifid. Drupe 4-lobed and 4-seeded, red, breaking into 4 one-seeded pyrenes.

Distr. India to the Philippines and the East Indies.

Ecol. In forests in rocky areas along streams where the plants are inundated during the wet season.

Note. *Rotula aquatica* is presently known from a single collection from Ceylon: Trimen (1895), under *Rhabdia lycioides* Mart., lists *C.P. 3491*. In notes I made in Sri Lanka (June–July 1973) I apparently saw *C.P. 3491* and agreed that it was this taxon. The above species description was taken in part from specimens collected in India, Burma and Thailand. The second species, *R. lycioides* (Mart.) I.M. Johnston, has been treated as conspecific with *R. aquatica* by some authors, but it differs in its soft brown indument, style noticeably but shortly once cleft, and longer leaves.

Specimen Examined. LOCALITY UNKNOWN: *s. coll. C.P. 3451* (PDA).

3. CORDIA

L., Sp. Pl. 190. 1753. Type species: *C. sebestena* L. (lectotype).

Trees or shrubs. Leaves alternate, deciduous or persistent, petiolate, the petioles usually canaliculate on the adaxial surface, stipules absent. Inflorescences variable, cymose, paniculate, spicate, capitate, or glomerate. Flowers perfect and often distylous or unisexual by abortion, the plants then dioecious, androdioecious, or subdioecious; calyx 3–5(–10) lobed or rarely circumscissile; corolla funnelform, campanulate, or tubular with reflexed or spreading lobes, (4–)5(–18) lobed, or sometimes the lobes nearly lacking and the corolla undulate or frilled at the apex or nearly truncate; stamens the same number as the corolla lobes, the lower part of the filaments adnate to the corolla tube, often with trichomes at or near the point of insertion, the anthers oblong to ellipsoid; ovary entire, disc annular to crateriform if distinct, style terminal, twice bifid, the four stigma lobes clavate, filiform, or discoid. Fruits variable, dry with a fibrous wall and capped by the persistent cartilaginous base of the style in sect. *Gerascanthus,* dry and bony walled in sect. *Rhabdocalyx,* or with a thin exocarp, juicy to mucilaginous mesocarp, and bony endocarp in sect. *Varronia, Myxa* and *Cordia,* usually 1-locular and 1-seeded.

Cordia comprises about 300 species that are widespread in tropical and subtropical regions. Although present in both tropical Asia and Africa, the genus is strongly centred in the Neotropics with centres of diversity in Mexico, the Greater Antilles, the Andes, and southeastern Brazil. The Asian tropics have the smallest number of species and only eight species are known from Ceylon.

KEY TO THE SPECIES

1 Corolla large, funnelform, orange; fruiting calyx accrescent .... **1. C. subcordata**
1 Corolla small, tubular with spreading or reflexed lobes, white; fruiting calyx not accrescent
  2 Inflorescence spicate; leaf margin serrate; corolla lobes broader than long ....
  ....**2. C. curassavica**
  2 Inflorescence cymose; leaf margin entire; corolla lobes longer than broad
    3 Upper leaf surface scabrous
      4 Leaves ovate to elliptic, alternate; calyx greater than 4 mm long .... **3. C. monoica**
      4 Leaves oblong to elliptic-oblong or rarely oblanceolate, opposite or subopposite; calyx less
        than 4 mm long
        5 Leaf undersurfaces with dense tufts of hairs in the axils of secondary veins...**4. C. sinensis**
        5 Leaf undersurfaces lacking dense tufts of hairs in the axils of secondary veins...**5. C. nevillii**
    3 Upper leaf surface glabrous or nearly so
      6 Lower leaf surface lightly villous to pubescent; leaf base truncate to subcordate; petioles thick
        ....**6. C. myxa**
      6 Lower leaf surface glabrous or with only small tufts of hairs in the axils of the major veins;
        leaf base rounded to obtuse or acute; petioles moderate to slender
        7 Leaf blade narrowly oblong-ovate to ovate or oblong; fruits greater than 1.7 cm long
        ....**7. C. oblongifolia**

7 Leaf blade elliptic to widely elliptic or occasionally ovate or obovate; fruits less than 1.3 cm
    long .... **8. C. dichotoma**

1. Cordia subcordata Lam., Tab. Enc. 1: 421. 1792; Trimen, Handb. Fl. Ceylon
3: 195. 1895.

Cordia orientalis R. Br., Prod. 498. 1810.
Cordia campanulata Roxb., Hort. Beng. 17. 1814.
Cordia hexandra Roem. & Schult., Syst. Veg. 4: 799. 1819.
Cordia rumphii Blume, Bijdr. 14: 843. 1826.
Cordia muluccana Roxb., Fl. Ind. 2: 337. 1824.

Tree 3–10 m tall, the twigs glabrous or with widely scattered hairs. Leaves
persistent, on short spurs, these often barely evident; petioles 20–42(−52) mm
long, sparsely strigillose to nearly glabrous; leaf blades widely ovate, (6.5–)8.5–16
cm long, (3.6–)5–11 cm wide, the apex acute to obtuse or rarely rounded and
abruptly acuminate, the base obtuse to rounded, the margin entire but often unevenly undulate, the upper surface glabrous or with sparse, very short, appressed hairs
from a multicellular cystolith, these becoming mineralized and easily visible with
age, the lower surface essentially glabrous but with dense tufts of hairs in the
axils of the veins, these often extending along the major veins. Inflorescence subterminal, a sparsely branched, loose cyme to 10 cm broad, the branches glabrous
or with very sparse appressed hairs. Flowers distylous; calyx tubular, 14–17.5
mm long, 6–8 mm wide at the mouth, essentially glabrous but with a few widely-
scattered, appressed hairs, strigose to strigillose near the apex on the interior surface, unevenly 2–3-lobed, the lobes 2–4 mm long; corolla orange-red, funnelform,
36–43 mm long, 5-merous, the lobes depressed ovate to widely depressed ovate,
13–15 mm long, 12–18 mm wide, frilled along the apical margin, the tube 17–27
mm long; stamens 5, the filaments 24–29 mm long, the upper 3–7 mm free,
glabrous, the anthers oblong to oblong-lanceoloid, 3–3.5 mm long; ovary ovoid
to conical-ovoid, the lower portion adnate to the calyx, 2–3 mm long, 1.4–1.5
mm broad, the disc not evident, the style 20–29 mm long, the stylar branches
3.5–6.5 mm long, the stigma lobes clavate. Fruits drupaceous, completely enclosed
in the accrescent calyx, the end projecting in a short tube beyond the fruit, the
exocarp smooth, lustrous, the mesocarp corky, the stone globose to ellipsoid,
21–31 mm long, 15–24 mm broad, the endocarp bony.

Distr. *Cordia subcordata* ranges from the east coast of Africa through India,
southeast Asia, the islands of the Malay Archipelago, to the Pacific Islands and
Hawaii.

Ecol. This is a common strand plant of dry areas surrounding the Indian Ocean
and Pacific islands that has apparently achieved its distribution by water dispersal of its fruits. The fruiting calyx is thickened and the mesocarp is corky making
the fruits buoyant.

Note. This is the only species of sect. *Cordia* that occurs in the Old World. The section consists of perhaps 15 species with the majority found on the Greater Antilles. This species has been reported for Ceylon (Trimen, 1895; Alston, 1931). According to notes taken by Nowicke in 1973, specimens of *C. subcordata* do exist at PDA. But I have not seen any collections and the above description is based upon plants from other areas.

Specimens Examined. According to Dr. Dassanayake (pers. comm.) two specimens of *C. subcordata* are represented at PDA: TRINCOMALEE DISTRICT: Norway Point, at high water, 30–11–1939, *Worthington 660*; Foul Point, Dec. 1885, *W.F(erguson) s.n.*

2. Cordia curassavica (Jacq.) Roem. & Schult., Syst. Veg. 4: 460. 1819.

Varronia curassavica Jacq., Enum. Pl. Carib. 14. 1760; Jacq., Select. Stirp. 40. 1763.

Shrub to 2(–4) m tall, rarely very short and densely branched, the bark dark brown, coarsely lenticellate, the twigs glabrous to strigillose or puberulent or rarely hirsute but always with small globose wax particles. Leaves deciduous, on short spurs to 1 mm long; petioles 1–8(–21) mm long, or rarely the leaves nearly sessile, canaliculate on the adaxial surface, strigillose or puberulent to hirsute; leaf blades lanceolate to narrowly-elliptic or ovate to elliptic-ovate, 2–10(–16) cm long, 0.5–4(–7.3) cm wide, the apex acute, the base cuneate to acute and sometimes decurrent along the petiole, the margin serrate, occasionally merely undulate, the upper surface scabrous to merely papillose, the lower surface strigillose with the majority of the trichomes restricted to the major veins to tomentulose. Inflorescence terminal or very occasionally subterminal, a dense spike or often with some of the flowers aborted and those remaining being scattered along the rachis, 1.5–8.8(–15) cm long, (4–)5–8(–9) mm broad, the peduncle (0.4–)1.8–7.8(–9.9) cm long, puberulent or strigillose to nearly glabrous, usually with small globose wax particles. Flowers distylous, sessile; calyx campanulate, 2.1–3.2(–3.8) mm long, (1.6–)2–2.8(–3) mm wide at the mouth, ribs absent, strigillose or puberulent with small globose wax particles, 5(–6) lobed, the lobes deltate, (0.4–)0.6–0.9(–1.3) mm long; corolla white, tubular with reflexed to spreading lobes, (3.8–)4.8–6.8(–8.3) mm long, 5(–6) lobed, the lobes ovate to depressed-ovate, (0.8–)1.2–1.8(–2.8) mm long, 1.0–1.8(–3.7) mm wide, the tube (1.6–)2.4–3.0(–3.4) mm long, puberulent to pubescent just beneath the point of staminal insertion; stamens 5(–6), the filaments (2.3–)3.2–5.0(–6.0) mm long, the upper (0.3–)0.8–1.8(–2.0) mm free, the free portion glabrous, puberulent to pubescent beneath the point of insertion, the anthers ellipsoid, (0.3–)0.7–1.0 mm long; ovary ovoid to broadly ovoid, (0.8–)1.0–1.2(–1.6) mm long, 0.5–0.8(–1.0) mm broad, glabrous, the style (1.4–)2–4(–5.7) mm long, the stigma lobes clavate, occasionally very narrow; fruit ½ – ¾ enclosed in the slightly accrescent calyx, bright red, the stone ovoid, (3.7–)4.0–4.5(–5.9) mm long, 2.2–3.1 mm broad, the surface ruminate, the endocarp bony, 1-locular.

Distr. *Cordia curassavica* is native to the Neotropics where it ranges from northern Mexico to South America and the West Indies. However, it has been introduced, escaped, and become widely established as a common weed in Ceylon as well as other parts of the Paleotropics. It was recorded by Petch (as *C. aubletii* DC.) from near the Anuradhapura Hotel (Alston, 1931) and probably was introduced into the country for the Botanic Garden which formerly existed there.

Ecol. This weedy shrub is characteristically found in young secondary growth, often along roadsides. Its small, bright red fruits are easily dispersed by birds and it often becomes a weed when introduced.

Note. This is an extremely variable species and many of the variants have been named as distinct species by authors in the past. Throughout its natural range it is quite variable but most of the variation is in overall size of the plants, leaves, and inflorescences. Plants from drier regions tend to be smaller in all aspects but most of these differences disappear when seed from various localities is collected and plants are raised in the greenhouse under constant conditions (Miller, 1988). In addition, *Cordia curassavica* is known to produce natural hybrids with several other Neotropical species (Miller, 1988). The appearance of hybrids coupled with a large amount of phenotypic plasticity has led to a great deal of taxonomic and nomenclatural confusion in the past. As this is the only species of sect. *Varronia* in Ceylon, there is no opportunity for hybridization and the species is morphologically more constant than in the New World.

Specimens Examined. ANURADHAPURA DISTRICT: Mihintale, collina templorum, alt. 250 m, *Bernardi 14227* (US); road between Anuradhapura and Habarana, alt. 100 m, *Comanor 1228* (US); 1 mile south of Anuradhapura, *Cooray 69093001R* (US); Dambulla-Anuradhapura Road, *Cooray 70013108R* (US); Anuradhapura, *Fosberg and Balakrishnan 53430* (US); *Fosberg and Balakrishanan 53432* (US); Galkulama, *Jayasuriya and Sumithraarachchi 1594* (US); Anuradhapura-Dambulla Road, *Kundu and Balakrishnan 267* (US); Anuradhapura, west shore of Nuwara Wewa, behind the Nuwara Wewa rest house, *Sohmer 8103* (US); 1 mile towards Turuwile Tank on the road leaving the Dambulla-Anuradhapura Road, at milestone c. 71, *Townsend 73/187* (US); between Wilpattu National Park and Anuradhapura, *van Beusekom and van Beusekom 1633* (US). KURUNEGALA DISTRICT: Wenduru Wewa, along banks of tank, alt. sea level, *Cramer 3056* (MO, US); Gongama, along roadside by Wilgamdematuwa Road, alt. 140 m, *Cramer 3592* (US); Elephant Rock, *Fosberg and Jayasuriya 52718* (US). TRINCOMALEE DISTRICT: Morawewa Tank Bund, *Fosberg and Jayasinghe 57134* (MO). AMPARAI DISTRICT: on the road from Akkaraipattu to Thottame, *Balakrishnan NBK381* (US); Tirukkovil, *Jayasuriya 2042* (MO); vicinity of Malvatai, between Amparai and coast, alt. 10 m, *Stone 11191* (US).

3. Cordia monoica Roxb., Pl. Corom. t. 58. 1796; Trimen, Handb. Fl. Ceylon 3: 193. 1895.

Tree 5(–10) m tall, the twigs densely brown puberulent with two types of trichomes, some short, stiff, and spreading, the others brown, erect, and dendritic. Leaves persistent, on very short spurs; petioles 5–18(–24) mm long, densely brown puberulent; leaf blade ovate to nearly elliptic, (2.4–)3.6–6.8(–10.3) cm long, (1.5–)2.3–5.1(–5.8) cm wide, the apex acute to obtuse, the base obtuse to rounded or rarely approaching acute, the margin undulate to unevenly serrate or denticulate, the upper surface scabrous, the trichomes erect to spreading and from a multicellular cystolith, the lower surface loosely to evenly brown pubescent. Inflorescence terminal, cymose-paniculate, 1–3.5 cm long, 1–3.5 cm broad, the branches densely brown puberulent. Flowers bisexual or male, sessile, the species androdioecious; calyx tubular, 4–7 mm long, 1–3 mm wide at the mouth, densely brown tomentulose, unevenly 3–4-lobed, the lobes to 1.7 mm long; corolla white, tubular with spreading lobes, (4–)5-lobed, the lobes oblong to straplike, 2.3–5.0 mm long, 0.8–2.0 mm wide, the tube 2.7–6.0 mm long; stamens (4–)5, the filaments 4–9 mm long, the upper 1.3–3.0 mm free, glabrous, the anthers oblong to ellipsoid, 0.6–1.6 mm long; ovary globose, 0.3–0.8 mm long, 0.3–0.8 mm broad. Fruits drupaceous, orange at maturity, the stone ovoid, 9–12 mm long, 5–8 mm broad, the endocarp bony.

Distr. *Cordia monoica* is known only from India and Ceylon.

Vern. Naruvili, Ponnaruvili.

Specimens Examined. MANNAR DISTRICT: Akattikulam, 4 miles south of Madhu Road, *Meijer 799* (US). ANURADHAPURA DISTRICT: Wilpattu National Park, Periya Naga Villu, *Fosberg et al. 50886* (US); Wilpattu National Park, Occapu Kallu in plot W34, *Mueller-Dombois et al. 69042818* (US). AMPARAI DISTRICT: West side of Ulpasse Wewa, 1.5 miles north of Panama, *Fosberg and Sachet 52933* (US); inter Arugam Bay et Panama, circa templum vetum deletumque, alt. 10 m, *Bernardi 15579* (US). HAMBANTOTA DISTRICT: Ruhuna National Park, Block 1, near Patanagala camp, *Cooray and Balakrishnan 69012201R* (US); Bundala, *Hepper and de Silva 4756* (US); Yala, where River Menik Ganga flows into the sea, *Nowicke and Jayasuriya 411* (US); Ruhuna National Park, Block 1, Palatupana, *Cooray 69111705R* (US); Ruhuna National Park, Andunoruwa Wewa, alt. 3–5 m, *Fosberg and Mueller-Dombois 50152* (US); Ruhuna National Park, near Yala Bungalow, *Mueller-Dombois 67082603* (US); Ruhuna National Park, Block 2, outside plot R15 but in scrub, *Mueller-Dombois 67083013* (US); Ruhuna National Park, Block 1, between Andunoruwa Wewa and Komawa Wewa, in plot R13, *Mueller-Dombois et al. 67093018* (US).

4. Cordia sinensis Lam., Tabl. Enc. 1: 423. 1792; Meikle, Israel J. Bot. 20: 22. 1971.

Cordia rothii Roem. & Schult., Syst. Veg. 4: 798. 1819; Trimen, Handb. Fl.
   Ceylon 3: 194. 1895.
Cornus gharaf Forssk., Fl. Aegypt.-Arab. 95. 1775, nomen invalidum.
Cordia gharaf (Forssk.) Ehrenb. ex. Asch., Sitzungsber. Ges. Naturf. Freunde
   Berlin 1879: 46. 1879; Alston in Trimen, Handb. Fl. Ceylon 6: 199. 1931.

   Shrub or small tree to 4 m tall, the branches glabrescent. Leaves persistent,
on short discoid spurs; petioles 5–10 mm long, canaliculate on the adaxial surface, sparsely strigillose to nearly glabrous; leaf blade oblong to elliptic-oblong
or rarely oblanceolate, (2.4–)3.0–7.2 cm long, 1.3–3.3 cm wide, the apex obtuse to rounded, the base obtuse, the margin entire, the upper surface scabrous,
the trichomes very short, stout, appressed and arising from a large, many-celled
cystolith, the lower surface sparsely strigillose to sparsely pubescent but with
dense tufts of trichomes in the axils of the lower secondary veins. Inflorescence
terminal, cymose-paniculate, small, 2–5 cm long, 2–5 cm broad, the branches
densely puberulent to tomentulose. Flowers monomorphic, sessile; calyx tubular-
campanulate, 3.3–3.5 mm long, 2.0–2.3 mm wide at the mouth, puberulent to
tomentose, very unevenly lobed, the lobes 0.7–1.0 mm long; corolla white, tubular
with reflexed lobes, 5–5.7 mm long, 4(–5)-lobed, the lobes oblong to oblong-
obovate, 2.3–2.6 mm long, 1.1–1.6 mm wide, the tube 2.8–3.2 mm long; stamens
4(–5), the filaments 4.0–4.5 mm long, the upper 1.3–1.5 mm free, glabrous, the
anthers oblong, c. 1 mm long; ovary ovoid to narrowly ovoid, 1.0–1.2 mm long,
0.6–0.9 mm broad, the style 2–5 mm long, the stigma lobes filiform to narrowly
clavate, the disc not evident. Fruits borne in the saucer-shaped calyx, drupaceous,
orange at maturity, mesocarp mucilaginous, endocarp bony, the stone ovoid, 6–9
mm long, 4–6 mm broad.

   Distr. *Cordia sinensis* ranges from Ethiopia and Northeast Africa through
the Middle East to India and Ceylon.
   Ecol. *Cordia sinensis* is a shrub of dry forests and deserts throughout the
Middle East and south-central Asia.
   Specimens Examined. HAMBANTOTA DISTRICT: near western boun-
dary of Ruhuna National Park, *Meijer 208* (US); Ruhuna National Park, Block
1 at Patanagala, *Cooray 68102205R* (US); Ruhuna National Park, Block 1, at
Kohombagaswala, *Cooray 69120305R* (US); Ruhuna National Park in An-
dunoruwa Wewa in plot R3, alt. 2 m, *Mueller-Dombois 67082613* (US); Ruhuna
National Park, Block 1, in Komawa Wewa, plot R17, *Mueller-Dombois 67093028*
(US); Ruhuna National Park, Block 1, Kohombagaswala, *Wirawan 730* (US).

5. Cordia nevillii Alston in Trimen, Handb. Fl. Ceylon 6: 199. 1931.

   Shrub to 3 m tall, the twigs densely pubescent, later glabrous. Leaves persistent, subopposite or opposite, the axillary buds pubescent; petioles 1.5–5(7.5)
mm long, narrowly canaliculate on the adaxial surface, strigose; leaf blade elliptic-

oblong to obovate, 1.9–6.5(–9) cm long, 1–4.5 cm wide, the apex obtuse, the base acute to obtuse, the margin entire or unevenly undulate, the upper surface scabrous, the hairs appressed and usually arising from a multicellular cystolith, the lower surface stiffly pubescent. Inflorescence terminal or axillary, cymose, less than 3 cm long, the branches scabrous. Flowers bisexual or male sterile, sessile; calyx tubular-campanulate, 3–4 mm long, 1.8–2.3 mm wide at the mouth, puberulent to tomentose, very unevenly 3–4-lobed, the lobes 0.5–1 mm long; corolla white, tubular with reflexed lobes, 4.5–6.5 mm long, 4(–5) lobed, the lobes oblong, 2–2.5 mm long, 1.2–1.6 mm wide, the tube 3–4 mm long; stamens 4(–5), the filaments 4–5.5 mm long, the upper 1.7–2.5 mm free, pubescent at the point of insertion, the anthers oblong, 1–1.5 mm long; ovary ovoid, 1.2–1.7 mm long, 0.6–1 mm broad, the style 3–5.5 mm long, the stigma lobes narrowly clavate to filiform, the disc not evident. Fruits borne in the saucer-shaped calyx, drupaceous, orange at maturity, mesocarp mucilaginous, endocarp bony, the stone ovoid, 7–11 mm long, 5–6.5 mm broad.

Distr. *Cordia nevillii* ranges from Ceylon and India west through Pakistan and the Middle East to eastern Africa.

Ecol. *Cordia nevillii* is a shrub of dry forests and deserts, often in rocky areas.

Specimens Examined. TRINCOMALEE DISTRICT: Kuchchaveli, on exposed rocks near the sea, *Nevill 578* (K, type).

6. Cordia myxa L., Sp. Pl. 190. 1753; Trimen, Handb. Fl. Ceylon 3: 193. 1895.

Tree to 8(–13) m tall, the twigs sparsely pubescent, the hairs c. 1 mm long. Leaves persistent, alternate, borne on short but prominent spurs, the axillary buds pubescent; petioles 13−32(−37) mm long, narrowly canaliculate on the adaxial surface, glabrous but curly brown pubescent in the groove; leaf blade widely ovate to nearly orbicular, 4.0–10(–11.3) cm long, 4.2–8.4(–9.2) cm wide, the apex obtuse to rounded, the base truncate to subcordate, the upper surface nearly glabrous but with a few very short, widely scattered appressed hairs mostly along the major veins, the lower surface lightly villous, the hairs simple, erect, curly. Inflorescence terminal, cymose-paniculate, 5–9 cm long, (2.0–)4.0–7.5 cm broad, the branches with a few, widely-scattered, curly brown hairs. Flowers bisexual, sessile; calyx tubular-campanulate, 4.7–7.0 mm long, 3.5–5.0 mm wide at the mouth, granular-puberulent on the upper half, densely white strigose on the interior surface, unevenly 4-lobed, the lobes ovate, c. 3 mm long; corolla white to cream-coloured, tubular with spreading lobes, 8.0–9.5 mm long, 5-merous, the lobes ovate, 4.5–5.0 mm long, 1.5–3.0 mm wide, the tube 3.5–4.5 mm long; stamens 5, the filaments 6.5–8.0 mm long, the upper 1.7–2.9 mm free, puberulent at and just above the point of insertion, the anthers oblong, 1.0–1.5 mm long; ovary broadly ovoid, 2.5–3.0 mm long, 1.0–1.4 mm broad, glabrous, the disc not evident, the style 3.0–4.0 mm long, the stigma lobes fan-shaped. Fruits drupaceous, yellow, globose, 19–24 mm long, 19–21 mm broad, the mesocarp mucilaginous, the endocarp bony.

Distr. *Cordia myxa* is probably native only in northeastern Africa and the adjacent Middle East although it is occasionally cultivated in parts of the Asian tropics. The single collection known from Ceylon is probably a cultivated or adventive plant.

Ecol. There is some question about the origin of this species and Hutchinson (1918) raises the possibility that it is simply a cultigen that has escaped and become adventive in areas surrounding where it is cultivated.

Vern. Lolu, Naruvili, Vidi.

Note. There has long been a great amount of confusion as to the application of the name *Cordia myxa*. In publishing the name, Linnaeus (1753) cited, as synonyms, the earlier works of Alpini, Bauhin, Rheede, Commelyn, and his own Hortus Cliffortianus. Hutchinson (1918) has shown that these earlier works include descriptions of three species, *C. myxa, C. obliqua,* and *C. crenata.* The limited description that is given in the first edition of Species Plantarum is of little help in resolving the problem. Hutchinson felt that based on material in the Linnaean herbarium, the name *C. myxa* should be applied to the species with more or less orbicular leaves and fan-shaped stigma lobes and this is the way in which the name has been traditionally used. However, he stated that the type was in the Linnaean herbarium in London. The only collection there labelled *C. myxa* is of a plant now treated as *C. dichotoma* and there are no collections of *C. myxa.* There is however, a collection of the plant that Hutchinson describes in the Linnaean herbarium at Stockholm and it is clearly labelled *Cordia myxa.*

It seems logical that the collection in the Linnaean herbarium at Stockholm should be treated as the lectotype for several reasons. Although the first edition of Species Plantarum provides little information as to what Linnaeus intended, the second edition (1760) provides a description that more clearly describes the plant and states that it is from Egypt. As *Cordia dichotoma* reaches the western geographic limit of its distribution in Pakistan and does not reach Africa, it seems that Linnaeus' concept of the species was the same as our current concept. Secondly, although Hutchinson (1918) cited the type as being in the Linnaean herbarium in London, this seems to be a clerical error as there is no material of it there and the plant he has illustrated (Kew Bull. 1918: 220. 1918) is clearly based on the Stockholm specimen.

Cordia myxa and its allies (*C. obliqua, C. crenata, C. domestica,* and *C. dichotoma*) are a very confusing group of species and in desperate need of serious monographic study. As Ivan Johnston (1951) has indicated, they are not nearly as distinct as Hutchinson (1918) has indicated. While the two members of this group, *C. myxa* and *C. dichotoma,* that occur in Ceylon are easily distinguished from one another, there are serious problems in other geographic regions. The problem is compounded by the fact that many of the species are commonly cultivated.

Specimens Examined. HAMBANTOTA DISTRICT: Ruhuna National Park, Block 2, plot R-16, *Fosberg et al. 51083* (MO, US).

7. Cordia oblongifolia Thw., Enum. Pl. Zeyl. 214. 1860; Trimen, Handb. Fl. Ceylon 3: 194. 1895.

Scrambling shrub to 3 m tall, the twigs glabrous or very sparsely strigillose. Leaves persistent, on short lateral spurs, the axillary bud with tufts of hairs on the upper and lower side; petioles (8–)10–30(–35) mm long, deeply canaliculate on the adaxial surface and spreading to slightly winged near the base of the lamina; leaf blade narrowly oblong-ovate to ovate or oblong, 4.3–9(–14) cm long, (1.5–)2–5(–7) cm wide, the apex acuminate to acute, the base rounded to obtuse or rarely acute or approaching cordate, the margin entire, the upper surface glabrous but papillose with small cystoliths of 5–9 cells, the lower surface glabrous but usually with small tufts of trichomes in the axils of the lowest secondary veins. Inflorescence terminal, a small, sparsely branched cyme seldom more than 4 cm long and 4.5 cm broad, the peduncle usually short, but sometimes elongating to 4 cm, glabrous or sparsely strigillose. Flowers apparently monomorphic, sessile, the buds obovoid; calyx tubular-campanulate, 9–12.5 mm long, 5.5–8 mm wide at the mouth, the outer surface glabrous to very sparsely strigillose, the inner surface densely sericeous, unevenly 3–4 lobed, the lobes 2.5–3.5 mm long, ovate to widely ovate; corolla white, tubular with spreading lobes, 13–16.5 mm long, 6–8 lobed, the lobes oblong to narrowly obovate, 5–7 mm long, 3–4 mm wide, the tube 7–9.5 mm long; stamens 6–10, the filaments 9.5–12 mm long, the upper 3–6 mm free, puberulent to pubescent above and below the point of insertion, the anthers oblong, 1.8–3.3 mm long; ovary globose, 0.7–1.5 mm long, 0.7–2 mm broad, the style short and lacking an evident stigma. Fruits drupaceous, symmetrically ovoid, the mesocarp mucilaginous, the stone 1.7–3.3 cm long, 1–2.1 cm broad, the wall bony, surrounded by a fibrous layer, often with several locules.

Distr. *Cordia oblongifolia* ranges from Africa to India and Ceylon.

Ecol. In Ceylon, this is an uncommon plant of sand dunes known mostly from Ruhuna National Park and adjacent forest in Hambantota District.

Specimens Examined. JAFFNA DISTRICT: Pallavarayankaddu-Mankulum Road, *Kundu and Balakrishnan 644* (US). AMPARAI DISTRICT: Pottuvil, ad oram sabulosam oceani, alt. 0 m, *Bernardi 16007* (MO, US). HAMBANTOTA DISTRICT: Ruhuna National Park, Block 1, at Bambawa, *Cooray 69111714R* (US); Ruhuna National Park, Block 1, north of Buttawa, in dune forest-scrub near ocean in plot R27, *Mueller-Dombois and Cooray 67121027* (US); Ruhuna National Park, Block 1, north of Buttawa, *Mueller-Dombois and Cooray 68022806* (US); near Hambantota Salterns, *Sumithraarachchi and Davidse*

DBS574 (US); Ruhuna National Park, Uraniya, in plot R33, *Wirawan 729* (US); Ruhuna National Park, Block 1, Buttawa, *Wirawan 808* (MO, US).

8. Cordia dichotoma Forst. f., Prod. 18. 1786; I.M. Johnston, J. Arnold Arbor. 32: 8. 1951.

Varronia sinensis Lour., Pl. Cochinch. 138. 1790; not *Cordia sinensis* Lam.
Cordia indica Lam., Tab. Enc. 1: 422. 1791.
Cordia suaveolens Blume, Bidjr. 14: 843. 1826.
Cordia brownii DC., Prod. 9: 499. 1845; not *Cordia brownei* (Friesen) I.M. Johnston, J. Arnold Arbor. 31: 177. 1950.
Cordia ixiocarpa F. Muell., Fragm. 1: 59. 1858.

Tree to 5(–10) m tall, the twigs glabrous. Leaves persistent, alternate or rarely subopposite, borne on short spurs, the axillary bud tomentulose; petioles slender, (7–)13–34(–42) mm long, canaliculate on the adaxial surface, glabrous or with a few scattered hairs in the groove; leaf blade elliptic to widely elliptic or occasionally ovate or obovate, (2.5–)5–10(–12) cm long, (1.8–)2.5–6(–8) cm wide, the apex obtuse to acute, rarely approaching rounded, occasionally abruptly acuminate, the base obtuse to rounded or less commonly acute, the margin entire or occasionally undulate to crenate near the apex, the upper surface essentially glabrous but with numerous small, few-celled cystoliths, the lower surface sparsely puberulent when young, later glabrous except for small tufts of hairs in the axils of the major veins. Inflorescence terminal, cymose-paniculate, 3.5–13 cm long, 3–10 cm broad, the branches glabrous or with a few, widely-scattered, short hairs. Flowers either bisexual or male, sessile, these similar in form; calyx tubular-campanulate, 4.2–6.0 mm long, 2.8–5.5 mm wide at the mouth, glabrous to sparsely short strigillose, densely strigillose on the interior surface, unevenly 3–4-lobed or raggedly fringed, the lobes to nearly 1 mm long; corolla white, tubular with spreading lobes, 6.2–10 mm long, 5(–6)-merous, the lobes oblong, 3.4–6.0 mm long, 1.3–2.0 mm wide, the tube 2.8–4.7 mm long; stamens 5(–6), the filaments 4.5–9.5 mm long, the upper 1.8–5.0 mm free, glabrous or sparsely puberulent at the point of insertion, the anthers oblong, 1.2–2.1 mm long; ovary broadly ovoid, 1.0–1.5 mm long, 0.8–1.0 mm broad, glabrous, the disc not evident, the style 6.0–6.5 mm long, absent in male flowers, the stylar branches 2.0–4.0 mm long, the stigma lobes filiform to clavate. Fruits drupaceous, yellow to orange, borne in the persistent saucer-shaped calyx, ovoid to globose, 8–13 mm long, 6–8 mm broad, the mesocarp mucilaginous, the endocarp bony.

Distr. *Cordia dichotoma* ranges from Pakistan, southern India, and Ceylon east through tropical Asia and the Malay Archipelago to Queensland and New Caledonia.

Ecol. According to F.R. Fosberg (pers. communication) *Cordia dichotoma* occurs in semi-deciduous and mature secondary forest in northern and eastern areas.

Note. This species has often been nomenclaturally confused with *Cordia myxa* but Hutchinson (1918) and Johnston (1951) have both shown that the two are not only distinct, but quite different. *Cordia dichotoma* is a species of tropical Asia and *C. myxa* is a species of northeast Africa and the adjacent Middle East although it may be cultivated outside of this range and is known in Ceylon from a single collection. Throughout its range, *C. dichotoma* is a somewhat variable species but it can be recognized by its more or less elliptic leaves, slender petioles, and filiform to clavate stigma lobes.

Specimens Examined. VAVUNIYA DISTRICT: Cheddikulam, *Kundu and Balakrishnan 566* (US). ANURADHAPURA DISTRICT: Wilpattu National Park, south of W14 at entrance to forest near Erige Ara, *Mueller-Dombois 69042823* (US); Vavuniya-Kebithigollawa Road, *Sumithraarachchi DBS832* (US). POLONNARUWA DISTRICT: Polonnaruwa Sacred Area, elev. 61 m, *Dittus WD700772804* (US); Polonnaruwa Sacred Area, *Dittus WD700880301* (US); Polonnaruwa Sacred Area, *Fosberg 51874* (US); Polonnaruwa Sanctuary, elev. 60 m, *Huber 431* (US); Polonnaruwa, near lake, Smithsonian Primate Research Camp site near old ruins, *Meijer and Balakrishnan 108* (US, WIS); Polonnaruwa Sacred Ruin Area, *Meijer and Balakrishnan 115* (US, WIS); Polonnaruwa Sacred Area, *Ripley 145* (US); Polonnaruwa Sacred Area, elev. 61 m, *Ripley 162* (US); Polonnaruwa Sacred Area, elev. 61 m, *Ripley 228* (US); Polonnaruwa Sacred Area, wooded section east of Watadage, elev. 61 m, *Ripley 300* (US); Polonnaruwa Sacred Area, Archaeological Quarters Yard, elev. 61 m, *Ripley 306* (US); Polonnaruwa Sacred Area, section 1C on bund, *Ripley 431* (US); Polonnaruwa Sacred Area, section 1C on bund, *Ripley 440* (US); Polonnaruwa-Batticaloa Road 53/3 mile marker, *Waas 629* (US); ad templum Medirigiriya, elev. 75 m, *Bernardi 16111* (US). PUTTALAM DISTRICT: on road Wanatavillu-Weerakuti Villu, *Cooray 69100609R* (US). MATALE DISTRICT: Ridiella, 10 miles north of Naula-Elahera Road, off mile post 5, *Jayasuriya 311* (US); Sigiriya Tank, *Nowicke et al. 367* (US). TRINCOMALEE DISTRICT: Trincomalee, China Bay, *Sumithraarachchi DBS5857* (US); Monkey Bridge, *Waas 1265* (US); Great Sober Island, *Wheeler and Balakrishnan 12577* (US). AMPARAI DISTRICT: near Kalmunai, *Kostermans 24344* (US). HAMBANTOTA DISTRICT: Ruhuna National Park, Block 2, 300 m west of Kumbukkan Oya in plot R16, *Mueller-Dombois and Comanor 67083018* (US).

4. TOURNEFORTIA

L., Sp. Pl. 140. 1753. Type species: *T. hirsutissima* L. (lectotype).

Shrubs, sometimes decumbent, vines, or small trees. Leaves alternate, sometimes opposite. Inflorescences cymose, scorpioid, sometimes irregular panicles, the bracts seemingly absent. Flowers perfect, 5-merous, rarely 4-merous; sepals free or weakly connate basally; corolla salverform or funnelform, the lobes frequently keeled; stamens mostly 5, borne on the corolla throat, included, the filaments very short, the anthers elongate, free or + or − connate apically; ovary 2-carpellate, each carpel with two ovules, disc + or − distinct, sometimes cup-shaped, style long and slender or + or − absent, the stigma peltate or conic. Fruit fleshy, + or − globose and unlobed to distinctly 4-lobed, separating into 2–4 nutlets at maturity.

Tournefortia is a large genus of more than 100 poorly-defined species in the warmer regions of the world, but centering in the neotropics. Two species occur in Ceylon.

KEY TO THE SPECIES

1 Plants densely silk-pubescent; leaves oblanceolate; flowers 5-merous .... **1. T. argentea**
1 Plants sparsely strigose; leaves ovate, ovate-elliptic; flowers 4-merous....... **2. T. walkerae**

1. Tournefortia argentea L. f., Suppl. Pl. 133. 1781; Trimen, Handb. Fl. Ceylon 3: 198. 1895.

Shrubs, to 5 m tall, or rarely small trees; densely silk-pubescent on leaves, young stems, inflorescences, and calyx. Leaves spiral, crowded near tips of branches, oblanceolate, to 17 cm long and 7 cm wide, acute, entire, the bases long-attenuate; sessile, subsessile, or shortly petiolate, to 1 cm long. Inflorescences conspicuous cymes, terminal, compact in bud, open and lax in fruit, the peduncle stout, to 14 cm long. Flowers small, sessile, crowded and two-ranked; calyx of 5 sepals, slightly connate basally, spatulate, c. 2.0 mm long, densely pubescent without, glabrous within; corolla shallowly salverform, white, to 4 mm across, the tube wide, 1.0 mm, and shallow, 1.2 mm long, 5-lobed, the lobes ovate-orbicular, longer than wide, pubescent down the middle outside; stamens 5, mounted low in tube, anthers subsessile, 0.8–1.0 mm long, slightly mucronate; ovary globose, stigma sessile, consisting of annular ring with conical apex cleft into two stout branches. Fruit globose, 5–6 mm in diam, consisting mostly of aerenchymatous mesocarp, 2–4 seeds near the apex surrounded by a hard endocarp, the exocarp thin.

Distr. Common along shorelines in tropical areas of Pacific and Indian Oceans.

Ecol. Sandy, saline soils and coastlines.

Vern. Karan.

Note. Johnston (1951) treated this species as one of three in the genus *Messerschmidia* L., the other two being *M. sibirica* and *M. gnaphalodes*. He believed that these species were closer to *Heliotropium* than to *Tournefortia*. In a palynological study of the latter genus, Nowicke and Skvarla (Amer. J. Bot. 61: 1021–1036. 1974) concluded that "the palynological evidence does not support the generic status which Johnston accorded this small complex but does not deny the validity of the taxon. When herbarium material of all three species is compared, however, it is difficult to perceive of *T. argentea* and *T. gnaphalodes*, with their woody habit and dense, silky pubescence as being more closely related to the herbaceous and sparsely pubescent *T. sibirica* than to many of the remaining species of *Tournefortia*.

The success of *Tournefortia argentea* is surely due to the adaptation of the seeds to ocean dispersal. According to Johnston (J. Arnold Arbor. 16: 164. 1935) the original description of *T. argentea* is based on material collected on the coast of Ceylon by König. Despite the extensive collecting done by various collaborators on the Flora of Ceylon, the U.S. National Herbarium has no material of *T. argentea* from Ceylon.

Specimens Examined. TRINCOMALEE DISTRICT: Foul Point, *Simpson 9678* (PDA); N. of Trincomalee, 1 March 1892, *Nevill s.n.* (PDA). HAMBANTOTA DISTRICT: Tangalla (as Tangalle), December 1882, *s. coll. s.n.* (PDA). LOCALITY UNKNOWN: *s. coll.*, Herb. Wright prop. (PDA).

2. Tournefortia walkerae Clarke in Hook. f., Fl. Br. Ind. 4: 147. 1883; Trimen, Handb. Fl. Ceylon 3: 198. 1895.

Vines or trailing shrubs. Leaves alternate, elongate-lanceolate, to 11 cm long and 4 cm wide, chartaceous, sparsely pubescent on both surfaces, acute-acuminate, entire, the bases rounded; petioles 8–10 mm long. Inflorescences terminal cymes, with slightly irregular dichotomous branching, peduncles to 4 cm long, flowers crowded, two-ranked, small, sessile; calyx of 4 sepals, deltoid, c. 2 mm long, slightly connate basally, persistent, sparsely strigose on both surfaces; corolla of four united petals, elongate-salverform, the tube c. 4–5 mm long, 4-lobed and 4-ribbed, the lobes small, c. 1 mm, not spreading; stamens 4, attached low in the corolla tube, the anthers 1.7–2 mm long; ovary at anthesis very small, c. 1.5 mm long, the style and stigma scarcely·differentiated. Fruit subglobose, c. 5 mm in diam, dry, hard, not separating into nutlets.

Distr. Known only from Ceylon and very rare even there.

Note. *Tournefortia walkerae* is known from a single specimen, identified by *C.P. 2697*. Although Johnston (1935) reduced this taxon to *T. tetrandra* Blume var. *walkerae* (Clarke) it is doubtful that he actually saw the above specimen. The collections of *T. tetrandra* Bl. at the US herbarium have very different leaves—ovate, thicker and more prominently veined. According to Trimen (1895) *T. walkerae* is yellow-flowered and found in low moist country.

Specimens Examined. LOCALITY UNKNOWN: *s. coll. C.P. 2697* (PDA).

5. EHRETIA

L., Syst. ed. 10, 936. 1759. Type species: *E. tinifolia* L.

Trees or shrubs, pubescent or glabrous. Leaves alternate, petiolate, estipulate, entire or serrate. Inflorescences terminal, cymose to paniculate. Flowers perfect; sepals 5, imbricate or open in bud; corolla white (or rarely blue in some African species), tubular with 5 spreading lobes; stamens 5, usually exserted, the anthers oblong to ellipsoid; ovary ovoid, 2- or 4-locular, the style terminal, bifid, the stigmas 2, clavate or capitate. Fruits drupaceous, ovoid to nearly spherical, the stone separating into 2, 2-seeded or 4, 1-seeded pyrenes.

The genus *Ehretia* comprises about 40 species from tropical and subtropical Asia, Africa, and the Neotropics. The majority of the species occur in Africa, only 3–5 species occur in the Neotropics, and Johnston (1951) reported 12 species for eastern Asia. Only a single species is known from Ceylon.

Ehretia laevis Roxb., Pl. Corom. 1: 42, t. 56. 1796; Trimen, Handb. Fl. Ceylon 3: 195. 1895.

Ehretia laevis var. *platyphylla* Merr., Lingnan Sci. J. 14: 55. 1935.

Shrub or small tree to 8 m tall, the twigs glabrous, the axillary buds usually drying black, resinous. Leaves deciduous; petioles 4–13(–19) mm long, sulcate on the adaxial surface; leaf blade elliptic or narrowly-elliptic to narrowly ovate or obovate, 3.0–9.2 cm long, (1.2–)1.7–4.2 cm wide, the apex acute or occasionally obtuse, the base acute to cuneate, the margin entire, the upper surface glabrous, lustrous, the lower surface glabrous but occasionally with widely scattered, very small, appressed trichomes when young. Inflorescence terminal, a small cyme, 2–5(–11) cm broad, the inflorescence branches somewhat scorpioid, glabrous. Flowers perfect, sessile; calyx campanulate, 1.2–2.0 mm long, 1.2–2.0 mm wide at the mouth, the 5 sepals free to nearly the base, glabrous, but shortly ciliate on the margins; corolla white, short tubular with spreading lobes, 3.2–4.6 mm long, (4–)5-lobed, the lobes oblong, 1.8–2.5 mm long, 0.7–1.6 mm wide, the tube 1.4–2.1 mm long; stamens (4–)5, the filaments (3.5–)4.0–5.0 mm long, the upper 1.8−3.0 mm free, glabrous, the anthers oblong, 0.9−1.2 mm long; ovary ovoid to globose, 0.6–1.2 mm long, 0.5–1.3 mm broad, the style 2.3–3.2 mm long, bifid or rarely unbranched, the branches to 1.7 mm long when present, the stigmas 2, discoid. Fruits drupaceous, orange to red at maturity, 4–6 mm long, 4–10 mm broad, with four pyrenes.

Distr. This species ranges from Hainan through India and Ceylon.

Vern. Addula, Chirupulichchul.

Note. *Ehretia laevis* is extremely variable in leaf shape and most plants of the Asian mainland have larger, broader leaves. Ceylon plants are much more uniform in having elliptic leaves that are usually acute at the apex and base and are generally glabrous on both surfaces. Plants of the Asian mainland often have broader leaves that range from acute to obtuse to rounded at the apex and base and more often have some trichomes on the lower leaf surface. *Ehretia laevis* var. *platyphylla* is based on one of these broader leaved populations from Hainan. This description applies only to the populations from Ceylon and does not include much of the variation found throughout the rest of the range of this species.

Specimens Examined. ANURADHAPURA DISTRICT: Wilpattu National Park, in nemore sicco arboribus 12–15 metralibus ut summum, *Bernardi 15374* (MO); Wilpattu National Park, 2.5 miles south of Maduru Odai, *Mueller-Dombois 68123001* (US); Wilpattu National Park, between Kokkare Villu and Kurutu Pandi Villu, *Wirawan et al. 875* (MO, US); Yakkalla Colony, *Balakrishnan and Jayasuriya NBK1064* (US). POLONNARUWA DISTRICT: open dry land east of Welikanda, alt. 30 m, *Bremer and Bremer 926* (US). MATALE DISTRICT: Dambulla, in colle rupestre ad templum, *Bernardi 15244* (MO); Dambulla, *Kostermans 26847* (US). AMPARAI DISTRICT: ad templum Magul Maha Vihara, prope Lahugala, alt. 10 m, *Bernardi 15559* (US). MONERAGALA DISTRICT: Udawalawe area, *Balakrishnan and Jayasuriya NBK891* (US); inter Timbolkety et Tanamalwila, alt. 110 m, *Bernardi 15498* (MO). HAMBANTOTA DISTRICT: Ruhuna National Park, Block 2, at Walaskema rock outcrop area, *Mueller-Dombois et al. 67100104* (US); Ruhuna National Park, Block 1, main road, alt. 4–8 m, *Comanor 844* (US); Ruhuna National Park Area, before gate, alt. 5 m, *Comanor 856* (US); *Comanor 857* (US); Ruhuna National Park, Block 1, *Cooray 67102202R* (US); Ruhuna National Park, Block 1 at Patangala, opposite Smithsonian Camp, *Cooray 68102206R* (MO, US); Ruhuna National Park, Patangala, in front of the Smithsonian Carnp, alt. 0 m, *Cooray 69091001R* (US); Yala (Buttawa), alt. sea level, *Cramer 3336* (US); Ruhuna National Park, Block 1, about 2 miles north of Yala Bungalow at N-road, alt. 2 m, *Mueller-Dombois 67093016* (US); Ruhuna National Park, Block 1, opposite Karaugaswala, *Mueller-Dombois and Cooray 67121007* (US); Ruhuna National Park, Block 1200 m west of Karaugaswala Junction, *Mueller-Dombois and Cooray 67121086* (US); Ruhuna National Park, Block 1, mile 11 at Yala Road, *Mueller-Dombois 68101838* (US); Ruhuna National Park, Block 1, Patangala, in front of the Smithsonian Camp, alt. 0 m, *Wirawan 682A* (US); Ruhuna National Park, Block 1, Patangala in front of the Smithsonian Camp, *Wirawan 802* (MO, US).

6. COLDENIA

L., Sp. Pl. 125. 1753. Type species: *C. procumbens* L.

Herbs, prostrate, strigose. Leaves numerous, small, markedly asymmetric, the veins ending in the sinuses. Inflorescences extra-axillary, sometimes glomerate, bracts absent. Flowers perfect, 4-merous; sepals persistent; stamens included; style terminal on ovary, bifid, stigmas 2, poorly differentiated. Fruit dry, breaking into four bony nutlets.

Coldenia now consists of a single species in Asia: the American species have been segregated as *Tiquilia* Persoon (Richardson, 1976). The similarity of *C.*

procumbens to the species of *Tiquilia* is apparently superficial, and *Coldenia* differs in its asymmetrical leaves, the veins of which end in the sinuses, annual habit, and pollen morphology.

Richardson, A.T. 1976. Reinstatement of the genus *Tiquilia* (Boraginaceae: Ehretioideae) and descriptions of four new species. Sida 6: 539–673.

Coldenia procumbens L., Sp. Pl. 125. 1753; Trimen, Handb. Fl. Ceylon 3: 197. 1895.

Herbs, prostrate and spreading; densely strigose, some trichomes 1–2 mm long. Leaves alternate, small, markedly asymmetrical, to 2 cm long and 1 cm wide, the widest part above the middle, coarsely dentate, densely strigose between the secondary veins with the free tips of trichomes forming a crest down the centre of areoles, the trichomes along the margins with conspicuous multicellular bases; subsessile to shortly petiolate, to 5 mm long. Inflorescences extra-axillary, few-flowered scorpioid cymes, or sometimes flowers seemingly solitary. Flowers small, 4-merous; calyx of 4 sepals, the sepals slightly connate basally, the lobes lanceolate, to 1.5 mm long, sparsely strigose, persistent at fruiting; corolla salverform?, white, 1.5–2.0 mm long; stamens 4, attached in mid-throat, the anthers included; ovary 4-locular, pyramid-shaped, styles 2, stigmas not differentiated. Fruit a 4-seeded nutlet, the styles persistent, 3–4 mm at widest point, the seed coat with tubercles, breaking into two 2-seeded halves, and then into four 1-seeded nutlets.

Distr. Common in S.E. Asia.

Ecol. *Coldenia* occupies dry, sun-baked wastelands.

Vern. Chirupaddi.

Note. The distinctive leaves—small, asymmetric, and the organization of the trichomes—would separate this species from all other Boraginaceae in Ceylon or elsewhere. *Coldenia procumbens* is most common in habitat extremes, particularly in very xeric conditions, where it may be the only species to occupy such sites. Its growth habit, prostrate and appressed, and densely strigose leaves may be adaptations against herbivory.

Specimens Examined. ANURADHAPURA DISTRICT: Between Wilpattu National Park and Anuradhapura, alt. low, *C.F. & R.J. Beusekom 1635* (US). POLONNARUWA DISTRICT: Polonnaruwa Sacred Area, on dry tank edge, alt. 61 m, *Ripley 202* (US). MATALE DISTRICT: Ereula Tank, c. 5 miles ESE of Dambulla, *Davidse 7397* (MO); road around Sigiriya Rock, *Nowicke et al. 353* (MO). PUTTALAM DISTRICT: Wilpattu National Park, Atha Villu, *Wirawan et al. 939* (US). TRINCOMALEE DISTRICT: prope Trincomalee, ditione Pereyakulam, *Bernardi 15285* (MO); Kantalai Tank, *Nowicke & Jayasuriya 300* (US). BATTICALOA DISTRICT: Batticaloa, *Kundu & Balakrishnan 205* (US). AMPARAI DISTRICT: Panama-Okanda Road, *Cooray 69073008R* (US); East shore of Senanayake Samudra S. of dam, alt. 36 m, *Mueller-Dombois &*

Comanor 67072613 (US). HAMBANTOTA DISTRICT: Yala National Park on road to Hinwewe Tank, *Nowicke & Jayasuriya 405* (US).

7. TRICHODESMA

R. Br., Prod. 496. 1810. Type species: *T. zeylanicum* (Burm.f.) R. Br.

Borago zeylanica Burm.f.

Herbs, erect, strigose. Leaves alternate or decussate, margins entire, chartaceous. Inflorescences raceme-like or flowers solitary, nodding; bracts mostly absent. Flowers perfect, 5-merous, distinctly pedicellate; calyx of 5 sepals, connate basally, elongating in fruit; corolla salverform-campanulate, the lobes shallow; stamens attached low in the corolla throat, the anthers connivent, the connectives elongated and connivent; ovary 4-lobed, style simple, stigma minute. Fruit of 4 nutlets, the exterior surface smooth, the inner tuberculate.

Trichodesma comprises about 35 species in the tropics and subtropics of Africa, Asia and Australia. The androecium is very unusual—the rather large anthers are connivent and the connectives are elongated, exserted, and also connivent.

Two species are found in Ceylon.

KEY TO THE SPECIES

1 Calyx lobes hastate at the bases, uniting with adjacent ones to appear as spurs; leaf bases auriculate
.... **1. T. indicum**
1 Calyx lobes with rounded bases; leaf bases acute .... **2. T. zeylanicum**

1. Trichodesma indicum (L.) Smith in Rees, Cyclop. 36(1), n.1. 1817; Trimen, Handb. Fl. Ceylon 3: 201. 1895.

Borago indica L., Sp. Pl. 137. 1753.

Herbs, wiry and erect, stems sparsely scabrate. Leaves opposite, spathulate to lanceolate, to 5 cm long and 1 cm wide, acute, entire, the bases auriculate, uniformly scabrate above but only on the veins beneath; sessile. Inflorescences racemose, lax, few-flowered. Flowers with slender pedicels to 1.5 cm long; calyx of five sepals, sagittate with acuminate tips, scabrate and one-ribbed, in fruit appearing as spurs; corolla salverform, the tube c. 5–7 mm long, the lobes c. 5 mm, abruptly acuminate; stamens five, conspicuous, anthers almost sessile, 4.5–5 mm long, densely pubescent on outer surface, the connectives elongated and twisted into a cone, sometimes exserted as much as 6 mm (past the throat); ovary subglobose, 4-celled, style elongate to 7–8 mm, stigma small, papillose. Fruit dry, separating into four nutlets, 5 mm by 3 mm, smooth on the outer surface, roughened on the inner.

Distr. Indian subcontinent, Burma, Philippines, Mauritius.

Ecol. Dry sites, fallow fields.

Vern. Kavil-tumpai.

Specimens Examined. JAFFNA DISTRICT: Elephant Pass, *Simpson 9310* (PDA); Jaffna, Feb 1890, *s. coll.* (PDA). HAMBANTOTA DISTRICT: Buddegiriya, *Cramer 2825* (US); Ruhuna National Park, Sithul Pahuwa, *Fosberg and Sachet 52912* (US, PDA).

2. Trichodesma zeylanicum (Burm.f.) R. Br., Prod. 496. 1810; Trimen, Handb. Fl. Ceylon 3: 202. 1895.

Borago zeylanica Burm. F., Fl. Ind. 41. 1768.

Herbs, coarse and wiry, to 1 m or more; vestiture compound, strigose with stiff, finely barbed trichomes with conspicuous cystolith-like bases, and villous with trichomes much smaller and appressed. Leaves opposite, lanceolate to oblanceolate, to 7(–10) cm long and 1.5(–3.0) cm wide, acute, entire, the bases attenuate, uniformly scabrate above, but only on the veins beneath; petioles indistinct. Inflorescences irregular, nodding, racemose and 5–10-flowered, or solitary and axillary; bracts mostly absent. Flowers pedicellate, the pedicels 3–4(–5) cm long; calyx of 5 prominent leaflike sepals, elongate-deltoid, 3-ribbed, to 1.5 cm long in fruit, scabrate without; corolla salverform-campanulate, the cylindrical part of tube c. 3–3.5 mm long, the widened throat c. 3.0 mm long, lobes shallow except for long acuminate tips; stamens five, attached low in throat, the anthers densely pubescent on outer surface, the connectives greatly elongated and connivent forming a spiral cone, total stamen length 6–6.5 mm long; ovary 4-lobed, style slender and elongated, stigma minute. Fruit of 4 nutlets, 4 mm by 3 mm, exterior surface smooth, the inner ridged and tuberculate.

Distr. India, Mascarene Isls., Malaya and Australia.

Ecol. Wastelands and dry sites.

Specimens Examined. PUTTALAM DISTRICT: Puttalam, *Kundu & Balakrishnan 394* (US); Wilpattu National Park, *Mueller-Dombois et al. 69043031* (MO, US). TRINCOMALEE DISTRICT: Foul Point, *Jayasuriya et al. 669* (US). BADULLA DISTRICT: Pallewela, *Balakrishnan & Jayasuriya NBK808* (US); Dowa, *Cramer 5183* (US). NUWARA ELIYA DISTRICT: Mulhalkelle, *Cramer 4523* (US).

8. CYNOGLOSSUM

L., Sp. Pl. 134. 1753. Type species: *C. officinale* L. (lectotype).

Herbs, mostly perennial, sometimes annual or biennial, slender to very robust, glabrous to coarsely pubescent. Leaves alternate, simple, pinnately-nerved, entire, the basal ones long petiolate. Inflorescences raceme-like or irregular panicles, scorpioid, the bracts mostly absent. Flowers perfect, actinomorphic; calyx of 5 sepals + or − connate basally, enlarging somewhat in fruit; corolla salverform, funnelform or + or − campanulate, blue, violet, or reddish, 5-lobed, the lobes + or − spreading, 5 faucal appendages generally well developed; stamens 5, included or barely exserted at the corolla throat, subsessile or filaments very short, the anthers oblong; ovary 4-lobed, distinctly so, the gynobase disc-like, the style gynobasic, the stigma 1 and + or − capitate. Fruit of 4 nutlets, spreading at maturity, adnate apically to the gynobase, the scar not extending below the middle on the ventral side, the surface covered with stout, glochidiate spines.

. A cosmopolitan genus of about 80 species. The Ceylon collections easily segregate into two taxa that Trimen distinguished primarily by the base of the stem leaves. *Cynoglossum micranthum* Desf. has subsessile leaves, while the variety "decurrens" is a much coarser plant having stem leaves with decurrent/clasping bases to some extent. Wallich (1824) described *C. furcatum* as having "Cauline leaves scattered, acute, slightly stem-clasping sometimes obscurely decurrent on one side." This species is 919 in Wallich's catalogue where he cites two collections, one from "Hindustania" collected in 1825 and the other from "Nepala" collected in 1821. In the microfiche of the Wallich Herbarium, however, there are six specimens of 919 on four herbarium sheets: one sheet with a single plant is labelled 919.1 and represents the 1825 collection, a second sheet with a single plant is labelled 919.2 and represents the 1821 collection, a third sheet with two plants is labelled 919.D, and the fourth sheet has two plants labelled 919.E and 919.F. As best I can determine from the microfiche, 919.1 and 919.2 have decurrent/clasping leaves to some extent. The only difference is that some of the Ceylon collections have longer inflorescences. Whether Wallich had all the above collections when he established *C. furcatum* is unknown. Later Clarke (Fl. Br. Ind. 155. 1885) reinforced this distinction when he described stem leaves for *C. furcatum* as (p. 156) "frequently subdecurrent." Most of the collections that I have referred to *C. furcatum* have leaf bases that are conspicuously decurrent—at least one cm long.

Although I have applied the name *Cynoglossum zeylanicum* to the other taxon found in Ceylon, I do so with reservation since I can find no distinctive characteristics either in the original description or in the actual plants. After examining all the Asian collections at the US and MO herbaria, I find the Ceylon material most similar to those collections identified as *C. zeylanicum*.

KEY TO THE SPECIES

1 Robust, coarse herbs; at least some stem leaves and/or bracts decurrent; nutlets sometimes with marginal spines slightly united basally forming a rim but not a crest .... **1. C. furcatum**
1 Slender herbs; stem leaves petiolate or subsessile, but not decurrent; nutlets with glochidiate spines closest to gynophore united basally, forming a crest .... **2. C. zeylanicum**

1. Cynoglossum furcatum Wall. in Roxb., Fl. Ind. 2: 6. 1824; Alston in Trimen, Handb. Fl. Ceylon 6: 201. 1931.

Cynoglossum micranthum Desf. var. *decurrens* Trimen, Handb. Fl. Ceylon 3: 203. 1895.

Herbs, erect, coarse, robust, to 1.5 m, the stems mostly strigose. Stem leaves alternate, lanceolate, sometimes oblanceolate, to 15 cm long and 5 cm wide, acute, entire, the bases decurrent, sometimes clasping, rarely petiolate; basal leaves appearing to form a rosette, lanceolate, 8–10 cm long, petioles to 5 cm long. Inflorescences axillary or terminal, one-sided, raceme-like, to 40 cm long in fruit, bracts with decurrent bases. Flowers crowded, later becoming remote, sessile or with pedicels 4–5 mm long; calyx of five sepals, ovate, c. 2.5 mm long, slightly connate basally, persistent and slightly enlarging in fruit; corolla campanulate, blue, the tube short, 1.5–2 mm long, the throat constricted by bilobed appendages at the base of each lobe, the lobes rounded, imbricate in bud; stamens inserted below appendages, subsessile, the anthers 0.7–0.8 mm long; ovary strongly 4-lobed at anthesis, the style stout, c. 0.5 mm long, the stigma scarcely differentiated. Fruit of 4 nutlets, 3.5–5 mm long, adherent to the central gynobase, covered with glochidiate spines, those along the margin sometimes united basally forming a ridge.

Distr. Ceylon and Burma.

Ecol. Roadsides and waste areas in central highlands, particularly Horton Plains.

Vern. Bu-katu-henda.

Note. After examining the microfiche of Wallich's Herbarium, I have concluded that *C. furcatum* should not be treated as conspecific with *C. zeylanicum*. There are only two Asian collections at US, both from Burma, that match the above concept of *C. furcatum*.

Specimens Examined. NUWARA ELIYA DISTRICT: Nuwara Eliya, surrounding hills, alt. 2000 m, *C.F. & R.J. van Beusekom 1422* (US); Horton Plains, fork at road to Diyagama west side, alt. 2181 m, *Comanor 447* (US); Horton Plains, Diyagama Road, alt. 2140 m, *Comanor 973* (US); Nuwara Eliya, beside stream on the climb to Single Tree, alt. 2165 m, *Cramer 2932* (US); Horton Plains, in grassy slopes by Ohiya Road, below Farr Inn, alt. 2333 m, *Cramer 2872* (US); Slopes of Mt. Pidurutalagala directly north of Nuwara Eliya, alt. 2250-2485 m, *Davidse 8068* (MO, US); 1 km N.W. of Ambawela Railroad Station, 6 mi. SSE, Nuwara Eliya, alt. 1940 m, *Fosberg 49972* (US); Horton Plains, near "World's End", *Fosberg & Jayasuriya 53248* (US); Horton Plains, forest back of Farr Inn, *Fosberg & Sachet 53273* (US); Horton Plains, about 1½ miles S. of Farr Inn on Ohiya Road, alt. 2400 m, *Gould & Cooray 13857* (US); Horton Plains, trail W. of Farr Inn, so-called "Primate Study Trail", alt. 7200 ft, *Koyama et al. 14090* (US); Horton Plains, alt. 2000 m, *Larsen 29502* (MO); Horton Plains, alt. 2000 m, *Larsen 29624* (MO); Mipalana patana (near Kande Ela Reservoir) c. 3 miles S. of Nuwara Eliya, on trackside in patana, alt. 1850 m, *Hoogland 11524* (US); about 2 miles N. of Horton Plains toward Agrapatana, in forest, alt. 2000 m, *Gould 13579* (US); Kandapola Forest Reserve, common weed in field cleared and planted in *Pinus, Maxwell & Jayasuriya 867* (MO, US); Horton Plains, below Resthouse at Ohiya Road, alt. 2175 m, *Mueller-Dombois & Comanor 67070916* (US); Horton Plains, Rd. 6 approach from Pattipola, *Nowicke & Jayasuriya 218* (MO, US); Horton Plains, road from Horton Plains to Ohiya, *Nowicke & Jayasuriya 273* (MO, US); Kandapola FR, along loop road, *Sohmer et al. 8352* (US); World's End, Horton Plains, alt. 7200 ft, *Sumithraarachchi et al. DBS47* (US); Horton Plains, along path to Greater World's End, *Tirvengadum & Cramer 282* (US); Hakgala, in secondary montane forest top of Hakgala hill, *Waas 833* (US).

2. Cynoglossum zeylanicum Thunb. ex Lehm., Neue Schriften Naturf. Ges. Halle 3(2): 20. 1817; Alston in Trimen, Handb. Fl. Ceylon 6: 201. 1931.

Anchusa zeylanica Vahl ex Hornem., Hort. Hafn. 1: 176. 1813, non Jacq. f. 1812.

Herbs, wiry and suffrutescent, to 2 m tall, scabrate. Leaves alternate, lanceolate to elliptic, to 14 cm long and 5 cm wide, entire, acute, the base obtuse to attenuate, scabrate above, the trichomes with cystolith-like bases, mostly strigose below; subsessile to petioles 15 mm long. Inflorescences mostly terminal, one-sided scorpioid racemes, to 26 cm long in fruit, the bracts with obtuse or attenuate bases, rarely clasping. Flowers with pedicels 2–5 mm long, crowded in bud, becoming remote; calyx of 5 sepals, lanceolate or spathulate, 3–4 mm long, the outer surface scabrate, densely so at the base; corolla campanulate, white with blue centre, the tube short, c. 1–1.3 mm long, the throat constricted by bilobed appendages at the base of each lobe, the lobes ovate, c. 1–1.2 mm wide; stamens 5, inserted below the appendages, subsessile, the anthers c. 0.6 mm long; ovary 4-lobed, the style stout, the stigma weakly capitate. Fruit of 4 nutlets, 3.2–3.8 mm long, covered with glochidiate spines which appear to form a row or crest at the area of attachment to the gynophore.

Distr. According to Alston (1931) this species extends into India and Malaya.
Ecol. Roadsides and waste ground.
Specimens Examined. KANDY DISTRICT: Udawatte, off Hanguranketa, alt. 875 m, *Jayasuriya & Balasubramaniam 461* (US); 8/12 road marker along Peradeniya-Galaha Road, alt. 1900 ft, *Sumithraarachchi DBS87* (US); Mount Pleasant, alt. 760 m, *Wheeler 12907* (US). NUWARA ELIYA DISTRICT: Horton Plains, forest back of Farr Inn, *Fosberg & Sachet 53273* (US); Kandapola Forest Reserve, *Maxwell & Jayasuriya 867* (US); Kandapola FR, along loop road, *Sohmer et al. 8352* (US); in gravelly slope of clearing close to mile 35, Maturata-Mulhalkelle Road, alt. 1000 m, *Cramer 4521* (US). RATNAPURA DISTRICT: Koslanda-Ratnapura Road, *Sumithraarachchi & Waas DBS302* (US). BADULLA DISTRICT: Ella, alt. 3400 ft, *Hepper & de Silva 4783* (US); Namunukula Kanda, *Waas 465* (US); Namunukula Kanda, *Waas 478* (US); Nikapatha, Koslanda–

Beragala road 120/19 mile marker, *Waas 485* (US). LOCATION UNKNOWN: *Fraser 126* (US).

9. HELIOTROPIUM

L., Sp. Pl. 130. 1753. Type species: *H. europaeum* L. (lectotype).

Herbs, sometimes suffrutescent, rarely shrubs; scabrous, the trichomes coarse, stiff, slightly barbed, sometimes with cystolith-like bases, or rarely + or − glabrous. Leaves alternate, sometimes + or − fasciculate, pinnately-veined, small to large. Inflorescences spike-like or raceme-like, scorpioid, solitary, paired or ternate, or the flowers solitary; bracts present or absent. Flowers perfect, 5-merous, sometimes weakly zygomorphic, pedicellate or subsessile; bracteoles mostly absent; calyx of 5 sepals, + or − connate basally, persistent or deciduous, the lobes linear or lanceolate; corolla mostly salverform to funnelform, white, yellow or blue, the limb 5-lobed, the lobes + or − spreading, the throat without appendages; stamens 5, included, borne on the corolla throat, the filaments short or absent, the anthers linear; ovary 4-loculed, lobed or unlobed, a glandular ring at the base, the style apical, the stigmas sessile or not, peltate or conic. Fruit of 4 nutlets, free, or combined in pairs, nutlets 1–2-seeded.

Heliotropium includes 200 species in the warmer regions of both hemispheres. Five species are found in Ceylon.

KEY TO THE SPECIES

1 Plants densely tomentose and woolly in appearance...................... **1. H. supinum**
1 Plants glabrous, or if pubescent, then not conspicuously woolly
  2 Leaves + or − ovate, at least some > 4 cm long and > 3 cm wide...... **2. H. indicum**
  2 Leaves narrowly linear, spatulate, or + or − elliptic, < 3.5 cm long and < 1 cm wide
    3 Inflorescences obscure, < 3 cm long, and < 7-flowered............... **3. H. scabrum**
    3 Inflorescences conspicuous or at least easily recognized, some > 3 cm, and 7–10-flowered
      4 Plants glabrous; leaves spatulate, at least some 3–4 mm wide and > 2 cm long; flowers sessile
        and crowded...**4. H. curassavicum**
      4 Plants scabrate; leaves narrowly linear, 3 mm wide and < 1.5 cm long; older flowers remote,
        not touching .... **5. H. zeylanicum**

1. Heliotropium supinum L., Sp. Pl. 130. 1753; Trimen, Handb. Fl. Ceylon 3: 199. 1895.

Annual, prostrate, less than 30 cm in diam, densely pubescent. Leaves alternate, ovate, small, 10–12 mm long and < 10 mm wide, the orientation of the trichomes giving a plicate appearance; petiole slender, to 5 mm long. Inflorescences spikelike cymes, axillary or terminal, to 6 cm long. Flowers small, crowded and sessile; calyx of five sepals, ovate, densely pubescent without, sparsely pubescent within, 3–3.5 mm long in fruit and enclosing it; corolla salverform, the tube 2 mm long, the lobes c. 1 mm long; stamens five, anthers c. 1 mm long;

ovary subglobose, style and stigma poorly differentiated. Fruit of three–four nutlets, c. 1 mm in diam, the surface tuberculate.

Distr. Warmer regions of the Old World.

Note. *Heliotropium supinum* is known from Ceylon from a single collection.

Specimens Examined. JAFFNA DISTRICT: Delft Island, *Gardner s.n., C.P. 2854* (PDA).

2. Heliotropium indicum L., Sp. Pl. 130. 1753; Trimen, Handb. Fl. Ceylon 3: 200. 1895; I.M. Johnston, J. Arnold Arbor. 32: 111. 1951.

Herbs, coarse, sometimes suffrutescent, to 1.1 m tall; stems and young leaves pubescent, the trichomes either short, < 0.2 mm, or long, > 0.8 mm. Leaves alternate, ovate or deltoid, to 12 cm long and 7 cm wide, sparsely pubescent on both surfaces, acute, the margins undulate, the bases rounded to obtuse and long-attenuate; petioles winged, indistinct, sometimes to 8 cm long. Inflorescences scorpioid cymes, unilateral, to 12 cm long, mostly terminal; bracts absent. Flowers small, sessile; calyx of 5 sepals, subulate, c. 1.5–2.0 mm long, sparsely and coarse-ly pubescent; corolla salverform, blue, violet, or rarely white, the tube 3–4 mm long, slightly constricted distally, 5-lobed, each lobe c. 1 mm long; stamens subsessile, mounted in middle of the tube; ovary weakly 4-lobed, style stout, the stigma conical. Fruit deeply 4-lobed, the lobes horizontally divergent, the nutlets in pairs, angulate, each nutlet 2–3 mm long.

Distr. Widely distributed in the warmer regions of the world.

Ecol. Wasteland and roadsides.

Vern. Et-setiya, Et-honda, Dimi-biya, Tedkodukku.

Specimens Examined. JAFFNA DISTRICT: *Dassanayake 409* (US). ANURADHAPURA DISTRICT: W. shore of Nuwara Wewa behind the Nuwara Wewa Rest House, *Sohmer 8094* (MO, US). POLONNARUWA DIS-TRICT: Polonnaruwa Sacred Area Section 4A, alt. 61 m, *Ripley 304* (US). KURUNEGALA DISTRICT: Ganewatte, alt. + or − 120 m, *Cramer 4099* (US). MATALE DISTRICT: near Wewala Tank, *Sumithraarachchi DBS 480* (MO). TRINCOMALEE DISTRICT: Mahaweli Ferry, Mahaweli Ganga about 20 km from mouth, *Fosberg 56397* (MO, US); Kantalai Tank, *Nowicke & Jayasuriya 299* (MO, US). AMPARAI DISTRICT: Crossing of Heda Oya, 4 miles S. of Pottuvil, *Fosberg & Sachet 53020* (US). COLOMBO DISTRICT: Negombo, *Sumithraarachchi et al. DBS 83* (US). GALLE DISTRICT: Bona Vista, *Balakrishnan NBK 1154* (US); Wakwelle, beside Ukwatte Road, alt. sea level, *Cramer 2656* (MO, US). HAMBANTOTA DISTRICT: Ruhuna National Park, Block 1, at N/S boundary NE. of Katagamuwa Tank, *Wirawan 68102015* (US); Ruhuna National Park, Block 1, Kosgastotupola, *Cooray 69120715R* (US).

3. Heliotropium scabrum Retz., Obs. Bot. 2: 8. 1781; Trimen, Handb. Fl. Ceylon 3: 200. 1895.

Herbs, spreading or procumbent; scabrate, the trichomes stiff, sometimes with small cystolith-like bases. Leaves alternate, fasciculate, or rarely subopposite, frequently crowded, lanceolate to elliptic, to 12 mm long and 5 mm wide, the margins entire; subsessile. Inflorescences obscure, few-flowered, short. Flowers obscure, small; calyx of 5 sepals, elongate-deltoid, c. 2.0 − 2.5 mm long; corolla salverform, white?, the tube c. 1.2–1.4 mm long, pubescent within at the throat, the lobes shallow; stamens subsessile, the anthers deltoid and connivent, c. 0.6 mm long; ovary 4-lobed, style short, c. 0.4 mm long, the stigma conical, c. 0.5 mm long. Fruit 4-lobed, eventually separating into four nutlets, each nutlet slightly pubescent near the style.

Distr. Southern India and Ceylon.

Ecol. Dry sites, wasteland and roadsides.

Note. *Heliotropium scabrum* is easily recognized by the scabrate pubescence and obscure inflorescences.

Specimens Examined. ANURADHAPURA DISTRICT: North of Akirikanda m. p. 98, *Fosberg & Balakrishnan 53475* (US); Mihintale Hill, alt. 260 m, *Balakrishnan & Jayasuriya NBK 1121* (US). POLONNARUWA DISTRICT: Attampitiya Hill, Polonnaruwa, *Kundu & Balakrishnan 221* (US). PUTTALAM DISTRICT: 2.5 miles S. Maduru Odai, 5 miles S. of Marai Villu, plot W18, *Fosberg et al. 50845* (US). TRINCOMALEE DISTRICT: Eight mile-post, N. of Trincomalee, just S. of Nilaveli, *Fosberg & Jayasinghe 57073* (US). AMPARAI DISTRICT: East side of Ulpasse Wewa, 2 mi. N. of Panama, *Fosberg & Sachet 5296* (US). HAMBANTOTA DISTRICT: Ruhuna National Park, Buttawa Plain in plot R35, *Cooray 69120202R* (US); Ruhuna National Park, Block 1, Karaugaswala, *Wirawan 809* (US); Ruhuna National Park, Block 1 near Bambawa Junction, *Cooray 68113011R* (US); Ruhuna National Park, Block 1 at Karaugaswala in Plot R28, *Mueller-Dombois & Cooray 69011702R* (MO, US); Kirinda, alt. 5 m, *Fosberg 50381* (US); Ruhuna National Park, Block 1, *Cooray & Balakrishnan 69011014R* (US); Ruhuna National Park, Block 1, 200 m N. of Karaugaswala Junction, plot R28, *Mueller-Dombois & Cooray 67121068* (US); Ruhuna National Park, Block 1 near Bambawa Junction, *Cooray 68113011R* (US); Ruhuna National Park, Karaugaswala, plot R28, *Cooray 69121204R* (US); Ruhuna National Park at Uraniya in plot R35, *Mueller-Dombois 68090101* (US); Ruhuna National Park, Block 1 in plot R35 in Buttawa Plain, *Cooray 69120202R* (US); Ruhuna National Park, Block 1, Uraniya in plot R36, *Mueller-Dombois 68101914* (US).

4. Heliotropium curassavicum L., Sp. Pl. 130. 1753; Alston in Trimen, Handb. Fl. Ceylon 6: 200. 1931.

Herbs, prostrate and spreading, glabrous and succulent, the younger parts sometimes glaucous. Leaves alternate or somewhat fasciculate, oblanceolate,

lanceolate, or spathulate, to 3.5 cm long and 0.6–0.7 cm wide, obtuse, the margins entire, the bases attenuate; petioles obscure. Inflorescences spike-like, scorpioid when immature, unilateral, mostly terminal, c. 2–4.5 cm long; bracts absent. Flowers small, sessile, crowded; calyx of 5 sepals, slightly connate basally, elongate-deltoid, c. 1.2–1.3 mm long; corolla salverform, white, the tube somewhat saccate, c. 1.2 mm long, the lobes c. 0.8 mm long; stamens 5, sessile, the anthers 0.6 mm long; ovary 4-lobed, the disc obscure, the style absent, the stigma conical, annular pubescent at the widened base. Fruit 4-lobed, 2.3 × 3.8 mm, the 4 nutlets frequently remaining together at maturity, the nutlets glabrous, wedge-shaped, c. 1.6–1.7 mm long.

Distr. Cosmopolitan in distribution.

Ecol. Dry sites, roadsides, wastelands.

Specimens Examined. JAFFNA DISTRICT: Between Pallavarayan Kaddu & Vellankulam, alt. 10 ft, *Grierson 1127* (US); Jaffna lagoon, *Waas 246* (US). MANNAR DISTRICT: Mantai-Pooneryn Road, 11/1 Road marker, *D.B. Sumithraarachchi & D. Sumithraarachchi DBS 779* (US); between Mannar & Thirukatheeswaram, *Kundu & Balakrishnan 615* (US). PUTTALAM DISTRICT: Puttalam, by lagoon, alt. sea level, *Jayasuriya 1441* (US); Puttalam, turf just back of shore of lagoon, *Fosberg & Jayasuriya 52738* (US); Wilpattu National Park, Pukulam Beach, *Cooray 70040404R* (US).

5. Heliotropium zeylanicum (Burm.f.) Lam., Enc. 3: 94. 1789; Alston in Trimen, Handb. Fl. Ceylon 6: 200. 1931.

Heliotropium curassavicum L. var. *zeylanicum* Burm.f., Fl. Ind. 41. t. 16, f. 2. 1768.

Herbs, semi-erect, spreading; scabrate, the trichomes coarse and stiff, cystolith-like bases if present with only one or two rings of cells. Leaves mostly alternate to subopposite, remote or at least not crowded, lanceolate, to 2 cm long and 0.4–0.5 cm wide, the margins entire; petioles indistinct or absent. Inflorescences scorpioid cymes, unilateral, mostly terminal, to 3 cm long; bracts present. Flowers at maturity with pedicels to 2.5 mm long, remote; bracts usually present, lanceolate, to 4 mm long; calyx of 5 sepals, elongate-deltoid, to 1.6 mm long, in fruit expanding laterally near the base and curving around nutlet; corolla salver-form, white, the tube 1.5 mm long, constricted below and above the attachment of the stamens; stamens subsessile, the anthers deltoid, connivent, 0.6–0.7 mm long; ovary 4-lobed, the style c. 0.2 mm long, the stigma conical, c. 0.2 mm long. Fruit 4-lobed, the lobes hemispherical, separating into 4 nutlets at maturity, each nutlet pubescent near the attachment to the style.

Distr. Southern India and Ceylon.

Ecol. Dryer sites, wastelands and roadsides.

Note. *Heliotropium zeylanicum* occurs sympatrically with *H. scabrum* and both have very similar stamens and nutlets. However, they can be distinguished by the inflorescences and habit. In *H. scabrum* the inflorescences are obscure, very short, with crowded flowers whereas in *H. zeylanicum* they are, if not conspicuous, at least easily recognized, two to four cm long, with older flowers remote. The leaves of *H. scabrum* are crowded and almost always have sand grains attached reflecting the prostrate habit. In *H. zeylanicum* the leaves are remote, and the absence of sand reflects the more erect habit.

Specimens Examined. HAMBANTOTA DISTRICT: Ruhuna National Park, Karaugaswala, *Cooray 69121204R* (US), *Wirawan 809* (US); 200 m north of Karaugaswala Junction, *Mueller-Dombois & Cooray 671210668* (US).

The US collection of *Cooray 6811301* from Ruhuna National Park, near Bambawa Junction, is mixed: *Heliotropium scabrum* and *H. zeylanicum*.

FABACEAE (LEGUMINOSAE)

Subfamily CAESALPINIOIDEAE

(by Velva E. Rudd*)

(R. Br.) DC., Prod. 2: 473. 1825; DC., Leg. Mem. 453. 1826 or 1827, as suborder Caesalpinieae; Cowan in Polhill & Raven, Adv. Leg. Syst. 57. 1981.

Caesalpinieae R. Br. in Flinders, Voy. Terra Austr. 2, append. 3: 551. 1814.
   Type: *Caesalpinia* L.

Trees or shrubs, sometimes scandent, sometimes herbaceous, armed or unarmed; leaves usually alternate, pinnate, sometimes bipinnate, rarely simple or unifoliolate; stipules in pairs, usually caducous; stipels present or absent; inflorescences spicate, racemose, or paniculate, rarely capitate, terminal or axillary, rarely leaf-opposed, sometimes cauliflorate; flowers usually zygomorphic, sometimes actinomorphic, 5-merous, sometimes large and showy, sometimes small; calyx with sepals usually imbricate, rarely valvate, free or somewhat connate; corolla with petals usually 5, sometimes fewer, or absent, imbricate in bud with the uppermost (adaxial) petal enveloped by the others, the bases free or sometimes connate; stamens 10, sometimes fewer, sometimes many, the filaments free or somewhat united; anthers various, sometimes with apical pores, usually dehiscing lengthwise; fruit various, dehiscent or indehiscent; seeds bilaterally symmetrical, sometimes with pleurograms and areoles, sometimes fracture lines, rarely winged, occasionally arillate, the hilum basal, small, the seed coat lignified to coriaceous, rarely thin, the embryo straight, endosperm usually absent but present in some genera, the cotyledons fleshy or foliaceous.

KEY TO THE NATIVE OR NATURALIZED GENERA

1 Leaves simple, usually bilobed, sometimes 2-foliolate .... **1. Bauhinia**
1 Leaves simple-pinnate or bipinnate
 2 Plants commonly armed; leaves bipinnate
  3 Rachis of pinnae flattened, narrowly marginate; sepals valvate in bud .... **2. Parkinsonia**
  3 Rachis of pinnae not flattened or marginate; sepals imbricate in bud ...... **3. Caesalpinia**
 2 Plants unarmed; leaves simple-pinnate or bipinnate

*Smithsonian Institution, Washington, D.C., U.S.A.

4 Leaves bipinnate
  5 Fruit winged along both margins; flowers yellow; stigma peltate .... **4. Peltophorum**
  5 Fruit not winged or with a wing along only one margin; flowers yellow to red; stigma truncate
    or oblique, not peltate .... **3. Caesalpinia**
4 Leaves simple-pinnate
  6 Petals present
    7 Calyx tube short or lacking
      8 Segments of calyx 5; anthers mostly basifixed, opening by terminal or basal pores; flowers
        yellow or, less commonly, pink or white .... **5. Cassia**
      8 Segments of calyx 4, rarely 5; anthers dorsifixed, opening by longitudinal slits; flowers white
        or pinkish .... **6. Cynometra**
    7 Calyx turbinate, elongated, 3–5 mm long
      9 Perfect stamens 5; petals 5; leaflets 3–5 pairs .... **7. Humboldtia**
      9 Perfect stamens 3; petals 3; leaflets 10–20 pairs .... **8. Tamarindus**
  6 Petals lacking
    10 Stamens 2(–3); anthers basifixed or sub-basifixed; leaves commonly 5-foliolate; leaflets alter-
      nate or subopposite .... **9. Dialium**
    10 Stamens 3–10; anthers dorsifixed
      11 Leaves 2–6-foliolate; leaflets alternate; fruit pubescent .... **10. Crudia**
      11 Leaves 8–12-foliolate; leaflets opposite; fruit glabrous .... **11. Saraca**

KEY TO THE GENERA CASUALLY INTRODUCED BUT NOT NATURALIZED

1 Plants armed, with thorns 5–15 mm long; fruit dehiscing longitudinally in centre of valves, not
  along the margins .... **19. Haematoxylum**
1 Plants unarmed; fruit indehiscent or dehiscing along one or both margins
  2 Leaves bipinnate
    3 Flowers relatively smaller; petals 1 cm long or less; leaflets predominantly ovate-oblong to ovate-
      elliptic; fruit about 16 cm long or less
      4 Petals yellowish, 2–3 mm long; leaves with 2–4 pairs of pinnae, each pinna 7–14-foliolate;
        leaflets alternate .... **18. Erythrophleum**
      4 Petals red, 6–10 mm long; leaves with 3–5 pairs of pinnae, each pinna 8–18-foliolate; leaflets
        opposite .... **12. Acrocarpus**
    3 Flowers larger; petals about 1.5–7 cm long; leaflets small and numerous, elliptic to linear-oblong;
      fruit linear-oblong, to 80 cm long .... **17. Delonix**
  2 Leaves simple-pinnate
    5 Flowers large and showy; petals about 3.5–21 cm long
      6 Bracteoles valvate, enclosing the flowers in bud; petals red, or mostly so, about 3.5–7.5 cm long
        7 Inflorescences capitate; petals red, about 3.5–6.5 cm long .... **15. Brownea**
        7 Inflorescences racemose; petals red with yellow markings, about 4.5–7.5 cm long
          .... **13. Amherstia**
      6 Bracteoles imbricate; almost valvate, not enclosing the flowers in bud; petals white or cream,
        to 21 cm long .... **14. Baikiaea**
    5 Flowers smaller; petals to about 2 cm long, or absent
      8 Petals present
        9 Leaflets 1 pair, pellucid-punctate; petals 5, white, sometimes purplish; fruit thick ......
          .... **20. Hymenaea**
        9 Leaflets 2–4 (–6) pairs, not pellucid-punctate; petals 3, purplish, and 2 small, rudimentary;
          fruit compressed .... **21. Lysidice**
      8 Petals absent; leaflets often pellucid-punctate .... **16. Copaifera**

Note. In the following text the specimens cited as in Paris, Biblio. Inst. France, are at the Bibliothéque de l'institut de France, Paris, in the herbarium of Paul Hermann, basis of the Thesaurus Zeylanicus of J. Burman (see A. Lourteig in Taxon 15: 23–33. 1966).

1. BAUHINIA

L., Sp. Pl. 374. 1753; L., Gen. Pl. ed. 5, 177. 1754. Lectotype species: *B. divaricata* L.

Phanera Lour., Fl. Cochinch. 37. 1790. Type species: *P. coccinea* Lour.
Piliostigma Hochst., Flora 29: 598. 1846. Type species: *P. reticulata* (DC.)
    Hochst. based on *Bauhinia reticulata* DC.
Lasiobema (Korth.) Miq., Fl. Ind. Bat. 1 (1): 71. 1855. Type species: *L. anguinum*
    (Roxb.) Korth. ex Miq. = *L. scandens* (L.) de Wit.
Alvesia Welw., Apont. 47: 587. 1858. Type species: *A. bauhinioides* Welw.

Additional generic synonyms are given by de Wit, Reinwardtia 3: 390. 1956, Hutchinson, Gen. Fl. Pl. 1: 242. 1964, and by Wunderlin in Polhill and Raven, Adv. Leg. Syst. 114–116. 1981.

Trees or shrubs, sometimes scandent; stems armed or unarmed, sometimes with tendrils; leaves alternate, simple, entire or bilobed, palmate with 3-many veins, sometimes 2-foliolate; stipules usually small, caducous; flowers in terminal or, rarely, in axillary racemes, panicles, or corymbs, or, sometimes solitary; bracts and bracteoles linear; flowers somewhat zygomorphic, large or small; calyx turbinate or elongate, 5-lobed or dentate, or spathaceous; petals 5, unequal or subequal, minute to large, white to pink, red, or purple; perfect stamens 10 or less with some reduced to staminodes; filaments free or shortly connate; anthers dorsifixed, ovate to linear, dehiscing longitudinally; ovary stipitate or subsessile, 2-many-ovulate, the stipe sometimes adnate to the calyx tube; style filiform or short; stigma small to peltate, terminal or oblique; fruit 2-valved, dehiscent or indehiscent, oblong to linear or, rarely, falcate, membranaceous to coriaceous or subligneous, sometimes septate within; seeds orbicular or ovate, compressed, with endosperm, the cotyledons flat, carnose; radicle straight or somewhat oblique; hilum crescentic.

A genus of about 200–500 or more species, depending on circumscription.

KEY TO THE SPECIES

1 Plants erect, shrubs or small trees; flowers 1.5–6.5 cm long; calyx spathaceous; fertile stamens
   10; leaves bilobed for ⅓ to ½ of their length
2 Flowers large; petals 3.5–6.5 cm long
  3 Petals white; lobes of leaf usually acute or acuminate, rarely rounded; flower buds narrowly
     ellipsoid, acuminate .... **1. B. acuminata**

3 Petals yellow, 4–7 cm long, one or more often with a dark spot near the base; lobes of leaf rounded; flower buds ovoid or fusiform .... **2. B. tomentosa**
2 Flowers smaller; petals 1.5–2 cm long .... **3. B. racemosa**
1 Plants scandent; flowers less than 1 cm long; calyx campanulate; fertile stamens 3; staminodes 2; leaves variable, shallowly bilobed to entire or the 2 lobes nearly free .... **4. B. scandens**

1. Bauhinia acuminata L., Sp. Pl. 375. 1753; Burm. f., Fl. Ind. 94. 1768; Wight & Arn., Prod. 295. 1834; Baker in Hook. f., Fl. Br. Ind. 2: 276. 1878; Trimen, Handb. Fl. Ceylon 2: 116. 1894; Brandis, Indian Trees 256. 1906; de Wit, Reinwardtia 3: 393. 1956; Backer & Bakh. f., Fl. Java 1: 532. 1963; Verdcourt, Man. New Guinea Leg. 114. 1979. Lectotype: Ceylon, *P. Hermann,* Herm. Herb. 1: 42, no. 148 (BM).

Shrub, to about 3 m tall; young stems puberulent; stipules linear, puberulent, 1–2 cm long, caducous; leaves subcoriaceous, ovate to subrotund, (3–)7–11(–20) cm long, (3–)6–10(–20) cm wide, bilobed for about 1/3 the length, lobes acute to acuminate or, rarely, rounded, the base rounded to subcordate, 7–11-nerved, glabrous above, puberulent, glabrescent, somewhat glaucous beneath; bracts and bracteoles linear; inflorescences lateral, few-flowered, racemose or cymose; flowers about 3.5–6 cm long; buds narrowly ellipsoid with 5 apical teeth about 3 mm long; calyx spathaceous, puberulent, about 3 cm long; petals white, elliptic, obtuse, 3.5–5(–6) cm long, (1.5–)2–3 cm wide; stamens 10, perfect, the filaments connate at the base, otherwise free, 1.5–2.5 cm long, pubescent at the base; anthers 4–7 mm long with puberulent connective; ovary stipitate, subglabrous or lightly puberulent; style glabrous; stigma peltate; fruit linear-oblong, somewhat septate within, 3–11-seeded, 7–15 cm long, 1–2 cm wide with sharp-rimmed, raised margins; seeds about 8–10 mm long, 6 mm wide, compressed, brown.

D i s t r. Believed to be native in Southeast Asia; introduced elsewhere in the Old World tropics, including Ceylon.

V e r n. Kobo-leela (S).

S p e c i m e n s   E x a m i n e d. WITHOUT EXACT LOCALITY: *Macrae 337* (BM, K); *Walker s.n.* (K). PUTTALAM DISTRICT: Chilaw, cultivated, *Simpson 8207* (BM). KEGALLE DISTRICT: About 2 mi. S. of Kitulgala, *Maxwell 941* (CAS, NY, SFV, US).

Note. According to de Wit (l.c. p. 397), *B. candida* Ait. has often been treated as a synonym of *B. acuminata,* but he prefers to consider it a synonym of *B. variegata* L. although, apparently, Aiton's type is not extant. Macmillan (Trop. Pl. & Gard., ed. 5, 103. 1962) cites *B. candida* and *B. variegata* but not *B. acuminata,* suggesting that he may have considered *B. candida* a synonym of *B. acuminata.*

De Wit also well explains his typification and the exclusion of Linnaeus's erroneous citation of American material under *B. acuminata.*

2. Bauhinia tomentosa L., Sp. Pl. 375. 1753; Burm. f., Fl. Ind. 94. 1768; Moon, Cat. 33. 1824; Wight & Arn., Prod. 295. 1834; Thw., Enum. Pl. Zeyl. 98. 1859; Baker in Hook. f., Fl. Br. Ind. 2: 275. 1878; Trimen, Handb. Fl. Ceylon 2: 116. 1894; Brandis, Indian Trees 256. 1906; de Wit, Reinwardtia 3: 409. 1956; Roti-Michelozzi, Webbia 13: 153. 1957; Worthington, Ceylon Trees 199. 1959; Macmillan, Trop. Pl. & Gard., ed. 5, 196. 1962; Backer & Bakh. f., Fl. Java 1: 532. 1963; Brenan in Fl. Trop. East Africa, Caesalp. 209. 1967; Ali, Fl. W. Pakistan, no. 54, Caesalp. 8. 1973; Verdcourt, Man. New Guinea Leg. 118. 1979. Lectotype: Burman, Thes. Zeyl. t. 18. 1837. Syntype: *P. Hermann,* Ceylon, Herm. Herb. 1: 42, no. 147 (BM).

Alvesia bauhinioides Welw., Apont. 47: 587. 1859 ('1858').
Alvesia tomentosa (L.) Britt. & Rose, No. Am. Fl. 23: 208. 1930.

Shrub or small tree, to about 8 m tall; young stems tomentulose, glabrescent; stipules linear, pubescent; leaves conduplicate, about (1-)2-10 cm long, 1.5-7(-11) cm wide, bilobed for about ⅓ to ½ of their length, lobes rounded at the apex, cordate at the base, glabrous above, puberulent beneath, sometimes glabrescent, 5- or 7-nerved; inflorescences terminal or axillary, racemose, commonly 1-3-flowered; bracts and bracteoles linear, subulate, puberulent, caducous; flowers 4-6.5 cm long; calyx spathaceous, about 2 cm long, puberulent, bifid at the apex; petals 4-7 cm long, 4-5 cm wide, glabrous, bright or pale yellow, commonly with 1 or more petals bearing near the base a maroon or purplish spot about 1.5 cm long and 1 cm wide; stamens 10, perfect, the filaments free, about 2-2.5 cm long, puberulent at the base; anthers linear, 3-4 mm long; ovary tomentulose, the stipe connate with the calyx; style 2-3 cm long, glabrous above, puberulent toward the base; stigma peltate, asymmetrical; fruit dehiscent, tomentulose, septate within, 6.5-15 cm long, 1-3, commonly about 1.5-2 cm wide, many-seeded; seeds dark brown or black, sublenticular, 7-10 mm long, 5-6 mm wide, asymmetrical at the hilar end.

Distr. Ceylon, India, tropical Africa; introduced elsewhere as an ornamental.
Vern. Kaha-petan, petan (S); tiruvathi, tiruvatti (T).
Specimens Examined. WITHOUT LOCALITY: *Stadhouder herb. p. 16* (P), *p. 23* (P); *Macrae 338* (BM); *ex herb. Pallas 70* (BM). MANNAR DISTRICT: Mannar, *Petch s.n.* (A); between Pooneryn and Mannar, *Rudd 3303* (K, NY, PDA, US). ANURADHAPURA DISTRICT: Anuradhapura, *Maxwell & Jayasuriya 794* (NY, SFV, US); near Tirappane, *Koyama & Koyama 13940* (K, NY, US). POLONNARUWA DISTRICT: Between Habarane and Kantalai, mi. marker 133, *Rudd 3127* (K, PDA, US); between Polonnaruwa and Habarane, *Ripley 370* (US). PUTTALAM DISTRICT: Wilpattu Natl. Park, Maduru Odai, Mannar-Puttalam Rd., *Wirawan et al. 922* (US); Anamaduwa, *Amaratunga 108* (PDA). KURUNEGALA DISTRICT: Wanduragala, quarry near Kurunegala, *Rudd 3328* (PDA, US). MATALE DISTRICT: Nalanda, *de Silva 15* (A); near

Matale, *Simpson 8059* (BM); Dambulla, *Cramer 3532* (US), *Kostermans 23539* (US), *Amaratunga 1065* (PDA), *Bernardi 15236* (US). KANDY DISTRICT: Between Hunnasgiriya and Madugoda, 1000 m elev., *Kostermans 25423* (K); Kandy, Hillcrest, *Worthington 5496* (K); Peradeniya, *de Silva 218* (NY); Gannoruwa, *Wirawan s.n.* (US). TRINCOMALEE DISTRICT: Kuchchaveli ferry crossing, *Comanor 796* (GH, PDA, US), *797* (K, NY, PDA, US); Kantalai, 6.5 mi. NE., *Fosberg & P. Jayasinghe 57139* (CAS, SFV, US). BADULLA DISTRICT: Mahiyangana, *Stone 11152* (US); Uma Oya ('Oma Oya'), *Thwaites C.P. 1498* (PDA); Kuruminiya Kandura, Uma Oya, *de Silva 266* (PDA). MONERAGALA DISTRICT: Between Wellawaya and Moneragala, mi. 142, *Rudd & Balakrishnan 3211* (PDA, US); near Kadu Oya, *Grierson 1146* (US). HAMBANTOTA DISTRICT: Ruhuna Natl. Park, 200 m W. of Karaugaswala Junction, *Mueller-Dombois & Cooray 67212085* (PDA, US); near Patanagala, *Wirawan 790* (US); Patanagala Rocks, *Fosberg 50131* (US); on road to Katagamuwa Tank, 1 mi. from Yala Bungalow, *Mueller-Dombois et al. 67093001* (PDA, US); near Katagamuwa Tank, *Wirawan 790A* (SFV, US); Hambantota-Tissamaharama Rd., *Comanor 837* (K, PDA, US); Bambowa, *Cooray 69111718 R* (K, NY, UC, US); Yala, *Worthington 7104* (K); Tissamaharama, *Soejarto & Balasubramaniam 4962* (US).

Note. The citation of a neotype by de Wit was superfluous in view of the existence of authentic syntypes.

As has been noted by various authors, this species is somewhat polymorphic. De Wit has indicated formas to distinguish specimens with spotted petals from those that are concolorous.

3. Bauhinia racemosa Lam., Enc. 1: 390. 1785, non Vahl 1794; Wight & Arn., Prod. 295. 1834; Hook., Ic. Pl. 141. 1837; Beddome, Fl. Sylv. t. 182. 1872; Baker in Hook. f., Fl. Br. Ind. 2: 276. 1878; Trimen, Handb. Fl. Ceylon 2: 116. 1894; Brandis, Indian Trees 256. 1906; de Wit, Reinwardtia 3: 537. 1956; Worthington, Ceylon Trees 198. 1959; Macmillan, Trop. Pl. & Gard., ed. 5, 408. 1962; Ali, Fl. W. Pakistan, no. 54, Caesalp. 4. 1973. Type: "Indies orientales", *Sonnerat s.n.* (P).

Bauhinia parviflora Vahl, Symb. Bot. 3: 55. 1794; Moon, Cat. 33. 1824. *Piliostigma racemosum* (Lam.) Benth. in Miq., Pl. Jungh. 262. 1852; Thw., Enum. Pl. Zeyl. 98. 1859.

Small tree, to about 7 m tall; young stems puberulent, glabrescent; stipules foliaceous, caducous; leaves conduplicate, about 2–4 (–5) cm long, 2–5 (–7.5) cm wide, bilobed from apex to about ⅓ of way down, the lobes rounded at the apex, the base somewhat cordate, 7- or 9-nerved, glabrous on both surfaces, or moderately appressed-pubescent beneath, moderately reticulate beneath; inflorescences racemose, terminal on new growth, later leaf-opposed, pseudo-axillary; bracts and bracteoles minute, acicular, caducous; flowers about

1.5–2 cm long; calyx spathaceous, puberulent, reflexed, 2- or 3-toothed; petals yellowish-white or greenish-white, linear, about 1.2–1.5 mm long; 1.5 mm wide, pubescent on the outer face; stamens 10, fertile, slightly longer than the petals; filaments villous, essentially separate to the base; anthers about 5 mm long; ovary glabrous; style lacking; stigma small, sessile; fruit indehiscent, glabrous, about 12–25 cm long, 2 cm wide, many-seeded; seeds 7–8 mm long, oblong, compressed, black.

Distr. Ceylon, India, Pakistan, Southeast Asia, and China.

Uses. Strong fibre from the inner bark is used in rope-making; gum and leaves are used medicinally.

Vern. Maila, Mayila (S); Atti (T).

Specimens Examined. WITHOUT EXACT LOCALITY: *Macrae 312* (BM), *382* (BM); *Col. Walker 148* (K); "Damboul, Jaffna, & Gardner, Lower Badulla", *Thwaites C.P. 1497* (PDA); without locality, *s. coll. s.n.* (BM, K, US?). JAFFNA DISTRICT: Talaiyadi, S. of Jaffna, *Simpson 8007* (BM). MANNAR DISTRICT: Near Mannar, *Kostermans 24874* (GH, K) Madhu forest, road to Mannar, *Kostermans 24914* (GH, K). VAVUNIYA DISTRICT: Along road to Mannar about 3 mi. W. of Cheddikulam, near mi. marker 113, 40 m elev., *Davidse & Sumithraarachchi 9181* (SFV, US). ANURADHAPURA DISTRICT: Horowpothana, tank near Rest House, *Sumithraarachchi & Sumithraarachchi 837* (CAS, US); Mihintale, *Kundu & Balakrishnan 342* (US). POLONNARUWA DISTRICT: Polonnaruwa, Sept. 1885, *Trimen? s.n.* (PDA), *Kostermans 24287 A* (K, US); Yard of Development Circuit Bangalow, *Ripley 325* (US), *422* (K, NY, US); Uradiwetti Kandakaduwa, *Waas 624* (SFV, US); Manampitiya, *Waas 2098* (US), *Worthington 5855* (K); about 3 mi. N. of Madirigiriya, *Davidse 9497* (SFV). PUTTALAM DISTRICT: Wilpattu Natl. Park, crossing of the Modaragam Aru below Marichchukkaddi, *Davidse & Sumithraarachchi 8236* (SFV, US); Manikepola Uttu Park Bungalow, *Mueller-Dombois 67051803* (US); at W. end of Mail Villu, *Mueller-Dombois 69043029* (GH, K, NY, UC, US); between Kokkare Villu and Kurutu Pandi Villu, *Wirawan et al. 876* (GH, K, SFV, UC, US). KURUNEGALA DISTRICT: Dekanduwela, 3 mi. W. of Rambe, *Maxwell & Jayasuriya 782* (NY, SFV, US); Wenduru Wewa, *Cramer 3055* (US). MATALE DISTRICT: Dambulla, *Simpson 8090* (BM); Etabendi Wewa (Dambulla 39 mi.), 700 m elev., *Worthington 4800* (BM); between Dambulla and Galewela, *Rudd 3313* (K, NY, PDA, US). NUWARA ELIYA DISTRICT: Padinawela, c. 1300 m elev., *Cramer 4365* (US). TRINCOMALEE DISTRICT: Trincomalee, *Waas 1242* (US); Trincomalee-Habarana Rd., near mi. marker 152, culvert 4, *Sohmer 8172* (GH). BATTICALOA DISTRICT: Batticaloa, *Petch s.n.* (A). AMPARAI DISTRICT: Arugam Bay, sea level, *Maxwell & Jayasuriya 755* (NY, SFV, US). BADULLA DISTRICT: Uma Oya ("Ooma Oya"), June 1881, *Trimen?* (PDA). MONERAGALA DISTRICT: Bibile, Uraniya Rd., *Worthington 4077* (K). HAMBANTOTA DISTRICT: Bundala, Dec. 1882, *Trimen?* (PDA); Ruhuna

Natl. Park, Buttawa Bungalow, *Comanor 362* (PDA, US), *1157* (K, US); Yala, *Cooray 68052908 R* (K, NY, PDA, UC, US); beach E. of Buttawa Modera, *Fosberg 50312* (PDA, US); Palatupana, *Fosberg 50379* (K, US).

4. Bauhinia scandens L., Sp. Pl. 374. 1753, non Roxb. 1832, nec Blanco 1837; Burm. f., Fl. Ind. 94. 1768. Type: Malabar, Rheede, Hort. Mal. 8: 57, t. 29. 1688.

Bauhinia anguina Roxb., Fl. Ind. ed. 2, 2: 328. 1832; Baker in Hook. f., Fl. Br. Ind. 2: 284. 1878; Trimen, Handb. Fl. Ceylon 2: 117. 1894; Macmillan, Trop. Pl. & Gard. ed. 5, 129. 1962. Type: Malabar, Rheede, Hort. Mal. 8: 30, 31. 1688.
Phanera scandens (L.) Raf., Sylv. Tell. 122. 1838.
Lasiobema anguinum (Roxb.) Korth. ex Miq., Fl. Ind. Bat. 1: 71. 1855.
Lasiobema horsfieldii Miq., Fl. Ind. Bat. 1: 71. 1855. Type: Java, *Horsfield s.n.* ("L. 169") (K).
Bauhinia anguina Roxb. var. *horsfieldii* (Miq.) Watt ex Prain, J. Asiat. Soc. Bengal 66 (2): 194. 1897.
Bauhinia horsfieldii (Miq.) Macb., Contr. Gray Herb. ser. 2, no. 59: 23. 1919.
Lasiobema scandens (L.) de Wit, Reinwardtia 3: 427. 1956.
Lasiobema scandens (L.) de Wit var. *horsfieldii* (Miq.) de Wit, Reinwardtia 3: 427. 1956; Backer & Bakh. f., Fl. Java 1: 536. 1963.

Additional synonymy is cited by de Wit, l.c.

Woody liana with compressed, serpentine stems, climbing to about 10–30 (–50) m; young stems glabrous, bearing pairs of opposite, stiff, flattened, circinately curved, prehensile tendrils; stipules small, linear-ovate, mucronate, caducous; leaves ovate, about (5–) 12–15 cm long, 5–10 cm wide, entire or shallowly bilobed, or sometimes the 2 lobes nearly free, cordate at the base, the surfaces glabrous, nitid, 7-nerved; inflorescences racemose or paniculate, terminal, many-flowered; bracts linear, 2 mm long; bracteoles linear, 1 mm long, puberulent; flowers small, about 3 mm long; calyx campanulate, 1.5–2 mm long with 5 deltoid teeth; petals yellowish-white, obovate, puberulent; ovary stipitate, glabrous; style glabrous; stigma small, indistinct; fruit elliptic or oblong, indehiscent, glabrous, (3–)4–5 cm long, 1.5–2 cm wide, 1- or 2-seeded; seeds compressed, about 6 mm long and wide.

Distr. India, Indochina, Malaysia, and in Ceylon, possibly as an introduction.
Uses. According to de Wit, l.c., this species is said to produce strong ropes; juice from the stems is used to cure severe coughs.
Vern. Snake climber (E).
Specimens Examined. KURUNEGALA DISTRICT: "Foot of Doluwe Kande", 19 Dec. 1893, *Trimen ? s.n.* (PDA). KANDY DISTRICT: Peradeniya, Royal Botanic Gardens, *Trimen ? s.n.* (PDA).

Note. Although de Wit recognizes two varieties he states that "differences (except the size of the pods) do not really exist."

In addition to the four preceding species, several more have been introduced as ornamentals, some, mentioned by Macmillan in Tropical Planting and Gardening, ed. 5. 1962, apparently are only found in the Botanic gardens. For a few there are herbarium vouchers.

Bauhinia diphylla Buch.-Ham. in Symes, Embassy, ed. 2, 3: 311. 1811. This is included by Macmillan, p. 124, in his list of "Ornamental-leaved Climbers" as a straggling climbing shrub from Malaya with small, 2-lobed leaves. De Wit, in Reinwardtia 3: 535. 1956, however, states that it is an Indian species, and that the specimens referred to it by J.G. Baker in Fl. Br. Ind. 2: 278. 1878 "belonged to two other species (cf. Ridley, Fl. Malay Penins. 1: 634. 1922)"

Bauhinia galpinii N.E. Br., Gard. Chron. 9: 728. 1891. *Bauhinia punctata* Bolle in Peters, Reise Mossamb. Bot. 1: 23. 1862, non Jacq. 1780. "Red Bauhinia, nasturtium Bauhinia" is a South African species with showy, brick-red flowers, cited by Macmillan (p. 103) as a flowering shrub. There are two specimens from Ceylon: Belihuloya, 1870 ft elev., *Worthington 4849* (K); Peradeniya, 1575 ft elev., *Worthington 4991* (K).

Bauhinia heterophylla Kunth, Mim. 157, t. 46. 1821. "Bejuco de corona; crown vine; wreath vine", a Venezuelan climber with white, purple-veined flowers, is represented in Ceylon by a specimen from Peradeniya, *F.W. de Silva 36* (K). According to Pittier (Las Plantas Usuales de Venezuela 130. 1926) antisyphilitic properties have been attributed to this species. It is also used as cordage.

Bauhinia monandra Kurz, J. Asiat. Soc. Bengal 42 (2): 3. 1873. *Bauhinia krugii* Urban, Ber. Deutsch. Bot. Ges. 3: 83. 1885. This species, characterized by having only one fertile stamen, is a widely cultivated ornamental shrub or small tree not known in the wild state. It was first described from Burma but is now believed to be of tropical American origin (de Wit, Reinwardtia 3: 402. 1956; Verdcourt, Man. New Guinea Leg. 118. 1979). Macmillan, p. 80, includes it in his list of flowering trees suitable for low or medium elevations. There is a specimen as "Kobonila" (S), Anuradhapura, near the temple at Isurumuniya, *Simpson 8125* (BM).

Bauhinia purpurea L., Sp. Pl. 375. 1753. *Bauhinia triandra* Roxb., Fl. Ind. ed. 2, 2: 320. 1832. "Butterfly tree; orchid tree", is a native of Southeast Asia, introduced widely in the tropics and warm temperate regions. The shrubs, or small trees, bear showy pinkish to purple flowers with three fertile stamens. Macmillan (p. 80) includes this, also, in his list of flowering trees suitable for low or medium elevations. There are several voucher specimens: "Ceylon, *Col. Walker s.n.* (K);

Peradeniya, *Macrae 336* (BM, K), *de Silva 8* (K), *9* (A, K), *10* (A, K), *11* (K), *12* (K).

Bauhinia semibifida Roxb., Fl. Ind. ed. 2, 2: 330. 1832; Wight, Ic. Pl. Ind. Or. 263. 1840. *Phanera semibifida* (Roxb.) Benth. in Miq., Pl. Jungh. 263. 1852; de Wit, Reinwardtia 3: 465, fig. 15. 1956. This is a white-flowered scandent shrub native to Malaysia, originally described from a plant introduced into the Botanic Garden at Calcutta, India. The one collection from Ceylon is presumably an introduction: Nuwara Eliya District, culvert 29/11 along A5, *Maxwell & Jayasuriya 858* (US).

Bauhinia vahlii Wight & Arn., Prod. 297. 1834. "Maloo, Malu creeper, camel's foot climber" is a huge, white or yellowish-flowered climber from India. Macmillan (p. 127) mentions that the large, roundish leaves, 12 inches wide or more in diameter, are sometimes used as plates in India. Santapau (Common Trees, 36: 1966) states that this very elegant plant, because of its shading the growing tops of useful forest trees, "suffers much persecution from foresters, who try to protect their trees from this plant, and from hill tribes who collect seeds as article of food". The plants also are said to have medicinal uses.

Bauhinia variegata L., Sp. Pl. 375. 1753; Wight & Arn., Prod. 296. 1834. *Bauhinia candida* Roxb., Fl. Ind. ed. 2, 2: 319. 1832, non Ait. 1789. *Bauhinia variegata* L. var. *candida* Voight, Hort. Suburb. Calcutta 253. 1845. *Bauhinia variegata* L. var. *alboflava* de Wit, Reinwardtia 3: 412. 1956. Macmillan (p. 80) includes this species in his list of flowering trees suitable for low or medium elevations. It is a widely cultivated native of eastern Asia, commonly known in English as "orchid tree; mountain ebony". The flowers can be distinguished from those of the similar *B. purpurea* by the five perfect stamens.

2. PARKINSONIA

L., Sp. Pl. 375. 1753; L., Gen. pl., ed. 5, 177. 1754; Brenan, Kew Bull. 17: 203. 1963; Polhill & Vidal in Polhill & Raven, Adv. Leg. Syst. 94. 1981; Allen & Allen, Legum. 495. 1981. Type species: *P. aculeata* L.

Peltophoropsis Chiov., Ann. Bot. (Genoa) 13: 385. 1915; Allen & Allen, Legum. 502. 1981. Type species: *P. sciona* Chiov.

Small trees or shrubs, usually armed; stipules minute or spinescent; leaves bipinnate but sometimes appearing to be simple-pinnate; pinnae 1–5, usually 2, each with many leaflets; inflorescences racemose, axillary; bracts minute, caducous; bracteoles lacking; flowers of medium size; calyx with 5 sepals separate nearly to the base, valvate or imbricate; petals 5, yellow, subequal; stamens 10 with filaments alternately longer and shorter; anthers dorsifixed; ovary stipitate or sessile, 2–6-ovulate; style usually glabrous, sometimes puberulent at the base; stigma terminal, truncate, glabrous or ciliolate; fruit indehiscent, 2-valved, usually constricted between the seeds, terete or compressed, commonly 1–5-seeded; seeds oblong, longitudinal; endosperm present.

A genus of 2–14 species depending on circumscription. *Cercidium* Tul. and *Cercidiopsis* Britt. & Rose are considered to be synonyms by some authors. Only one species, *P. aculeata* L., is known from Ceylon.

Parkinsonia aculeata L., Sp. Pl. 375. 1753; Moon, Cat. 34. 1824: Baker in Hook. f., Fl. Br. Ind. 2: 260. 1878; Trimen, Handb. Fl. Ceylon 2: 102. 1894; Alston in Trimen, Handb. Fl. Ceylon 6: 90. 1931; Worthington, Ceylon Trees 187. 1959; Macmillan, Trop. Pl. & Gard., ed. 5, 71, 203. 1962; Saldanha & Nicolson, Fl. Hassan 225. 1976. Type: Represented by L., Hort. Cliff. t. 13. 1737. There appears to be no specimen in BM-Cliff.

Small trees, to about 10 m tall; young stems greenish, finely appressed-pubescent, glabrescent and darkening with age, armed with straight, acicular spines to about 3 cm long; stipules acicular, spinescent; leaves bipinnate with (1–) 2(–4) pinnae, when 1-pinnate appearing to be simple-pinnate; primary rachis very short; secondary rachis to about 50 cm long, 1–2 mm wide, flattened and narrowly marginate; leaflets about 20–50, alternate or subopposite, linear to oblong or obovate, 2–6 mm long, 1–2 mm wide, obtuse at apex, acute or rounded at base, puberulent to glabrous on both surfaces, usually deciduous; inflorescences axillary, racemose, few-flowered; bracts minute, caducous; flowers about 1.5–2 cm long on pedicels (0.5–)1–2 cm long; calyx 6–8 mm long, glabrous or nearly so, the lobes deciduous leaving a persistent basal tube; petals bright yellow, villous at the base; fruit pendent, glabrous, longitudinally striate, about 5–15 cm long, 6–8 mm wide, 3–5 mm thick, about (1–)2–5-seeded, constricted between the seeds, narrowed at apex and base; seeds mottled olive and brown, oblong, about 1 cm long, 5 mm wide, 3–4 mm thick.

Distr. Probably native to dry areas of Mexico but widely distributed in tropical and subtropical America; introduced and escaped in dry tropical areas of the Old World, including Ceylon.

Uses. The plants are used as ornamental street trees and as hedges; the wood is burned as fuel or made into charcoal; leaves, young branches, and fruit are eaten by livestock; in Mexico an infusion of the leaves is sometimes used as a febrifuge and sudorific, as a remedy for epilepsy, and as an abortifacient (Standley, Trees and Shrubs of Mexico, Contr. U.S. Natl. Herb. 23: 428. 1922).

Vern. Mulvakai (T); Jerusalem thorn, Mexican palo verde, retama (E). There also are a number of local names used in Latin America.

Specimens Examined. JAFFNA DISTRICT: Jaffna, *Trimen ? s.n.* (PDA); near Kayts, *Rudd 3277* (PDA, SFV, US). MANNAR DISTRICT: Illuppaik-kadavai (as Illapai Kaduvai), Feb. 1890, *Trimen ?* (PDA); Mannar Isl., *Koster-*

mans 25265 (K, US); Murunkan, Giant's Tank, *Worthington 6717* (K). POLON-
NARUWA DISTRICT: Mannampitiya, culvert 52/3, *Cramer 4382* (US). PUT-
TALAM DISTRICT: Along road to Parukkuwattan, *Sumithraarachchi 700* (US);
Puttalam, near seashore, *Meijer 362* (US); Puttalam, along roadside near mi. post
2, Puttalam-Chilaw Rd., *Cramer 3043* (US). TRINCOMALEE DISTRICT: Kan-
niyai, *Cramer 5109* (K, MO); Trincomalee, *Worthington 221* (K). HAMBAN-
TOTA DISTRICT: Yala, *Waas 70* (K, NY, US); Ruhuna Natl. Park, between
gate and office, *Comanor 1054* (GH, K, NY, PDA, SFV, UC, US); on road
to Elephant Rock opposite Buttawa Rd., *Mueller-Dombois 67120901* (PDA, US);
"Hambantota, in garden, also semi-wild about H'tota & Kirinda", 31 Dec. 1926,
Alston s.n. (PDA).

3. CAESALPINIA

L., Sp. Pl. 380. 1753; L., Gen. Pl. ed. 5, 178. 1754 (as "Caesalpina"); Hattink, Reinwardtia 9: 9. 1974; Polhill & Vidal in Polhill & Raven, Adv. Leg.
Syst. 93. 1981; Allen & Allen, Legum. 119. 1981. Type species: *C. brasiliensis* L.

Poinciana L., Sp. Pl. 380. 1753; L., Gen. Pl. ed. 5, 178. 1754. Type species:
   P. pulcherrima L.
Guilandina L., Sp. Pl. 381. 1753; L., Gen. Pl. ed. 5, 179. 1754. Type species:
   G. bonduc L.
Bonduc Mill., Gard. Dict., abridg. ed. 1754; ed. 8, 1768, referred to *Guilandina.*
Mezoneuron Desf., Mém. Mus. Natl. Hist. Nat. 4: 245. 1818 (as *Mezonevron*);
   Allen & Allen, Legum. 432. 1981. Type species: *M. glabrum* Desf.

Additional synonyms are given by Hattink, l.c., and by Polhill & Vidal in
Polhill, l.c.

Trees or shrubs, erect or sometimes scandent or scrambling, usually armed
with spines or prickles or sometimes unarmed; stipules present, minute to
foliaceous; stipels usually lacking; leaves bipinnate or rarely reduced to scales;
leaflets usually opposite, rarely alternate, sometimes glandular; inflorescences
terminal or axillary, racemose or paniculate, (1–)few- to many-flowered; bracts
usually caducous; bracteoles lacking; flowers small and inconspicuous to large
and showy; calyx campanulate with 5 imbricate sepals, sometimes valvate; petals
5, imbricate, subequal, cream or yellow to red; stames 10, usually all fertile;
filaments alternately long and short, pubescent, sometimes glandular; anthers
dorsifixed; ovary sessile or short-stipitate, (1–)2–10(–13)-ovulate; style glabrous
or sometimes pubescent at base; stigma terminal, truncate or oblique, glabrous
or ciliolate; fruit dehiscent or indehiscent, usually 2-valved, glabrous to spiny,
coriaceous to ligneous, sometimes with a marginal wing, compressed or sometimes
cylindrical, 1–10-seeded; seeds transverse, lenticular to subreniform or
subglobose, endosperm present or absent.
   A pantropical genus of about 100–200 species.

KEY TO THE SPECIES

1 Shrubs or trees, unarmed or nearly so
  2 Flowers with petals flame-red to yellow,15–25 mm long on elongate pedicels 7–8 cm long; stamens exserted, 5–6.5 cm long; fruit dehiscent, straight, 6–12 cm long, 2 cm wide . . . .
     . . . .**1. C. pulcherrima**
  2 Flowers with petals yellow, 4–10 mm long on pedicels about 3–6 mm long; stamens not long-exserted; fruit indehiscent, straight or contorted, 3–8 cm long, 1–2 cm wide
    3 Fruit about 3–5 cm long, 1–2 cm wide, usually contorted; flowers about 4 mm long in condensed racemes; leaflets small, linear-oblong . . . . **C. coriaria**
    3 Fruit 7–8 cm long, 2 cm wide, straight; flowers about 8–10 mm long in elongate racemes or panicles; leaflets elliptic to obovate . . . . **C. punctata**
1 Shrubs, small trees, often scandent, scrambling, or prostrate, armed with recurved prickles
  4 Fruit spiny
    5 Stipules about 5–20 mm long, pinnate, foliaceous with 3–5 "leaflets"; leaves with 3–9(–11) pairs of pinnae, each pinna with about (3–)6–12 pairs of leaflets; flowering pedicels 2–6 mm long; ovules 2; seeds grey at maturity . . . . **2. C. bonduc**
    5 Stipules 1–3 mm long, subulate or absent; leaves with 3–8 pairs of pinnae, each pinna with about (3–)5–7 pairs of leaflets; flowering pedicels 6–12 mm long; ovules 4; seeds yellow to greenish grey at maturity . . . . **3. C. major**
  4 Fruit not spiny
    6 Dorsal suture of fruit not conspicuously winged
      7 Leaves with 2–5 pairs of pinnae, each pinna with 2–3(–5) pairs of leaflets; stipules lacking; fruit 1 (or 2)-seeded . . . . **4. C. crista**
      7 Leaves with 3–14 pairs of pinnae, each pinna with 5–20 pairs of leaflets; stipules present but often caducous; fruit 1–9-seeded
        8 Pinnae with 5–12 pairs of leaflets; leaflets essentially symmetrical with central costa
          9 Fruit 6–12 cm long, 4–9-seeded, dehiscent; petals 10–15 mm long. . . . .
           . . . . **5. C. decapetala**
          9 Fruit 3–5 cm long, 1–3(–4)-seeded, indehiscent; petals 5 mm long. . . . .
           . . . . **6. C. digyna**
        8 Pinnae with 10–20 pairs of leaflets; leaflets asymmetrical with excentric costa. . . . .
         . . . . **7. C. sappan**
    6 Dorsal suture of fruit with longitudinal wing about 6 mm wide or wider
      10 Leaflets opposite or alternate, 10–18 per pinna, 1.1–2.8 cm long, 0.5–1.6 cm wide; fruit 6–15 cm long, 2–4 cm wide including wing 6–8 mm wide . . . . **8. C. hymenocarpa**
      10 Leaflets mostly alternate, 7–9 per pinna, 2–7 cm long, 1.5–5.5 cm wide; fruit 10–17 cm long, 3–6 cm wide including wing about 10–20 mm wide . . . . **C. sumatrana**

1. Caesalpinia pulcherrima (L.) Sw., Obs. Bot. 166. 1791; Willd., Sp. Pl. 2: 531. 1799; Baker in Hook. f., Fl. Br. Ind. 2: 255. 1878; Trimen, Handb. Fl. Ceylon 2: 99. 1894; Alston in Trimen, Handb. Fl. Ceylon 6: 89. 1931; Macmillan, Trop. Pl. & Gard., ed. 5, 103, 104. 1962; Backer & Bakh. f., Fl. Java 1: 544. 1963; Brenan in Fl. Trop. East Africa, Caesalp. 31. 1967; Hattink, Reinwardtia 9: 50. 1974; Verdcourt, Man. New Guinea Leg. 27. 1979.

Poinciana pulcherrima L., Sp. Pl. 380. 1753; Moon, Cat. 24. 1824; Roxb., Fl. Ind., ed. 2, 2: 355. 1832; Wight & Arn., Prod. 282. 1834. Syntypes: Herb. Linn. 529.1 (LINN); Herb. Hermann 4: 50, 5: 26, 5: 436 (BM-Herm.);

isosyntypes: Herb. Hermann fol. 96, 114, 116, 117 (L); Herb. Burman p. 99 (Paris, Biblio. Inst. France). There appears to be no specimen in BM-Cliff.
Caesalpinia pulcherrima var. *flava* Hort. ex Bailey & Rehder in Bailey, Cycl. Amer. Hort. 1: 206. 1900.
Poinciana pulcherrima forma *flava* Degener, New Illust. Hawaii Flora, 169b, 1938. Type: Hawaii, Waialua, Oahu, cultivated, *Degener & Salucop 11319.*
Caesalpinia pulcherrima forma *flava* (Degener) St. John, Pacific Trop. Bot. Gard. Mem. 1: 179. 1973.

Shrubs or small trees, to about 5 or 6 m tall; stems and leaf axes glabrous, unarmed or sometimes with a few or numerous short, straight prickles; stipules subulate, about 2 mm long, caducous; leaves with rachis about (10–)20–30(–40) cm long; pinnae (3–)5–10 pairs, each pinna with about 6–10(–13) pairs of leaflets; leaflets glabrous, elliptic to obovate, about 1–2(–3.5) cm long, (3–)5–15 mm wide, rounded or emarginate at the apex, rounded at the base; bracts lanceolate-attenuate, about 7 mm long, caducous; inflorescences glabrous, racemose, terminal or axillary, to about 20–50 cm long; flowers showy, on pedicels (4–)5–8(–10) cm long; calyx about 1–1.5 cm long, the outer lobe cucullate, enclosing the flower in bud; petals 1.5–2.5 cm long, subequal, commonly flame-red but sometimes yellow or orange; stamens 10 with filaments reddish, exserted to about 5(–7.5) cm long; fruit dehiscent, glabrous, compressed, obliquely oblong, short-stipitate, about 6–12 cm long, 1.5–2 cm wide, narrowed at the base, beaked at the apex, 8–10-seeded; seeds dark brown or black, obovate, compressed, 8–10 mm long, 6–8 mm wide, 2–3 mm thick.

Distr. Presumably a native of tropical America but is widely distributed pantropically as an ornamental and as an escape. Apparently introduced into Ceylon at an early time.

Uses. Chiefly planted as an ornamental but sometimes reputed to have medicinal properties such as a vermifuge, a laxative, and for coughs and catarrh.

Vern. Peacock flower, Barbados' pride (E). There are also innumerable local names in the Americas.

Specimens Examined. WITHOUT EXACT LOCALITY: *Ripley 34* (US). ANURADHAPURA DISTRICT: Tirappane, *Rudd 3312* (K, PDA, US). PUTTALAM DISTRICT: 0.5 mi. beyond Kalpitiya, *Fosberg & Jayasuriya 52753* (K, MO, SFV, US), *52754* (GH, K, MO, SFV, UC, US). MATALE DISTRICT: Dambulla Rest House, *s. coll. s.n.* (PDA). KANDY DISTRICT: Peradeniya, Botanic Garden, 6 Oct. 1930, *de Silva s.n.* (PDA); Gonawatte, 15 Oct. 1920, *"J.M.S." (Silva?)* (K, PDA, US), Dec. 1887, *s. coll. s.n.* (PDA). HAMBANTOTA DISTRICT: Ruhuna Natl. Park, Kambukkan Oya, *Comanor 424* (US), *425* (GH, K, MO, NY, UC, US); near shore of Katagamuwa Tank, *Mueller-Dombois & Comanor 67062521* (US).

Note. Moon, l.c., mentioned both red (ratu) and yellow (kaha) under this species. Two collections cited above, *Fosberg & Jayasuriya 52754* and the specimen *s. coll.* from the Dambulla Rest house, correspond to variety, or forma *flava*.

2. Caesalpinia bonduc (L.) Roxb., Fl. Ind., ed. 2, 2: 362. 1832, emend. Dandy & Exell, J. Bot. 76: 179. 1938; Macmillan, Trop. Pl. & Gard., ed. 5, 364, 439. 1962; Backer & Bakh. f., Fl. Java 1: 545. 1963; Hattink, Reinwardtia 9: 17. 1974.

Guilandina bonduc L., Sp. Pl. 381. 1753. Lectotype: Ceylon, Herb. Hermann 3: 35 (BM-Herm.). Syntype: Herb. Hermann 2: 17 (BM-Herm.).

Caesalpinia crista L., Sp. Pl. 380. 1753, pro parte, non emend. Dandy & Exell; Gamble, Fl. Pres. Madras 1: 393. 1919; Saldanha & Nicolson, Fl. Hassan 218. 1976.

Guilandina bonducella L., Sp. Pl., ed. 2, 1: 545. 1762; Moon, Cat. 34. 1824. Type: Same as for *G. bonduc* L.

Bonduc minus Medik., Theod. Sp. 41. 1786, based on *G. bonducella* L.

Caesalpinia bonducella (L.) Fleming, Asiat. Res. 11: 159. 1810; Trimen, Handb. Fl. Ceylon 2: 99. 1894; Alston in Trimen, Handb. Fl. Ceylon 6: 88. 1931.

Additional synonymy is given by Dandy & Exell, cited above.

Climbing or prostrate shrubs with stems to about 20 m long; stems tomentulose to subglabrous, armed with recurved prickles; stipules 5–20 mm long, foliaceous with 3–5 "leaflets", usually persistent; leaves bipinnate with main axis to about 1.3 m long, puberulent and armed with recurved prickles like the stems; pinnae 3–9(–11) pairs, each pinna with about (3–)6–12 pairs of leaflets; leaflets ovate to elliptic, about 1–6.5 cm long, 0.5–2(–3) cm wide, obtuse to subacute, mucronate at the apex, rounded at the base, puberulent to glabrous on both surfaces; inflorescences racemose, supra-axillary or terminal, many-flowered; bracts lanceolate, attenuate, to about 8 mm long, 1–2 mm wide, caducous; flowers unisexual on pedicels about 2–6 mm long; calyx with 5 sepals separate almost to the base, about 5–8 mm long, minutely tomentulose; petals yellow to orangish, reflexed at anthesis, about as long as the sepals, or shorter; ovules 2; fruit elliptic, tardily dehiscent, about 4.5–9 cm long, 3.5–4.5 cm wide, 1.5–2 cm thick, densely covered with straight, puberulent spines to 10 mm long, 1 mm wide at base, 1- or 2-seeded; seeds globose or subglobose, about 1.5–2 cm in diameter, grey at maturity, lustrous, finely lined horizontally.

Distr. A pantropical species, chiefly in coastal areas.

Uses. The hard seeds are used ornamentally as beads in necklaces, rosaries, etc., also medicinally. According to Macmillan, l.c., tender leaves are applied for toothache and for worms in children.

Vern. Kumburu-wel, wael-kumburu (S); panaikkalaichi (T); grey nicker (E).

Specimens Examined. JAFFNA DISTRICT: Smallpox Island, Jaffna, Feb. 1890, *Trimen ? s.n.* (PDA). MANNAR DISTRICT: South shore of Mannar Island SW. of Pesalai, *Fosberg & Balakrishnan 53628* (GH, K, MO, SFV, UC, (US).

ANURADHAPURA DISTRICT: Anuradhapura, *Alston 1240* (PDA); Wilpattu Natl. Park, near Kokkare Villu, *Maxwell & Jayasuriya 817* (MO, NY, SFV, US). PUTTALAM DISTRICT: Island in Puttalam Lagoon, *Alston 1239* (PDA). TRINCOMALEE DISTRICT: Vakarai, *Worthington 6533* (K); beach S. of Elizabeth Point, 5 mi. marker on Trincomalee-Nilaveli Rd., *Davidse 7537* (MO). BATTICALOA DISTRICT: Kalkudah, *Jayasuriya et al. 687* (PDA, US). KALUTARA DISTRICT: Biyagama, 18 Feb. 1922, *J.M.S. s.n.* (PDA). HAMBANTOTA DISTRICT: Between Tissamaharama and Hambantota, *Cooray 69091404 R* (GH, K, MO, PDA, SFV, UC, US); Ruhuna Natl. Park, N. of Buttawa, *Mueller-Dombois & Cooray 68022803* (PDA, SFV, US); Buttawa beach area, *Comanor 897* (K, MO, PDA, SFV, UC, US).

Note. In trying to distinguish *C. bonduc* from *C. major* it is difficult to know if foliaceous stipules of *C. bonduc* were present but have fallen off or if the specimen is truly *C. major*. For example, some sheets of *Fosberg & Balakrishnan 53628* show well developed stipules but in others they are lacking.

Because of the confusion of names, specimens annotated as *C. bonduc, C. bonducella, C. crista,* and *C. major* should be reexamined for correct identification, as indicated in the preceding key to species.

3. Caesalpinia major (Medik.) Dandy & Exell, J. Bot. 76: 180. 1938; Fosberg, Taxon 22: 162. 1973; Hattink, Reinwardtia 9: 39. 1974; Verdcourt, Man. New Guinea Legum. 26. 1979.

Guilandina bonduc L., Sp. Pl. ed. 2, 1: 545. 1762, pro parte, non L., 1753.
Bonduc majus Medik., Theod. Spec. 43, t. 3, sup. 1786, excl. syn. L. Type: *Frutex globulorum* Rumph., Herb. Amboin. 5: 89, t. 48. 1747.
Caesalpinia bonduc sensu auctt., Baker in Hook. f., Fl. Br. Ind. 2: 255. 1878; Trimen, Handb. Fl. Ceylon 2: 98. 1894; Alston in Trimen, Handb. Fl. Ceylon 6: 89. 1931 (non L., 1753).
Guilandina major (DC.) Small, Fl. Southeast U.S. 591, 1331. 1903; Skeels, Science, new ser. 37: 922. 1913.
Caesalpinia globulorum Bakh. f. & van Royen, Blumea 12: 62. 1963; Backer & Bakh. f., Fl. Java 1: 545. 1963.

Additional synonymy is given by Dandy & Exell, l.c.

Shrubby vine with stems to about 15 m long; stems puberulent to subglabrous, armed with recurved prickles; stipules subulate, 1–3 mm long, or absent; leaves bipinnate with main axis to about 75 cm long, puberulent or subglabrous and armed like the stems, with 3–8 pairs of pinnae, each pinna with 3–7 pairs of leaflets; leaflets ovate-oblong to suborbicular, 3–9 cm long, 1.5–5 cm wide, acute or acuminate to obtuse, mucronate at the apex, rounded at the base, puberulent to glabrous on both surfaces; inflorescences racemose, terminal or supra-axillary, sometimes branched, many-flowered; bracts lanceolate-attenuate, about 8 mm long, 1 mm wide at the base, caducous; flowers mostly unisexual, on pedicels 6–12 mm long; calyx with 5 sepals separate almost to the base, about 7 mm long, minutely tomentulose; petals yellow, about as long as the sepals; ovules 4; fruit tardily dehiscent, elliptic, 5–13 cm long, 4–6 cm wide, densely covered with straight, puberulent spines to about 10 mm long, 1 mm wide or less at the base, 2–4-seeded; seeds globose, 1.5–2.5 cm in diameter, yellow to greenish grey at maturity, lustrous, finely lined horizontally.

Distr. A pantropical species growing along beaches and in forest up to about 1400 m elevation.

Vern. Yellow nicker (E).

Specimens Examined. WITHOUT EXACT LOCALITY: *Walker 1866* (K), *Thwaites C.P. 1524* (BM).

Note. In addition to stipule and seed characters, this species appears to differ from *C. bonduc* in having coarser leaflets, and the spines on the fruit finer and more numerous.

4. Caesalpinia crista L., Sp. Pl. 380. 1753, emend. Dandy & Exell, J. Bot. 76: 179. 1938; Skeels, Science, new ser. 37: 922. 1913; Backer & Bakh. f., Fl. Java 1: 545. 1963; Hattink, Reinwardtia 9: 20. 1974. Lectotype: Ceylon, Herb. Hermann 1: 68 (BM-Herm.).

Guilandina nuga L., Sp. Pl., ed. 2, 1: 546. 1762. Type: *Nugae silvarum,* Rumph., Herb. Amboin. 5: t. 50. 1747.
Guilandina paniculata Lam., Enc. Meth. 1: 434. 1785; Willd., Sp. Pl. 2: 535. 1799. Type: India, Rheede, Hort. Mal. 6: t. 19. 1686.
Caesalpinia nuga (L.) Ait. f., Hort. Kew., ed. 2, 3: 32. 1811; Baker in Hook. f., Fl. Br. Ind. 2: 255. 1878; Trimen, Handb. Fl. Ceylon 2: 99. 1894; Alston in Trimen, Handb. Fl. Ceylon 6: 88. 1931; Macmillan, Trop. Pl. & Gard. ed. 5, 127. 1962.
Caesalpinia paniculata (Lam.) Roxb., Hort. Beng. 32. 1814; Moon, Cat. 34. 1824; Roxb., Fl. Ind. ed. 2, 2: 364. 1832; Wight & Arn., Prod. 281. 1834; Wight, Ic. Pl. Ind. Or. 36. 1838; Thw., Enum. Pl. Zeyl. 95. 1859.

Additional synonymy is given by Hattink, l.c.

Woody climber, to about 15 m high; stems glabrous, armed with recurved prickles; stipules lacking; leaves bipinnate with main axis 10–30 cm long, glabrous, armed or unarmed; pinnae 2–5 pairs, each pinna with 2–3(–5) pairs of leaflets; leaflets ovate-elliptic, 2–10 cm long, 1–5 cm wide, usually obtuse, sometimes acute to acuminate at the apex, acute at the base, glabrous on both surfaces; inflorescences racemose or paniculate, axillary or terminal, many-flowered; bracts about 1 mm long, caducous; flowers sweet scented, about 10 mm long on pedicels 7–15 mm long; calyx with 5 glabrous sepals 6–8 mm long, the lowest one cucullate; petals 8–10 mm long, yellow, the standard sometimes with reddish lines; ovules 1 or 2 (−3); fruit glabrous, indehiscent, elliptic, narrowed at the base, 4−7 cm long, 2.5–3.5 cm wide, 1- or 2-seeded; seeds black, orbicular to subreniform, 1.5–2 cm long, 0.5−1 cm wide.

Distr. Coastal areas of Asia from India and Ceylon to the Ryukyu Isl., Palau Is., Queensland, Australia, and New Caledonia, on sandy and limestone soil at elevations up to about 350 m.

Vern. Diya-wawuletiya (S).

Specimens Examined. WITHOUT EXACT LOCALITY: *Kelaart s.n.* (K); *Walker 68* (K); *s. coll. Walker ? s.n.* (GH); "Van luce Bay & Hantane, *Gardner,* Galle, Dec. 1863, Peradenia Aug. 1858", *Thwaites C.P. 1525* (PDA); without data, *C.P. 1525* (BM, PDA). KANDY DISTRICT: Peradeniya, Royal Botanic Gardens, 9.8.17, *s. coll. s.n.* (PDA).

Note. Moon, l.c., also cites "Caltura".

5. Caesalpinia decapetala (Roth) Alston in Trimen, Handb. Fl. Ceylon 6: 89. 1931; Brenan in Fl. Trop. East Africa, Caesalp. 36. 1967; Backer & Bakh. f., Fl. Java 1: 545. 1963; Hattink, Reinwardtia 9: 24. 1974; Saldanha & Nicolson, Fl. Hassan 218. 1976.

Reichardia decapetala Roth, Nov. Pl. Sp. Ind. Or. 212. 1821. Type: India, *Heyne* (K, isotype).

Caesalpinia sepiaria Roxb., Fl. Ind., ed. 2, 2: 360. 1832; Wight & Arn., Prod. 282. 1834; Wight, Ic. Pl. Ind. Or. 37. 1838; Thw., Enum. Pl. Zeyl. 95. 1859; Baker in Hook. f., Fl. Br. Ind. 2: 256. 1878; Trimen, Handb. Fl. Ceylon 2: 100. 1894; Macmillan, Trop. Pl. & Gard., ed. 5, 71. 1962. Type: India, Mysore, *Roxburgh* (holotype BM; isotype or syntype K, Roxb. in Wallich 5834a).

Caesalpinia sepiaria var. *auricoma* Trimen, Handb. Fl. Ceylon 2: 100. 1894. Type: Ceylon, "Amherst, Uda Pusselawe, Sept. 1885, W.F." (*Ferguson*) (PDA).

Additional synonyms are given in Hattink, l.c.

Shrub or woody climber, to about 25 m high; stems densely pubescent with brownish hairs, glabrescent with age, armed with recurved prickles; stipules lunate or obliquely semi-cordate, (6–)8–13 mm long, 3–7 mm wide, acuminate, caducous; leaves bipinnate with main axis to about 47 cm long, armed with recurved prickles; pinnae 3–10 pairs, each pinna with 5–12 pairs of leaflets; leaflets elliptic to elliptic-oblong, essentially symmetrical with central costa, about 1–2.5 cm long, 5–10 mm wide, rounded at apex and base, finely puberulent on both surfaces; inflorescences paniculate, axillary and terminal, many-flowered; bracts ovate-deltoid to lanceolate, acuminate, 4–8 mm long, 2–2.5 mm wide, pubescent, caducous; flowers 1.5–2 cm long on pedicels (1.5–)2–3(–3.5) cm long, tomentulose; petals 10−15 mm long, bright yellow to yellowish-white; ovules

8–10; fruit dehiscent, puberulent, elliptic-oblong, sessile, 6–12 cm long excluding beak 1–2 cm long, about 2–3 cm wide, 1–1.5 cm thick, 4–9-seeded; seeds black, or brown and tan mottled, ellipsoid, about 8–12 mm long, 6–8 mm wide, 3–4 mm thick.

Distr. Native to tropical and subtropical Asia but now widely introduced and naturalized in Africa and the Pacific area.

Uses. The plants are sometimes used as ornamental hedges; the bark has been used for tanning, and the roots reported to be a purgative.

Specimens Examined. WITHOUT EXACT LOCALITY: *Walker s.n.* (PDA); *Thwaites C.P. 2784* (BM, K). MATALE DISTRICT: Matale, Mar. 1868, *Thwaites C.P. 3602* (PDA). KANDY DISTRICT: Peradeniya, *Gardner 251* (BM, GH, K); between Madugoda and Hunnasgiriya, mi. marker 24/19, *Rudd & Balakrishnan 3246* (GH, K, NY, PDA, SFV, US). NUWARA ELIYA DISTRICT: Rikiligaskade, *Amaratunga 879* (PDA). BATTICALOA DISTRICT: Batticaloa, Mar. 1858, *Thwaites C.P. 3602* (PDA). BADULLA DISTRICT: S. of Hakgala, A5, *Maxwell & Jayasuriya 886* (NY, PDA, SFV, US).

Note. As described by Trimen, *C. sepiaria* var. *auricoma* is a robust plant differing from typical specimens by its dense pubescence. It is known only from the holotype at PDA, a flowering specimen, and needs to be studied further.

Thwaites customarily gave one *C.P.* number to all collections of a given species. In the case of *C.P. 2784* he found that to be a mixture. In his Enum. Pl. Zeyl., p. 95, he indicated a correction, giving the two elements new numbers: *3601* (*2784* partim), cited in this paper as *Caesalpinia hymenocarpa*, and *3602* (*2784* partim), cited above as *C. decapetala*. The collections of *C.P. 2784* at BM and K apparently predate the division of the material and, presumably, are also referrable to *C.P. 3602*.

6. Caesalpinia digyna Rottl., Ges. Naturf. Freunde Berlin Neue Schriften 4: 200, t. 3. 1803; Wight & Arn., Prod. 281. 1834; Thw., Enum. Pl. Zeyl. 95. 1859; Baker in Hook. f., Fl. Br. Ind. 2: 256. 1878; Trimen, Handb. Fl. Ceylon 2: 100. 1894; Alston in Trimen, Handb. Fl. Ceylon 6: 88. 1931; Macmillan, Trop. Pl. & Gard., ed. 5, 431. 1962; Backer & Bakh. f., Fl. Java 1: 546. 1963; Hattink, Reinwardtia 9: 28. 1974. Type: S. India, Marmelon, *Rottler* (K).

Caesalpinia oleosperma Roxb., Fl. Ind. ed. 2, 2: 357. 1832. Type: India, *Roxburgh*.

Woody climber or small tree, to about 10 m high; stems pilose, glabrescent, armed with recurved prickles 4–5 mm long; stipules subulate, to about 3 mm long, caducous; leaves bipinnate with main axis about 15–23 cm long, pubescent and armed with recurved prickles; pinnae 8–13 pairs, each pinna with 9–12 pairs of leaflets; leaflets closely spaced, essentially symmetrical with central costa, oblong or linear-oblong, 5–11 mm long, 2.5–4.5 mm wide, truncate or obtuse

at apex, obliquely truncate at base, glabrous above or puberulent on both surfaces; inflorescences racemose, axillary and terminal, many-flowered; bracts
puberulent, setaceous, about 4 mm long, caducous; flowers on pedicels 1.5–2.5
cm long; calyx glabrous with lowest sepal 6–8 mm long, the others 3.5 mm long;
petals yellow, the standard about 5 mm long, the others 6–8 mm long; stamens
exserted to about 12 mm long; ovules 2 – 4; fruit indehiscent, glabrous, oblong,
3 – 5 cm long, 1.5 – 2 cm wide, thickened along the sutures, somewhat constricted
between the seeds. 1–3(–4)-seeded; seeds subglobose, dark brown, about 10–12
mm in diameter.

Distr. In southern Asia from India and Ceylon to Malaysia.

Uses. Macmillan, l.c., mentions that the fruit is used in tanning. Watt states
that Roxburgh named the bush *C. oleosperma* in allusion to the oil in the seeds
which was used in some parts of the country as lamp oil. (The commercial products of India, reprint ed., 193. 1966.) The roots are reputed to have properties
as a febrifuge and as a narcotic.

Vern. Hingourou (fide Jonville); tari pods.

Specimens Examined. WITHOUT EXACT LOCALITY: *Jonville s.n.*
(BM); *Beddome 2590* (BM); *Macrae 626* (BM); *Walker 1490* (K), *s.n.* (GH).
KURUNEGALA DISTRICT: Kurunegala Garden, Sept. 1866, *Thwaites C.P.*
1527 (PDA), without data, *C.P. 1527* (BM, GH); Sept. 1888, Kurunegala,
Trimen ? (PDA); without locality *s. coll. s.n.* (K).

7. Caesalpinia sappan L., Sp. Pl. 381. 1753; Moon, Cat. 34. 1824; Wight. &
Arn., Prod. 281. 1834; Beddome, Fl. Sylv. t. 90. 1871; Baker in Hook. f., Fl.
Br. Ind. 2: 255. 1878; Trimen, Handb. Fl. Ceylon 2: 99. 1894; Macmillan, Trop.
Pl. & Gard., ed. 5, 417. 1962; Backer & Bakh. f., Fl. Java 1: 546. 1963; Hat-
tink, Reinwardtia 9: 51. 1974. Type: Ceylon, Herb. Hermann 4: 31 (BM-Herm).

Small trees or shrubs, to about 10 m high; stems minutely tomentulose, glabrescent, sparsely armed with short, straight or recurved prickles; stipules (?)
spiniform, usually recurved, to about 3–5 mm long, 3 mm wide at the base; leaves
bipinnate with main axis 25–40 cm long; pinnae 9–14 pairs, each pinna with 10–20
pairs of leaflets; leaflets asymmetrically oblong with excentric costa, 1–2.5 cm
long, (3–)5–8(–11) mm wide, obtuse at apex and base, glabrous or nearly so on
both surfaces; inflorescences racemose, supra-axillary and terminal, many-
flowered; bracts lanceolate, attenuate, about 5–12 mm long, puberulent, caducous;
flowers about 1 cm long on pedicels 1–2 cm long; calyx with sepals ciliate, otherwise glabrous, the lowest one about 10 mm long, the others 7 mm long; petals
yellow, about 1 cm long; stamens exserted to about 1.5 cm long; ovules 3–6;
fruit dehiscent, puberulent when young but essentially glabrous, black at maturity, obliquely oblong, somewhat beaked, 7–10(–12.5) cm long, 3–4 cm wide,
2–4-seeded; seeds brown to black, oblong to ellipsoid, 15–8 mm long, 8–11 mm
wide, 5 – 7 mm thick.

Distr. Known from southern and southeastern Asia to Malaysia.

Uses. The dark red heartwood yields a red dye. As shrubs, the plants can form good barrier hedges.

Vern. Pattangi (S); sappan-wood (E).

Specimens Examined. KANDY DISTRICT: Peradeniya, Royal Botanic Gardens, *C.F. Baker 117* (CAS, GH, NY, PDA, UC, US); Gampola Rd., river bank, *Cooray 70013002R* (K, PDA, SFV, US); Ulapane, *Maxwell 929* (NY, PDA, SFV, US); Kandy catchment, 2150 ft elev., *Worthington 5495* (K); Yattawatta, *Amaratunga 899* (PDA).

8. Caesalpinia hymenocarpa (Prain) Hattink, Reinwardtia 9: 35. 1974.

Mezoneuron hymenocarpum Jacks. ex Prain, J. Asiat. Soc. Bengal 2, 66: 233, 472. 1897. Type: Burma, Tanog Doung, Herb. Wallich *5832* (holotype K; isotypes BM, K).

Shrubs or climbers; stems puberulent or glabrous, armed with recurved prickles; stipules minute, scale-like, about 0.5 mm long, 0.5–1 mm wide; leaves bipinnate with main axis 20–40 cm long; pinnae 6–10 pairs, each pinna with about 10–18 leaflets; leaflets opposite or alternate, ovate to obovate-oblong, 1.1–2.8 cm long, 0.5–1.6 cm wide, rounded to retuse at the apex, asymmetrically rounded to cuneate at the base, puberulent to glabrous on both surfaces; inflorescences racemose or paniculate, axillary or terminal, many-flowered; bracts minute, caducous; flowers about 7–10 mm long on pedicels 8–15 mm long; calyx with sepals 6–10 mm long, ciliate; petals yellow, about 7–8 mm long; stamens exserted to about 7–17 mm long; ovules 4–6; fruit indehiscent, oblong or elliptic, compressed, 6–15 cm long, 2–3(–4) cm wide including a marginal wing about 6–8 mm wide along the dorsal suture, (1–)3–6-seeded; seeds ellipsoid, compressed, about 5–10 mm long, 3–5 mm wide, 1 mm thick.

Distr. Southern and eastern Asia from Ceylon to Yunnan, China.

Vern. Godawawuletiya (S).

Specimens Examined. RATNAPURA & KANDY DISTRICTS: Ratnapura, Sept. 1858, Peradeniya, July 1858, *Thwaites C.P. 3601* (PDA); without data, *C.P. 3601* (BM, GH, PDA). KANDY DISTRICT: Ginigathena, *Simpson 8548* (BM).

Note. The above-mentioned collections of *Thwaites, C.P. 3601,* were cited as *Mezoneurum enneaphyllum* Wight & Arn. by Thwaites, Enum. Pl. Zeyl. 414. 1864 and by Trimen, Handb. Fl. Ceylon 2: 102. 1894. Alston in Trimen, Handb. Fl. Ceylon 6: 90. 1931, referred them to *M. furfuraceum* Prain, with *M. enneaphyllum* as a synonym. Those two *Caesalpinia* spp. are treated as distinct by Hattink, op. cit.

Additional species:

Caesalpinia coriaria (Jacq.) Willd., Sp. Pl. 2: 532. 1799, based on *Poinciana coriaria* Jacq., Select. 123, t. 175, f. 36. 1763. "Siliqua arboris Guatapana ex Coraçao insula. Breyn. Cent. p. 58, f. 5." Divi-divi (Carib); vanni (T).

This tropical American species was introduced into India about 1834 and into Ceylon probably about the same time according to Macmillan, Trop. Pl. & Gard., ed. 5, 419. 1962.

The trees are unarmed, to about 10 m tall; leaves are bipinnate with about 3–9 pairs of pinnae, each pinna with 15–28 pairs of small, glabrous, linear-oblong leaflets; flowers are small, yellowish in condensed, axillary or terminal racemes; fruits are indehiscent, oblong, contorted or flexuous at maturity, 1–10-seeded. The fruits are valuable for their yield of tannin, and sometimes used medicinally.

Specimens Examined. "Planted in R.B.G. and in Trincomalee rest house compound", *Worthington 1313* (BM). "Colombo, Queen's Road, trees planted either side of University library gates (since mutilated)", *Worthington 1816* (K).

Caesalpinia mimosoides Lam., Enc. l: 462. 1785, Ill. t. 335, f. 2. 1797; Rheede, Hort. Mal. 6: 15, t. 8. 1686; Moon, Cat. 34. 1824; Wight & Arn., Prod. 281. 1834; Wight, Ic. Pl. Ind. Or. 392. 1840; Thw., Enum. Pl. Zeyl. 95. 1859; 414. 1864; Baker in Hook. f., Fl. Br. Ind. 2: 256. 1878; Trimen, Handb. Fl. Ceylon 2: 100. 1894; Alston in Trimen, Handb. Fl. Ceylon 6: 88. 1931; Saldanha & Nicolson, Fl. Hassan 218. 1976.

This species may not actually occur in Ceylon. It apparently was first attributed to the country by Moon, cited as Goda-wawul-aetiya, from Colombo. Thwaites included it, in 1859, as Goda-wawool-atteya, *C.P. 3601* (*2784* partim), but in 1864, he rescinded that: "Dele '3. *C. mimosoides*, Lam.', cum descriptione, et in ejus loco insere: '3. *C. Glenieii*, Thw.' . . . . *C.P. 3815*". The latter was not intended to be a synonym but a new species based on a different collection, and is now considered a synonym of *Peltophorum pterocarpum*. The collection *Thwaites C.P. 3601* (*2784* partim) was assigned to *Mezoneurum enneaphyllum*, which is cited in this paper, following Hattink, in Reinwardtia 9: 36. 1974, as *Caesalpinia hymenocarpa*. However, *C. mimosoides* and *C. hymenocarpa* are very distinct species. The confusion seems to relate to the original mixture of *C.P. 2784*, one part of which was later separated, correctly, as *C.P. 3602, C. sepiaria* (=*C. decapetala*), the other as *C.P. 3601*. Authors subsequent to Thwaites seem not to have been alerted to his amendment. Trimen noted that "*C. mimosoides*, Lam. is recorded for Ceylon in Fl. B. Ind. II. 256, but I am not aware upon what authority". At BM there is a specimen of *Gardner 251*, now identified as *C. decapetala* (= *C. sepiaria*), that was originally annotated as *C. mimosoides*. That might be the authority on which *C. mimosoides* was recorded for Ceylon.

Caesalpinia punctata Willd., Enum. Hort. Berol. 1: 445. 1809; Urban, Symb. Antill. 2: 284. 1900. Lectotype: Herb. Willdenow 8022 (B-Willd.).*Caesalpinia paucijuga* Benth. ex Oliver, Hooker's. Ic. Pl. 20: t. 1977. 1891. Type: Trinidad Bot. Gard., "Only known to us from the Botanic Garden; sent by Mr. Prestoe." (K). "It occurs also in St. Thomas; introduced from Trinidad, *Eggers* (No. 134)", (K).

There are several specimens collected by Worthington, originally identified as *Peltophorum linnaei* Benth., that appear to be referable to *Caesalpinia punctata* Willd., a native of the Caribbean area. Macmillan, in Trop. Pl. & Gard., ed. 5, 215. 1962, presumably was referring to this taxon which he included as "Brazilleto-Wood (Peltophorum Linnaei. Leguminosae) W. Indies" in his list of "Important timbers and cabinet woods of the tropics."

Trees, unarmed; leaves with 2–4 pairs of pinnae, each pinna with about 4–7 pairs of elliptic to obovate, obtuse leaflets 8–20 mm long, 5–12 mm wide, glabrous on both surfaces; flowers yellow, about 8–10 mm long in elongate racemose inflorescences; pedicels about 5 mm long; fruit indehiscent, 7–8 cm long, 2 cm wide, about 10-seeded; seeds dark brown, ellipsoid, slightly compressed, about 10 mm long, 5 mm wide, 3 mm thick.

Specimens Examined. KANDY DISTRICT: Andiatenne, Kadugannawa, (*W.R. Hancock*) *Worthington s.n.* (K); Udawattakelle, Kandy, (*W.R. Hancock*) *Worthington s.n.* (K). TRINCOMALEE DISTRICT: Trincomalee, 'Spyglass', Welcombe Hotel Garden, planted, seed brought from Jamaica 1938, nurseried in Kadugannawa, *Worthington 1273* (K), *1276* (K); Trincomalee, seed brought from Trinidad R.B.G. in 1938, *Worthington 3096* (A, BM, K). BADULLA DISTRICT: Badulla esplanade, *Worthington 4957* (K). COLOMBO DISTRICT: Heneratgoda Garden, 16 May 1922, *Stockdale s.n.* (K).

Caesalpinia sumatrana Roxb., Fl. Ind. ed. 2, 2: 366. 1832; Hattink, Reinwardtia 9: 55. 1974. *Mezoneurum sumatranum* (Roxb.) Wight & Arn. ex Miq., Fl. Ind. Bat. 1: 105. 1855; 1081. 1858.

This climbing species, a native of Sumatra, has been included in the key because it has been introduced into Ceylon. One voucher specimen so identified is, Peradeniya, Royal Botanic Gardens, 20 Oct. 1926, *Jayasinghe 358* (PDA).

4. PELTOPHORUM

(Vog.) Benth., J. Bot. (Hooker) 2: 75. 1840, nom. cons.; Polhill & Vidal in Polhill & Raven, Adv. Leg. Syst. 90. 1981; Allen & Allen, Legum. 503. 1981. Type species: *P. vogelianum* Walp., nom. illeg., a synonym of *P. dubium* (C. Spreng.) Taub. based on *Caesalpinia dubia* C. Spreng.

Baryxylum Lour., Fl. Cochinch. 266. 1790, nom. rejic. Type species: *B. rufum* Lour.
Caesalpinia sect. *Peltophorum* Vog., Linnaea 11: 406. 1837.

Trees, unarmed; leaves bipinnate, deciduous; leaflets small, numerous, opposite; stipules small, caducous; inflorescences simple racemose or paniculate, terminal or lateral; bracts lanceolate, sometimes caducous; bracteoles lacking; flowers showy; petals 5, yellow, crinkled, subequal; stamens 10, filaments free, pilose at the base; anthers uniform, dorsifixed, dehiscent longitudinally; stigma peltate; fruit elliptic-oblong, samaroid, essentially sessile, narrowed at apex and base, winged along the margins, compressed, indehiscent, usually longitudinally striate, 1–6-seeded; seeds compressed, transverse, lenticular.

A genus of about 7–9 tropical species.

KEY TO THE SPECIES

1 Leaflets about 10–25 mm long, 3–10 mm wide, obtuse to retuse; flowers 15–25 mm long
  2 Inflorescences paniculate, terminal and lateral; flowering pedicels about as long as the calyx, 7–10
    mm long; stipules small, unbranched; fruit striate...................**1. P. pterocarpum**
  2 Inflorescences simple-racemose, lateral; flowering pedicels about 3 or more times the calyx, 30–40
    mm long; stipules pinnately divided; fruit not striate or indistinctly so.....**2. P. dasyrachis**
1 Leaflets about 3–12 mm long, 1–4.5 mm wide, subacute, mucronate; flowers about 10 mm long;
  fruit striate...**3. P. africanum**

1. Peltophorum pterocarpum (DC.) Backer ex K. Heyne, Nutt. Pl. Ned.-Ind., ed. 2, 2: 755. 1927; Backer & Bakh. f., Fl. Java 1: 547. 1963; Ali, Fl. W. Pakistan, no. 54, Caesalp. 42. 1973.

Inga pterocarpa DC., Prod. 2: 441. 1825. Type: "Timor. Fl. ign. (v. s. comm. a Mus. Par.)".
Caesalpinia inermis Roxb., Fl. Ind., ed. 2, 2: 367. 1832. "Type: A native of the Moluccas".
Poinciana roxburghii G. Don, Gen. Hist. 2: 433. 1834. Type: Based on "*Caesalpinia inermis* Roxb., non L."
Caesalpinia ferruginea Decne., Nouv. Ann. Mus. Hist. Nat. 3: 462. 1834.
Peltophorum ferrugineum (Decne.) Benth., Fl. Austral. 2: 279. 1864; Baker in Hook. f., Fl. Br. Ind. 2: 257. 1878; Trimen, Handb. Fl. Ceylon 2: 101. 1894, pl. 32. 1894; Macmillan, Trop. Pl. & Gard., ed. 5, 87, 88, 196, 209. 1962.
Caesalpinia gleniei Thw., Enum. Pl. Zeyl. 414. 1864. Type: Ceylon, Trincomalee, *Glenie s.n.* June 1863, *Thwaites C.P. 3815* (PDA holotype; BM, GH, K isotypes).
Peltophorum inerme (Roxb.) Naves ex F.-Vill., Nov. App. 69. 1880; Alston in Trimen, Handb. Fl. Ceylon 6: 89. 1931; Worthington, Ceylon Trees 186. 1959.
Peltophorum roxburghii (G. Don) Degener, Fl. Hawaii, Fam. 169b. 1938.

Trees, to about 24 m tall; young stems brown-tomentulose; stipules small, unbranched; leaves with about 4 – 14 pairs of pinnae, each pinna with

(8–)10–15(–22) pairs of leaflets; leaflets oblong, 10–20 mm long, 3–10 mm wide, obtuse to retuse at the apex, asymmetrically acute and rounded at the base, glabrous above, glabrous to puberulent beneath; inflorescences paniculate, terminal and lateral; bracts minute, deltoid, attenuate, caducous; flowers fragrant, on pedicels about 7–10 mm long; calyx lobes acute, about 7–10 mm long, minutely rufo-tomentulose; petals bright yellow, about 17–25 mm long; fruit oblong-elliptic, 5–11.5 cm long, 1.7–2.8 cm wide including marginal wings about 4 mm wide, minutely tomentulose, glabrescent, longitudinally striate, 1–4-seeded; seeds lenticular, compressed, light brown, about 10 mm long, 5 mm wide.

Distr. Ceylon, Malaysia, and northern Australia; introduced elsewhere in the tropics.

Uses. Planted as an ornamental, especially as a street tree. Burkill, Dict. Econ. Prod. Malay Penins. 2: 1716. 1966, states that the bark is used for colouring cotton a yellow-brown, and mentions that the astringent bark has various medicinal uses.

Vern. Iya-vakai (T); copper-pod, yellow-flame, yellow poinciana (E).

Specimens Examined. WITHOUT LOCALITY: *Willis s.n.* (A). ANURADHAPURA DISTRICT: Tissa Wewa rest house, *Ripley 91* (US). POLONNARUWA DISTRICT: Polonnaruwa sacred area, *Ripley 128* (US). KANDY DISTRICT: Peradeniya, Royal Botanic Gardens,*Worthington 2869* (BM, K); University grounds, *Comanor 1170* (GH, K, NY, PDA, SFV, UC, US); Haragama, on river bank, *Amaratunga 877* (PDA). TRINCOMALEE DISTRICT: Trincomalee, Welcombe Hotel, *Worthington 2666* (BM). BATTICALOA DISTRICT: Batticaloa, *Walker 186* (PDA).

2. Peltophorum dasyrachis (Miq.) Kurz ex Baker in Hook. f., Fl. Br. Ind. 2: 257. 1878; Backer & Bakh. f., Fl. Java 1: 547. 1963.

Caesalpinia dasyrachis Miq., Fl. Ind. Bat., Suppl. 292. 1961.

Trees; young stems brown-tomentulose; stipules pinnately divided; leaves with about 5–9 pairs of pinnae, each pinna with 6–16 pairs of leaflets; leaflets elliptic-oblong, finely pubescent on both surfaces, glabrescent, 15–25 mm long, 4–7 mm wide, rounded at apex and base, inflorescences lateral, simple-racemose; bracts linear, 10–12 mm long, subpersistent; flowers on pedicels about 20–40 mm long; calyx lobes 10–15 mm long, tomentulose; petals 15–25 mm long, light yellow; fruit brown-puberulent, glabrescent, not or indistinctly striate, 9–13 cm long, 2.5–3.5 cm wide, 1–6-seeded.

Distr. A native of Sumatra and the Malay Peninsula; introduced elsewhere in the tropics.

Uses. Planted as a shade tree; in Java in coffee and cacao plantations according to Backer and Bakh. f., l.c.

Specimens Examined. WITHOUT LOCALITY: *Worthington 1062* (K).

KANDY DISTRICT: Kandy, *Worthington (Mrs. Epps) 4968* (K); Hillcrest, *Worthington 6910* (K); Peradeniya, Royal Botanic Gardens, *Worthington 4102* (K), *5276* (BM, K), *6448* (K); University grounds, lower Hantane Rd., *Comanor 493* (GH, K, NY, PDA, SFV, UC, US). TRINCOMALEE DISTRICT: Trincomalee, Welcombe Hotel, *Worthington 2665* (BM, K). RATNAPURA DISTRICT: Avisawella-Ratnapura Road, about 42 mi., *Worthington 4866* (K).

3. Peltophorum africanum Sond., Linnaea 23: 35. 1850; Palgrave, Trees Southern Africa 291, tab. 90. 1977. Type: Africa, "Am Ufer des Crocodillrivier und in den Hainer an der nördlichen Seite des Magalishberges, *Zeyher 554.*"

Trees, to about 10 m high; young stems brown-puberulent; stipules foliaceous, compound, caducous; leaves with about 4–7(–10) pairs of pinnae, each pinna with 10–23 pairs of leaflets; leaflets elliptic-oblong, 3–12 mm long, 1–4.5 mm wide, averaging 7 mm long, 2 mm wide, subacute, mucronate, puberulent-tomentulose on both surfaces; inflorescences paniculate or clusters of racemes, terminal; bracts linear-subulate; flowers on pedicels about 5–11 mm long; calyx lobes 4–6 mm long, brown-tomentulose; petals bright yellow, about 10 mm long; fruit striate, 5–10 cm long, 1.5–2 cm wide, 1 or 2-seeded.

Distr. A native of southern Africa but introduced elsewhere.

Uses. The bark is said to be used medicinally; the wood is soft with black heartwood suitable for carving.

Vern. Weeping wattle, African wattle (E).

Specimens Examined. KALUTARA DISTRICT: Allapella bungalow-garden, 1,800 ft alt., *Worthington 1424* (K). KANDY DISTRICT: Peradeniya, Royal Botanic Gardens, "planted from seed sent by Forest Director Rhodesia", *Worthington 2063* (BM).

Peltophorum linnaei Benth., J. Bot. (Hooker) 2: 75. 1840, Brazilletto-wood, was included by Macmillan, op. cit. p. 215, in his list of "important timbers and cabinet woods of the tropics." Presumably he was referring to the species cited earlier in this treatment as *Caesalpinia punctata* Willd. Any material at PDA should be checked for correct identification. Some specimens referred to *P. linnaei* are of another species, also known as *Caesalpinia violacea* (Mill.) Standl.

5. CASSIA

L., Sp. Pl. 376. 1753; L., Gen. Pl., ed. 5, 178. 1754; de Wit, Webbia 11: 198. 1955; Irwin & Barneby in Polhill & Raven, Adv. Leg. Syst. 104. 1981; Allen & Allen, Legum. 140. 1981; Irwin & Barneby, Mem. New York Bot. Gard. 35: 4. 1982. Lectotype species: *C. fistula* L.

Senna Miller, Gard. Dict., Abridg. ed., 3. 1754, ed. 8, *Senna no. 1.* 1768; Gaertner, Fruct. 2: 312, t. 146. 1791; Irwin & Barneby, Mem. New York Bot.

Gard. 35: 64. 1982. Type species: *S. alexandrina* Miller.

Chamaecrista Moench, Meth. 272. 1794; Irwin & Barneby, Brittonia 28: 28–36. 1976; Irwin & Barneby, Mem. New York Bot. Gard. 35: 636. 1982. Type species: *C. nictitans* (L.) Moench.

Additional synonymy and references are given by de Wit, l.c. and by Irwin & Barneby, 1982, l.c.

Trees, shrubs, or herbs, usually unarmed; stipules present; leaves paripinnate or rarely reduced to phyllodes; leaflets 2–many; petiole and rachis often glanduliferous; inflorescences racemose or paniculate, terminal or axillary, sometimes cauliflorous; (1–)few–many-flowered; bracts present; bracteoles present or absent; flowers of moderate size to small; sepals 5, imbricate; petals 5, imbricate, subequal, commonly yellow to orange, sometimes white or pink to reddish; stamens usually 10, sometimes fewer, some reduced to staminodes; filaments similar or unequal in length, sigmoidally curved, simply curved, or straight; anthers basifixed or dorsifixed, opening longitudinally or by terminal or basal pores or slits; stigma terminal, small; ovules few–many; fruit dehiscent or indehiscent, large and woody to small and membranaceous, terete or compressed; seeds few–many, compressed, transverse, sometimes with areoles.

A pantropical to warm-temperate genus, considered sensu latior, of about 500–600 species.

The recent treatments of the tribe Cassieae subtribe Cassiinae by Irwin and Barneby, l.c., give rather compelling reasons for separating *Cassia* L. into three genera: *Cassia* L., sensu strictior, *Senna* Miller, and *Chamaecrista* Moench. However, to refrain from departing too drastically from previous treatments of the Ceylon species, and because some of the Asian taxa have not yet been transferred to segregate genera, the more conservative approach is here maintained. Alternative names are provided for use if so desired. The warning and plea of Brenan (Fl. Trop. East Africa, Caesalp. 48. 1967) that ''the keys and descriptions... be read with caution, and I hope also with charity...'' is appropriate in this presentation.

KEY TO THE SPECIES

1 Stamens 10, unequal, with filaments of 3 lowest longer, sigmoidally curved and anthers longer, dehiscent by longitudinal slits; 4 or 5 intermediate stamens with straight or lightly curved filaments and smaller anthers dehiscent by basal pores; 2 or 3 uppermost stamens small, abortive; bracteoles present at base of pedicels; fruit indehiscent, woody, terete; leaves eglandular . . . .(*Cassia* L. sensu strictior)

 2 Flowers yellow

  3 Leaves with 2–8 pairs of leaflets; leaflets 6–20 cm long, 3.5–9 cm wide, acute to subacuminate at the apex, glabrous or nearly so on both surfaces; petals 2–3.5 cm long . . . . .1. **C. fistula**

  3 Leaves with 10–20 pairs of leaflets; leaflets 2.5–3 cm long, 0.8–2 cm wide, obtuse, emarginate, or subacute at the apex, puberulent on both surfaces; petals 1–1.5 cm long . . . . .2. **C. moschata**

2 Flowers pink or reddish to white
  4 Petals 1–1.5 cm long; fruit 20–80 cm long, 2–5 cm wide; flowers on pedicels 1–2 cm long
    5 Fruit (25–)45–80 cm long, 2–5.5 cm in diameter, transversely rugose, one margin with a prominent rib, the other with 2 ribs; seeds imbedded in pulp; leaflets glabrous or subglabrous above, puberulent beneath, glabrescent.....................................**3. C. grandis**
    5 Fruit 20–32 cm long, 2 cm in diameter, seeds imbedded in dry pith; leaflets puberulent on both surfaces or glabrate above.................................**4. C. roxburghii**
  4 Petals 1.2–3.5(–5) cm long; fruit 20–60(–70) cm long, 1–1.5(–2) cm in diameter; flowers on pedicels (2.5–)3–6 cm long.................................**5. C. javanica**
    6 Leaflets mostly rounded or obtuse; calyx dark red; petals 2.5–3.5 cm long, pink to dark red
....subsp. **javanica**
    6 Leaflets mostly acute; calyx green; petals 1.2–2.5 cm long, pink to nearly white, later turning yellowish...subsp. **nodosa**
1 Stamens 10 or fewer, with all filaments essentially straight or only lightly curved, subequal in length; anthers dehiscent apically; 2 or 3 uppermost stamens sometimes abortive; bracteoles near or above the middle of pedicel or lacking; fruit dehiscent or indehiscent, terete or compressed; leaves with glands on petiole and/or rachis, or eglandular
  7 Fruit indehiscent or dehiscent with valves not elastic, terete or compressed; leaves eglandular or bearing glands on petiole or rachis; stamens 10, usually with 7 perfect, 3 reduced or staminodial
....*(Senna* Moench)
    8 Leaves eglandular
      9 Valves of fruit with a longitudinal wing...............................**6. C. alata**
      9 Valves of fruit without wings
        10 Fruit terete or subquadrangular...............................**7. C. spectabilis**
        10 Fruit compressed
          11 Shape of fruit elliptic to oblong or falcate
            12 Leaflets narrow, 3–15 mm wide, acute; fruit oblong to subfalcate, valves not crested
....**8. C. senna**
            12 Leaflets wider, 5–27 mm wide, obtuse; fruit oblong-falcate; valves with a longitudinal ridge of raised·crests...**9. C. italica**
          11 Shape of fruit linear-oblong
            13 Stipules small, subulate, caducous; fruit 15–30 cm long, 1–2 cm wide............
....**10. C. siamea**
            13 Stipules larger, foliaceous; fruit 7.5–15 cm long
              14 Flowers on pedicels 20–30 mm long; fruit glabrous, nitid; trees.................
....**11. C. timoriensis**
              14 Flowers on pedicels 5–7 mm long; fruit short-pilose, glabrescent; shrubs.........
....**12. C. didymobotrya**
    8 Leaves with glands on petiole or rachis
      15 Foliar gland near the base of petiole
        16 Plants glabrous or subglabrous
          17 Gland clavate, about 3–10 mm above base of petiole; leaves with 6–10 pairs of leaflets
....**13. C. sophera**
          17 Gland hemispherical, at base of petiole; leaves with 4–6 pairs of leaflets..........
....**14. C. occidentalis**
        16 Plants hirsute..**15. C. hirsuta**
      15 Foliar glands on rachis
        18 Fruit terete or subterete, about 10–15 mm in diameter
          19 Leaves with 2 pairs of leaflets; gland on rachis between leaflets of lower pair; petals (1–)2–3.3 cm long.......................................**16. C. bacillaris**
          19 Leaves with 3–5 pairs of leaflets; petals 1–2 cm long

20 Leaflets rounded or subacute; gland on rachis between leaflets of lowest pair only; shrubs, sometimes weak and sprawling.............................**17. C. bicapsularis**

20 Leaflets acuminate; glands on rachis often between leaflets of each pair; shrubs or small trees...**18. C. septemtrionalis**

18 Fruit compressed or, if terete, less than 7 mm in diameter

21 Stipules reniform-rotund, persistent............................**19. C. auriculata**

21 Stipules linear, caducous or subpersistent

22 Leaves with about 12–40 pairs of leaflets......................**20. C. multijuga**

22 Leaves with 2–11 pairs of leaflets

23 Fruit 1–2 cm wide

24 Leaflets glabrous above, appressed-pubescent beneath; flowers on pedicels 2.5–4 cm long; fruit glabrous.......................**21. surattensis** subsp. **glauca**

24 Leaflets tomentose; flowers on pedicels 1.5–2.5 cm long; fruit thinly tomentose ....**22. C. tomentosa**

23 Fruit less than 1 cm wide

25 Leaves with 6–11 pairs of leaflets; fruit 6–8 mm wide.......**23. C. divaricata**

25 Leaves with 2–4 pairs of leaflets; fruit 2.5–6 mm wide

26 Seeds 4–5 mm long, 2–3 mm wide, with a lighter coloured areole 3.5–4.5 mm long, 1.5–2 mm wide on each face; pedicels in flower (4–)6–10 mm long, in fruit 12–15 mm long; petals 8–12 mm long; anthers of 3 largest stamens blunt at the apex..**24. C. tora**

26 Seeds 4–5 mm long, 1.5–2 mm wide, with a concolorous areole 2.5–4 mm long, 0.25–0.5 mm wide on each face; pedicels in flower 9–23 mm long, in fruit to 30 mm long; petals (9–)12–15 mm long; anthers of 3 largest stamens slightly narrowed at the apex..**25. C. obtusifolia**

7 Fruit elastically dehiscent, compressed; leaves bearing glands on petiole or rachis; stamens 10 or fewer, similar but usually subequal, sometimes some reduced to staminodes....... ....*(Chamaecrista* Moench)

27 Leaves with 2 pairs of leaflets, bearing a linear gland on the rachis between the leaflets of one or both pairs; leaflets elliptic to elliptic-obovate, 1–3 cm wide.......**26. C. absus**

27 Leaves with more than 2 pairs of leaflets, bearing glands on the petiole below the leaflets; leaflets linear, less than 1 cm wide

28 Glands on petioles stalked or obconoid

29 Foliar gland stalked, peltate, occurring slightly below the lowest leaflets; midrib of leaflets marginal or submarginal.......................................**27. C. kleinii**

29 Foliar gland obconoid, occurring about midway between the base of the petiole and the lowest leaflets; midrib of leaflets excentric but not marginal..**28. C. aeschinomene**

28 Glands on petioles sessile, annulate, occurring just below the lowest pair of leaflets

30 Fruit appressed- or subappressed-pubescent; leaves with 16–60(–85) pairs of leaflets; rachis usually serrate- or crenate-crested above, between the leaflets; plant generally appressed- to crisp-pubescent, or subglabrous................**29. C. mimosoides**

30 Fruit villous; leaves with about 12–36 pairs of leaflets; rachis sometimes slightly crenate above, often ciliate.......................................**30. C. auricoma**

1. **Cassia fistula** L., Sp. Pl. 377. 1753; Wight & Arn., Prod. 285. 1834; Thw., Enum. Pl. Zeyl. 95. 1859; Baker in Hook. f., Fl. Br. Ind. 2: 261. 1878; Trimen, Handb. Fl. Ceylon 2: 103. 1894; Alston in Trimen, Handb. Fl. Ceylon 6: 91. 1931; De Wit, Webbia 11: 207. 1955; Worthington, Ceylon Trees 188. 1959; Macmillan, Trop. Pl. & Gard., ed. 5. 82, 356, 365, 437. 1962; Ali, Fl. West

Pakistan, no. 54, Caesalp., 12. 1973; Saldanha & Nicolson, Fl. Hassan 219. 1976; Irwin & Barneby, Mem. New York Bot. Gard. 35: 14. 1982. Type material: Herb. Hermann 2: 29, 2: 84; 3: 20; 4: 7 (BM-Herm.); Herb. Hermann, fol. 122 (L-Herm); Herb. Linnaeus 528. 15 (LINN).

Cathartocarpus fistula (L.) Pers., Syn. Pl. 1: 459. 1805; Moon, Cat. 34. 1824. *Cassia rhombifolia* Roxb., Fl. Ind., ed. 2, 2: 334. 1832; Wight, Ic. Pl. Ind. Or. 269. 1840. Type: "A native of Ceylon, from thence General Hay Macdowall sent seeds to the Botanic garden at Calcutta in 1802."

Trees, to about 20 m tall; stems glabrous; stipules deltoid to linear-lanceolate, setaceous, 1–2 mm long; caducous; leaves 10–40(–60) cm long with (2–)3–8 pairs of leaflets, eglandular, deciduous; leaflets ovate to ovate-oblong or oblong-lanceolate, 6–10(–20) cm long, 3.5–9 cm wide, acute to subacuminate at the apex, acute to cuneate at the base, glabrous or nearly so on both surfaces; inflorescences racemose, pendent, axillary, many-flowered, 10–40 cm long; bracts and bracteoles ovate to lanceolate-subulate, caducous; flowers on pedicels 3–5(–7.5) cm long; sepals glabrous or minutely puberulent, ovate to oblong, about 6–9(–10) mm long, 5 mm wide, obtuse, deciduous; petals yellow, broadly obovate to ovate-elliptic, 2–3.5 cm long, 1–2 cm wide; stamens 10, unequal, with 3 lower filaments 2.5–4 cm long, anthers about 3.5–4.5 mm long, the other stamens smaller, the 3 uppermost abortive; fruit pendulous, black, terete, glabrous, smooth, indehiscent, 20–60 cm long, 1.5–2 cm in diameter, 40–100-seeded, septate within; seeds brown, glossy, subellipsoid, about 7.5 mm long, 6–7 mm wide, 2.5–3 mm thick with a raised raphe along one side, transverse, imbedded in blackish pulp.

Distr. Probably native to Southeast Asia but distributed pantropically as an ornamental.

Uses. The trees are grown as ornamentals; the flowers are used to decorate temples for religious ceremonies. According to Thwaites, l.c., "every part is used medicinally by the Singhalese as a purgative." The heart of the tree yields a good timber.

Vern. Ahalla-gass, aehaela-gaha, ekela (S); tirukkontai, kavani, konnei (T); showers of gold (E); caña fistula (Latin America).

Specimens Examined. POLONNARUWA DISTRICT: Polonnaruwa, sacred area, *Ripley 66* (US), *74* (US); Habarana, along Habarana-Kantalai Rd., *Cramer 2988* (US); 3 mi. SW. of Elahera, near mi. marker 12/5, *Davidse 7336* (MO). PUTTALAM DISTRICT: Wilpattu Natl. Park, at margin of Periya Naga Villu, *Wirawan et al. 878 A* (K, MO, PDA, SFV, UC, US). KURUNEGALA DISTRICT: Kurunegala, Ooma Oya, July 1853, *Thwaites C.P. 1502* (PDA); without data, *C. P. 1502* (BM); Kurunegala, Yakdessakande, *Worthington 4006* (K). MATALE DISTRICT: Sigiriya, Kibissa, 600', *Worthington 4357* (K). NUWARA ELIYA DISTRICT: Watagoda, Medacombra, 3300', ("planted in garden and reported not to have flowered, no doubt too high elevation"),

Worthington (H. Andrews) 4315 (K). TRINCOMALEE DISTRICT: Trincomalee, *Worthington 182* (K). AMPARAI DISTRICT: Pottuvil, *Simpson 8277* (BM). HAMBANTOTA DISTRICT: Rugamtota, on Menik Ganga, *Fosberg 50185* (PDA, SFV, US); near shore of Katagamuwa, *Mueller-Dombois & Comanor 67062522* (PDA, US). KANDY DISTRICT: Kandy Lake, *Worthington 6833* (K); Galphele, Panwila, 2300', *Worthington 1047* (K).

2.   Cassia moschata H.B.K., Nov. Gen. & Sp. 6: 338. 1824; Macmillan, Trop. Pl. & Gard., ed. 5, 82, 83. 1962; Irwin & Barneby, Mem. New York Bot. Gard. 35: 33. 1982. Type: Colombia, Bolívar, Mompós, "Cresit in sylvis, ad fluvium Magdalenae, prope Mompox", *Humboldt & Bonpland 1479* (P-Humb.).

Trees, to about 20 m tall; young stems pilose, glabrescent; stipules 1.5–4.5 mm long, auriculate below the point of attachment, the upper lobes attenuate, usually longer than the acute lower lobe, caducous; leaves eglandular, with about 10–20 pairs of opposite or subopposite leaflets, deciduous; leaflets lance-oblong to ovate-oblong, 2.5–5 cm long, 0.8–2 cm wide, obtuse, emarginate or subacute at the apex, rounded to subcordate at the base, puberulent on both surfaces, the margin somewhat revolute; inflorescences racemose, terminal or axillary, about (6–)9–32 cm long, 25–70-flowered; bracts and bracteoles lance-subulate, to about 3 mm long, caducous; flowers on pedicels (5–)6–12(–14) mm long; sepals ovate-orbicular, obtuse, (5.5–)6–8(–8.5) mm long; petals yellow, often with red venation, 10–15.5 mm long; stamens 10, unequal with 3 filaments 1.5–2.5 cm long, sigmoid and anthers 3–4 mm long, opening by terminal pores; the other stamens with shorter filaments, 4 with anthers opening by basal pores, 3 more or less rudimentary; fruit terete, puberulent, glabrescent, 35–50 cm long, 1.2–2 cm in diameter, about 75–100-seeded, septate and pulpy within; seeds brown, lustrous, ellipsoid, 7–8 mm long, 5–6 mm wide.

Distr. Native to the American tropics, distributed pantropically as an ornamental.

Uses. Planted as an ornamental, and the musky-smelling pulp reputedly has medicinal properties.

Vern. Cañafistolo (Colombia, Venezuela).

Specimens Examined. KANDY DISTRICT: Peradeniya, *Worthington 824* (K), *4272* (K); Kandy, planted from Peradeniya, *Worthington 7181* (K).

3.   Cassia grandis L.f., Suppl. 230. 1781; de Wit, Webbia 11: 212. 1955; Macmillan, Trop. Pl. & Gard. ed. 5, 82. 1962; Irwin & Barneby, Mem. New York Bot. Gard. 35: 30. 1982. Syntypes: Breyne, Exot. Pl. Cent., t. 14. 1678; Surinam, *Dahlberg s.n.* (cf. Herb. Linn. 528.29, LINN).

Trees, to about 30 m tall; young stems tomentulose; stipules falcate or subulate, 0.5–1.5 mm long, caducous; leaves 15–25 cm long, short-petiolate, eglandular, with 8−20 pairs of leaflets, deciduous; leaflets elliptic-oblong, 2.5−6 cm long,

1.4–2.2 cm wide, rounded at apex and base, glabrous or with sparse, minute hairs above, puberulent or tomentulose, glabrescent beneath; inflorescences simple-racemose, lateral, 10–20 cm long, usually appearing when the tree is leafless or nearly so, many-flowered; bracts and bracteoles ovate, acute, tomentose, about 5 mm long, caducous; flowers on pedicels 1–2 cm long; sepals tomentulose, obovate, obtuse, 6–9 mm long; petals pink or white, orbicular to obovate, about 1–1.5 cm long; stamens 10, unequal with 3 lower filaments about 2 cm long, sigmoid, the others shorter; anthers pilosulous, the three larger opening by apical and basal pores or slits, 5 on shorter filaments opening by apical and basal pores, 2 smaller and abortive; fruit indehiscent, terete or subterete, transversely rugose, one margin with a prominent obtuse rib, the other with 2 ribs; (25–)45–80 cm long, 2–5.5 cm wide, 70–80-seeded; seeds brown, lustrous, obovoid-ellipsoid, 14–16 mm long, 9–10 mm wide, transverse, imbedded in pulp.

Distr. A native of tropical America but distributed pantropically as an ornamental, sometimes becoming naturalized.

Uses. An ornamental tree with strong, handsome wood useful for various purposes; the musky-smelling pulp of the fruit reportedly is of medicinal value.

Vern. Pink shower, pink coral shower, horse cassia, stinking toe (E); marimari, cañaflote, etc. (Latin America).

Specimens Examined. WITHOUT LOCALITY: *Worthington 7010* (K). KANDY DISTRICT: Kandy, Hillcrest, *Worthington 7171* (K); Peradeniya, Royal Botanic Gardens, *Worthington 4268* (K); Attabagie, Gampola, *Worthington 2831* (BM, K).

4. Cassia roxburghii DC., Prod. 2: 489. 1825: Wight, Ic. Pl. Ind. Or. 83. 1838; Thw., Enum. Pl. Zeyl. 95. 1859; Ali, Fl. W. Pakistan, no. 54, Caesalp., 23. 1973; De Wit, Webbia 11: 226. 1955; Irwin & Barneby, Mem. New York Bot. Gard. 35: 51. 1982. Type: based on *C. marginata* Roxb., non Willd. 1809.

Cassia marginata Roxb., Fl. Ind. ed. 2, 2: 338. 1832, non Willd., 1809; Wight & Arn., Prod. 286. 1834; Baker in Hook. f., Fl. Br. Ind. 2: 262. 1878; Trimen, Handb. Fl. Ceylon 2: 104. 1894; Alston in Trimen, Handb. Fl. Ceylon 6: 91. 1931; Worthington, Ceylon Trees 189. 1959; Macmillan, Trop. Pl. & Gard., ed. 5, 82. 1962. Type: "A native of Ceylon introduced into the Botanic Gardens at Calcutta by General Macdowall in 1802", *Wallich 5308* (K).

Trees, to about 10 m tall; young stems tomentulose, glabrescent; stipules pubescent, caducous, falcate, produced below the point of insertion, the upper lobe acuminate, to about 10 mm long, the lower lobe sagittate, acute or acuminate, about 5 mm long; leaves 15–25 cm long, eglandular, with about 8–15 pairs of leaflets; leaflets ovate to elliptic-oblong, 1.5–4.5 cm long, 1–2 cm wide, obtuse or retuse at the apex, asymmetrically rounded at the base, puberulent or glabrate above, appressed pubescent with minute hairs beneath; inflorescences 5–8 cm long, racemose, somewhat corymbose, axillary; bracts sericeous, ovate, acute, about 3–8 mm long, 4–5 mm wide, caducous; bracteoles, at base of pedicel, oblong, acute, 3–4 mm long, 1 mm wide; flowers on pedicels 10–14(–19) mm long; sepals puberulent, ovate, obtuse, 3–6 mm long, 3–4 mm wide; petals pink to salmon or rose, turning yellowish or orange, ovate to obovate, obtuse or subacute, 10–15 mm long, 5–8 mm wide, puberulent; stamens 10, the 3 lower filaments 2.5–3 cm long, the others shorter; 3 stamens with reduced filaments and anthers; fruit terete, indehiscent, smooth, black, 20–32 cm long, 2 cm in diameter, sometimes slightly constricted between the seeds, septate within, many-seeded.

Distr. A native of Ceylon and the Indian peninsula.

Uses. Cultivated as an ornamental; the wood is used for handles of tools, hubs of wheels, etc.

Vern. Ratoo-waa, ratu-wa (S); vakai (T).

Specimens Examined. WITHOUT EXACT LOCALITY: *Macrae 375* (BM); *Col. Walker 55* (K); *1493* (K); *Worthington 7046* (K); Matale & Jaffna, *Gardner,* Hayama, July 1852, Ooma Oya, July 1853, *Thwaites C.P. 1505* (PDA); without data, *C.P. 1505* (BM, GH). KANDY DISTRICT: in jungle near Nishaoya Estate, Haragama, *Alston 459* (PDA). ANURADHAPURA DISTRICT: Madatugama to Kekirawa, *Hepper & Jayasuriya 4610* (K, US); Kekirawa, along road to Anuradhapura, *van Beusekom 1599* (US). POLONNARUWA DISTRICT: Polonnaruwa, Mutugale, *Worthington 6410* (K); Polonnaruwa sacred area, *Dittus WD 70012104* (US), *Sohmer & Sumithraarachchi 10821* (GH). KURUNEGALA DISTRICT: Dekanduwela, 3 mi. W. of Rambe, *Maxwell & Jayasuriya 781* (MO, PDA, SFV, US). MATALE DISTRICT: Dambulla, along road to Kandalama tank, *Sumithraarachchi 446* (CAS, SFV, US); Kandy-Dambulla Rd., 36 mi. marker, culvert 4, *Sohmer 8059* (GH, MO, US); Nalanda, 1000', *Worthington 6354* (K); Inamaluwa, turn-off to Sigiriya on road to Trincomalee, *Fosberg 56376* (GH, MO, US); Sigiriya, *Reitz 30024* (US); Galewela, 650', dry zone, *Worthington 4258* (K); road from Naula to Dambulla, mi. marker 36/4, *Nowicke et al. 322* (K, MO, US). TRINCOMALEE DISTRICT: N. shore of Kinniya Ferry, entrance to Tambalagam Bay, SW. of Trincomalee, *Davidse 7575* (MO); Trincomalee, Clappenberg Bay, *Worthington 4374* (K); China Bay, *Worthington 504* (BM, K); Trincomalee, *Worthington 2059* (BM, K), *Alston 544* (PDA); near Kuchavelli, *Kostermans 24832* (BM, K, US). BATTICALOA DISTRICT: Batticaloa, 16-8-26, *Petch s.n.* (A); dunes, *Simpson 8262* (BM, PDA). AMPARAI DISTRICT: Pottuvil, 2 mi. toward Lahugula, *Mueller-Dombois & Comanor 67072563* (PDA, US); Maha Oya, 1 mi. S., *Mueller-Dombois 67081422* (PDA, US). BADULLA DISTRICT: Near Mahiyangana, *Kostermans 25334* (K). MONERAGALA DISTRICT: Bibile, 12-7-24, *Silva s.n.* (PDA). HAMBAN-TOTA DISTRICT: Ruhuna Natl. Park, along Menik Ganga NW. of Yala,

Davidse 7797 (US); Yala camp, *Wirawan 707* (GH, K, MO, PDA, SFV, UC, US); about 1 mi. above mouth of Menik Ganga, *Fosberg 50217* (PDA, SFV, US); Situlpahuwa Rd., *Cooray 70032703 R* (K, MO, PDA, SFV, US); Kotabandu Wewa, *Comanor 415* (BM, GH, K, PDA, SFV, US); Kumbukkan Oya, *Comanor 424* (K, PDA, SFV, US).

5. Cassia javanica L., Sp. Pl. 379. 1753.

Trees, to about 30–40 m tall; young stems glabrous or sparsely puberulent, sometimes armed with thorn-like remnants of branches; stipules variable, foliaceous, broadly elliptic to falcate or lunate to ovate, acuminate, 12–25 mm long, 1.5–10 mm wide, often caducous; leaves 15–40 cm long, eglandular, with (4–)8–17(–20) pairs of leaflets; leaflets elliptic or elliptic-oblong, 2.5–5.5(–10) cm long, 1.5–3.5 cm wide, obtuse, retuse, or acute at the apex, rounded to acute at the base, essentially symmetrical, subnitid but puberulent with minute hairs on both surfaces, glabrescent above; inflorescences racemose-subcorymbose, to about 18 cm long, terminal or lateral, few- to many-flowered; bracts and bracteoles at base of pedicel, ovate-oblong, long-acuminate, 10–12(–18) mm long, to about 5 mm wide; flowers on pedicels (2–)4–5(–6) cm long; sepals puberulent, ovate, obtuse to subacute, about 6 mm long, 3.5 mm wide; petals puberulent, pale pink to dark reddish, (1.5–)2–3.5 cm long; stamens 10, unequal, 3 with longer filaments, recurved, about 3–4 cm long, glabrous, enlarged at about mid-point, the others shorter; anthers dorsally pubescent, those with longest filaments dehiscing longitudinally, the others by basal pores and apical slits, 3 much reduced, abortive; fruit pendulous, terete, glabrous, black, lustrous, indehiscent, septate, 20–70 cm long, 1–2 cm in diameter, 50–80-seeded; seeds transverse, lenticular to orbicular, 6.5–8 mm in diameter, 4–5 mm thick, light brown, lustrous, imbedded in dry pith.

Distr. Native to tropical Asia but introduced elsewhere in the tropics.

1. subsp. javanica

Cassia javanica L., Sp. Pl. 379. 1753; de Wit, Webbia 11: 214. 1955; Macmillan, Trop. Pl. & Gard., ed. 5, 82. 1962; Backer & Bakh. f., Fl. Java 1: 537. 1963; Ali, Fl. W. Pakistan, no. 54, Caesalp. 23. 1973; K. & S. Larsen, Nat. Hist. Bull. Siam Soc. 25: 205. 1974; Irwin & Barneby, Mem. New York Bot. Gard. 35: 46. 1982. Type: Commelin, Hort. Med. Amst. 1: 217, t. 111. 1697.

Cassia bacillus Gaertn., Fruct. 2: 313. 1791; Wight, Ic. Pl. Ind. Or. 252. 1840. Type: ''Cassia fistula indica, flore carneo, *Jaegeri,* Breyn. prodr. 2: 51, Ab amiciss meo Hudsono.''

The leaflets of this typical subspecies are rounded or obtuse; rachis of the inflorescence stout; flowers with calyx dark red and petals pink, turning dark red, 2.5–3.5 cm long.

Uses. Planted as an ornamental.

Vern. Java cassia (E).

Specimens Examined. KANDY DISTRICT: Peradeniya, cultivated, *Alston 614* (K), *1402* (UC). "This is quite distinct from *C. nodosa* with which it has been confused as specimens of *nodosa* have often been distributed as *javanica*."

2. subsp. **nodosa** (Buch.-Ham. ex Roxb.) K. & S. Larsen, Nat. Hist. Bull. Siam Soc. 25: 205. 1974.

Cassia nodosa Buch.-Ham. ex Roxb., Fl. Ind., ed. 2, 2: 336. 1832; Wight, Ic. Pl. Ind. Or. 410. 1840; de Wit, Webbia 11: 223. 1955; Macmillan, Trop. Pl. & Gard., ed. 5, 83. 1962; Backer & Bakh. f., Fl. Java 1: 537. 1963; Ali, Fl. W. Pakistan, no. 54, Caesalp. 21. 1973. Type: India, Chittagong (K; isotype BM).

Cassia javanica L. var. *indochinensis* Gagnep., Fl. Gén. Indochine 2 (2): 158. 1913; Irwin & Barneby, Mem. New York Bot. Gard. 35: 50. 1982. Syntypes: According to Irwin & Barneby, l.c., some 13 collections were cited, from Annam, Laos, Siam, Cambodia, and Cochinchina.

This subspecies is characterized by acute leaflets; rachis of the inflorescence slender; flowers with calyx green and petals pink to nearly white, turning to yellowish pink, 1.5−2 cm long.

Uses. Planted as ornamentals.

Vern. Wesak gaha (S); pink cassia (E).

Specimens Seen, but should be checked. *Worthington 6282* (K). Colombo, in garden, *Worthington (Mrs. Wignaraja) 4176* (K). Kadugannawa, garden, *Worthington s.n.* (K); Getambe, Peradeniya, *Worthington 4267* (K); Gampola, Attabagie Estate, *Worthington 2829* (BM, K); Kandy, Haloya Estate, Bungalow Garden, *Worthington 3017* (BM, K); Kandy, "Rest Harrow", *Worthington 6283* (K), *6831* (K), *6832* (K).

Note. As I recall there are specimens at PDA that I did not have time to examine.

6.   Cassia alata L., Sp. Pl. 378. 1753; Wight, Ic. Pl. Ind. Or. 253. 1840; Thw., Enum. Pl. Zeyl. 97. 1859; Baker in Hook. f., Fl. Br. Ind. 2: 264. 1878; Trimen, Handb. Fl. Ceylon 2: 108. 1894; Alston in Trimen, Handb. Fl. Ceylon 6: 92. 1931; de Wit, Webbia 11: 231. 1955; Macmillan, Trop. Pl. & Gard. 104. 1962; Ali, Fl. W. Pakistan, no. 54, Caesalp. 25, f. 6, 1973. Syntype: "America calidiore" (BM-Cliff).

Senna alata (L.) Roxb., Fl. Ind., ed. 2, 2: 349. 1832; Irwin & Barneby, Mem. New York Bot. Gard. 35: 460. 1982.

Shrubs, to about 5 m tall; stems puberulent to subglabrous; stipules deltoid, acute to acuminate at the apex, auriculate at the base, 1–2 cm long; leaves eglandular, 30–60 cm long with 6–14 pairs of leaflets; leaflets oblong to obovate-oblong, 5–15(–19) cm long, 2–7(–12) cm wide, rounded to retuse at the apex, rounded to subcordate at the base, puberulent to glabrous on both surfaces; inflorescences terminal or axillary, erect, spicate-racemose, many-flowered, 50–70 cm long; bracts petaloid, yellow or orange, 1–2.5 cm long, ovate-elliptic to oblong or suborbicular, subacute; bracteoles lacking; flowers on pedicels 4–10 mm long; sepals orange-coloured, 1–1.5 cm long, puberulent; petals yellow or yellow-orange, 1.5–2 cm long, obovate to spatulate; stamens 10; filaments essentially straight; anthers 2 large, 5 medium-sized, and 3 small; fruit dehiscent, chartaceous, linear, 10–15 cm long, 1.5–2 cm wide, septate, many-seeded, appearing 4-alate, each valve with a longitudinal wing about 5–10 mm wide, glabrous or puberulent; seeds light or dark brown, rhomboid, compressed, transverse, 5.5–7 mm long, 4.5–5.5 mm wide, bearing on each face an elliptic to linear-elliptic areole, 2–3 mm long, 0.5–0.9 mm wide.

Distr. Native to tropical America but introduced pantropically.

Uses. An ornamental shrub, also used medicinally, especially for ringworm and as a purgative.

Vern. Candle-bush, candle-stick, ringworm shrub (E) and various local names in tropical America.

Specimens Examined. ANURADHAPURA DISTRICT: Nuwara Wewa, *Simpson 93599* (BM, PDA). PUTTALAM DISTRICT: About 14 mi. from Kalpitiya along road to Palavi, *Sumithraarachchi 707* (SFV, US). KURUNEGALA DISTRICT: Near Ganewatta, *Simpson 8211* (BM, PDA). KANDY DISTRICT: Kadugannawa, *J.M. de Silva 128* (NY).

7. Cassia spectabilis DC., Cat. Hort. Monsp. 90. 1813; de Wit, Webbia 11: 267. 1955; Worthington, Ceylon Trees 192. 1959. Type: "Hab. ad Caracas", Venezuela, "cult. in h. m. oct. 1809", Hort. Montpelier (G-DC).

Senna spectabilis (DC.) Irwin & Barneby, Mem. New York Bot. Gard. 35: 600. 1982.

Trees, to about 18 m tall; young stems pilosulous; stipules linear, about 10 mm long, subpersistent; leaves eglandular, 20–45(–50) cm long with (8–)10–16 pairs of leaflets; leaflets lanceolate to oblong-elliptic, about 3–9 cm long, 2(–3) cm wide, acute to acuminate, glabrous or subglabrous above, sparsely pilosulous to tomentulose beneath; inflorescences axillary, racemose, few-many-flowered or terminal panicles to 30 cm long; bracts lanceolate, caducous; flowers on pedicels 2.5–3 cm long; sepals ovate-orbicular, 5–10 mm long, glabrous to puberulent;

petals yellow, (1.5–)2–2.5(–3.5) cm long, unequal; stamens 10, 1 large, 6 medium-sized, 3 reduced to staminodes; fruit linear, terete or subquadrangular, indehiscent or tardily dehiscent along one margin, coriaceous, glabrous, 15–30 cm long, 1 cm in diameter, septate, many-seeded, sometimes irregularly constricted between the seeds; seeds olive, suborbicular, 5 mm long, 4 mm wide bearing an ovate areole about 2 mm long, 1.5 mm wide on each face.

Distr. A native of South America, introduced elsewhere.

Uses. Planted as an ornamental.

Vern. Kaha-kona (S); munjal-kona (T); various local names in tropical America.

Specimens Examined. KANDY DISTRICT: Kandy, "from seed brought over in 1934 and planted at Kadugannawa", *Worthington 3039* (BM, K); Kadugannawa, *Worthington 558* (BM, K), *644* (BM), "introduced from E. Africa by collector, 1934", *s. coll. 1210* (K).

8. Cassia senna L., Sp. Pl. 377. 1753, in part, excluding β *Senna italica*; Brenan, Kew Bull. 13: 243. 1958; Brenan in Fl. Trop. East Africa, Caesalp. 65. 1967; Ali, Fl. W. Pakistan, no. 54, Caesalp. 12, fig. 2, C, E. 1973. Type: Uncertain, fide Brenan 1967, l.c. Lectotype: Morison, Pl. Hist. Univ. Oxon. 2(2): t. 24, fig. 1. 1715, vide Irwin & Barneby, l.c.

Senna alexandrina Mill., Gard. Dict., ed. 8, *Senna* no. 1. 1768; Irwin & Barneby,
    Mem. New York. Bot. Gard. 35: 481. 1982.
Cassia angustifolia Vahl, Symb. 1: 29. 1790. Type: Yemen, *Forsskal* (C holotype);
    (BM isotype).
Cassia acutifolia Delile, Fl. Aegypt. Ill. 61, t. 27. 1813. Type: Egypt, *Delile* ? (P).

Shrubs or suffruticose perennials, to about 3 m tall; stems puberulent to glabrous; stipules linear-deltoid, subulate, 1.5–5 mm long, caducous; leaves eglandular, 5–16 cm long with (3–)5–9(–12) pairs of leaflets; leaflets lanceolate to elliptic or ovate, 1.2–5(–6) cm long, 3–10(–15) mm wide, usually acute at the apex, rounded at the base, pubescent to glabrous on both surfaces; inflorescences terminal or axillary, racemose, (7–)10–30-flowered; bracts cupuliform, 5–10 mm long, 3.5–5 mm wide; flowers on pedicels 3–5 mm long; sepals 7–11 mm long; petals yellow, 10–17 mm long, 7–9 mm wide; stamens 10, 2 with large anthers, 5 medium, 3 small; fruit oblong, subfalcate, compressed, dehiscent, chartaceous, 3–7 cm long, 1.6–2.6 cm wide, about 4–10-seeded, pubescent, the surface of the valves not crested; seeds compressed, 6–7 mm long, 4 mm wide, reticulate or rugose, bearing an areole 1–2 mm long, 0.5–0.7 mm wide on each face.

Distr. Africa, West Pakistan, India and Ceylon.

Uses. Drug, used as a cathartic.

Vern. True senna (E).

Specimens Examined. WITHOUT LOCALITY: *Walker s.n.* (K, PDA).
ıNURADHAPURA DISTRICT: Anuradhapura, Oct. 1853, *s. coll. s.n.* (PDA).

ı. **Cassia italica** (Mill.) Spreng., Nachtr. Bot. Gart. Univ. Halle. 1801; F.W.
ındr., Fl. Pl. Anglo-Egypt. Sudan 2: 117. 1952; Brenan, Kew Bull. 13: 239.
958; Brenan in Fl. Trop. East Africa, Caesalp. 65. 1969; Ali, Fl. W. Pakistan,
ıo. 54, Caesalp. 15, f. 3 A–C, 16. 1973.

ıenna italica Mill., Gard. Dict. ed. 8, *Senna* no. 2. 1768; Irwin & Barneby,
    Mem. New York. Bot. Gard. 35: 482. 1982.
ıassia obovata Collad., Hist. Cass. 92, t. 15A. 1816, nom. illegit; Baker in Hook.
    f., Fl. Br. Ind. 2: 264. 1878.
ıassia obtusa Roxb., Hort. Beng. 31. 1814, nom. nud.; Wight, Ic. Pl. Ind. Or.
    757. 1843.

Perennial herbs, usually suffrutescent, to about 1 m high, sometimes prostrate;
tems puberulent to glabrous; stipules linear-lanceolate, acute to acuminate, 3–9
ım long, persistent; leaves 3–12 cm long, eglandular, with 3–7(–7) pairs of
ıaflets; leaflets elliptic to obovate-oblong, about 1–4 cm long, 0.5–2(–2.7) cm
vide, obtuse at the apex, mucronulate, rounded at the base, glabrous or
ubglabrous on both surfaces; inflorescences axillary, racemose, 2–25 cm long;
ıracts 3–5 mm long, caducous; flowers on short pedicels; sepals subequal, 8–13
ım long; petals subequal, pale to bright yellow, 9–20 mm long; stamens 9 or
0, 2 with large anthers, 4 or 5 medium, and 3 small; fruit compressed, oblong-
ıalcate, (2.5–)3–6 cm long, 1–2 cm wide, rounded at each end, dark when mature,
ıalves glabrous, chartaceous, strongly veined with a longitudinal ridge of raised
ırests, 6–12-seeded; seeds dark brown, wedge-shaped, 6–7 mm long, 3–4 mm
vide, with an areole 1–1.2 mm long, 0.5 mm wide on each face.

Distr. Middle East, Africa, Pakistan, India and northern Ceylon; also inroduced in the West Indies.

Uses. The leaves serve as a cathartic and are sometimes substituted for *C.*
ıenna.

Vern. Nilavakai (T); Italian senna (E).

Specimens Examined. JAFFNA DISTRICT: Between Jaffna and
ıankesanturai, Feb. 1890, *Trimen ?* (PDA). MANNAR DISTRICT: Mannar,
890, *M.S. Crawford 138* (PDA). TRINCOMALEE DISTRICT: "Trincomalie,
ıev. S.O. Glenie", and *C.P. 3843* (PDA).

ı0. **Cassia siamea** Lam., Enc. 1: 648. 1783; Baker in Hook. f., Fl. Br. Ind.
ı: 264. 1878; Trimen, Handb. Fl. Ceylon 2: 108. 1894; Alston in Trimen, Handb.
ıl. Ceylon 6: 91. 1931; de Wit. Webbia 11: 263. 1955; Worthington, Ceylon
ırees 191. 1959; Ali, Fl. W. Pakistan, no. 54, Caesalp. 25, f. 7. 1973; Saldanha
ı Nicolson, Fl. Hassan 221. 1976. Type: *Commerson s.n.* (P-Lam).

Cassia sumatrana Roxb., Hort. Beng. 31. 1814, nom. nud; Moon, Cat. 33. 1824; DC., Prod. 2: 506. 1825.

Senna sumatrana Roxb., Fl. Ind. ed. 2, 2: 347. 1832.

Cassia florida Vahl, Symb. Bot. 3: 57. 1794; Thw., Enum. Pl. Zeyl. 96. 1859.

Senna siamea (Lam.) Irwin & Barneby, Mem. New York Bot. Gard. 35: 98. 1982.

Trees, to about 6–12 m tall; young stems puberulent to glabrous; stipules small, subulate, caducous; leaves 12–18 cm long, eglandular, with 6–10(–14) pairs of leaflets; leaflets elliptic-oblong to oblong-lanceolate, (3–)5–8 cm long, 1–2 cm wide, obtuse, mucronate or sometimes retuse, glabrous above, glabrous or minutely pubescent with raised or appressed hairs beneath, glaucous; inflorescences 15–30(–60) cm long; corymbose-racemose or paniculate, many-flowered, terminal or axillary; bracts linear, curved, dilated at the middle, 3–6 mm long; flowers on pedicels 2.5–3.5 cm long; sepals suborbicular, subequal, greenish-yellow, 5–7 mm long, puberulent; petals yellow, 10–16(–20) mm long; stamens 10, 2 or 3 with long anthers, 4 or 5 with medium, 3 small; fruit linear with somewhat thickened margins, nearly straight, velutinous or puberulent, glabrescent, compressed, coriaceous, (15–)20–30 cm long, 1–2 cm wide, 8–20(–25)-seeded; seeds light brown, ovoid, glossy, about 6.5–8 mm long, 5.5–6 mm wide, with an areole about 3–4.5 mm long, 0.9–1.2 mm wide on each face.

Distr. Apparently native to India and Pakistan to Malaysia; in Ceylon may be native or introduced; introduced in other tropical areas of the new and old worlds.

Uses. The wood is used commercially for small articles and for construction; the trees are planted as coffee shade; the roots presumably are medicinal; the flowers are sometimes eaten in curries.

Vern. Wa, aramana (S); manga konnei, vakai (T); kassod tree (E).

Specimens Examined. WITHOUT EXACT LOCALITY: "Batalagolla Veva," *Simpson 8854* (BM, NY). KURUNEGALA DISTRICT: Kurunegala, *Worthington 2706* (K). KANDY DISTRICT: Kandy, "rest house, planted 10 years ago and felled", *Worthington 7209* (K); Kandy catchment, *Worthington 6896* (K), *7218* (K); Peradeniya, *Gardner 260* (K, NY), *Gardner, Thwaites C.P. 3485* (BM, PDA). TRINCOMALEE DISTRICT: Trincomalee, Nov. 1890, *Nevill s.n.,* (PDA). KEGALLE DISTRICT: Yatiyantota, *Worthington 407* (BM, K).

11.   Cassia timoriensis DC., Prod. 2: 499. 1825; Thw., Enum. Pl. Zeyl. 96. 1859; Baker in Hook. f., Fl. Br. Ind. 2: 265. 1878; Trimen, Handb. Fl. Ceylon 2: 108. 1894; Alston in Trimen, Handb. Fl. Ceylon 6: 91. 1931; de Wit, Webbia 11: 273. 1955.

Senna glauca Roxb., Fl. Ind. ed. 2, 2: 351. 1832, non *Cassia glauca* Lam.

Senna timoriensis (DC.) Irwin & Barneby, Mem. New York Bot. Gard. 35: 98. 1982.

Trees, to about 20(–30) m tall; young stems pubescent; stipules foliaceous, broadly falcate with a filiform tip, often lobate or dentate, about 9 mm long; semipersistent; leaves 14–20 cm long, eglandular, with 10–20 pairs of leaflets; leaflets narrowly oblong, 1.5–5.5 cm long, 0.8–1.7 cm wide, obtuse, apiculate, pubescent on both surfaces; inflorescences corymbose-paniculate, axillary or terminal; bracts linear, acute, 3–4 mm long, pubescent, caducous; flowers on pedicels about 2–3 cm long; sepals unequal, 5–10 mm long, ovate, obtuse, puberulent; petals yellow, obovate, about (1.5–)2–3 cm long; stamens 10, anthers long in 2, medium in 5, and reduced in 3; fruit dehiscent, glabrous, nitid, about 6–15 cm long, 1–1.5 cm wide, not raised along the margins, septate, 10–20-seeded; seeds brown, suborbicular, beaked at the hilum, bearing areoles on both faces.

Distr. Malaysia, Indonesia, introduced elsewhere.
Uses. Planted as shade trees and said to be medicinal.
Vern. Limestone Cassia (E); various local names in Malaysia and Indonesia.
Specimens Examined. WITHOUT LOCALITY: *Walker s.n.* (GH). KEGALLE DISTRICT: "Four Korales Distr., common, *Gardner,* Sitawaka, Sept. 1857", *Thwaites C.P. 1504* (BM, K, PDA); "Allegalle", 3000 ft, *Gardner 257* (K). COLOMBO DISTRICT: Heneratgoda, *Beddome 2521* (BM).

12. Cassia didymobotrya Fresen., Flora 22: 53. 1839; de Wit, Webbia 11: 241. 1955; Brenan in Fl. Trop. East Africa, Caesalp. 66, f. 12. 1967; Ali, Fl. W. Pakistan, no. 54, Caesalp. 25. 1973. Type: Ethiopia, *Rueppell.*

Senna didymobotrya (Fresen.) Irwin & Barneby, Mem. New York Bot. Gard. 35: 467. 1982.

Shrub, to about 4.5(–9) m tall; young stems puberulent; stipules cordate-ovate, acuminate, about 1.5–2.5 cm long, 8–12 mm wide, pubescent or subglabrous, ciliate, persistent; leaves 10–30(–36) cm long, eglandular, with 7–18 pairs of leaflets; leaflets elliptic-oblong, 2–6.5 cm long, 0.6–2.5 cm wide, obtuse or subacute, mucronate at the apex, rounded at the base, sparsely pubescent with subappressed hairs on both surfaces; inflorescences racemose, subspicate, axillary, to about 40 cm long; bracts ovate, 9–27 mm long, 5–14 mm wide; flowers on pedicels 5–7 mm long; sepals puberulent, 12 mm long; petals yellow with darker veins, 1.8–3 mm long; stamens 10, 2 anthers large, 5 medium, and 3 small, staminodial; fruit oblong, compressed, short-pilose, glabrescent, 7.5–12 cm long, 1.5–2.5 cm wide, dehiscent, septate, 9–16-seeded; seeds compressed, sublenticular, apiculate at the apex, 8–9 mm long, 4–5 mm wide, 2.5 mm thick, with in oblong areole 4 mm long, 1.5 mm wide on each face.

Distr. Native to Africa, introduced into Ceylon, India, Malaysia and some areas of tropical America.
Uses. Planted as an ornamental and as green manure; leaves are boiled and drunk for constipation and for gonorrhoea.

Vern. At tora, rata tora (S).

Specimens Examined. KANDY DISTRICT: Madulkelle-Katooloya Estate, planted for nitrogen, *Worthington 1973* (BM, K); near Pussellawa, hwy 5, mi. 22, *Mueller-Dombois 67082405* (PDA, US), between hwy 23/11 and 23/13, *Maxwell & Jayasuriya 857* (MO, NY, SFV, US); Kandy, *Alston s.n.* (PDA); Moray Tea Estate, *Fosberg 58137* (US); Hewaheta Rd., *Sumithraarachchi 109* (SFV, US); Oodoowela Estate, *Worthington 1771* (BM); Kadugannawa, *Worthington 4999* (K). BADULLA DISTRICT: Bandarawela Hotel, *Worthington 5832* (K).

13. Cassia sophera L., Sp. Pl. 379. 1753; Moon, Cat. 33. 1824; Thw., Enum. Pl. Zeyl. 95. 1859; Baker in Hook. f., Fl. Br. Ind. 2: 262. 1878; Trimen, Handb. Fl. Ceylon 2: 105. 1894; Alston in Trimen, Handb. Fl. Ceylon 6: 91. 1931; de Wit, Webbia 11: 265. 1955; Brenan in Fl. Trop. East Africa, Caesalp. 78. 1967; Ali, Fl. W. Pakistan, no. 54, Caesalp. 27. 1973. Type: Ceylon, *P. Hermann*, Herb. Herm. 4: 79 (holotype BM-Herm.); isotype ?, Herb. Burman 45 (Paris, Biblio. Inst. France).

Senna sophera (L.) Roxb., Fl. Ind., ed. 2, 2: 347. 1832, as *S. "sophora"*; Irwin & Barneby, Mem. New York Bot. Gard. 35: 440. 1982.

Shrubs or suffrutescent herbs, to about 3 m high; young stems glabrous or subglabrous; stipules ovate, subglabrous, caducous, about 6 mm long; leaves about (7–)15–25 cm long with 6–10 pairs of leaflets, bearing a clavate gland about 2 mm long, 0.5 mm in diameter, 3–10 mm above the base of the petiole; leaflets lanceolate, (2–)3–9 cm long, 1–2.5 cm wide, acute, mucronate at the apex, rounded at the base, ciliolate, otherwise glabrous or subglabrous; inflorescences simple or compound corymbose-racemose, axillary or terminal, about 4–10-flowered; bracts ovate, 4–5 mm long, acute or obtuse, pubescent; flowers on pedicels (8–)10–15(–18) mm long; sepals ovate, obtuse, 5–6.5 mm long, puberulent or glabrous; petals yellow, orbicular to obovate, obtuse, 10–17 mm long; stamens 10, the 3 lower with longer filaments and anthers, 3 or 4 medium, 3 or 4 reduced; fruit lightly compressed, somewhat inflated, oblong, puberulent but glabrous at maturity, about 5–10 cm long, 0.5–1 cm wide, 30–40-seeded; seeds compressed, ovate or suborbicular, pointed at the hilum, 4–4.5 mm long, 3.5–4.5 mm wide, brownish-grey or brownish-olive with an elliptical areole 2–2.5 mm long, 1–1.5 mm wide on each face.

Distr. A pantropical weed.

Vern. Uru-tora (S); ponaverai, takarai (T).

Specimens Examined. WITHOUT EXACT LOCALITY: *Worthington 1342* (K). ANURADHAPURA DISTRICT: Kekirawa on road from Kandy to Anuradhapura, *van Beusekom 1589* (US). KURUNEGALA DISTRICT: Maho, rest house, *Maxwell & Jayasuriya 787* (MO, NY, PDA, SFV, US). TRINCOMALEE DISTRICT: Trincomalee, gravel pits, *Worthington 2053* (A, K).

COLOMBO DISTRICT: Negombo, on sand just above beach, *Rudd 3067* (K, MO, PDA, SFV, US in part). HAMBANTOTA DISTRICT: Tissamaharama, *Soejarto & Balasubramaniam 4959* (US). RATNAPURA DISTRICT: Godakawela, *Simpson 9965* (PDA); Madampe, *Worthington 2132* (K). JAFF-NA DISTRICT: Jaffna, Feb. 1890, *Trimen* ? *s.n.* (PDA). JAFFNA & BAT-TICALOA DISTRICTS: "Jaffna & *Gardner* Batticaloa, Mar. 1858", *Thwaites C.P. 1507* (PDA).

14. Cassia occidentalis L., Sp. Pl. 377. 1753; Thw., Enum. Pl. Zeyl. 95. 1859; Baker in Hook. f., Fl. Br. Ind. 2: 262. 1878; Trimen, Handb. Fl. Ceylon 2: 105. 1894; Alston in Trimen, Handb. Fl. Ceylon 6: 91. 1931; de Wit, Webbia 11: 256. 1955; Brenan in Fl. Trop. East Africa, Caesalp. 78. 1967; Ali, Fl. W. Pakistan, no. 54, Caesalp. 17. 1973; Saldanha & Nicolson, Fl. Hassan 220. 1976. Type: America, Hort. Cliff. 159, *Cassia* no. 7 (BM-Cliff.).

Senna occidentalis (L.) Link, Handb. 2: 140. 1831; Roxb., Fl. Ind. ed. 2, 2: 343. 1832; Irwin & Barneby, Mem. New York Bot. Gard. 35: 436. 1982.

Herbs, sometimes suffrutescent, to about 3 m high; stems glabrous or subglabrous; stipules linear-lanceolate, semi-sagittate, 4–8 mm long, acuminate, caducous; leaves to about 25 cm long, with 4–6 pairs of leaflets, bearing a sessile, hemispherical gland at the base of the petiole; leaflets ovate or lance-ovate, (1–)3–7(–12) cm long, 1.5–4 cm wide, acute or attenuate at the apex, rounded at the base, ciliolate at the margin, minutely appressed-pubescent when young, glabrate; inflorescences racemose, axillary, few-flowered; bracts lanceolate, acute, caducous; flowers on pedicels about 10–12 mm long; sepals glabrous or subglabrous, obtuse, 6–10 mm long; petals orange-yellow, sometimes tinged with purple, often white when dry, 10–15 mm long; stamens 9 or 10, 1 sterile or lack-ing, 2 large, 4 medium, and 3 staminodes; fruit compressed, linear, often slight-ly curved, 5.5–12 cm long, 6–9 mm wide, glabrous, 20–30-seeded; seeds com-pressed, brown, ovate or suborbicular, about 4.5–5 mm long, 3.5–4.5 mm wide, with an elliptic areole 2.5–3.5 mm long, 1.5–2 mm wide on each face.

Distr. A pantropical weed.

Uses. Seeds are used as a substitute for coffee, and as a purgative; leaves and other parts are used medicinally and sometimes in insecticides.

Vern. Peni-tora (S); ponnantakarai (T); coffee-senna, coffee-weed (E).

Specimens Examined. DISTRICT UNKNOWN: Illegible, *Thwaites, C.P. 1269* (PDA). ANURADHAPURA DISTRICT: Anuradhapura, W. shore of Nuwara Wewa, *Sohmer 8084* (US). POLONNARUWA DISTRICT: Polon-naruwa, *Ripley 282* (US), *345* (US), *Hladik 1204* (US), *Sohmer 8230* (GH, MO). KURUNEGALA DISTRICT: Wanduragala, *Rudd 3329* (NY, PDA, SFV, US); Millawa, 2.5 mi. from Kurunegala on road to Kandy, *Rudd 3333* (K, MO, NY, PDA, US). KANDY DISTRICT: Kandy, *Worthington 6920* (K); between

Madugoda and Hunnasgiriya, mi. marker 24/19, *Rudd & Balakrishnan 3244* (NY, PDA, SFV, US). TRINCOMALEE DISTRICT: Trincomalee, *Comanor 771* (K, MO, PDA, SFV, US), *Theobald & Grupe 2321* (US). COLOMBO DISTRICT: Ja-Ela-Gampaha Road, *Comanor 1014* (GH, K, NY, PDA, SFV, US); Negombo, *Rudd 3067* (US in part). HAMBANTOTA DISTRICT: Tangalla, *Rudd 3091* (US); Ruhuna Natl. Park, near Patangala Beach, *Mueller-Dombois 67083112* (US); Rakina Wewa, near Gonalabbe Lewaya, *Fosberg 50246* (GH, PDA, UC, US); Uraniya, *Mueller-Dombois & Cooray 68020305* (PDA, US); Patanagala, *Fosberg 50355* (PDA, SFV, US).

Note. There is a specimen identified, perhaps correctly, as "? *C. occidentalis* × *C. hirsuta*": Urugalla, roadside with *C. hirsuta*, "a single bush", *Simpson 9220* (BM).

15. Cassia hirsuta L., Sp. Pl. 378. 1753; Trimen, Handb. Fl. Ceylon 2: 106. 1894; Alston in Trimen, Handb. Fl. Ceylon 6: 91. 1931; de Wit, Webbia 11: 250. 1955; Brenan in Fl. Trop. East Africa, Caesalp. 80. 1967; Saldanha & Nicolson, Fl. Hassan 220. 1976. Type: America, Hort. Cliff. 159, *Cassia* no. 4 (BM-Cliff.).

Senna hirsuta (L.) Irwin & Barneby, Phytologia 44: 499. 1979; Irwin & Barneby, Mem. New York Bot. Gard. 35: 425. 1982.

Herbs, sometimes suffrutescent, to about 2.5 m tall; stems hirsute; stipules linear, acuminate, 7–10 mm long, sometimes persistent; leaves to about 22 cm long, usually with 3–5 pairs of leaflets, bearing an antrorse, subclavate gland near the base of the petiole; leaflets elliptic to ovate-elliptic or ovate-lanceolate, acute or subacuminate at the apex, rounded at the base, hirsute on both surfaces; inflorescences racemose, short, subumbellate, axillary, few-flowered; bracts linear, attenuate, about 13 mm long; flowers on pedicels 12–15 mm long; sepals hirsute at the base, glabrous above, 7–8 mm long; petals yellow, 1–1.7 cm long; stamens 10, 2 large, 5 medium, and 3 small; fruit linear, curved, densely hirsute, septate, compressed, many-seeded, (8–)12–15(–20) cm long, 3–7 mm wide; seeds orbicular, light brown, 3 mm long, 2.5 mm wide, compressed laterally, with a line from hilum to base on each face and an areole 1.5–2 mm long, 1 mm wide at each side.

Distr. A pantropical weed.

Uses. According to Burkill, Econ. Prod. Malay Penins. 1: 482. 1966, the leaves are used for treating herpes, and are eaten steamed with food.

Specimens Examined. POLONNARUWA DISTRICT: Polonnaruwa, *Cramer 3680* (US), *Ripley 293* (US), *342* (US). KANDY DISTRICT: Madugoda, *Simpson 9130* (BM, PDA); Kandy, *van Beusekom 1649* (US); Pussellawa, Sangilipalama, *Cramer 3857* (K, US); Moray Tea Estate, *Fosberg 58141* (US); Hatale, Madulkelle, *Worthington 4988* (BM, K). NUWARA ELIYA DISTRICT:·

MacDonald's Valley, below Hakgala, *Rudd & Balakrishnan 3169* (PDA, SFV, US). COLOMBO DISTRICT: Colombo, Maitland Place, *Cooray 70011302R* (US). BADULLA DISTRICT: Canyon just below Ella rest house, *Fosberg & Sachet 53179* (PDA, SFV, US); Uma Oya off Ettampitiya, *Jayasuriya & Townsend 1176* (US); N. of Welimada, near Ancient Site, Ft. MacDonald, near culvert 8/8, *Maxwell & Jayasuriya 892* (MO, US).

16. Cassia bacillaris L.F., Suppl. 231. 1781. Type: Surinam, *Dahlberg s.n.*, Herb. Linn. no. 528. 2 (LINN).

Senna bacillaris (L.F.) Irwin & Barneby, Mem. New York Bot. Gard. 35: 111. 1982.

Cassia fruticosa sensu auth.; Benth., Trans. Linn. Soc. London 27: 521. 1871; de Wit, Webbia 11: 247. 1955; Brenan in Fl. Trop. East Africa, Caesalp. 70. 1967; non Mill., 1768, fide Irwin & Barneby, l.c.

Shrubs or trees about 5–8 m tall; young stems minutely appressed-pubescent; stipules linear to linear-oblanceolate, straight or falcate, acute, 4.5–20 mm long; leaves 12–30 cm long, with 2 pairs of leaflets, bearing an ovate-elliptic gland on rachis between the lower pair of leaflets; leaflets elliptic or ovate, (2.5–)4–16(–22) cm long, 1.7–7.2(–11) cm wide, acute to subacuminate, sometimes obtuse at the apex, minutely appressed-pubescent on both surfaces; inflorescences short-racemose, (1.5–)2–7(–8.5) cm long, 5–35-flowered; bracts ovate to lanceolate, 1–3 mm long, caducous; flowers on pedicels 2–5.5 cm long; sepals 8–12.5 mm long, rounded at the apex; petals pale or dull yellow, obovate, puberulent, (1–) 2–3.3 cm long; stamens 9 or 10 with 2 or 3 large, 4 medium, and 3 reduced to staminodes; fruit terete or subterete, (10–)15–27(–36) cm long, 9–15 mm in diameter, slightly thickened along the margins, dehiscent along one suture, not septate within, many-seeded; seeds dark brown, subreniform, embedded in sticky pulp, 3–5.5 mm long, 2–3.5 mm wide, with a faint areole, 2.6–4.4 mm long, 1.7–2.4 mm wide on each face.

Distr. Native to American tropics but introduced elsewhere.

Vern. Wal ehela.

Specimens Examined. MATALE DISTRICT: Belligama Estate, *Worthington 4448* (K). NUWARA ELIYA DISTRICT: Maddecombra, Watagoda, 3300' alt, *Worthington & H. Andrews 4221* (BM, K). KANDY DISTRICT: Kandy, Hantane Estate, 2400', *Worthington 4260* (K); Kadugannawa, *Worthington & Banda 7060* (K); Poilakanda, Kadugannawa, *Worthington 1209* (K).

17. Cassia bicapsularis L., Sp. Pl. 376. 1753; Baker in Hook. f., Fl. Br. Ind. 2: 263. 1878; Alston in Trimen, Handb. Fl. Ceylon 6: 92. 1931; de Wit, Webbia 11: 235. 1955; Backer & Bakh. f., Fl. Java. 1: 539. 1963; Brenan in Fl. Trop. East Africa, Caesalp., 71. 1967. Syntype: Herb. Linnaeus 528.10 (LINN). There appears to be no specimen in BM-Cliff.

Senna bicapsularis (L.) Roxb., Fl. Ind., ed. 2, 2: 342. 1832; Irwin & Barneby,
Mem. New York Bot. Gard. 35: 399. 1982.

Shrubs, sometimes weak and sprawling; stems glabrous, about 3 ın long;
stipules linear, acute, about 4 mm long, 1 mm wide, caducous, leaves (2.5–)3–6(–9)
cm long, with 3–5 pairs of leaflets, bearing a clavate gland between the leaflets
of the lowest pair; leaflets obovate to oblong or suborbicular, 1–4 cm long, 1–2
cm wide, rounded or subacute at the apex, rounded or acute at the base, glabrous
on both surfaces; inflorescences racemose, axillary, to about 15 cm long; bracts
lance-subulate, 1–3 mm long, caducous; flowers on pedicels 2–8 mm long; sepals
glabrous, often ciliolate, ovate, 8–12 mm long, rounded at the apex; petals yellow,
1–1.5 cm long; stamens 10, 3 with large anthers, 4 medium, and 3 reduced; fruit
short-stipitate, terete, straight, 8–20 cm long, 1–1.5 cm thick, glabrous, lustrous,
septate, tardily dehiscent along one suture or rupturing irregularly, many-seeded;
seeds grey-brown, suborbicular, compressed, 5–6(–8) mm long, 3.5–4.5 mm wide
with no apparent areole.

Distr. Native to tropical America; introduced elsewhere in the tropics.
Uses. Often planted for hedges.
Specimens Examined. JAFFNA DISTRICT: Jaffna, *Rudd 3282* (K, NY,
PDA, SFV, US). KANDY DISTRICT: Katukelle, *de Silva s.n.* (PDA); Piachaud
Gardens, Katukelle, *Rudd 3322* (PDA, SFV, US); Trinity College Estate, 16.1.26,
Alston s.n., (PDA); Haragama, roadside, escape, 4.9.27, *Alston s.n.*, (PDA);
Peradeniya, *Simpson 8960* (BM); between Madugoda and Hunnasgiriya, mi. 26/4,
Rudd 3240 (K, MO, PDA, SFV, US); road from Teldeniya to Mahiyangana,
mi. 26/5, *Comanor 559* (K, MO, PDA, SFV, UC, US); Hindagala on Galaha
Road, *Cooray 70012801R* (K, PDA, SFV, US); Kandy-Mahiyangana Road, near
mi. 27 from Kandy, *Sohmer 8265* (MO, NY, GH); Kadugannawa, Poilakanda
garden, *Worthington 1205* (K). BADULLA DISTRICT: N. of Welimada, Am-
bagasdowa, *Maxwell & Jayasuriya 898* (MO, NY, PDA, SFV, US).

18.   Cassia septemtrionalis Viviani, Elench. Pl. Hort. J. Car. Dinegro 14. 1802.
Type: Not known.

Cassia laevigata Willd., Enum. Pl. Hort. Berol. 441. 1809; Alston in Trimen,
Handb. Fl. Ceylon 6: 92. 1931. Type: Cultivated, Berlin, *Willdenow no 7952*
(B-Willd).
Cassia floribunda sensu auth.; de Wit, Webbia 11: 245. 1955; Brenan, Fl. Trop.
East Africa, Caesalp., 70. 1967, non *C. floribunda* Cav. 1802, nec *Senna*
× *floribunda* (Cav.) Irwin & Barneby, 1982.
Senna septemtrionalis (Viviani) Irwin & Barneby, Mem. New York Bot. Gard.
35: 365. 1982.

Shrubs or small trees, to about 6 m high; stems glabrous; stipules linear, attenuate, 3–8 mm long, caducous; leaves 8–25 cm long, with 3–5 pairs of leaflets,

bearing oblong-clavate glands on rachis often between each pair of leaflets; leaflets ovate to elliptic-oblong, 3–10 cm long, 1–4.5 cm wide, acuminate, glabrous on both surfaces, the margin sometimes ciliolate; inflorescences racemose, axillary, 1.5–8 cm long, 3–13-flowered; bracts linear-lanceolate, caducous; flowers on pedicels 5–30 mm long; sepals suborbicular, glabrous, rounded at apex, 6–10 mm long, often yellowish; petals yellow, 1–2 cm long; stamens 10, 2 large, 1 medium, 4 smaller, and 3 reduced; fruit subterete, short-stipitate, 6–9 cm long, 1–1.2 cm wide, rounded and short-rostrate or apiculate at the apex, indehiscent or tardily dehiscent, glabrous, the valves thin, brittle; seeds numerous, dark brown, transverse, surrounded by pulp, lustrous, 4–5 mm long, 3–3.5 mm wide, without apparent areole.

Distr. Native of tropical America; introduced elsewhere in the tropics.

Uses. Planted for hedges; sometimes used as a substitute for coffee; used medicinally especially as a purgative and as an emmenagogue.

Vern. Various names used in Latin America.

Specimens Examined. KANDY DISTRICT: Near Maskeliya, *van Beusekom 1532* (US); Madugoda, *Simpson 8834* (BM); Hatale-Madulkele, *Worthington 4989* (K). NUWARA ELIYA DISTRICT: Hakgala, *s. coll. s.n.* (PDA), *s. coll. 180* (PDA); Nuwara Eliya, *Petch s.n.* (K); between Horton Plains and Nuwara Eliya, mi. 20/2, *Meijer et al. 643* (US); Nuwara Eliya to Talawakele, 2 mi. E. of Talawakele, *Maxwell & Jayasuriya 907* (MO, NY, PDA, SFV, US). BADULLA DISTRICT: Near road just above Bandarawela station, 26.3.06, *A. M. S. s.n.* (PDA); N. of Welimada, near Ft. MacDonald, culvert 8/8, *Maxwell & Jayasuriya 893* (MO, NY, PDA, SFV, US). MONERAGALA DISTRICT: Govindahela, roadside, 27.2.1906, *s. coll. s.n.* (PDA).

19.   Cassia auriculata L., Sp. Pl. 379. 1753; Moon, Cat. 33: 1824; Thw., Enum. Pl. Zeyl. 96. 1859; Baker in Hook. f., Fl. Br. Ind. 2: 263. 1878; Trimen, Handb. Fl. Ceylon 2: 106. 1894, pl. 33. 1894; Alston in Trimen, Handb. Fl. Ceylon 6: 91. 1931; de Wit, Webbia 11: 234. 1955; Macmillan, Trop. Pl. & Gard. ed. 5, 29, 104, 134, 302, 364, 421. 1962; Brenan in Fl. Trop. East Africa, Caesalp. 76. 1967; Ali, Fl. W. Pakistan, no. 54, Caesalp. 27. 1973; Saldanha & Nicolson, Fl. Hassan 219. 1976. Syntypes: Ceylon, *P. Hermann,* 4: 20, 4: 79 (BM-Herm).

Senna auriculata (L.) Roxb., Fl. Ind., ed. 2, 2: 349. 1832.

Shrubs or small trees, to about 7 m tall; stems puberulent; stipules large, 7–22 mm wide, reniform-rotund, produced at base next to the petiole into a filiform point to 7 mm long, persistent; leaves (5–)10–12 cm long, with 6–13 pairs of leaflets, bearing erect linear glands on the rachis usually between leaflets of each pair; leaflets elliptic-oblong, rounded at apex and base, mucronate at the apex, 1–3.5 cm long, 1–2 cm wide, glabrous above, puberulent or glabrous beneath;

inflorescences corymbose, few-flowered but aggregated into large, terminal panicles in the axils of the upper leaves; bracts linear, attenuate, 5–7 mm long; flowers on pedicels 1.5–3 cm long; sepals yellow, rounded at the apex, pubescent at the base; petals bright yellow, 2–2.5 cm long; stamens 10, 3 with long filaments and anthers, 4 medium, and 3 reduced to staminodes; fruit linear-oblong, compressed, undulate between the seeds, pilose or minutely crisp-pubescent; 7.6–10 cm long with a stipe about 5 mm long and a persistent style 1.5 cm long, about 1–2 cm wide, 12–20-seeded; seeds brown, ovate-oblong, 7–9 mm long, 4–5 mm wide, with an areole 3–3.5 mm long, 0.5–0.75 mm wide on each face.

Distr. Native to Ceylon and the Indian peninsula; introduced in Southeast Asia and Africa.

Uses. Leaves form "Ceylon tea", or "Matara tea"; the bark is "Avaram bark" important in tanning; the plants are sometimes used medicinally, and as ornamentals.

Vern. Ranawara (S); avarai (T).

Specimens Examined. WITHOUT LOCALITY: *Jonville s.n.* (BM). JAFFNA DISTRICT: Jaffna, *Gardner, Thwaites C.P. 1503* (PDA); between Kayts and causeway to Jaffna, *Rudd 3275* (PDA, SFV, US); Kurikadduvan, beside Bettu Kulam, *Cramer 3359* (US); Ponneryn, *Bernardi 14281* (US). MANNAR DISTRICT: Vankalai, *Kundu & Balakrishnan 629* (US); Madu Rd. to Mannar Isl., *Kostermans 25226* (K), *24876* (A, K, US); Mannar, *Petch s.n.* (A). ANURADHAPURA DISTRICT: Mihintale Sanctuary, 55.5 mi. marker, *Sohmer 8111* (GH, MO); between Dambulla and Kekirawa, mi. 52, *Rudd 3252* (K, PDA, SFV, US); near Kokkare Villu, *Maxwell & Jayasuriya 818* (MO, NY, PDA, SFV, US). PUTTALAM DISTRICT: Wilpattu Natl. Park, Kali Villu, *Mueller-Dombois 67051816* (PDA, US); before Erige Ara, *Mueller-Dombois & Wirawan 68091306* (K, PDA, US); Nelun Villu, *Mueller-Dombois 67051830* (US); Atha Villu, *Wirawan et al. 942* (K, MO, US); 2.5 mi. S. of Maduru Odai, 5 mi. S. of Marai Villu, *Fosberg et al. 50858* (K, PDA, SFV, US); Smithsonian Camp, *Hladik 822* (US). KANDY DISTRICT: Peradeniya, Royal Botanic Gardens, *Kostermans 24543* (A, BM, K, US); Kadugannawa, *Worthington 1207* (BM), *7136* (K), *7168* (K). TRINCOMALEE DISTRICT: Trincomalee, *Worthington 2061* (BM, K), *2692* (A, BM, K), *Davidse 7532* (MO); Punani Wewa, *Simpson 9234* (BM). MONERAGALA DISTRICT: NW. of "Juda Oya" [Kuda ?], N. of Tanamalwila, *Meijer 197* (US). HAMBANTOTA DISTRICT: Yala, *Hepper & de Silva 4747* (K, NY, US), *Cramer 3333* (US), *Kostermans 25447* (K, US), *Comanor 634* (GH, K, PDA, SFV, UC, US), *635* (GH, K, UC, US); 1 mi. E. of Bundala, *Davidse 7764* (MO, US); Hambantota, 14-10-1955, *Senaratna s.n.* (PDA), *Soejarto & Balasubramaniam 4948* (US); Ruhuna Natl. Park, near Katupila Ara, *Comanor 348* (PDA, US); Patanagala, *Fosberg 50370* (GH, K, PDA, SFV, US); 1 mi. N. of Yala camp, *Mueller-Dombois & Comanor 67062506* (PDA, US); Padikema, *Cooray 69120311R* (K, MO, US).

20.   Cassia multijuga L.C. Rich., Actes Soc. Hist. Nat. Paris 1: 108. 1782; Bentham in Martius, Fl. Bras. 15 (2) : 123, tab. 37. 1870; de Wit, Webbia 11: 253. 1955; Worthington, Ceylon Trees 190. 1959; Macmillan, Trop. Pl. & Gard., ed. 5, 83. 1962; Backer & Bakh. f., Fl. Java 1: 541. 1963. Type. Cayenne, French Guiana, *Le Blond s.n.* (P, P-Lam).

Senna multijuga (L.C. Rich.) Irwin & Barneby, Mem. New York Bot. Gard. 35: 492. 1982.

Trees, to about 40 m tall; young stems puberulent or subglabrous; stipules linear-acuminate, somewhat falcate, 4–12 mm long, 0.8–2.5 mm wide, caducous; leaves 12–30(–35) cm long, with about 12–40 pairs of leaflets, bearing a cylindric or subconate gland on rachis between the leaflets of the lowermost pair and sometimes between others; leaflets oblong, oblong-elliptic to linear-oblong, about 2–5(–5) cm long, 4–12 mm wide, glabrous or puberulent on both surfaces, rounded, mucronulate at the apex, asymmetrically rounded at the base; inflorescences paniculate, terminal or subterminal; bracts linear, acuminate, caducous; flowers on pedicels 1.5–2.5 cm long; sepals ovate, 4–7 mm long, 5 mm wide, subglabrous; petals yellow, unequal, ovate to obovate or suborbicular, the longest 16–26 mm long; stamens 10, 3 largest, 4 medium, and 4 reduced; fruit compressed, broadly linear, (4–)10–20 cm long, 1.5–2(–2.5) cm wide, septate, the margin somewhat sinuate by constriction between some seeds, tardily dehiscent, lustrous, many-seeded; seeds light brown, oblong, 4.5–8(–9) mm long, 1.5–2 mm wide, with a linear areole 2–4.5 mm long, 0.25–0.5 mm wide on each face.

Distr. A native of tropical America, widely planted elsewhere as an ornamental; introduced to Ceylon in 1851.

Specimens Examined. KANDY DISTRICT: Kandy, Hillcrest, *Worthington 7082* (K); Kadugannawa, *Worthington 1208* (K), *Worthington & Banda 4278* (BM, K).

21.   Cassia surattensis Burm. f. subsp. **glauca** (Lam.) K. & S. Larsen, Fl. Cambodge, Laos & Viêt-Nam, Legum. Caesalp. 102. 1980.

Cassia glauca Lam., Enc. 1: 647. 1785; Thw., Enum. Pl. Zeyl. 96. 1859; Baker in Hook. f., Fl. Br. Ind. 2: 235. 1878; Trimen, Handb. Fl. Ceylon 2: 109. 1894. Type: India, *Sonnerat* (P-Lam), non *Senna glauca* Roxb. 1832.
Cassia sulphurea DC. ex Collad., Hist. Cass. 84. 1816.
Senna sulphurea (Collad.) Irwin & Barneby, Mem. New York Bot. Gard. 35: 78. 1982.
Cassia surattensis sensu auth., Alston in Trimen, Handb. Fl. Ceylon 6: 93. 1931; de Wit, Webbia 11: 269. 1955; Backer & Bakh. f., Fl. Java 1: 538. 1963; Ali, Fl. W. Pakistan, no. 54, Caesalp., 29. 1973; Saldanha & Nicolson, Fl. Hassan 221. 1976.

Shrubs or small trees, to about 7 m tall; young stems puberulent or glabrous;

stipules linear-falcate, puberulent, 6–15 mm long, subpersistent; leaves 14–30 cm long, with 4–7 pairs of leaflets, bearing clavate glands between the leaflets of the 2–4 lowermost pairs; leaflets ovate to ovate-oblong, 3–9 cm long, 1.2–5 cm wide, obtuse, emarginate at the apex, rounded to cuneate at the base, glabrous above, sparsely appressed-pubescent beneath; inflorescences racemose, 5–13 cm long, 10–20-flowered; bracts ovate-acute, 2.5–8 mm long, caducous or subpersistent; flowers on pedicels 2.5–4 cm long; sepals unequal, to 8 mm long, ciliolate, otherwise glabrous; petals yellow, 2–3 cm long, puberulent on the outer face; stamens 10, all fertile, 3 with longer filaments, 7 shorter; fruit compressed, glabrous, dehiscent, straight or curved, 15–20 cm long with stipe 1–2 cm long, 1.2–2 cm wide, septate, 20–35-seeded; seeds transverse, blackish, glossy, compressed, 6–10 mm long, 3.5–4 mm wide, oblong-elliptic, with an areole 4.5–5 mm long, 1.2–1.5 mm wide on each face.

Distr. Native to southeast Asia and introduced into other tropical areas.
Uses. Planted as an ornamental; roots and other parts used medicinally.
Specimens Examined. WITHOUT LOCALITY: "Comm. Mrs. C. 1851" (BM); *Macrae 9* (BM), *s.n.* (K); "Manpalakouni", *Jonville s.n.* (BM). KANDY DISTRICT: Kadugannawa, Poilakande, *Worthington 1206* (BM, K).

22. Cassia tomentosa L. f., Suppl. 231. 1781; Thw., Enum. Pl. Zeyl. 95. 1859; Baker in Hook. f., Fl. Br. Ind. 2: 263. 1878; Trimen, Handb. Fl. Ceylon 2: 106. 1894; de Wit, Webbia 11: 275. 1955; Macmillan, Trop. Pl. & Gard., ed. 5, 302. 1962. Type: America meridionali, presumably Colombia, *Mutis*, herb. Linnaeus 528.23 (LINN), non *Senna tomentosa* Batka, 1854.

Cassia multiglandulosa Jacq., Ic. Pl. Rar. 1 (3): 8, t. 72. 1783; Jacq., Collect.
    1: 42. 1787. Type: Cult. hort. Schoenbrun (W).
Senna multiglandulosa (Jacq.) Irwin & Barneby, Mem. New York Bot. Gard.
    35: 357. 1982.

Shrubs to about 4(–7) m tall; young stems velutinous; stipules linear, acute, about 5–7 mm long, tomentose, caducous; leaves 7–17 cm long, with 4–8(–9) pairs of leaflets, bearing small, sessile, conical glands on the rachis between some or all pairs of leaflets; leaflets oblong to elliptic, 2–3 cm long, 1–1.5 cm wide, rounded, or subacute, mucronate at the apex, rounded to acute or cordate at the base, lightly tomentulose above, densely so beneath; inflorescences racemose, terminal, usually compound into corymbose panicles, few-flowered; bracts ovate, acute, 5 mm long, tomentulose, caducous; flowers on tomentose pedicels, 1.5–2.5 cm long; sepals unequal, 7–9 mm long, pubescent; petals yellow, 10–13(–19) mm long; stamens 10, 2 large, 1 smaller, 4 medium, 3 reduced; fruit compressed, straight, thinly tomentose, about 10–14 cm long, 1–1.5 cm wide, septate, 50–90-seeded; seeds brown, obovoid, 4–5 mm long, 2.5–3.5(–4) mm wide, without areoles.

Distr. A native of tropical America, introduced into tropical areas elsewhere.

Vern. Various local names in Latin America.

Specimens Examined. NUWARA ELIYA DISTRICT: Nuwara Eliya ("N. Ellia"), *Gardner 256* (K), "*Gardner & 1857*", *Thwaites C.P. 2410* (PDA, K). KANDY DISTRICT: "RBG, in Magnoliaceae", 26.2.30, *de Silva s.n.* (PDA). BADULLA DISTRICT: Ambawela, *Worthington 5719* (K).

23. Cassia divaricata Nees & Blume, Syll. Pl. Nov. Ratisb. 1: 94. 1825; de Wit, Webbia 11: 242, fig. 2. 1955; Backer & Bakh. f., Fl. Java 1: 538. 1963. Type: Java, *Blume s.n.* (L).

Shrubs 2–5 m tall; stipules linear-falcate, about 8–10 mm long, pubescent, subpersistent; leaves about 10–14 cm long, with 6–11 pairs of leaflets, bearing a doliiform or clavate gland on the rachis between the leaflets of the lowermost pair, and sometimes a similar but smaller gland between the uppermost; leaflets elliptic-oblong, 1.5–4.5 cm long, 5–15 mm wide, obtuse, mucronulate, rounded to acute at the base, glabrous above, glabrous or appressed-pubescent beneath; inflorescences racemose, axillary, 2- or 3-flowered; bracts boat-shaped, glabrous, 1.5 mm long; flowers on pedicels 1.5–2.5 cm long; sepals unequal, to 1 cm long, glabrous; petals bright yellow, 1.5–2.5 cm long; unequal; stamens 10, all fertile, 3 with large, crested anthers and longer filaments and 7 smaller; fruit compressed, straight or slightly curved, glabrous, 11–22 cm long including stipe 5–7 mm long, 6–8 mm wide, 15–50-seeded; seeds compressed, about 5 mm long.

Distr. Malaysia and the Philippines; introduced in Ceylon.

Vern. False ranawara, fide Worthington.

Specimens Examined. KANDY DISTRICT: Roseneath, Hermitage, *Worthington s.n.* (K); Gampola, *Worthington 6538* (K). NUWARA ELIYA DISTRICT: Lindula, Lippakelle Estate, "Planted in tea", *Worthington 5357* (BM, K); Nildandahinna, *Worthington 5647* (BM, K).

Note. de Wit suggests that this might actually be an American species because of its similarity to *Cassia biflora* L. Irwin & Barneby do not include *C. divaricata* in their treatment of the New World species and have placed *C. biflora*, with a question, under *C. pallida* Vahl (*Senna pallida* (Vahl) Irwin & Barneby, Mem. New York Bot. Gard. 35: 531. 1982).

24. Cassia tora L., Sp. Pl. 376. 1753; Thw., Enum. Pl. Zeyl. 96. 1859; Trimen, Handb. Fl. Ceylon 2: 106. 1894; Prain, J. Asiat. Soc. Bengal 66 (2): 475. 1897; Alston in Trimen, Handb. Fl. Ceylon 6: 91. 1931; de Wit, Webbia 11: 276. 1955; Brenan, Kew Bull. 13: 248. 1958; Backer & Bakh. f., Fl. Java 1: 539. 1963; Saldanha & Nicolson, Fl. Hassan 221. 1976. Syntype: *P. Hermann,* Ceylon, herb. Hermann no. 4: 79 (BM-Herm).

Senna tora (L.) Roxb., Fl. Ind. ed. 2, 2: 340. 1832.

Herbaceous annual about 1–1.2 m high; stems glabrous or puberulent; stipules linear-subulate, somewhat falcate, 10–15 mm long, 1 m wide, subpersistent; leaves with 2−4 pairs of leaflets, bearing a slender cylindric gland, 2 mm long, on the rachis between the leaflets of the 2 lower pairs; leaflets obovate, 1.5–5 cm long, 1.5–2.5 cm wide, obtuse to subacute at the apex, often mucronate, asymmetrically rounded to acute at the base, glabrous above, subappressed-pubescent or glabrous beneath; racemes short, axillary, few-flowered; bracts 2–4 mm long, subulate; flowers on pedicels (4–)6–10 mm long; fruiting pedicels to 12–15 mm long; sepals unequal, oblong to rounded, 5–6 mm long; petals yellow, 8–12 mm; stamens 7–10, 3 large, 4 medium, 3 staminodial or absent, rarely perfect; fruit linear or subtetragonous, 10–15 cm long, 3–6 mm wide, falcate or almost straight, margins thickened, indehiscent, 20–30-seeded; seeds chestnut-brown, lustrous, 4–5 mm long, 2–3 mm wide, compressed, with an areole lighter coloured, 3.5–4.5 mm long, 1.5–2 mm wide on each face.

Distr. A pantropical or Asian weed, origin debatable.

Uses. The leaves are sometimes eaten when young and are used medicinally; the seeds are sometimes used as a coffee substitute and as a mordant in dyeing.

Vern. Peti-tora (S); vaddutakarai (T).

Specimens Examined. JAFFNA DISTRICT: Between Jaffna causeway and Kayts, *Rudd 3271* (PDA, SFV, US). PUTTALAM OR ANURADHAPURA DISTRICT: Puttalam-Anuradhapura Road, jungle margin, *Simpson 9163* (BM, PDA); Wilpattu Natl. Park, near Eerige Ara confluence with Moderagam Ara, *Fosberg et al. 50783* (GH, K, MO, PDA, UC, US). POLONNARUWA DISTRICT: Between Habarane and Kantalai, mi. 123, *Rudd & Balakrishnan 3128* (PDA, SFV, US); Polonnaruwa, sacred area, *Ripley 288* (US), *239* (US), *344* (US). KANDY DISTRICT: Kandy, *van Beusekom 1652* (US); Peradeniya, *Thwaites, C.P. 2785* (PDA), *J.M. de Silva 122* (NY); in 1916, *s. coll. s.n.* (PDA). TRINCOMALEE DISTRICT: Kantalai-Trincomalee Road, mi. 140–144, *Comanor 770* (K, MO, PDA, SFV, US). MONERAGALA DISTRICT: Between Wellawaya and Moneragala, mi. 142, *Rudd & Balakrishnan 3212* (K, PDA, SFV, US).

Notes. Most of the specimens known as *C. tora* L. have been correctly identified. However, a few in Ceylon are actually *C. obtusifolia* L., a very similar species treated by Bentham as conspecific with *C. tora* (Trans. Linn. Soc. London 27: 535–6. 1871) and followed by many authors. The differences between the two species have been recognized by others, as indicated in the key and under the following species in this treatment, *C. obtusifolia*.

25. Cassia obtusifolia L., Sp. Pl. 377. 1753; Prain, J. Asiat. Soc. Bengal 66 (2): 475. 1897; de Wit, Webbia 11: 254. 1955; Brenan, Kew Bull. 13: 248. 1958. Type: Dill., Hort. Elth. 71, t. 62, f. 72. 1732; typotype, OXF.

Cassia toroides Roxb., Hort. Beng. 31. 1814, nom. nud.; Raf., Med. Bot. 96. 1828.

Senna toroides Roxb., Fl. Ind., ed. 2, 2: 341. 1832. "The seeds of this plant were sent from Mysore to the Botanic garden at Calcutta by Dr. Buchanan in 1800."

Cassia tora L. var. β Wight & Arn., Prod. 291. 1834.

Cassia tora L. var. *obtusifolia* (L.) Haines, Bot. Bihar & Orissa 304. 1922.

Senna obtusifolia (L.) Irwin & Barneby, Mem. New York Bot. Gard. 35: 252. 1982.

Annual, sometimes suffrutescent, to 2 m tall, glabrous or pubescent; stipules linear, falcate, 15–20 mm long, 1–1.5 mm wide, caducous or subpersistent; leaves with 3 pairs of leaflets, bearing a cylindric gland 2–3 mm long on the rachis between the lowermost 1 or 2 pair of leaflets; leaflets obovate, the largest 1.5–6.5 cm long, 0.5–4 cm wide, glabrous above, glabrous or subappressed-pubescent beneath, rounded to subacute, often mucronulate at the apex, rounded to acute at the base; inflorescences short-racemose, axillary, few-flowered; bracts linear-acute, 4–8 mm long; flowers on pedicels about 9–23 mm long; fruiting pedicels to about 30 mm long; sepals 5.5–9.5 mm long, glabrous, ciliolate, ovate, acute or obtuse; petals yellow, unequal, (9–)12–15 mm long; stamens 10, 3 with longest filaments and anthers slightly narrowed at the apex, 4 smaller, 3 reduced to staminodes; fruit terete or subtetragonous, straight or curved, (6–)10–18 cm long, 2.5–6 mm wide, 20–50-seeded, pubescent when young, glabrous at maturity; seeds elongate-rhomboidal, lustrous, dark brown or chestnut-brown, 4–5 mm long, 1.5–2 mm wide, with an areole about the same colour, 2.5–4 mm long, 0.25–0.5 mm wide, on each face.

Distr. Presumably of tropical American origin; an introduced weed in parts of Asia.

Uses. The leaves are eaten and are used medicinally.

Specimens Examined. GALLE DISTRICT: Galle, *Rudd 3083* (K, PDA, SFV, US); Induruwa, *Rudd 3082* (PDA, SFV, US). HAMBANTOTA DISTRICT: Ruhuna Natl. Park, road to Katagamuwa Tank, *Comanor 641* (K, NY, PDA, SFV, US).

Note. This species has been characterized as differing from *C. tora* L. as slightly larger plants having leaves with 3 pairs of leaflets, with a gland on the rachis only between the leaflets of the lowermost pair, longer pedicels, slightly larger flowers, anthers of the 3 longer stamens narrowed at the apex, and seeds with very narrow areoles. In the case of the specimens from Ceylon, only the last character can be readily ascertained, if the material includes seeds. The collections of *C. tora* and *C. obtusifolia* seem to be uniformly 3-jugate and bear 2 glands; the pedicels vary somewhat, but not decisively; and the differences in anther shape are often ambiguous.

26. Cassia absus L., Sp. Pl. 376. 1753; Moon, Cat. 33. 1824; Thw., Enum. Pl. Zeyl. 96. 1859; Baker in Hook. f., Fl. Br. Ind. 2: 265. 1878; Trimen, Handb. Fl. Ceylon 2: 109. 1894; Alston in Trimen, Handb. Fl. Ceylon 6: 91. 1931; de Wit, Webbia 11: 279. 1955; Brenan in Fl. Trop. East Africa, Caesalp. 81, f. 15. 1967; Ali, Fl. W. Pakistan, no. 54, Caesalp. 20, f. 2, A, B. 1973; Irwin & Barneby, Mem. New York Bot. Gard. 30: 277. 1978. Lectotype: Ceylon, *P. Hermann*, herb. Herm. 2: 4 (BM-Herm); isolectotypes: Herb. Burm. page 27 (Paris, Biblio. Inst. France); Herb. Herm. fol. 110 (L).

Chamaecrista absus (L.) Irwin & Barneby, Mem. New York Bot. Gard. 35: 664. 1982.

Plants annual; stems erect or sometimes decumbent, to about 1 m long, viscid-pilose with long and short hairs; stipules linear-attenuate, 2.5–4 mm long, 1 mm wide, persistent; leaves slender-petiolate with 2 pairs of leaflets, the rachis bearing a linear gland between the leaflets of 1 or both pairs; leaflets elliptic to elliptic-obovate, 1–4 cm long, 1–3 cm wide, rounded at the apex, asymmetrical at the base, glabrous or subglabrous above, puberulent or subappressed pubescent beneath; inflorescences viscid-pilose, racemose, leaf-opposed, 2–13 cm long, few-flowered; bracts ovate, about 2 mm long; bracteoles 1 mm long; flowers on pedicels 3–5 mm long; sepals 3–4 mm long, viscid-pubescent; petals yellow, sometimes lined with red, appearing pink or reddish-yellow; stamens 2–7, usually 5, subequal; fruit linear-oblong, elastically dehiscent, compressed, 2.5–4.5(–5.5) cm long, 4–8 mm wide, sparsely setulose-pilose, 5–8-seeded; seeds obovate or rhombic, black or dark brown, lustrous, 4–5.5 mm long, 3.5–4.5 mm wide, marked with several rows of minute puncta.

Distr. A paleotropic weed, introduced in some areas of the New World.

Uses. According to Ali, l.c., "chaksine, extracted from this plant, is reputed to have an action on central and peripheral nervous system." The seeds are used for eye problems, as a cathartic, and in cases of skin infection.

Vern. Bu-tora (S).

Specimens Examined. WITHOUT LOCALITY: *Walker s.n.* (K); *Thwaites, C.P. 1506* (BM, K). KANDY DISTRICT: Kolugala, Tumpane Valley, *Worthington 1467* (K). POLONNARUWA DISTRICT: Between Habarane and Kantalai, near mi. 123, *Rudd & Balakrishnan 3126* (K, MO, PDA, SFV, US). TRINCOMALEE DISTRICT: Near Agbopura, SW. of Kantalai, *Fosberg & P. Jayasinghe 57146* (SFV, US). AMPARAI DISTRICT: Padagoda, S. of Inginiyagala, *Fosberg & Sachet 53090* (K, MO, PDA, SFV, US). COLOMBO DISTRICT: Gampaha, *J.M. de Silva 181* (NY). HAMBANTOTA DISTRICT: Ruhuna Natl. Park, *Wirawan 810* (K, NY, PDA, SFV, US); Patanagala, *Wirawan 697* (US); Near Rakinawala, *Cooray 69120710R* (K, MO, NY, US); near Buttawa Bungalow, *Cooray & Balakrishnan 69011027R* (US), *Mueller-Dombois & Cooray 69010535* (US); 200 m W. of Karaugaswala, *Mueller-Dombois & Cooray 67121052* (US).

27. Cassia kleinii Wight & Arn., Prod. 293. 1834; Thw., Enum. Pl. Zeyl. 96. 1859; Baker in Hook. f., Fl. Br. Ind. 2: 266. 1878; Trimen, Handb. Fl. Ceylon 2: 110. 1894; Cooke, Fl. Pres. Bombay 1: 452. 1903; Alston in Trimen, Handb. Fl. Ceylon 6: 91. 1931. Type: India, Travancore, Herb. Wight 863 (K).

Cassia dimidiata Klein ex Wall. ex Wight & Arn., Prod. 293. 1834, in synon., non Buch.-Ham. ex Roxb., 1832, nec Buch.-Ham. ex G. Don, 1832. Type: Wall. Cat. *5328* (K).

Cassia kleinii var. *pilosa* Thw., Enum. Pl. Zeyl. 97. 1859; Trimen, Handb. Fl. Ceylon 2: 110. 1894. Type: Ceylon, Hantane [*Gardner*], *Thwaites C.P. 1509* (*1508* in part) (PDA, K, GH).

Perennial herb arising from a woody base; stems erect or procumbent, pilose, to about 1 m high; stipules linear-attenuate, persistent, 3–5 mm long; leaves about (1.5–)2.5–6 cm long, with (3–)10–25 pairs of leaflets, bearing a stalked, peltate gland almost 1 mm tall on the rachis, 0.5–1 mm below the lowermost pair of leaflets; leaflets linear-oblong, somewhat falcate, asymmetrical with the midrib marginal or submarginal, many secondary veins, commonly 5–10 mm long, 1–2 mm wide, mucronate at the apex, rounded at the base, glabrous to pilose on both surfaces; flowers 1–2(–3), supra-axillary; bracts linear-attenuate, about 3 mm long, at base of pedicel; bracteoles 1–2 mm long, below the calyx; pedicels 10–15 mm long; petals yellow, 8–10 mm long; sepals 6–8 mm long, pubescent; stamens 10, subequal; fruit oblong, elastically dehiscent, 4–5 cm long, 5 mm wide, about 10–12-seeded, sparsely pilose; seeds rhombic, 3 mm long, 2 mm wide, castaneous, marked with a few lines of minute puncta.

Distr. Apparently endemic to Ceylon and SW. India.

Vern. Bin-siyambala (S).

Specimens Examined. WITHOUT EXACT LOCALITY: *Walker 21* (K), *47* (K), *356* (K), *s.n.* (K); *Fraser 144* (US); *Macrae 177* (BM). BADULLA DISTRICT: In Diyatalawa Military shooting range, *Mueller-Dombois 68051911* (PDA, US). PUTTALAM DISTRICT: W. of Kala Oya, *Jayasuriya et al. 1438* (K, MO, US). MATALE DISTRICT: Rattota-Illukkumbura Rd., near mi. 27, *Jayasuriya 264* (NY, SFV, US). KANDY DISTRICT: Peradeniya, Hantane, *Mueller-Dombois 67110903* (PDA, US), *Mueller-Dombois & Cooray 67111323* (PDA, US); Peradeniya, above University Circuit Bungalow, near radio relay station, *Mueller-Dombois 67110903* (US), *Maxwell & Jayasuriya 851* (MO, NY, SFV, US); E. of Madugoda, *Maxwell & Jayasuriya 739* (US), *Simpson 8467* (BM); Nilambe, *Worthington 2861* (K). NUWARA ELIYA DISTRICT: Road from Kandy to Maturata, culvert 20/5, *Maxwell 999* (MO, NY, SFV, US); Fort Macdonald valley, *A.M.S., s.n.* (PDA). KALUTARA DISTRICT: Kalutara ?, "Kaltura, Govinna Estate", *Simpson 8841* (BM). RATNAPURA DISTRICT: Wewelketiya, April 1930, *de Alwis s.n.* (PDA). MONERAGALA DISTRICT: 7 mi. E. of Bibile, *Fosberg & Sachet 53156* (US).

Note. Noting some variation in degree of pubescence, Thwaites published the variety *pilosa*. However, because it is difficult to know where to draw the line between it and typical *C. kleinii*, the two varieties are being combined.

A similar species, *C. pumila* Lam., was reported from Ceylon but, according to Trimen, l.c., it has not been verified. It differs, particularly, in having 5 stamens in contrast to 10 in *C. kleinii*.

28. Cassia aeschinomene DC. ex Collad., Hist. Cass. 127. t. 17. 1816. Type: "Santo Domingo", "Brotero, misit Balbis" (G-DC).

Cassia lechenaultiana DC., Mem. Soc. Phys. Genève 2 (2): 132. 1824; Baker in Hook. f., Fl. Br. Ind. 2: 266. 1878; Petch, Ceylon J. Sci., Ann. Peradeniya 9: 229. 1924; Alston in Trimen, Handb. Fl. Ceylon 6: 93. 1931; de Wit, Webbia 11: 280. 1955; Backer & Bakh. f., Fl. Java 1: 536. 1963; Verdcourt, Man. New Guinea Leg. 48, f. 11. 1979. Type: Bengal, *Lechenault s.n.* (G).
Cassia wallichiana DC., Mem. Soc. Phys. Genève 2 (2): 133. 1824; Wight & Arn., Prod. 292. 1834; Thw., Enum. Pl. Zeyl. 96. 1859; Ali, Fl. W. Pakistan, no. 54: 21. 1973. Type: Wall. Cat. 5320.
Cassia patellaria Collad. var. *glabrata* Vog., Syn. Gen. Cass. 66. 1837. Type: Brazil, *Sieber*, Herb. Willd. 8000 (B-Willd.).
Cassia mimosoides L. var. *aeschinomene* (Collad.) Benth., as "*aeschynomene*", Trans. Linn. Soc. London 27: 579. 1871.
Cassia mimosoides L. var. *wallichiana* (DC.) Baker in Hook. f., Fl. Br. Ind. 2: 266. 1878; Trimen, Handb. Fl. Ceylon 2: 110. 1894.
Chamaecrista aeschinomene (Collad.) Greene, Pittonia 4: 32. 1899.
Cassia mimosoides L. subsp. *lechenaultiana* (DC.) Ohashi, J. Jap. Bot. 50: 308. 1975.
Chamaecrista nictitans (L.) Moench subsp. *patellaria* (Collad.) var. *glabrata* (Vog.) Irwin & Barneby, Mem. New York Bot. Gard. 35: 822. 1982.

Herb, sometimes suffrutescent at base, to 1.5 m tall; stems crisp-pubescent; stipules linear-attenuate, 10–15 mm long; leaves 7–8 cm long, with about 8–24 pairs of leaflets, bearing a slightly raised, obconoid, or cupuliform gland on the petiole about midway between the base of the petiole and the lowest pair of leaflets; leaflets linear, straight or slightly falciform, asymmetrical, 10–20 mm long, 2–2.5 mm wide, rounded or subacute, mucronate at the apex, obliquely rounded at the base, ciliolate, otherwise glabrous or nearly so, 7–8 nerved, the mid-rib excentric but not marginal; inflorescences short-racemose or fasciculate, axillary or supra-axillary, 3–4-flowered; bracts linear-lanceolate, 6–7 mm long; bracteoles ovate, acute, 2–5 mm long inserted on pedicels 1–2 mm below the calyx; pedicels in flower 5–7 mm long, in fruit to about 9 mm long; sepals ovate, obtuse to acute, 6–7 mm long, lightly crisp-pubescent; petals yellow, oblong or obovate, 7–9 mm long, usually slightly pubescent on the outer face toward the base; stamens 10, subequal; fruit (2.2–)3–5 cm long, 3.5–5 mm wide, 8–16-seeded, crisp-pubescent; seeds black, lustrous, 4 mm long, 3.5 mm wide.

Distr. A weedy plant of Asia and tropical America.

Specimens Examined. WITHOUT LOCALITY: *Mackenzie s.n.* (K). KANDY DISTRICT: Hantane, *Macrae,* Haragama ? (illegible), Aug. 1856, *Thwaites, C.P. 2786* (PDA, K); Peradeniya, Getambe, *Worthington 4266* (K); University, Peradeniya campus, *Jayasuriya 2223* (US), *Comanor 710* (K, MO, SFV, US); University between church and dagoba, *Comanor 701* (K, MO, SFV, UC, US); New Peradeniya-Oodawela Rd., before marker 2/14, *Comanor 500* (K, MO, SFV, US); above University to radio tower, *Maxwell & Jayasuriya 845* (MO, NY, PDA, SFV, US); Peradeniya, *J.M. de Silva 202* (NY), *263* (NY).

29. Cassia mimosoides L., Sp. Pl. 379. 1753; Moon, Cat. 33. 1824; Thw., Enum. Pl. Zeyl. 96. 1859; Baker in Hook. f., Fl. Br. Ind. 2: 266. 1878; Trimen, Handb. Fl. Ceylon 2: 110. 1894; Petch, Ceylon J. Sci., Ann. Peradeniya 9: 229. 1924; Alston in Trimen, Handb. Fl. Ceylon 6: 93. 1931; Steyaert, Bull. Jard. Bot. Etat. 20: 236, pl. 8. 1950; de Wit, Webbia 11: 283. 1955; Brenan in Fl. Trop. East Africa, Caesalp. 100. 1967; Ohashi, J. Jap. Bot. 50: 307. 1975; Saldanha & Nicolson, Fl. Hassan 220. 1976. Lectotype: Ceylon, *P. Hermann,* herb. Herm. 2: 13 (BM-Herm); syntypes or isotypes: herb. Herm. 2: 78 (BM-Herm), Herb. Herm., fol. 85 (L).

Cassia angustissima Lam., Enc. Meth. 1: 650. 1783; Wight & Arn., Prod. 292. 1834. Type: "Maderaspatana, Pluk., Almagest. 252, tab. 5, f. 2". 1694.

Annuals, sometimes woody at base; stems usually erect, to about 1 m high, pubescent; stipules linear, attenuate, to about 10 mm long, persistent; leaves subappressed, about 4–10 cm long, with 15–60(–85), commonly about 30–50, pairs of leaflets, bearing a sessile, annulate, concave gland just below the lowest pair of leaflets; upper surface of rachis usually serrate- or crenate-crested between the leaflets, glabrous or pubescent; leaflets crowded, overlapping, linear-oblong, 2–9 mm long, 0.5–2 mm wide, acute, mucronulate, the midrib excentric but not marginal, secondary veins (1–)2–3, glabrous or subglabrous on both surfaces; flowers 1–2(–3) axillary or supra-axillary; bracts linear-attenuate, about 2–5 mm long, at base of pedicels; bracteoles 1–2 mm long, about 1 mm below the calyx; pedicels 5–30 mm long; sepals about 8 mm long, sometimes reddish, subglabrous; petals yellow, 4–13 mm long; stamens 10, subequal; fruit (1.5–)5–6.5(–8) mm wide, linear or linear-oblong, moderately pubescent with appressed hairs, about 20-seeded; seeds brown, rhombic, 2–3.5 mm long, 1–2 mm wide.

Distr. A paleotropical weed.

Vern. Bin-siyambala (S).

Specimens Examined. WITHOUT LOCALITY: *Walker s.n.* (K), *23* (K), *39* (K). KANDY & NUWARA ELIYA DISTRICTS: Pussellawa, *Gardner,* Maturata, Oct. 1853, *Thwaites, C.P. 1510* (BM, K, PDA). POLONNARUWA DISTRICT: 4 mi. E. of Welikanda, *Jayasuriya et al. 708* (K, US). KANDY

DISTRICT: Kandy, Hillcrest, *Worthington 5012* (K); Kandy, *van Beusekom 1646* (US); Pussellawa, *Waas & Tirvengadum 826* (SFV, US); Urugala, *Cramer 3613* (US); Gampola, *Balakrishnan 621* (US). AMPARAI DISTRICT: Between Amparai and Maha Oya, mi. 34–35, *Rudd & Balakrishnan 3229* (PDA, SFV, US). HAMBANTOTA DISTRICT: Ruhuna Natl. Park, near Buttawa Bungalow, *Cooray & Balakrishnan 69011026R* (US). PUTTALAM OR ANURADHAPURA DISTRICT: Wilpattu Natl. Park, near Sadpuda Kallu (Occapu Junction), 2 mi. E. of Kattankandal Kulam, *Fosberg et al. 50820* (GH, PDA, SFV, US).

Note. The treatment of this and the related species is rather simplistic and based almost entirely on Ceylon specimens. It is hoped that the names used are reasonably correct. Considering collections from more extended areas makes the problems frustratingly complicated.

30. Cassia auricoma Grah. ex Steyaert, Bull. Jard. Bot. Etat. 20: 246. 1950. Type: "Gualpora ? Mont Sillet ? W. Gomez" (illegible fide Steyaert l.c.) , Wall. Cat. 5322 (K).

Cassia mimosoides L. var. γ *villosula* Thw., Enum. Pl. Zeyl. 96. 1859. Type: Ceylon, "Maturatte District", *Thwaites C.P. 3603* (holotype PDA; isotypes? BM, K).

Cassia mimosoides L. forma δ *auricoma* Grah. in Wall., Cat. ex Benth., Trans. Linn. Soc. London 27: 580. 1871.

Cassia mimosoides L. var. *auricoma* Grah. ex Baker in Hook. f., Fl. Br. Ind. 2: 266. 1878; Trimen, Handb. Fl. Ceylon 2: 110. 1894; Petch, Ceylon J. Sci., Ann. Peradeniya 9: 233. 1924.

Cassia lechenaultiana DC. var. *auricoma* Grah. ex de Wit, Webbia 11: 282. 1955.

Cassia mimosoides L. subsp. *lechenaultiana* (DC) Ohashi var. *auricoma* (Benth.) Steyaert; Ohashi, J. Jap. Bot. 50: 307. 1975.

Herb, sometimes woody at base, 7–70 cm tall; young stems villous, older somewhat glabrescent, crisp-pubescent; stipules lanceolate-attenuate, 8–15 mm long, with 12–36 pairs of leaflets, the rachis villous and crisp-pubescent, bearing an annulate gland just below the lowest pair of leaflets; leaflets somewhat falcate, 3–8 mm long, 1–1.5 mm wide, acute, mucronulate at the apex, obliquely rounded at the base, ciliate, otherwise glabrous above, subglabrous beneath, venation about 5- or 6-costate, with the midrib excentric or submarginal; inflorescences subfasciculate, axillary or supra-axillary, 1- or 2-flowered; pedicels villous and crisp-pubescent, about 10 mm long in flower, in fruit to 2 cm long; bracts lanceolate-attenuate, about 5 mm long; bracteoles similar, about 2 mm long, borne 1 mm below the calyx; sepals villous, attenuate, about (5–)6–7 mm long; petals yellow, (5–)6–6.5 mm long, glabrous; stamens commonly 7, subequal; fruit elastically dehiscent, septate within, linear-oblong, 3.5–4.5 cm long, 4–5 mm wide, villous, about 10–14-seeded; seeds light brown with lines of black puncta, 4 mm long, 1.5 mm wide.

Distr. Ceylon and India.

Specimens Examined. NUWARA ELIYA DISTRICT: Along A5 to Nuwara Eliya, culvert 32/2, *Maxwell & Jayasuriya 860* (MO, SFV, US); along Horton Plains-Hakgala Rd., near Warwick Tea Factory, *Sumithraarachchi et al. 70* (SFV, US); Macdonald's Valley, below Hakgala, *Rudd & Balakrishnan 3175* (PDA, US). BADULLA DISTRICT: Haputale, patana on hill above Monamaya, *Rudd & Balakrishnan 3205* (K, PDA, SFV, US).

Note. As a matter of convenience this taxon has been kept as a separate species. It appears to share characteristics of both *C. mimosoides* and *C. aeschinomene* (*lechenaultiana*) and has been treated as a variety of each, but it is beyond the scope of this study to decide on its proper placement. If recognized as subspecific, the correct epithet, because of priority, should be *villosula* rather than *auricoma*. The types of those taxa should be reexamined to decide if they are truly synonymous. Graham's name, *auricoma*, was published as a nomen nudum.

6. CYNOMETRA

L., Sp. Pl. 382. 1753; L., Gen. Pl., ed. 5, 179. 1754; Knaap-van Meeuwen, Blumea 18: 12. 1970; Cowan & Polhill in Polhill & Raven, Adv. Leg. Syst. 124. 1981; Allen & Allen, Legum. 208. 1981; Kostermans, Reinwardtia 10: 63. 1982. Type species: *C. cauliflora* L.

Shrubs to large trees, unarmed; leaves paripinnate with 1–6 pairs of leaflets; stipules caducous; stipels lacking; inflorescences usually axillary, racemose or paniculate, sometimes corymbose, fasciculate, or cauliflorate; bracts scale-like, broadly ovate, imbricate, usually persistent; bracteoles present but caducous; flowers small with 4(–5) imbricate sepals; petals 5, subequal; stamens (8–)10(–15), sometimes with filaments alternately longer or shorter; anthers uniform, dorsifixed; ovary 1(–4)-ovuled; style filiform; stigma terminal, truncate or capitate; fruit usually oblique, lunate or subreniform, compressed or turgid, dehiscent or indehiscent; 1 or 2(–4)-seeded; seeds compressed or subglobular.

A pantropical genus of about 70 species.

KEY TO THE SPECIES

1 Inflorescences axillary or terminal
  2 Leaflets 2 pairs; fruit rugose with a conspicuous lateral beak.................**1. C. iripa**
  2 Leaflets 1 pair; fruit minutely verrucose, not conspicuously beaked........ **2. C. zeylanica**
1 Inflorescences cauliflorate on trunk or roots. Introduced..................**3. C. cauliflora**

1. Cynometra iripa Kostel., Allg. Med. Pharm. Fl. 4: 1341. 1835; Knaap-van Meeuwen, Blumea 18: 21. 1970; Kosterm., Reinwardtia 10: 66. 1982.Type: Rheede, Hort. Mal. 4: t. 31. 1673.

Cynometra ramiflora var. *heterophylla* Thw., Enum. Pl. Zeyl. 97. 1859; Beddome, Fl. Sylv. t. 315. 1873; Baker in Hook. f., Fl. Br. Ind. 2: 267. 1878;
Trimen, Handb. Fl. Ceylon 2: 111. 1894; Alston in Trimen, Handb. Fl. Ceylon
6: 94. 1931 as syn. of *C. bijuga;* Knaap-van Meeuwen, l.c. as syn. of *C.
ramiflora.* Type: Ceylon, "Trincomalee, Gardner. Caltura District"
[Kalutara], *Thwaites C.P. 1500* (holotype PDA; isotypes BM, K).

Cynometra ramiflora var. *mimosoides* Wall. ex Baker in Hook. f., Fl. Br. Ind.
2: 267. 1878; Trimen, Handb. Fl. Ceylon 2: 112. 1894. Type: Wall. Cat.
5817.

Cynometra ramiflora ssp. *bijuga* var. *mimosoides* (Baker) Prain, J. Asiat. Soc.
Bengal 66 (2): 478. 1897.

Cynometra ramiflora ssp. *bijuga* var. *heterophylla* (Thwaites) Prain, J. Asiat.
Soc. Bengal 66 (2): 198. 1897.

Trees, to about 8 m tall; young stems glabrous; leaves with 2 pairs of leaflets,
the upper pair larger; leaflets obliquely obovate-oblong to oblong, 6–9 cm long,
2–4 cm wide, obtuse or breviacuminate at the apex, asymmetrically acute at the
base, glabrous, nitid on both surfaces; inflorescences corymbiform pseudoracemes;
calyx with 4 or 5 sepals 3–4 mm long; petals white, about 4–5 mm long; fruit
ellipsoid, compressed, about 3 cm long, 2 cm wide, lightly pubescent with short
crispate hairs, glabrescent, strongly rugose with a conspicuous lateral beak extending to about 6 mm long.

Distr. In coastal areas back of mangrove forest, Ceylon, India, Malaysia,
and northeastern Australia (Queensland).

Vern. Opulu (S); attukaddupulli, kadumpuli (T).

Specimens Examined. WITHOUT EXACT LOCALITY: *Walker s.n.* (K);
Ambagamawa, *Silva 6* (NY). PUTTALAM DISTRICT: Puttalam, Nov. 1881,
Ferguson s.n. (PDA). TRINCOMALEE DISTRICT: Sober Isl., Trincomalee har-
bour, Mar. 1892, *Nevill s.n.* (PDA); Niroddumunai, littoral scrub jungle, *Simpson 8496* (BM).

2. Cynometra zeylanica Kosterm., Reinwardtia 10: 63. 1982. Type: Ceylon,
Amparai District: Devulane forest, side road Amparai-Mahiyangane Road, near
stream, *Kostermans 25351* (holotype L; isotypes G, K, PDA, US).

Cynometra ramiflora sensu auctt., non L; Wight & Arn., Prod. 293. 1834; Thw.,
Enum. Pl. Zeyl. 97. 1859, excl. var. *heterophylla;* Baker in Hook. f., Fl.
Br. Ind. 2: 267. 1878 in part as to Ceylon; Trimen, Handb. Fl. Ceylon 2:
111. 1894; Alston in Trimen, Handb. Fl. Ceylon 6: 93. 1931; Knaap-van
Meeuwen, Blumea 18: 23. 1970 in part.

Cynometra longifolia Trimen ex Alston in Trimen, Handb. Fl. Ceylon 6: 93.
1931, nom. nud. in synon. under *C. "ramiflora".*

Trees, up to about 25 m tall; young stems glabrous; leaves with 1 pair of
leaflets; leaflets sessile, obliquely-oblong, about 5 – 13 cm long, 2 – 4 cm wide,

acute to subacuminate at the apex, asymmetrical at the base, glabrous, nitid on both surfaces; inflorescences axillary; bracts caducous; calyx 3 mm long; petals 3–4 mm long; stamens 10; fruit subglobose-ovoid, about 2–2.5 cm long, 1.5 cm wide, brown, verrucose and minutely and sparsely pilose, glabrescent, 1-seeded.

Distr. Endemic to Ceylon in gallery forest in the dry zone.

Vern. Trimen mentioned the name *gal mendora* but Kostermans, l.c., p. 65, says it "is certainly wrong."

Specimens Examined. POLONNARUWA DISTRICT: Alutoya-Kaudulla, *Jayasuriya & Sumithraarachchi 1622* (US); Mannampitiya, *Balakrishnan 601* (US), *Kundu & Balakrishnan 227* (US). BADULLA DISTRICT: Galbokka, between Uraniya and Ekiriyankumbura, *Jayasuriya 1929* (K, PDA); Passara-Yakkala Rd., *Worthington 5103* (BM, K). MONERAGALA DISTRICT: Bibile-Nilgala Rd., mile-marker 3, *Waas 656* (SFV, UC, US); between Nilgala and Pattipola-ara, Uva, Jan. 1888, *Trimen* ? (PDA); "Pattipola & Feb. 1856", *Thwaites C.P. 3604* (PDA); Moneragala Rd., *Worthington 2978* (BM); between Muppane and Indigaswela, *Silva 180* (K, PDA). KALUTARA DISTRICT: Nambapana, *Mackenzie s.n.* (K). KANDY DISTRICT: Weragamtota, *Alston 1672* (PDA).

3. Cynometra cauliflora L., Sp. Pl. 382. 1753; Wight & Arn., Prod. 283. 1834; Baker in Hook. f., Fl. Br. Ind. 2: 268. 1878; Trimen, Handb. Fl. Ceylon 2: 112. 1894; Macmillan, Trop. Pl. & Gard. ed. 5, 253, 254. 1962; Knaap-van Meeuwen, Blumea 18: 20. 1970; Kosterm., Reinwardtia 10: 68. 1982. Type: *Cynomorium* Rumph., Herb. Amboin. 1: t. 62. 1741.

Trees, to about 15 m tall; leaves with 1 pair of leaflets; leaflets obovate to obovate-oblong or obovate-lanceolate, about 5.5–16.5 cm long, 1.6–5.6 cm wide, obtuse or breviacuminate at the apex, glabrous on both surfaces, sometimes pubescent on the petiolules; inflorescences cauliflorous, usually on the trunk, sometimes on the roots; bracts 1–10 mm long; bracteoles 1.5 mm long inserted on the pedicels about 3–6.5 mm long; sepals 4(–5), 2–4 mm long; petals white or pinkish, 3–4 mm long; stamens 8–10, variable in number; fruit rugose, glabrescent, 2.7–3 cm long, 1.8–2 cm wide, 1 cm thick.

Distr. A native of Malaysia introduced into Ceylon at an early period, apparently before mid-18th century.

Uses. The fleshy fruit is eaten, usually cooked. The seeds yield a medicinal oil.

Vern. Nam-nam (Malaysia).

Specimens Examined. KANDY DISTRICT: Peradeniya Royal Botanic Gardens, Sept. 1887, *Trimen* ? (PDA). COLOMBO DISTRICT: Colombo, in a garden (Fraser home), *Worthington 1812* (K).

Note. Linnaeus included *Cynometra cauliflora* (no. 166) and *C. ramiflora* (no. 167) in his Flora Zeylanica, which was based on collections of Paul

Hermann. However, as pointed out by Trimen (J. Linn. Soc. Bot. 24: 141. 1888), there are no specimens in the Hermann herbarium at BM, "only drawings which are not determinable, and seem to have been partly made up from *Averrhoa*."

7. HUMBOLDTIA

Vahl, Symb. Bot. 3: 106. 1794, nom. cons., non Ruiz & Pavon, 1794; Cowan & Polhill in Polhill & Raven, Adv. Leg. Syst. 141. 1981; Allen & Allen, Legum. 335. 1981. Type species: *H. laurifolia* (Vahl) Vahl.

Batschia Vahl, Symb. Bot. 3: 39, t. 56. 1794, non Gmel., 1791, nec Mutis ex Thunb., 1792, nec Moench, 1794. Type species: *B. laurifolia* Vahl.

Trees, unarmed; stipules large, persistent; leaves paripinnate with short petiole; rachis sometimes winged; leaflets sessile, coriaceous, 3–5 pairs; stipels lacking; inflorescences racemose or paniculate, dense, axillary or supra-axillary, many-flowered; bracts and bracteoles ovate to lanceolate; calyx with 4 subequal lobes, imbricate in bud; petals 3 or 5, white or pinkish, long-clawed; perfect stamens 5, exserted, equal in length, alternating with 5 staminodes; anthers dorsifixed; style long, filiform; stigma terminal, truncate; ovary about 3- or 4-ovulate borne on a stipe adnate to the disc; fruit dehiscent, 2-valved, elliptic-oblong, compressed, 1–4-seeded; seeds transverse, ovate, exalbuminous.

A genus of about six species in Ceylon and India. Only one species is known from Ceylon.

Humboldtia laurifolia (Vahl) Vahl, Symb. Bot. 3: 106. 1794; Moon, Cat. 17. 1824; Wight & Arn., Prod. 285. 1834; Wight, Ic. Pl. Ind. Or. 1605. 1850; Thw., Enum. Pl. Zeyl. 97. 1859; Baker in Hook. f., Fl. Br. Ind. 2: 273. 1878; Trimen, Handb. Fl. Ceylon 2: 115. 1894; Alston in Trimen, Handb. Fl. Ceylon 6: 88. 1931; Worthington, Ceylon Trees 197. 1959; Macmillan, Trop. Pl. & Gard. ed. 5, 107, 441. 1962; Backer & Bakh. f., Fl. Java 1: 530. 1963.

Batschia laurifolia Vahl., Symb. Bot. 3: 39, t. 56. 1794. Type: "Habitat in Zeylona, König, Thunberg", *König s.n.* (C).

Trees, to about 10 m tall; young stems glabrous, often swollen and inhabited by ants; stipules leaflike above the point of insertion, about 2–3.5 cm long, 1–2 cm wide, acute, the lower portion falcate; leaves with 3–5 pairs of leaflets, the rachis about 1–20 cm long, glabrous or nearly so; leaflets ovate to ovate-oblong or lanceolate-oblong, about 4–15 cm long, 2–5.5 cm wide, acuminate at the apex, obliquely rounded to cuneate at the base, glabrous, nitid above, puberulent beneath, usually glabrous at maturity; inflorescences axillary, racemose or paniculate, 8–12 cm long; bracts deltoid, about 4 mm long; bracteoles oblong-spatulate, 6–7 mm long; flowers on pedicels 5–6 mm long; calyx usually reddish or orangish with tube about 3 mm long, segments 8–10 mm long, subequal; petals 5, white,

10–12 mm long; stamens exserted to about 12–20 mm long; fruit puberulent or subglabrous when young, glabrous at maturity, ellipsoid, compressed, 5–10 cm long, 2.5–3 cm wide, 3–4-seeded.

Distr. Native to Ceylon and India; introduced elsewhere as an ornamental.

Vern. Gal-karanda, ruan-karanda (S).

Specimens Examined. WITHOUT EXACT LOCALITY: *Moon s.n.* (BM); *Macrae 367* (BM), *s.n.* (BM); *Walker s.n.* (GH, PDA); Amadulla, Matale, & *Gardner,* Atagalla (?), *Thwaites C.P. 328* (PDA), without data, *C.P. 328* (BM, GH, NY). PUTTALAM DISTRICT: Kalu mookalana—off Lihiriyagama, *Sumithraarachchi 426* (PDA, SFV, US). KANDY DISTRICT: Peradeniya, *de Silva 215* (NY), *Kostermans 24557* (K, PDA, US). KALUTARA DISTRICT: Morapitiya, *Sumithraarachchi et al. 1007* (PDA, SFV, US); Kalugala Forest, *Sohmer & Waas 10253* (GH, NY, PDA, US); Mandagala, *Bernardi 15729* (NY, PDA, US). KEGALLE DISTRICT: Kitulgala, *Comanor 546* (NY, PDA, US), *Worthington 395* (BM, K); Pitawela, Kitulgala 57/12, *Worthington 2091* (BM). KANDY DISTRICT: Bambaragohakele, Kadugannawa, *Worthington 711* (A, BM, K). RATNAPURA DISTRICT: Heramitiyagala, near Rassagala, *Huber 539* (US); Gilimale Forest Reserve, *Nooteboom & Huber 3157* (US); *Sohmer et al. 8787* (GH, US), *Meijer & Gunatilleke 1356* (US), *Dassanayake 477* (US), *Bernardi 14114* (K, PDA, US), *Meijer 420* (US); Paragala along Pokonodole, *Bremer 796* (PDA, S, US); Delgoda, on Kalawana-Weddagala Rd. *Faden 76/475* (PDA, US); Sri Palabaddala, *Waas 256* (K, PDA, US); Karawitakanda, *Waas 20* (K, NY, PDA, US). GALLE DISTRICT: Galle, *Gardner 253* (BM); Kanneliya, *Nooteboom 3219* (US), *Sohmer et al. 8929* (GH), *Balakrishnan 260* (PDA, US); Bonavista, *Cooray 70020911 R* (PDA, SFV, US), *Kundu & Balakrishnan 492* (US); Bengamuwa, Ratmale Forest, *Comanor 1199* (US); Kottawa, *Worthington 622* (K). KURUNEGALA DISTRICT: Weudakanda below Galagedera, *Worthington 5545* (K).

8. TAMARINDUS

L., Sp. Pl. 34. 1753; L., Gen. Pl., ed. 5, 20. 1754; Cowan & Polhill in Polhill & Raven, Adv. Leg. Syst. 141. 1981; Allen & Allen, Legum. 641. 1981. Type species: *T. indica* L.

Trees, unarmed; stipules lanceolate, minute, caducous; leaves paripinnate; leaflets opposite, numerous; stipels lacking; inflorescences racemose; terminal or lateral; bracts ovate-oblong, fugacious; bracteoles ovate-oblong, valvate, enclosing the flowers in bud, usually reddish, caducous; flowers small; calyx 4-parted; imbricate in bud; petals 5, unequal with 3 larger, 2 smaller and narrower, white or yellowish, usually with some reddish venation; perfect stamens 3 with filaments connate alternating with 4 or 5 minute staminodes; anthers dorsifixed; ovary many-ovuled with stipe adnate to the calyx tube; style elongated; stigma minutely

capitate; fruit indehiscent, oblong, lightly compressed, 1–12-seeded, often constricted between the seeds, exocarp brittle, thin, scurfy outside; mesocarp pulpy;
seeds compressed, brown, lustrous, with a closed areole on each face,
exalbuminous.

A genus with only one species, native to the Old World tropics but widely
introduced and often naturalized in warm areas of the New World.

Tamarindus indica L., Sp. Pl. 34. 1753; Moon, Cat. 48. 1824; Wight & Arn.,
Prod. 285. 1834; Beddome, Fl. Sylv. t. 184. 1872; Baker in Hook. f., Fl. Br.
Ind. 2: 273. 1878; Trimen, Handb. Fl. Ceylon 2: 114. 1894; Alston in Trimen,
Handb. Fl. Ceylon 6: 88. 1931; Worthington, Ceylon Trees 196. 1959; Mac-
millan, Trop. Pl. & Gard., ed. 5, 29, 197, 216, 362, 366, 487. 1962; ·Backer
& Bakh. f., Fl. Java 1: 529. 1963; Brenan in Fl. Trop. East Africa, Caesalp.
153. 1967; Ali, Fl. W. Pakistan, no. 54, Caesalp. 35. 1973; Saldanha & Nicolson,
Fl. Hassan 224. 1976. Type material: Hort Cliff. 18 (BM-CLIFF, sterile); Herb.
Linn. 49: 1, 2, 3 (LINN); Herb. Hermann 2: 73, 2: 80 (BM-HERM); Hermann,
fl. 73, above (L-HERM); Hermann in Herb. Burm. p. 78 (Paris, Biblio, Inst.
France); Rheede, Hort. Mal. 1: t. 23. 1678.

Tamarindus occidentalis Gaertn., Fruct. 2: 310, t. 146, fig. 2. 1791.
Tamarindus officinalis Hook., Bot. Mag. t. 4563. 1851.

Additional synonymy is given by Roti-Michelozzi, Webbia 13: 134. 1957.

Trees, to about 15(–24) m tall; young stems puberulent, glabrescent; stipules
minute, lanceolate, caducous; leaves with about 8–20 pairs of leaflets; leaflets
opposite, oblong-elliptic, 12–25(–30) mm long, (3–)5–10 mm wide; rounded to
truncate or retuse at apex, rounded at the base, glabrous or nearly so on both
surfaces; inflorescences racemose, terminal or lateral, about 1–8-flowered; flowers
small, on pedicels 3–14 mm long; calyx reddish, glabrous, 8–12 mm long; petals
white or yellowish with reddish venation, about 10–13 mm long; fruit light brown,
scurfy outside, commonly 5–15 cm long including stipe about 5 mm long, 2 cm
wide, 1.5 cm thick, 2–4-seeded, indehiscent; seeds brown, lustrous, with a closed areole on each face, compressed, about 1–1.5 cm long, 5–8 mm wide.

Distr. Native to the Old World tropics; introduced into warm areas of the
New World.

Uses. The trees are useful as shade near houses and along roadsides; the wood
provides a beautiful red timber; the leaves, fruit, and seeds are reputed to have
medicinal properties and the fruit pulp is used in preparing refreshing beverages
and for chutneys and preserves. Beddome, 1.c., also states that a strong glue
can be made from finely ground seeds and that a red dye is made from the leaves.

Vern. Siyambala, maha-siyambala (S); puli (T); tamarind, Indian date (E).

Specimens Examined. WITHOUT EXACT LOCALITY: *Worthington
4558* (K). JAFFNA DISTRICT: Between Elephant Pass and Paranthan,

Rudd 3293 (K, PDA, US). ANURADHAPURA DISTRICT: N. bank of Rajangana Reservoir, 18 mi. S. of Anuradhapura, *Maxwell & Jayasuriya 793* (NY, PDA, SFV, US). POLONNARUWA DISTRICT: Polonnaruwa, sacred area, *Dittus WD 710900422* (PDA, US), *Hladik 693* (US). MATALE DISTRICT: Kandalama, *Amaratunga 1466* (PDA); Matale-Dambulla Rd., before mi. marker 44 on A9, *Comanor 725* (K, NY, PDA, US); Lenodora, near mi. marker 38, Kandy-Dambulla Rd., *Waas 708* (US). KANDY DISTRICT: Kandy, *Worthington 2120* (K), *4784* (BM, K), *5590* (K); Peradeniya, Royal Botanic Gardens, *Gardner, Thwaites C.P. 1501*(PDA), *Baker 131* (CAS, GH, NY, PDA, US). HAMBANTOTA DISTRICT: Ruhuna Natl. Park, near Yala Bungalow, *Mueller-Dombois & Comanor 67062210* (US). KURUNEGALA DISTRICT: Melsiripura, *Amaratunga 1808* (PDA). RATNAPURA DISTRICT: Galenda, *Worthington 4448* (K). BADULLA DISTRICT: 'Mutigalla Kulam', *Worthington 6414* (K).

9. DIALIUM

L., Mant. 3, 24. 1767; Irwin & Barneby in Polhill & Raven, Adv. Leg. Syst. 101. 1981; Allen & Allen, Legum. 234. 1981. Type species: *D. indum* L.

Trees or large shrubs, unarmed; stipules minute, caducous; leaves imparipinnate; leaflets few, alternate or subopposite; stipels lacking; inflorescences paniculate, terminal or lateral, many-flowered; bracts and bracteoles minute, fugacious; flowers small; calyx 5-parted, imbricate in bud; petals lacking; stamens 2(–10), with free filaments; anthers basifixed or subbasifixed; ovary sessile or subsessile, 2-ovuled; style short, filiform-subulate; stigma small, terminal; fruit indehiscent, subglobose, somewhat compressed, pulpy, 1- or 2-seeded; seeds compressed; endosperm present.

A genus of about 40 species, chiefly in the Old World tropics.

Dialium ovoideum Thw., Enum. Pl. Zeyl. 97. 1859; Beddome, Fl. Sylv. pl. 181. 1872; Baker in Hook. f., Fl. Br. Ind. 2: 269. 1878; Trimen, Handb. Fl. Ceylon 2: 112. 1894; Worthington, Ceylon Trees 194. 1959; Macmillan, Trop. Pl. & Gard., ed. 5, 254. 1962. Type: Ceylon, Kandy, *Thwaites C.P. 3149* (BM, K).

Trees, large, unarmed; stems glabrous; stipules minute, fugacious; leaves 4 or 5-foliolate, leaflets alternate or subopposite, lanceolate-ovate, about 2–9 cm long, 1–3.5 cm wide, bluntly acuminate to subacute at apex, rounded at base, glabrous, nitid on both surfaces; bracts and bracteoles minute, deltoid, caducous; flowers small, whitish, about 6 mm long on pedicels about 5 mm long; calyx lobes petaloid, lanceolate, subacute, ciliolate, about 6 mm long; petals lacking; stamens 2(–3); fruit orbicular, compressed, dark brown, velutinous, about 2 cm long, 1.5 cm wide, 5 mm thick, 1(–2)-seeded; seeds orbicular, compressed, light brown, striate, about 1–1.5 cm long.

Distr. Native to Ceylon but has been introduced into Southeast Asia.

Uses. Macmillan, 1.c. states that the fruit is edible and is used in the preparation of chutneys. The trees are valuable as timber which is used for furniture making.

Vern. Gal-siyambala (S); kaddupuli (T); velvet tamarind (E).

Specimens Examined. ANURADHAPURA DISTRICT: Ela at Ritigala W., *Worthington 5060* (K); Ritigala Strict Natl. Reserve, E. slope of Andikanda, *Jayasuriya 1062* (PDA, US), *Meijer & Jayasuriya 1284* (US). POLON-NARUWA DISTRICT: Gunners Quoin, *Worthington 5292* (K). KURUNEGALA DISTRICT: Kurunegala, *Kostermans 25240* (K, PDA, US); Dolukande, *Cramer 3047* (PDA, US), *Meijer 367* (K, PDA, US). MATALE DISTRICT: Nalanda, *Worthington 3062* (BM, K), *6389* (K), *1842* (BM); Lenadora Kanda, Nalanda hills W. of mi. post 38, *Worthington 6643* (K); Erawalagala Mtn., E. of Kandalama Tank, *Davidse & Sumithraarachchi 8109* (K). AMPARAI DISTRICT: Gal Oya Natl. Park, Inginiyagala, *Meijer & Balakrishnan 147* (PDA, US). MONERAGALA DISTRICT: Near Mahiyangane, road to Bibile, *Kostermans 24514* (K); Wellawaya, *Worthington 4748* (BM, K).

Note. In addition to the above species, Macmillan, op. cit., p. 255, mentions that *D. guineense* Willd., the West African velvet tamarind, was introduced to Ceylon in 1893.

10. CRUDIA

Schreber, Gen. 1: 282. 1789, nom. cons.; Cowan & Polhill in Polhill & Raven, Adv. Leg. Syst. 131. 1981; Allen & Allen, Legum. 199. 1981.Type species: *C. spicata* (Aubl.) Willd.

Trees; unarmed; leaves imparipinnate; stipules small and caducous or large and persistent; leaflets alternate or opposite; stipels lacking; inflorescences racemose, axillary or terminal; bracts and bracteoles deciduous or subpersistent; flowers small; calyx with short tube and 4 imbricate lobes reflexed at anthesis; petals lacking; stamens 8–10; filaments joined into a basal tube adnate to the calyx tube, free above; anthers dorsifixed; ovary short-stipitate, 1–6-ovuled; style filiform; stigma small, terminal; fruit suborbicular to oblong, coriaceous or woody, usually compressed, dehiscent, 1–2(–3) -seeded; seeds suborbicular to reniform.

A pantropical genus of about 55 species.

Crudia zeylanica (Thw.) Benth., Trans. Linn. Soc. London 25: 314. 1865; Beddome, Fl. Sylv. 190. 1872; Baker in Hook. f., Fl. Br. Ind. 2: 271. 1878; Trimen, Handb. Fl. Ceylon 2: 113. 1894.

Detarium zeylanicum Thw., Enum. Pl. Zeyl. 414. 1864. Type: Ceylon, "Galpaata, near Caltura" [Kalutara], *Thwaites C.P. 3714* (holotype PDA; isotypes BM, K).

Large trees; leaves imparipinnate with 2–6 leaflets; leaflets alternate, ovate to oblong, about 5–15 cm long, 2–5 cm wide, obtusely breviacuminate at the apex, rounded at the base, glabrous on both surfaces, minutely reticulate; flowers small in spicate terminal racemes, on short pedicels; calyx segments ovate, obtuse, persistent; petals lacking; ovary tomentose; fruit (immature) oblong, compressed, slightly falcate, apiculate, densely tomentose.

Distr. Apparently endemic to Ceylon.

Specimens Examined. In addition to the type material cited above, there is a specimen at PDA from the Royal Botanic Gardens, Peradeniya, collected by J.M. Silva, 20 Feb. 1911. Later someone wrote on the sheet "cut down". Apparently the species has not been recollected at the type locality and there are no other voucher specimens.

11. SARACA

L., Mant. 13. 1767; Zuijderhoudt, Blumea 15: 414. 1967; Cowan & Polhill in Polhill & Raven, Adv. Leg. Syst. 128. 1981; Allen & Allen, Legum. 592. 1981. Type species: *S. indica* L.

Jonesia Roxb., Asiat. Res. 4: 355. 1799. Type species: *J. asoca* Roxb.

Trees or shrubs, to about 10 m tall, unarmed; stipules relatively large, connate, enveloping the buds, caducous; leaves paripinnate; stipels lacking; leaflets 1–7 pairs; inflorescences short, paniculate, fasciculate, subglobular corymbs, terminal, lateral, or sometimes cauliflorous; bracts and bracteoles present or absent, persistent or fugacious; flowers of medium size; calyx with an elongated tube and 4(–6) petaloid lobes, yellow to orange or red, sometimes purplish; petals lacking; stamens 3–10; filaments free, exserted, elongated; anthers dorsifixed; style filiform; stigma minute, terminal; fruit oblong, compressed, dehiscent, 1–8-seeded; seeds ellipsoid, compressed, 1–4 cm long, exalbuminous.

A genus of eight species native to Asia.

Although the genus is generally said to have 4(–6) petaloid calyx lobes and no petals, studies of the vascular anatomy have shown that there are 10 vascular bundles, suggesting that both sepals and petals are represented in the apparent "calyx" lobes (Puri, Bot. Rev. 17: 492. 1951).

Saraca asoca (Roxb.) De Wilde, Blumea 15: 393. 1967; Zuijderhoudt, Blumea 15: 422. 1967; Saldanha & Nicolson, Fl. Hassan 224. 1976; Verdcourt, Man. New Guinea Leg. 88. 1979.

Jonesia asoca Roxb., Asiat. Res. 4: 355, f. 252, 253. 1799; Moon, Cat. 30. 1824; Wight & Arn., Prod. 284. 1834; Wight, Ic. Pl. Ind. Or. 266. 1840; Thw., Enum. Pl. Zeyl. 97. 1859 ("Asoka"). Type: Roxburgh, f. 252–253, fide De Wilde.

Jonesia pinnata Willd., Sp. Pl. 2: 287. 1799, nom. illeg.; Moon. Cat. 30. 1824. Type: Based on Roxburgh, f. 252–253 and Rheede, Hort. Mal. 5, t. 59.

Saraca indica sensu auctt., non L.; Beddome, Fl. Sylv., pl. 57. 1870; Baker in Hook. f., Fl. Br. Ind. 2: 271. 1878; Trimen, Handb. Fl. Ceylon 2: 113. 1894; Alston in Trimen, Handb. Fl. Ceylon 6: 94. 1931; Worthington, Ceylon Trees 195. 1959; Macmillan, Trop. Pl. & Gard., ed. 5, 90. 1962.

Trees, to about 9 m tall; stipules small, deciduous; leaves with (1–)4–6 pairs of leaflets, subsessile; leaflets subcoriaceous, oblong to lanceolate, 3.5–25 cm long, 1–9 cm wide, acute to acuminate at the apex, acute to rounded, rarely cordate at the base, glabrous on both surfaces; inflorescences compact corymbose panicles, axillary and terminal; bracts elliptic to ovate or obovate, 1–6 mm long, 1–3.5 mm wide, persistent or caducous; bracteoles elliptic to ovate or obovate, 2–7 mm long, 1.5–4 mm wide, persistent; flowers fragrant; calyx yellowish to red, with tube 10–17 mm long, 1–2 mm in diameter, lobes elliptic to obovate, 7–10 mm long, 5–9 mm wide, rounded at the apex; petals lacking; stamens (3–)6–8(–10); filaments reddish, exserted, 17–25 mm long; fruit elliptic to oblong, 4.5–25 cm long, 2–6 cm wide, about 1.5(–2) cm thick, glabrous, 4–8-seeded; seeds ovoid, compressed, about 3.5–4.5 cm wide.

Distr. Ceylon and southern India to western Burma; introduced elsewhere in Southeast Asia and Africa.

Uses. Planted as an ornamental and for its sacred nature. Buddha is said to have been born under an ashoka tree. The bark and flowers are reputed to have medicinal properties (Jain, Medicinal Plants 124, 125. 1968).

Vern. Ashoka, asoca, asoka (Sanskrit); diya-ratmal, diya-ratambala (S); asogam (T).

Specimens Examined. WITHOUT EXACT LOCALITY: *Walker s.n.* (PDA); *Moon s.n.* (BM); *Macrae 115* (BM); Matale, Gardner, Arabegama, Mar 1852, Bintenne, July 1853, *Thwaites C.P. 653* (PDA); without data, *C.P. 653* (BM, PDA). ANURADHAPURA DISTRICT: Mihintale, *Bernardi 14235* (PDA, US); Ritigala, *Cramer 4143* (K, PDA, US), *Jayasuriya 1337* (K, US), *Jayasuriya & Ashton 1343* (K, US), *Jayasuriya et al. 2270* (K). MATALE DISTRICT: Kalugaloya, *Waas 570* (K, PDA, US); Hunugalla, *Worthington 122* (K); road to Midlands, junction of road and river after Illukkumbura, *Nowicke & Jayasuriya 211* (US). KANDY DISTRICT: Hantane, *Gardner 252* (BM); Peradeniya, University grounds, *Comanor 1167* (GH, NY, PDA, SFV, US). BADULLA DISTRICT: 7 mi. on Uraniya Rd., from Bibile, *Worthington 4067* (BM, K); Mahiyangane, *Kostermans 24431* (K). MONERAGALA DISTRICT: W. of Wellawaya at 137 mi. from Colombo, *Townsend 73/152* (PDA, US); Wellawaya, culvert 130/6, *Worthington 5449* (BM, K).

Note. In addition to *S. asoca*, other species of *Saraca* have been introduced as ornamentals, including *S. declinata* (Jack.) Miq., *S. indica* L., and *S. thaipingensis* Cantley ex Prain, all natives of SE. Asia.

*   *   *

In addition to the preceding taxa many species, especially of ornamental plants, have been introduced into Ceylon, some apparently by the mid-17th century, others more recently. Some are known only from introductions into the botanic gardens, especially at Peradeniya; some have been more widely planted but not known to have become naturalized.

The following are a few taxa that are documented by voucher specimens and therefore have been included in the key to genera. There also are a number of additional specimens of cultivated plants at PDA which, because of limited time, could not be carefully examined and recorded for this study.

12. Acrocarpus fraxinifolius Wight & Arn. in Wight ex Arn., Mag. Zool. Bot. 2: 547. 1839; Wight, Ic. Pl. Ind. Or. 254. 1840; Beddome, Fl. Sylv. t. 44. 1870; Baker in Hook. f., Fl. Br. Ind. 2: 292. 1878; Worthington, Ceylon Trees 183. 1959; Macmillan, Trop. Pl. & Gard., ed. 5, 171, 214. 1962. Howligemara, malai-konnai (T); pink cedar, red cedar, shingle tree (E).

A large, unarmed tree native to India and Southeast Asia, grown as shade for tea and coffee; leaves bipinnate with 3–5 pairs of pinnae, each pinna with about 4–9 pairs of ovate to oblong leaflets; flowers in racemes; petals red, 6–10 mm long; fruit glabrous, compressed, stipitate, about 8–13 cm long, 1.5–2 cm wide with a wing about 4 mm wide along one margin.

Specimens Examined. NUWARA ELIYA DISTRICT: Labukellie, 5000', planted as shade in Estates (tea), *Worthington 1632* (BM, K); Agras, 5000', *Worthington 1753* (BM, K). KANDY DISTRICT: Carolina/Norton Rd. 4400', *Worthington 2771* (BM, K); Peradeniya, Royal Botanic Gardens, cult., *s. coll. s.n.* (K), *Worthington 6859* (K).

13. Amherstia nobilis Wall. in Taylor & Phillips, Philos. Mag. J. 68: 323. 1826; Wall., Pl. As. Rar. 1: 1, t. 1, 2. 1830; Baker in Hook. f., Fl. Br. Ind. 2: 272. 1878; Macmillan, Trop. Pl. & Gard., ed. 5, frontispiece, 79. 1962; Backer & Bakh. f., Fl. Java 1: 530. 1963; Verdcourt, Man. New Guinea Leg. 106, fig. 24. 1979.

"Queen of flowering trees". According to Stockdale, Petch & Macmillan, The Royal Botanic Gardens, Peradeniya, Ceylon, Trop. Agric. (Ceylon) reprint 9. 1922, this ornamental species was introduced to the Gardens from Burma in 1860.

Trees, unarmed, to about 20–30 m tall; young leaves brownish-bronze, pendulous; mature leaves abruptly pinnate with 5–8 pairs of oblong-lanceolate, acuminate leaflets 15–25 cm long, 5.5–7 cm wide; flowers showy, long-pedicelled, in pendulous racemes; bracteoles valvate, enclosing flowers in bud; bracts and sepals bright red, petals 4.5–7.5 cm long, mixed reddish and yellow; fruit oblong, brown, dehiscent, compressed, about 15–20 cm long, 3.5–4 cm wide, 3–6-seeded.

Specimens Examined. KANDY DISTRICT: Peradeniya, Royal Botanic Gardens, *Fairchild & Dorsett 416* (US), *Fosberg 56330* (GH, MO, NY, US), *Worthington 826* (K).

14. Baikiaea insignis Benth., Trans. Linn. Soc. London 25: 314. 1865; Hutch. & Dalz., Fl. W. Trop. Africa, ed. 2, 1, 456. 1958; Macmillan, Trop. Pl. & Gard., ed. 5, 80. 1962; Brenan in Fl. East Trop. Africa, Caesalp. 109. 1967.

This native of West Africa was introduced at Peradeniya in 1902 according to Macmillan, l.c.

Trees, to about 30 m tall, unarmed; leaves pinnate, commonly 3–8-foliolate; leaflets large, alternate, coriaceous, ovate or elliptic to oblong-lanceolate, to 40 cm long, 17 cm wide; bracteoles imbricate, almost valvate, but not enclosing flowers in bud; flowers large, to about 21 cm long in few-flowered racemes; petals white or cream; fruit oblong, compressed, dehiscent, 17–60 cm long, 5–12 cm wide.

Specimens Examined. KANDY DISTRICT: Peradeniya, University grounds, *Comanor 1172* (K, NY, PDA, SFV, US); Kadugannawa (planted from Peradeniya), *Worthington 1101* (K), *1111* (K). RATNAPURA DISTRICT: Balangoda Estate, *Worthington (Mrs. Hazel Thomas) s.n.* (K).

15. Brownea Jacq., Enum. Pl. Carib. 6, 26. 1760; Pittier, Contr. U.S. Natl. Herb. 18: 145–157. 1916; Macmillan, Trop. Pl. & Gard., ed. 5, 81. 1962; Allen & Allen, Legum. 111. 1981; Velásquez & Agostini, Ernstia, no. 5: 1–13. 1981.

Trees, unarmed; leaves paripinnate with about 2–15 pairs of fairly large, oblong, leaflets; bracteoles valvate, enclosing flowers in bud; flowers in showy red or pinkish capitate inflorescences; petals about 3.5–6.5 cm long; fruit oblong, compressed, dehiscent, few-seeded.

Macmillan, l.c., cited the following species that were introduced to Ceylon in the 19th century:

Brownea ariza Benth., Pl. Hartw. 171. 1857, a native of Colombia, South America, with bright red flowers in dense clusters, introduced to Ceylon in 1884.

Brownea coccinea Jacq., Select. 95, pl. 1788. 1788, a Venezuelan species with flowers in small, scarlet clusters, introduced to Ceylon in 1849.

Brownea grandiceps Jacq., Collect. 3: 287. 1789, "Rose of Venezuela", "palo de cruz", "rosa de montaña", originally described from Caracas, Venezuela, is a large tree with large dense, many-flowered inflorescences, introduced to Ceylon in 1870.

Specimens Examined. Peradeniya, Royal Botanic Gardens, *Fairchild & Dorsett 415* (UC, US).

Brownea macrophylla Linden., Cat. no. 18, 11. 1863; Linden, Gard. Chron. 1873: 777, f. 149. 1873. from Colombia, with large heads of flowers "bright rose" to "fire-red", introduced to Ceylon in 1894.

Specimen Examined: Peradeniya, Royal Botanic Gardens, *Worthington 822* (K).

Brownea × crawfordii, with salmon-pink flowers, cited by Macmillan as a hybrid between *B. grandiceps* and *B. macrophylla*.

16. Copaifera L., Sp. Pl., ed. 2, 557. 1762, nom. cons.; Record & Hess, Timbers of the New World, 248. 1943; Dwyer, Brittonia 7: 143–172. 1951; Macmillan, Trop. Pl. & Gard., ed. 5, 94, 393. 1962.

Trees or shrubs, unarmed; leaves pinnate; leaflets alternate or opposite, often pellucid-punctate; flowers small in terminal or axillary racemes; sepals 4; petals absent; fruit suborbicular or obliquely elliptic, 1- or rarely 2-seeded, usually dehiscent. Two species were mentioned by Macmillan.

Copaifera officinalis (Jacq.) L., Sp. Pl., ed. 2, 557. 1762.

Balsam copaiba, a native of the West Indies and South America, introduced to Ceylon in 1880.

Trees, to about 20 m tall, resiniferous; leaflets about 4–8, alternate or subopposite, obliquely ovate-oblong or subfalcate, 2.5–8 cm long, 2–4 cm wide, pellucid-punctate, glabrous; flowers in axillary panicles; sepals about 3–4.5 mm long; stamens exserted to about 7 mm long; fruit suborbicular, glabrous, smooth, 2–3 cm long, 1.5–2.5 cm wide. The exudate from this plant, an oleoresin, is used medicinally and in the manufacture of varnishes and lacquers. The timber is used in carpentry and general construction.

Specimens Examined. KANDY DISTRICT: Nawalapitiya, *Worthington 123* (K), *370* (K), *586* (BM, K).

Copaifera langsdorfii Desf., Mém. Mus. Hist. Nat. 7: 377. 1821.

From Brazil, is similarly used medicinally and in carpentry, and is also known as Balsam copaiba, or copaiba balsam.

Trees; leaflets 6–8(–12), alternate or subopposite, oblong-elliptic to ovate-oblong, 1.5–5.5 cm long, 1–3 cm wide, pellucid-punctate, glabrous above, pubescent or glabrous beneath; flowers in axillary panicles; sepals about 3–4.5 mm long; stamens 4–7 mm long; fruit somewhat rugose, elliptic-oblong to suborbicular, 1.5–2 cm long, 1.3 cm wide.

17. Delonix Raf., Fl. Tellur. 2: 92. 1836; Allen & Allen, Legum. 222. 1981.

Unarmed trees; leaves bipinnate; leaflets small, numerous; flowers showy, in terminal or axillary racemes; petals flame-red or white; fruit linear-oblong, dehiscent, many-seeded. Two species have been introduced into Ceylon.

Delonix regia (Bojer ex Hook.) Raf., Fl. Tellur. 2: 92.1836, based on *Poinciana regia* Bojer ex Hook., Bot. Mag., pl. 2884. 1829; Worthington, Ceylon Trees 185. 1959; Macmillan, Trop. Pl. & Gard., ed. 5, 89, 99, 210. 1962, as *Poinciana regia*. Mal-mara (S); mayaram, poo-vahai (T); flamboyant, flame-tree, royal poinciana, gold mohur (E).

A beautiful flowering tree, introduced into Ceylon before 1841 according to Macmillan; originally native to Madagascar but now widely planted worldwide in the tropics as an ornamental. The trees may be more than 12 m tall; leaves with 8–20(–25) pairs of pinnae, each pinna with 10–25(–40) pairs of elliptic to linear-oblong leaflets about 7 mm long, 3 mm wide; flowers in lateral or terminal corymbose racemes; petals 4–7 cm long, one petal white, the others flame-red; fruit linear-oblong, (30)40–60(–80) cm long, 4–7 cm wide.

Specimens Examined. KANDY DISTRICT: Kandy, Oodoowela, *Worthington 245* (K); The Maligawa, *Worthington 6853* (K); Kadugannawa, Poilakanda Garden, "planted from Colombo seed", *Worthington 886* (K), *649* (K).

Delonix elata (L.) Gamble, Fl. Pres. Madras 1(3): 396. 1919, based on *Poinciana elata* L., Cent. II Pl.: 16. 1756; Moon, Cat. 34. 1824, as *Poinciana elata*; Worthington, Ceylon Trees 184. 1959. Vatham-nairaini, vatham rasi (T); creamy peacock flower (E).

This is an African species; trees to about 15 m tall; leaves with 2–12 pairs of pinnae, each pinna with about 8–25 pairs of small, linear-oblong leaflets; flowers with an upper, smaller, pale yellow petal, the others white, about 2–4 cm long and wide, all petals withering to orangish or apricot colour; fruit 13–25 cm long, 2–3.7 cm wide.

Specimens Examined. MANNAR DISTRICT: Mannar, rest house, *Worthington 4495* (K), *6716* (K). TRINCOMALEE DISTRICT: Trincomalee, Main street, *Worthington 1154* (BM, K), *1176* (K). According to Worthington the trees are planted for shade, as ornamentals, for fence posts, and as green manure. He mentions that the Tamil name, "vatham-nairaini" suggests a medicine for rheumatism.

18. Erythrophleum suaveolens (Guill. & Perr.) Brenan, Taxon 9: 194. 1960; Brenan, Fl. Trop. East Africa, Caesalp. 18. 1967, based on *Fillaea suaveolens* Guill. & Perr.; *Erythrophleum guineënse* G. Don; Macmillan, Trop. Pl. & Gard., ed. 5, 372. 1962.

Sassy bark, red water tree of Sierra Leone. This was cited by Macmillan as a well-known poisonous tree of Sierra Leone whose bark was formerly used in ordeal trials and for poisoning arrows, introduced into Ceylon in 1888.

Trees 9–30 m high, unarmed; leaves bipinnate; pinnae 2–4 pairs; leaflets alternate, 7–14 pairs per pinna, ovate, ovate-elliptic, to lanceolate, slightly asymmetrical, 2.7–9 cm long, 1.3–5.3 cm wide, usually glabrous or nearly so; flowers small in panicles of spicate racemes; petals yellowish-white to greenish-yellow, 2–3 mm long; fruit woody, dehiscent, oblong, glabrous, compressed, 8–17 cm long, 2.5–5.3 cm wide, 6–11-seeded or less by abortion.

Specimens Examined. KANDY DISTRICT: Peradeniya, Royal Botanic Gardens, *Worthington 1793* (K), *5983* (K), *Meijer 691* (MO, PDA, US); Kandy, Barracks, "naturalized in Kandy jungles", *Worthington 746* (BM, K).

19. Haematoxylum campechianum L., Sp. Pl. 384. 1753; Record & Hess, Timbers of the New World 276. 1943; Macmillan, Trop. Pl. & Gard., ed. 5, 70, 416, 434. 1962; Backer & Bakh. f., Fl. Java 1: 543. 1963.

Logwood, a species introduced to Ceylon in 1845. Macmillan mentioned the small, spiny trees as suitable for boundaries or barriers. The heartwood and roots are the source of blue, violet, and purple dyes used for ink, and for dyeing woollen and silk goods. They also yield hematoxylin, a stain used in histological work.

Trees, up to about 8–15 m tall, somewhat gnarled, armed with thorns 5–15 mm long; leaves paripinnate, or sometimes bipinnate in part; leaflets about 2–4 pairs, cuneate-obovate to obcordate, glabrous, finely veined; flowers yellow, fragrant, about 5–7 mm long in subspicate racemes; fruit oblong, obtuse at the apex, acute at the base, 2–5 cm long, 8–12 mm wide, with thin valves dehiscent longitudinally in the centre rather than along the margins.

Specimens Examined. KANDY DISTRICT: Kandy, Hillcrest, 20-year old tree, "not very happy", sterile, *Worthington 7216* (K).

20. Hymenaea L., Sp. Pl. 1142. 1753. *Trachylobium* Hayne, Flora 10: 743. 1827; Record & Hess, Timbers of the New World 251. 1943; Lee & Langenheim, Univ. Calif. Publ. Bot. 69: 1–109. 1975; Allen & Allen, Legum. 337, 658. 1981.

Trees, unarmed, resiniferous; leaves 2-foliolate; leaflets obliquely oblong to oblong-lanceolate, pellucid-punctate, coriaceous; flowers mostly white in terminal panicles; fruit ellipsoid, subterete, indehiscent, few-seeded. Two species have been introduced into Ceylon, one from tropical America, the other from Africa.

Hymenaea courbaril L., Sp. Pl. 1192. 1753; Macmillan, Trop. Pl. & Gard., ed. 5, 394. 1962; Backer & Bakh. f., Fl. Java 1: 528. 1963.

West Indian locust; South American locust. There also are many local names used in Latin America. This species is native to the American tropics, widely distributed from southern Mexico and the West Indies to Brazil.

Trees up to about 25–30 m tall; leaflets obliquely oblong-ovate, 4–9 cm long, 2–5 cm wide, glabrous; flowers yellowish-white or slightly purplish, in short corymbose panicles; petals about 2 cm long; fruit lustrous, smooth, about 5–15 cm long, 5 cm wide, subterete; seeds usually 2, oblong, 2–3 cm long. The trees yield valuable timber. The resin is used in varnishes.

Specimens Examined. KANDY DISTRICT: Peradeniya, Royal Botanic Gardens, *Worthington 6679* (K); Kadugannawa, Andiatenne, *Worthington 6854* (K).

Hymenaea verrucosa Gaertn., Fruct. 2: 306, t. 137, fig. 7. 1791; Macmillan, Trop. Pl. & Gard., ed. 5, 394. 1962. *Trachylobium verrucosum* (Gaertn.) Oliver in Fl. Trop. Africa 2: 311. 1871; Macmillan, l.c.; Backer & Bakh. f., Fl. Java 1: 528. 1963; Brenan, Fl. Trop. East Africa, Caesalp. 132. 1967.

Zanzibar copal. This species is native to east tropical Africa, Madagascar, Mauritius, and the Seychelles.

Trees, to about 25 m tall, or more; leaflets asymmetrically ovate to elliptic or oblong, glabrous or subglabrous, 3.5–12 cm long, 2–5.5 cm wide; flowers white, in elongated panicles to about 35 cm long; petals 1.5–2 cm long; fruit about 2.5–5 cm long, 1.5–3 cm wide, resinous and warty on the surface, 1–2(–3)-seeded. The resin is hard and used commercially.

Specimens Examined. GALLE DISTRICT: Galle reservoir on road to Hiniduma, *Kostermans 28553* (K); Hiyare reservoir, *Balakrishnan 982* (K, PDA, US). COLOMBO DISTRICT: Colombo, Victoria Road, *Worthington 954* (K), *1149* (K). KANDY DISTRICT: Peradeniya, Royal Botanic Gardens, *Worthington 6680* (K), *Thwaites s.n.* (K), *Fairchild & Dorsett 231* (UC, US). WITHOUT EXACT LOCALITY: *Walker 1845* (K), s.n. (GH, K).

21. Lysidice rhodostegia Hance, J. Bot. 5: 298. 1867; Rock, Legum. Pl. Hawaii 63. 1920; Macmillan, Trop. Pl. & Gard., ed. 5, 86. 1962; Allen & Allen, Legum. 412. 1981.

A species of ornamental, unarmed trees native to southern China, introduced to Ceylon in 1882.

Leaves paripinnate with (2–)4–6 pairs of oblong to ovate or lanceolate, acuminate leaflets 4–12 cm long, 2.5–5 cm wide; flowers fragrant, in terminal or axillary panicles; bracts pink, persistent; petals 3, 1–2 cm long, purplish-red, and 2 rudimentary petals; fruit oblong, compressed, dehiscent, about 10–22 cm long, 3–4.5 cm wide; seeds brown, compressed, transverse, separated by spongy septae.

Specimens Examined. Peradeniya, Royal Botanic Gardens, near the big pond, *Worthington 1322* (K). Curator's Garden, *Worthington 3466* (BM, K). Matale, *Worthington 3148* (K). Palmadulla, *Worthington 6448* (K).

Note. Another taxon, not included in my generic key and of which I have seen no specimens from Ceylon, was included by Macmillan, Trop. Pl. & Gard., ed. 5, 123. 1962, as *Wagatea spicata,* a climber with spikes of scarlet flowers. Recently the genus has been recognized as synonymous with *Moullara* [Rheede] Adanson, Fam. Pl. 2: 318, 579. 1763. The specific name, accordingly, is *Moullava spicata* (Dalz.) Nicolson, in Manilal, ed., Botany and History of Hortus Malabaricus 184. 1980.

FABACEAE (LEGUMINOSAE)

Subfamily FABOIDEAE (PAPILIONOIDEAE) (continued)

(by Velva E. Rudd*)

REVISED KEY TO THE TRIBES

1 Flowers with stamens free, the filaments separate to the base; plants woody, trees or shrubs
 . . . .**Sophoreae** (vol. 1, p. 430)
1 Flowers with the stamens monadelphous, the 9 or 10 filaments all joined at the base, or diadelphous, either 5:5 with filaments joined in two fascicles of 5 stamens each, or 9:1 with the vexillar filament free, at least at the base, the others joined
  2 Fruit lomentaceous, 1–many-articulate, usually breaking into 1-seeded articles at maturity; plants herbaceous or woody
   3 Leaves with stipels present, pinnately 1-, 3-, 5-, or 7-foliolate; pubescence of hooked hairs and sometimes glochidiate or capitate glandular hairs; stamens monadelphous or diadelphous 9:1; anthers uniform . . . . **Desmodieae**
   3 Leaves without stipels, pinnately 3- or 5–many-foliolate or digitately paripinnate 2- or 4-foliolate or, sometimes, 1-foliolate; pubescence of straight hairs, sometimes with enlarged, glandular bases; stamens monadelphous or diadelphous 5:5, sometimes the filaments alternately long and short; anthers uniform or dimorphic . . . . **Aeschynomeneae** (p. 160)
  2 Fruit not articulated, dehiscent or indehiscent
   4 Leaflets gland-dotted
    5 Pubescence of medifixed (T-shaped) hairs. . . . .**Indigofereae** (p. 114)
    5 Pubescence of basifixed hairs
     6 Plants annual, herbaceous; leaves mostly 1-foliolate, rarely with 2- or 3-foliolate leaves on the same plant. . . . .**Psoraleeae** (p. 112)
     6 Plants mostly perennial, woody or suffrutescent; leaves 3-foliolate. . .**Phaseoleae** (p. 236)
   4 Leaflets not gland-dotted
    7 Leaves digitately (1–)3- or 5(–7)-foliolate
     8 Stamens diadelphous 9:1; anthers uniform; plants herbaceous; leaflets mostly with toothed margins. . . . . **Trifolieae** (vol. 1, p. 449)
     8 Stamens monadelphous; anthers dimorphic, alternately basifixed and dorsifixed, plants herbaceous or woody
      9 Plants shrubby, sometimes spiny; stamens in a closed tube; fruit compressed or turgid, not inflated . . . . **Genisteae** (p. 182)
      9 Plants herbaceous, annual or perennial, sometimes suffrutescent; stamen tube open on the upper side; fruit usually inflated. . . . . **Crotalarieae** (p. 183)

*Smithsonian Institution, Washington D.C., U.S.A.

7 Leaves pinnately 3–many-foliolate or sometimes simple or unifoliolate
  10 Fruit commonly samaroid or drupaceous, sometimes chartaceous, indehiscent; plants woody,
    trees, shrubs, or lianas . . . .**Dalbergieae** (p. 217)
  10 Fruit 2-valved, commonly dehiscent
    11 Leaflets 3, rarely 1.
      12 Plants erect, herbaceous or woody; stipels absent.
        13 Pubescence of medifixed (T-shaped) hairs; stamens diadelphous 9:1 . . . .
        . . . .**Indigofereae** (p. 114)
        13 Pubescence of basifixed hairs; stamens monadelphous with filaments united into a sheath
        split above . . . .**Bossiaeeae** (p. 217)
      12 Plants mostly twining, herbaceous or woody; stipels present . . . .**Phaseoleae** (p. 236)
    11 Leaflets 5 or more
      14 Plants erect, herbaceous or woody; leaves paripinnate or imparipinnate
        15 Pubescence predominantly of medifixed (T-shaped) hairs; anthers apiculate or appen-
        daged . . . .**Indigofereae** (p. 114)
        15 Pubescence of basifixed hairs; anthers not apiculate or appendaged
          16 Leaves epulvinate or the pulvinus reduced; fruit sometimes inflated, sometimes
          longitudinally septate . . . .**Galegeae** (p. 109)
          16 Leaves pulvinate; fruit compressed, sometimes transversely septate
            17 Flowers, with hypanthium, in axillary racemes . . . .**Robinieae** (p. 135)
            17 Flowers lacking hypanthium, in terminal pseudopanicles or axillary pseudoracemes
            . . . .**Tephrosieae** (p. 144)
      14 Plants twining; leaves paripinnate
        18 Stamens 9, the vexillar stamen lacking; leaves terminating in a bristle; plants woody
        or suffrutescent . . . .**Abreae** (vol. 1, p. 445)
        18 Stamens 10; leaves usually terminating in a tendril; plants herbaceous . . . .
        . . . .**Vicieae** (vol. 1, p. 457)

Tribe GALEGEAE

(Bronn) Torrey & Gray, Fl. North. Amer. 1: 292. 1838; Polhill in Polhill & Raven, eds. Adv. Leg. Syst. 357. 1981. Type genus: *Galega* L.

Herbs or shrubs, sometimes subscandent; leaves alternate, epulvinate or with the pulvinus reduced, paripinnate or imparipinnate, (1–3)–many-foliolate; stipules present, joined abaxially; stipels lacking; inflorescences racemose, axillary or terminal; bracts present; bracteoles present or lacking; calyx campanulate with 5 subequal lobes; petals 5, variously coloured; keel petals sometimes with claws joined; stamens 10, diadelphous 9:1 with the vexillar filament free; anthers uniform; ovary sessile or stipitate, few–many-ovulate; style slender, straight or curved; stigma terminal or lateral, capitate or minute; fruit variously compress-ed or inflated, dehiscent or indehiscent, sometimes longitudinally septate, 1–many-seeded; seeds oblong-reniform.

About 20 genera chiefly in the northern hemisphere but also in Africa, Australia, and South America. Only two genera, *Clianthus* and *Swainsona,* of the subtribe Coluteinae, are known from Ceylon, both as introductions from Australia.

KEY TO THE GENERA

1 Flowers large and showy, red or white, to about 7.5 cm long; keel exserted; fruit oblong, turgid or sometimes inflated...**1. Clianthus**
1 Flowers smaller, variously red, pink, purple, or white, to about 5 cm long; fruit inflated, sometimes almost 2-celled by intrusion of the upper suture...........................**2. Swainsona**

1. CLIANTHUS

Soland. ex Lindl., Trans. Hort. Soc. London ser. 2, 1: 519. 1835; Lindl., Edward's Bot. Reg. 21: t. 1775. 1835, nom. cons.; Polhill in Polhill & Raven, Adv. Leg. Syst. 360. 1981; Allen & Allen, Legum. 170. 1981. Type species: *C. puniceus* (G. Don) Soland. ex Lindl.

Donia Lam. ex G. Don, Gen. Hist. 2: 468. 1832, non R. Br. 1813, nec R. Br. 1819.

Herbs or shrubs, usually scandent or trailing; leaves imparipinnate with numerous leaflets; stipules present; stipels lacking; inflorescences racemose, axillary; bracts and bracteoles present, subpersistent; flowers relatively large and showy; calyx campanulate with 5 subequal lobes or the two vexillar lobes broader and connate in part; petals red or white, subequal; keel beaked, longer than the other petals; stamens diadelphous 9:1 with the vexillar filament free; ovary stipitate, many-ovuled; style curved, bearded above; stigma minute, terminal; fruit linear-oblong, turgid, 2-valved, many-seeded; seeds small, reniform.

Two species, one in Australia, the other nearly extinct, in New Zealand; introduced elsewhere as ornamentals, one in Ceylon.

Clianthus formosus (G. Don) Ford & Vickery, Contr. New South Wales Natl. Herb. 1: 303. 1950.

Donia formosa G. Don, Gen. Hist. 2: 468. 1832. Type: *Capt. King,* N.W. coast of New Holland [Australia], at the Curlew River.
Donia speciosa G. Don, Gen. Hist. 2: 468. 1832. Type: New Holland, at Regent's Lake (K).
Clianthus dampieri Cunn. ex Lindl., Trans. Hort. Soc. London ser. 2, 1: 522. 1835; Benth., Fl. Austral. 2: 214. 1864; Macmillan, Trop. Pl. & Gard. ed. 5, 176. 1962. Type: *Sturt 18,* near Darling, Australia (BM, K).
Clianthus speciosus (G. Don) Aschers. & Graebn., Syn. Mitteleurop. Fl. 6: 725. 1909, non Steud. 1840.

Herb, sometimes perennial; stems grey-tomentose or villous, erect, prostrate, or subscandent, about 1 m long; leaves 15–21-foliolate; leaflets sessile, oblong or elliptic, 1.5–2.5 cm long, 3–5 mm wide, grey-tomentose; bracts acuminate; bracteoles linear; flowers about 4–6 in upright racemes shorter than the leaves; calyx about 1 cm long with attenuate lobes, grey-villous; petals about 5–7 cm long, scarlet with a large purplish-black spot on the standard; fruit somewhat inflated, oblong, to about 4–6.5 cm long, 7 mm wide, grey-villous or tomentose.

Distr. A native of Australia; introduced elsewhere as an ornamental.

Vern. Desert pea, glory pea, lobster claw (E).

Specimens Examined. NUWARA ELIYA DISTRICT: Hagkala Botanical Garden, *Simpson 9021* (PDA).

Note. Macmillan (l.c.) includes this "beautiful flowering, straggling shrub, difficult to cultivate" in a list of plants suitable for up-country planting.

2. SWAINSONA

Salisb., Parad. Lond. t. 28. 1806; Hutchins., Gen. Fl. Pl. 1: 406. 1964; Allen & Allen, Legum. 633. 1981. Type species: *S. coronillaefolia* Salisb., a synonym of *S. galegifolia* (Andr.) R. Br.

Herbs or small shrubs; stems glabrous or puberulent with appressed hairs; leaves imparipinnate with numerous leaflets; stipules small; stipels lacking; inflorescences axillary, racemose; bracts small; bracteoles minute, appressed to the calyx or on the pedicel slightly below the calyx; flowers of medium size, ornamental; calyx campanulate with 5 subequal lobes; petals red, pink, purple, or white; stamens diadelphous 9:1 with the vexillar filament free; anthers uniform; ovary sessile or stipitate; style curved, longitudinally bearded along the inner edge; stigma small, terminal; pod inflated, oblong or elliptic, sometimes bladder-like, sometimes almost 2-celled by the intrusion of the upper suture, subindehiscent; seeds small, reniform.

About 50–60 species, chiefly in Australia, with one species in New Zealand; some species introduced elsewhere.

Swainsona galegifolia (Andr.) R. Br. in Ait., Hort. Kew, ed. 2, 4: 327. 1812.

Vicia galegifolia Andr., Bot. Repos. 5: t. 319. 1803.
Colutea galegifolia (Andr.) Sims, Bot. Mag. 7. 792. 1804.
Swainsona coronillifolia Salisb., Parad. Lond. t. 28. 1806; Curtis, Bot. Mag. t. 1725. 1815.
Swainsona galegifolia (Andr.) R. Br. var. *albiflora* Lindl., Bot. Reg. 994.1826.
Swainsona albiflora (Lindl.) G. Don, Gen. Syst. 2: 245. 1832.
Swainsona coronillifolia Salisb. var. *galegifolia* (Andr.) Maiden & Betche, Census New South Wales 107. 1916.
Swainsona coronillifolia Salisb. var. *albiflora* (Lindl., as "Benth.") Maiden & Betche, Census New South Wales 107. 1916.

Shrub about 1–1.5 m tall; stems glabrous, flexuous, semiscandent; leaves 11–21-foliolate; leaflets oblong, 5–20 mm long, 3–5 mm wide, obtuse or slightly emarginate; stipules ovate, about 3 mm long, 2 mm wide; flowers 1.5−2 cm long in axillary racemes usually longer than the leaves; calyx 4–5 mm long with short, deltoid lobes, villous-ciliolate, otherwise glabrous; petals deep red, rose-red, rose-violet, pink, or sometimes white; style pubescent; fruit glabrous, somewhat inflated, 2.5–5 cm long, 1–1.2 cm wide, with stipe about 0.75–1 cm long.

Distr. A native of New South Wales, Australia, introduced elsewhere as an ornamental.

Vern. Swan flower, winter sweet pea, small-leaved bladder-senna, Darling River Pea (E).

Specimens Examined. NUWARA ELIYA DISTRICT: Hagkala Botanical Garden, 11 May 1928, *de Sliva s.n.* (PDA).

Note. Macmillan (Trop. Fl. & Gard. ed. 5, 182. 1962): "Swainsona. Several species". There are a number of horticultural varieties in addition to var. *albiflora* Lindl., cited above.

Tribe PSORALEEAE

(Benth.) Rydb., North. Amer. Fl. 24: 1. 1919; Rydb., Amer. J. Bot. 15: 195. 1928 (as "Psoraleae"); Stirton in Polhill & Raven, Adv. Leg. Syst. 337. 1981. Type genus: *Psoralea* L.

Tribe Galegeae subtribe Psoralieae Benth. & Hook., Gen. Pl. 443. 1865.

Herbs, shrubs, or small trees; leaves alternate, pinnately or palmately (1–)3(–5)-foliolate, or sometimes reduced to scales; leaflets usually gland-dotted, entire or dentate; stipules present, more or less adnate to the petiole; stipels lacking; inflorescences indeterminate, paniculate, racemose, or spicate; bracts present; pedicels sometimes subtended by a cupulum; bracteoles usually lacking; calyx campanulate with 5 subequal lobes, the vexillar lobes partially fused, sometimes gland-dotted; petals 5, blue to purplish, sometimes white; stamens 10, monadelphous or diadelphous 9:1; anthers uniform, alternately basifixed and dorsifixed; ovary sessile to stipitate, 1-ovulate; style curved; stigma minutely capitate to penicillate; fruit indehiscent, 1-seeded, sometimes glandular; seeds suborbicular to reniform.

Six genera, widely dispersed in Old and New Worlds, chiefly in temperate areas, sometimes in tropics. Only *Cullen* Medik. known in Ceylon.

CULLEN

Medik., Vorles. Churpfälz. Phys.-Oekon. Ges. 2: 380. 1787; Stirton in Polhill & Raven, Adv. Leg. Syst. 342. 1981; Allen & Allen, Legum. 201. 1981. Type: *C. corylifolium* (L.) Medik., based on *Psoralea corylifolia* L.

Psoralea L., Sp. Pl. 762. 1753, in part, not as to lectotype, *P. pinnata* L.

Dorychnium Moench, Meth. 109. 1784, in part. Type not designated, non *Dorycnium* Mill. 1754.

Meladenia Turcz., Bull. Soc. Imp. Naturalistes Moscou 21 (1): 576. Type: *M. densiflora* Turcz.

Bipontinia Alef., Pollichia 22–24: 121. 1866. Type not designated.

Shrubs or herbs; stipules lanceolate to linear-subulate, more or less adnate to base of petiole; stipels lacking; leaves 1–3, or 5-foliolate; leaflets entire or dentate, ovate to lanceolate, elliptic or orbicular, gland-dotted; flowers in axillary spikes or clusters; bracts ovate; bracteoles lacking; calyx sometimes glandular-punctate, campanulate with subequal lobes, sometimes bilabiate; petals purplish to light violet; stamens diadelphous 9:1 with the vexillar filament free, at least above; ovary 1-ovulate, often glandular; style curved, sometimes swollen; stigma capitate, minute, sometimes penicillate; fruit 1-seeded, erect, beaked, usually glandular, indehiscent; seed obliquely reniform.

A genus of 35 species, in India and Ceylon to Burma, the Philippines, Papua New Guinea, Australia, and a few species in Africa.

Cullen corylifolium (L.) Medik., Vorles. Churpfälz. Phys.-Oekon. Ges. 2: 380. 1787.

Psoralea corylifolia L., Sp. Pl. 764. 1753; Moon, Cat. 55. 1824; Roxb., Fl. Ind. 3: 388. 1832; Thw., Enum. Pl. Zeyl. 84. 1859; Baker in Hook. f., Fl. Br. Ind. 2: 103. 1876; Trimen, Handb. Fl. Ceylon 2: 28. 1894; Alston in Trimen, Handb. Fl. Ceylon 6: 74. 1931; Macmillan, Trop. Pl. & Gard., ed. 5, 28. 1962; Backer & Bakh. f., Fl. Java 1: 593. 1963; Ali, Fl. W. Pakistan, no. 100, Papil. 215. 1963. Type: India, Herb. Linn. 928. 24 (holotype LINN).

Trifolium unifolium Forssk., Fl. Aegypt.-Arab. 140. 1775. Type: *Forsskal s.n.* (BM).

Annual herb, to about 1.25 m tall; stems angular-ribbed, puberulent and studded with reddish glandular pustules; stipules lanceolate, essentially free, 6–8 mm long, 2 mm wide, puberulent and glandular-punctate; leaves 1–3-foliolate; petiole about 1–3 cm long; blade of leaflets ovate to elliptic, 2.5–8 cm long, 2–6 cm wide, obtuse to subacute at the apex, subcordate to cuneate at the base, erose-dentate, gland-dotted, subglabrous, puberulent or subappressed-pubescent on both surfaces; inflorescences axillary, spicate; bracts like the stipules; peduncles 1.5–8.5 cm long, usually longer than the petioles; flowers about 5–6 mm long; calyx glabrous, puberulent, or appressed-pubescent, gland-dotted, 4–5 mm long, campanulate with 5 subequal lobes, the vexillar lobes shortest, joined in part; petals glabrous, light violet to purplish-blue, slightly longer than the calyx; ovary glandular-warty, puberulent toward the apex; style glabrous, curved; stigma capitate, papillose; fruit obliquely ellipsoid, mucronate, 4–5 mm long, glabrous, glandular-warty.

Distr. A native of Asia, locally naturalized in other areas.

Uses. Macmillan (l.c.) cites this species as suitable for a green-manure cover crop. Ali (l.c.) says the seeds are reputed to be medicinally important. Allen & Allen (l.c.), also, mention that "this plant shows promise as a green manure in Sri Lanka" and that the "aromatic bitter seeds. . . . are administered in India and Indochina as a laxative and as a tonic for stomach ache."

Vern. Bodi (S); kavoti, kavothi, karporgam (T).

Specimens Examined. WITHOUT EXACT LOCALITY: in 1819, *Moon s.n.* (BM), cited in Cat. (l.c.) as "Walpany" (Walapane ?). MATALE DISTRICT: Near Nalande, Nov. 1882, *Trimen* ? *s.n.* (PDA). JAFFNA DISTRICT: Delft Isl., *Gardner s.n., Thwaites, C.P. 1443* (PDA), without data (BM, K, P, PDA). MANNAR DISTRICT: Near Giant's Tank, Feb. 1890, *"Sayaneris coll."* *s.n.* (PDA); Nochchikulam, *Jayasuriya et al. 615* (US). ANURADHAPURA DISTRICT: Anuradhapura, Aug. 1885, *Trimen* ? *s.n.* (PDA). KANDY DISTRICT: Peradeniya, 12.4.16, *J. M. S. s.n.* (PDA). NUWARA ELIYA DISTRICT: Without exact locality, *Drieberg s.n.* (PDA). HAMBANTOTA DISTRICT: Tissamaharama, *Simpson 9914* (BM, PDA).

Note. This species is characterized by having one leaflet, in contrast to *C. americanum* (L.) Rydb. with three. However, in Ceylon, *C. corylifolium* has been found to bear some 3-foliolate leaves: the collection, *Jayasuriya et al. 615,* has leaves with 1, 2, and 3 leaflets on the same stalk. The two species, if not synonymous, appear to be very closely related.

Tribe INDIGOFEREAE

(Benth.) Rydb., North Amer. Fl. 24: 137. 1923; Polhill in Polhill & Raven, Adv. Leg. Syst. 289. 1981. Type genus: *Indigofera* L.

Herbs or sometimes shrubs or small trees; hairs typically medifixed (T-shaped); leaves generally imparipinnate, 1–many-foliolate, sometimes simple; stipules present; stipels present or absent; leaflets opposite or alternate, entire or toothed, sometimes glandular-punctate; flowers usually in axillary racemes, sometimes single; petals white or yellowish to red or purplish; calyx campanulate. 5-lobed; bracts small, caducous; bracteoles lacking; stamens 10, monadelphous or diadelphous 9:1 with the vexillar filament free; anthers essentially uniform, apiculate, gland-tipped or with expanded connective; ovules 1–many; fruit 1–many-seeded, 2-valved, generally septate, usually dehiscent, rarely jointed; endocarp often spotted with tannin deposits; seeds globular to rectangular; hilum small.

Four genera, principally in tropical Africa, South Africa, Madagascar, and Asia, but *Indigofera* widely distributed in tropical to warm temperate regions of Old and New Worlds.

KEY TO THE GENERA

1 Leaflets dentate; petals glabrous with evident venation; stamens monadelphous or submonadelphous
  with the vexillar filament lightly attached to the others; fruit compressed, erect, longitudinally ridged,
  6–10 mm wide...1. Cyamopsis
1 Leaflets entire; petals glabrous or pubescent with inconspicuous venation; stamens diadelphous with
  the vexillar filament free; fruit usually terete, torulose, or lunate, sometimes slightly ridged or with
  longitudinal wings..2. Indigofera

1. CYAMOPSIS

DC., Prod. 2: 215. 1825; DC., Mém. Leg. 77, 89, 230. 1826; Gillett, Kew Bull.
Addit. Ser. 1: 6. 1958; Allen & Allen, Legum. 202. 1981. Type species: *C.
psoraloides* (Lam.) DC., a synonym of *C. tetragonoloba* (L.) Taub.

Herbs, to about 3 m tall with pubescence of appressed, medifixed hairs; leaves
imparipinnate, (1–)3–7-foliolate; stipules small; stipels lacking; leaflets opposite,
linear-oblong to ovate or obovate, entire or dentate; inflorescences axillary,
racemose, sessile or short-pedunculate, usually few-flowered; flowers relatively
small; calyx obliquely campanulate with the two vexillar teeth shortest; petals
lilac to purple; stamens 10, monadelphous or submonadelphous with the vexillar
filament lightly attached to the others; anthers uniform, apiculate; style incurved
at apex; stigma capitate; fruit compressed, indehiscent or tardily dehiscent,
3-ridged on each valve, beaked, many-seeded, septate between the seeds, erect;
seeds compressed, quadrate, minutely tuberculate.

A genus of four species native to the tropics of Africa and Asia; *C.
tetragonoloba* introduced elsewhere.

As Gillett has indicated (l.c.) this genus is dubiously separable from *Indigofera*.

Cyamopsis tetragonoloba (L.) Taub. in Pflanzenfam. 3 (3). 259. 1894; Alston
in Trimen, Handb. Fl. Ceylon 6: 72. 1931; Gillett, Kew Bull., Addit. Ser. 1:
6. 1958; Backer & Bakh. f., Fl. Java 1: 589. 1963.

Psoralea tetragonoloba L., Mant. 2: 104. 1767. Type: Suratte, Herb. LINN
    928.23 (LINN).
Dolichos psoraloides Lam., Enc. 2: 300. 1786. From Ceylon, cultivated Jardin
    du Roi, Paris.
Dolichos fabaeformis L'Her., Stirp. t. 78. 1791, based on *Psoralea tetragonoloba*
    L.
Lupinus trifoliatus Cav., Ic. 1: t. 59. 1791. From Mexico, cultivated in Regio
    horto Matritense.
Cyamopsis psoraloides (Lam.) DC., Prod. 2: 216. 1825: Wight, Ic. Pl. Ind. Or.
    248. 1840; Baker in Hook. f., Fl. Br. Ind. 2: 92. 1876.

Herb, erect, to about 2 m tall; stems grooved, appressed-pubescent with medifixed hairs; stipules linear-subulate, about 6–10 mm long; leaves pinnately 3-foliolate with petioles 2.5–3.5 cm long; leaflets elliptic to ovate, acute, dentate, 3.5–7.5 cm long, 1.25–5 cm wide, appressed-pubescent on both surfaces; inflorescences densely racemose, 6–30-flowered; bracts linear-subulate, 4.5–5 mm long; flowers sessile, about 7–8 mm long; calyx pubescent, 5–6.5 mm long with subequal teeth, the carinal tooth longest; corolla slightly longer than the calyx, the standard and keel whitish, the wings pinkish-purple; stamens monadelphous; style bent; stigma terminal, capitate; fruit in clusters, erect, about 3.5–9 cm long, 6–10 mm wide, tetragonal, somewhat fleshy, beaked, 3–12-seeded, slightly constricted between the seeds; seeds white, grey, or black.

Distr. Possibly a native of India and Ceylon, cultivated and sometimes an escape in Pakistan, Afghanistan, Arabia, and other warm areas of the Old and New Worlds.

Uses. Cultivated as fodder and green manure; cooked as a vegetable.

Vern. Cluster bean, guar (E); koth-averay (T).

Specimens Examined. KANDY DISTRICT: Peradeniya, 9 Jan. 1901, *s. coll. s.n.* (PDA), *Alston 1082* (PDA).

2. INDIGOFERA

L., Sp. Pl. 751. 1753; Gillett, Kew Bull. Addit. Ser. 1: 1–139. 1958; Ali, Bot. Not. 111: 543. 1958; Hutchins., Gen. Pl. 1: 400. 1964; Allen & Allen, Legum. 341. 1981. Type species: *I. tinctoria* L.

A list of 12 generic synonyms is given by Hutchinson, l.c.

Herbs or shrubs, usually pubescent with characteristic medifixed hairs; leaves usually imparipinnate, or sometimes alternately pinnate, or sometimes simple; stipules present; stipels present or absent; leaflets usually opposite, sometimes alternate; margins entire; inflorescences usually axillary, sometimes terminal, racemose or sometimes spicate, or the flowers solitary; bracts small, caducous; bracteoles lacking; flowers relatively small; calyx campanulate with 5 subequal lobes or teeth; petals pink to red or purplish; stamens diadelphous 9:1 with the vexillar filament free; anthers dorsifixed, uniform, gland-tipped; style glabrous; stigma capitate, sometimes minutely penicillate; fruit usually dehiscent, septate within, linear, terete, sometimes tetragonous, torulose, arcuate, lunate, or globose, 1–many-seeded; seeds globose to cylindrical or quadrate, hilum lateral.

A genus with 400 or more species widely distributed in the tropics and subtropics of Old and New Worlds.

KEY TO THE SPECIES

1 Leaves simple; fruit usually 1-seeded
  2 Fruit lunate, about 5–7 mm long, beset with hooked spines; leaves suborbicular to obovate, moderately pubescent to subglabrous, minutely gland-dotted below...**1. I. nummulariifolia**

2 Fruit globose, 1.5–2 mm long, not spiny; leaves linear, acute, densely appressed-pubescent, not gland-dotted . . . .**2. I. linifolia**

1 Leaves compound, imparipinnate, alternately pinnate, or digitate, (1–)3–many-foliolate; fruit (1–)2–many-seeded

  3 Fruit globose, (1–)2(–3)-seeded, 3–5 mm long; leaflets mostly alternate or with some opposite . . . .**3. I. linnaei**

  3 Fruit oblong to linear, commonly 6–12-seeded; leaflets alternate or opposite

    4 Flowers solitary; leaves digitate, 1–5-foliolate; leaflets minute, 2–3 mm long . . . . **4. I. aspalathoides**

    4 Flowers in axillary racemes; leaves pinnate or digitate, (1–)3–many-foliolate; leaflets larger, mostly 5–50 mm long

      5 Leaflets predominantly alternate

        6 Leaves (1–)3–5(–7)-foliolate; leaflets pubescent on both surfaces; fruit terete, about 8–21 mm long, 1–1.5 mm in diameter, constricted between the seeds . . . . **5. I. oblongifolia**

        6 Leaves 5–11-foliolate; leaflets glabrous above, pubescent below; fruit somewhat tetragonous, 20–25 mm long, 2 mm in diameter . . . . **6. I. spicata**

      5 Leaflets opposite

        7 Leaves 3-foliolate

          8 Lower surface of leaflets gland-dotted; fruit 1–1.7 cm long

            9 Terminal leaflet sessile; leaflets appressed-pubescent on both surfaces; fruit 6–10-seeded . . . . **7. I. trifoliata**

            9 Terminal leaflet with petiole about 1 cm long; leaflets pubescent with minute lax hairs; fruit 2–4-seeded . . . . **8. I. barberi**

          8 Lower surface of leaflets not gland-dotted; fruit 2–3 cm long, spiniform at the apex . . . . **9. I. trita**

        7 Leaves 5–many-foliolate

          10 Pubescence generally appressed, strigose to sericeous

            11 Flowers about 4–8 mm long; leaflets 0.5–40 mm long

              12 Fruit essentially straight or slightly curved, 1.5–6 cm long

                13 Constrictions between seeds pronounced, fruit torulose

                  14 Racemes few-flowered, 3–6(–8) flowers; fruit 1.5–2.5 cm long . . . . **10. I. karnatakana**

                  14 Racemes many-flowered; fruit (3–)3.5–5 cm long . . . . **11. I. constricta**

                13 Constrictions between seeds not pronounced, fruit not torulose

                  15 Leaflets densely canescent-sericeous on both surfaces; flowers about 7–8 mm long; calyx 3–4 mm long; petals red or orange-red; fruit 2.5–3 mm in diameter, finely canescent; leaves 13–21-foliolate . . . . **12. I. wightii**

                  15 Leaflets glabrous on both surfaces; flowers about 5–7 mm long; petals salmon-pink or lilac; fruit 2 mm in diameter, appressed-pubescent to glabrous; leaves 5–15(–17)-foliolate

                    16 Flowers about 5–7 mm long, calyx 1–1.5 mm long; petals salmon-pink . . . . **13. I. tinctoria**

                    16 Flowers about 7 mm long; calyx 2–4 mm long; petals lilac . . . . **14. I. parviflora**

               12 Fruit falcate, 1–1.5 cm long . . . . **15. I. suffruticosa**

            11 Flowers 10–13 mm long; leaflets 10–75 mm long, 1.2–3.2 cm wide, acute or obtuse

              17 Fruit 5.5–7.5 cm long including beak to 1–1.3 cm long, 2 mm wide; leaflets 1–5 cm long . . . . **16. I. galegoides**

              17 Fruit 3.5–4.5 cm long including beak about 2 mm long, 4–5 mm wide; leaflets 2–7.5 cm long . . . . **17. I. zollingeriana**

          10 Pubescence generally spreading or with stalked glands, not strigose

Specimens cited as type material of *Aspalathus indica* L., *Hedysarum nummulariifolium* L., and *Indigofera hirsuta* L. At Biblio. Inst. France, Paris are in the Bibliothéque de l'Institute France, Paris, in the herbarium of Paul Hermann, basis of the Thesaurus Zeylanicus of J. Burman (see A. Lourteig in Taxon 15: 23–33. 1966).

1. Indigofera nummulariifolia (L.) Livera ex Alston in Trimen, Handb. Fl. Ceylon 6: 72. 1931; Gillett, Kew Bull., Addit. Ser. 1: 8. 1958; Gillett, Fl. Trop. East Africa, Papil. 217. 1971.

Hedysarum nummulariifolium L., Sp. Pl. 746. 1753, as *"nummularifolium"*; Moon, Cat. 54. 1824. Type: *P. Hermann*, Ceylon (Herb. no.) 3: 10 (BM). Probable isotypes, P-Lam; Paris, Biblio. Inst. France, Burmann Coll. Pl. Zeyl. 119.

Indigofera echinata Willd., Sp. Pl. 3: 1222. 1802; Wight. & Arn., Prod. 198. 1834; Wight, Ic. Pl. Ind. Or. 316. 1840; Baker in Hook. f., Fl. Br. Ind. 2: 92. 1876. Type: *Koenig,* India. Isotype ? (BM).

Hedysarum erinaceum Poir. in Lam., Enc. 6: 393. 1804. Type: *Sonnerat,* "Les Indies" (P-lam).

Onobrychis rotundifolia Desv., J. Bot. 3: 84. 1814. Type: India (P).

Acanthonotus echinatus (Willd.) Benth. in Hook., Fl. Nig. 293. 1849; Thw., Enum. Pl. Zeyl. 83. 1859.

Herbaceous annual; stems usually prostrate, spreading, glabrous or subglabrous, sometimes rooting at the nodes; stipules lanceolate-subulate, 2–3 mm long; leaves simple, suborbicular to obovate, 1.5–5 cm long, 1.5–2.5 cm wide, obtuse at the apex, cuneate at the base, subglabrous above, appressed-pubescent and gland-dotted below; inflorescences racemose, axillary, about 6–15-flowered, reflexed; bracts minute, subulate; flowers about 3–3.5 mm long; calyx 2.5 mm long, with attenuate lobes, strigillose; petals pink or reddish, the vexillum lightly strigillose on the outer face; ovary 2-ovulate; fruit lunate, about 5–7 mm long, indehiscent, beset with hooked spines, 1(–2)-seeded.

Distr. Widespread in India, Ceylon, Madagascar, and tropical Africa in dry, sandy, rocky, or cultivated areas.

Specimens Examined. LOCALITY ILLEGIBLE: *Gardner s.n.*, Batticaloa, Mar. 1858, Colombo, *Macrae 168* (BM, K); *Ferguson s.n.*, *Thwaites C.P. 1453* (PDA). WITHOUT DATA: *s. coll. C.P. 1453* (BM, GH, K, P, US). ANURADHAPURA DISTRICT: Galpitagala, Kekirawa, *Worthington 5057* (K);

Kotagala, *Jayasuriya et al. 624* (US); between Medawachchiya and Vavuniya, mi. 98/1, *Rudd 3265* (PDA, US). PUTTALAM DISTRICT: Wilpattu Natl. Park, Mail Villu, *Fosberg et al. 50940* (K, SFV, US). MATALE DISTRICT: Dambulla Rock, *Amaratunga 524* (PDA), *Rudd & Balakrishnan 3113* (K, PDA, SFV, US); 6.5 mi. N. of Dambulla, *Fosberg & Balakrishnan 53420* (SFV, US); Dambulla Rd., *Alston 1002* (PDA). AMPARAI DISTRICT: Between Amparai and Maha Oya, mi. 34–35, *Rudd & Balakrishnan 3230* (K, PDA, SFV, US); About 5 mi. W. of Amparai, *Fosberg & Jayasinghe 57211* (K, US). COLOMBO DISTRICT: Anguruwella, *Amaratunga 1128* (PDA); Colombo, *Macrae s.n.* (K), *Moon* (fide Moon, Cat.). HAMBANTOTA DISTRICT: "Near Kirinda, S. Prov.", 20 Dec. 1882, *Trimen ? s.n.* (PDA); Ruhuna Natl. Park, Padikema, near Patanagala, *Fosberg et al. 51175* (GH, K, SFV, US); Patanagala, *Cooray 69121608R* (US); Buttawa Bungalow, *Mueller-Dombois & Cooray 69010533* (CAS, K, US); Buttawa Plain, *Cooray 69121205R* (US). BADULLA DISTRICT: Nilgala, Uva, Jan. 1888, *Trimen ? s.n.* (PDA); Ekiriyankumbura, Jan. 1888, *Trimen ? s.n.* (PDA).

2. Indigofera linifolia (L. f.) Retz., Obs. Bot. 4: 29. 1786; 6: 33, t. 2. 1791; Wight & Arn., Prod. 198. 1834; Wight, Ic. Pl. Ind. Or. t. 313. 1840; Thw., Enum. Pl. Zeyl. 83. 1859; Baker in Hook. f., Fl. Br. Ind. 2: 94. 1876; Trimen, Handb. Fl. Ceylon 2: 22. 1894; Backer & Bakh. f., Fl. Java 1: 590. 1963; Saldanha & Nicolson, Fl. Hassan 257. 1976; Ali, Fl. West Pakistan, no. 100, Papil. 69. 1977; Verdcourt, Man. New Guinea Leg. 353. 1979.

Hedysarum linifolium L. f., Suppl. 331. 1781. Type: *Koenig s.n.* Ind. Orient, Herb. Linn. 921.5 (LINN); isotype (BM).
Sphaeridiophorum linifolium (L. f.) Desv., J. Bot. 1: 125, t. 6. 1813.

Herbaceous annual; stems prostrate to erect, numerous, appressed-pubescent with white or gray hairs; stipules subulate, 3–5 mm long; leaves simple, linear to elliptic, about (5–)10–35 mm long, 1–2.5 mm wide, acute, mucronate at the apex, acute at the base, densely appressed-pubescent on both surfaces; inflorescences racemose, axillary, subsessile, about 3–12-flowered; flowers about 5–8 mm long; calyx 2.5–5 mm long, with attenuate lobes, appressed-pubescent; petals pink to bright red; fruit 1-seeded, globose, 1.5–2(–3) mm long, apiculate, densely white- or greyish-appressed-pubescent.

Distr. Dry regions of Ceylon, India and Pakistan to Malesia, China, Australia, and in the Sudan and Ethiopia.
Specimens Examined. KANDY DISTRICT: Banks of the Mahaweli Ganga at Kurundu-Oya, May 1856, *Thwaites C.P. 3514* (PDA), without data, *C.P. 3514* (BM, GH, K, P, US). TRINCOMALEE DISTRICT: Kantalai, Aug. 1885, *Trimen ? s.n.* (PDA). POLONNARUWA DISTRICT: Polonnaruwa, Govt. Farm, *J.E. Senaratne 3499* (PDA).

3. Indigofera linnaei Ali, Bot. Not. 111: 549. 1958; Ali, Fl. West Pakistan, no. 100, Papil. 75. 1977; Backer & Bakh. f., Fl. Java 1: 591. 1963; Verdcourt, Man. New Guinea Leg. 353. 1979. Type: Coromandel, Herb. Sloane, vol. 95: 186 (BM).

Hedysarum prostratum L., Mant. 1: 102. 1767. Type: Burm., Fl. Ind. t. 55, f. 1. 1768; isotype ? *Koenig s.n.* (BM).

Indigofera enneaphylla L., Mant. 2: 272. 1771, Append. 571. 1771, nom. illeg., based on *Hedysarum prostratum* L.; Moon, Cat. 54. 1824; Wight, Ic. Pl. Ind. Or. 403. 1840–43; Thw., Enum. Pl. Zeyl. 83. 1859, as "*cuneaphylla*", 411. 1864; Baker in Hook. f., Fl. Br. Ind. 2: 94. 1876; Trimen, Handb. Fl. Ceylon 2: 22. 1894; Alston in Trimen, Handb. Fl. Ceylon 6: 72. 1931.

Indigofera caespitosa Wight in Wall., Cat. ex Wight & Arn., Prod. 199. 1834, nomen in synon.

Indigofera prostrata (Burm. f.) Domin, Bibl. Bot. Stuttgart 187. 1926, non Willd. 1803, nec Perr. ex DC. 1825, nec Roxb. 1832, nec Klein ex Wight & Arn. 1834.

Herbaceous annual or perennial from a woody root; stems prostrate or trailing, sparsely appressed-pubescent with white hairs; stipules attenuate, about 3 mm long, hyaline along the margins; leaves imparipinnate, 5–9(–11)-foliolate; leaflets alternate, or some opposite, obovate, subsessile, about 3–12 mm long, 1.5–5 mm wide, obtuse to emarginate at the apex, cuneate at the base, appressed-pubescent on both surfaces; inflorescences spicate, axillary, about 1 cm long; bracts deltoid, hyaline, about 2 mm long; flowers 3–4.5 mm long; calyx 3–4 mm long with attenuate lobes, strigillose; petals salmon-pink to reddish; fruit oblong-cylindrical, 3–4(–6) mm long, 1.5–2(–3) mm in diameter, pubescent, (1–)2(–3)-seeded.

Distr. Ceylon, India and Pakistan to Southeast Asia and Australia, usually in dry, barren places.

Uses. According to Verdcourt (l.c.) this plant has proved poisonous to horses in Australia, but in New Guinea is reported to be liked by cattle and is soon eaten out of pastures.

Vern. Bin avari, bin awari (S); Cheppunerenchi (T).

Specimens Examined. WITHOUT EXACT LOCALITY: *Gardner 209* (K); Puttalam, Galle, 1824 (or 34 ?) , *W. Brown,* Trincomalee, Nov. 1859, *Glenie, Thwaites C.P. 2775* (PDA), without data, *2775 ?* (US). JAFFNA DISTRICT: Keerimale ("Kirimalai"), Punakai, Feb. 1890, *Trimen ? s.n.* (PDA); between Jaffna and Elephant Pass, *Rudd 3292* (K, PDA, US). MANNAR DISTRICT: Causeway to Mannar, *Davidse & Sumithraarachchi 9150* (US). ANURADHA-PURA DISTRICT: Anuradhapura, W. shore of Nuwara Wewa, *Sohmer 8089* (US); between Medawachchiya and Vavuniya 98/1, *Rudd 3264* (GH, K, PDA,

SFV, US). PUTTALAM DISTRICT: Wilpattu Natl. Park, *van Beusekom 1630* (US). POLONNARUWA DISTRICT: Polonnaruwa, near Parakrama Samudra, *Townsend 73/234* (K, US); sacred area, *Ripley 352* (US), *Waas 370* (CAS, K, SFV, US); Minneriya (as "Mineri"), Aug. 1885, *Trimen* ? *s.n.* (PDA). TRIN-COMALEE DISTRICT: Trincomalee, *Worthington 1094* (K), *Amaratunga 564* (PDA); Fullerton Cove, *Worthington 681* (K); Kantalai, Aug. 1885, *Trimen* ? *s.n.* (PDA); fishing village 14 mi. NW. of Trincomalee, 4 mi. SE. of Kuchchaveli ferry, *Fosberg & Jayasinghe 57104* (K, US). COLOMBO DISTRICT: Negombo, *Simpson 8577* (BM), *Senaratne 3246* (PDA), 21 Oct. 1949, *s. coll. s.n.* (PDA), July 1930, *de Silva s.n.* (PDA); between ocean and railroad tracks, *Maxwell 1030* (US). RATNAPURA DISTRICT: near Embilipitiya, bund of Chandrika Wewa, *Sohmer 8840* (US). GALLE DISTRICT: Galle, *Gardner 209* (K). HAMBAN-TOTA DISTRICT: Kirinda, Dec. 1882, *Trimen*? *s.n.* (PDA); Ruhuna Natl. Park, between Buttawa Plain and Buttawa Bungalow, *Wirawan 673* (PDA, US); near Buttawa Bungalow, *Cooray & Balakrishnan 69010916* (CAS, PDA, US), *Mueller-Dombois, et al. 69010532* (BM, K, PDA, SFV, US), *Mueller-Dombois & Cooray 68040604* A (PDA, US); Patanagala, *Fosberg 50363* (PDA, SFV, US).

4. Indigofera aspalathoides Vahl ex DC., Prod. 2: 231. 1825; Wight & Arn., Prod. 199. 1834; Wight, Ic. Pl. Ind. Or. 332. 1840–43; Thw., Enum. Pl. Zeyl. 83. 1859; Baker in Hook. f., Fl. Br. Ind. 2: 94. 1876; Trimen, Handb. Fl. Ceylon 2: 23. 1894. Type: "Vahl in herb. Juss." (P).

Aspalathus indica L., Sp. Pl. 712. 1753; Moon, Cat. 52. 1824. Lectotype: *P. Hermann*, Herb. Herm. 4: 76, Ceylon (BM). Syntypes. *P. Hermann*, Herb. Herm. 3: 33, 4: 12, 5: 155, 5: 395 (ic), Ceylon (BM). Probable type material, Paris, Biblio. Inst. France, Burman coll. Pl. Zeyl. 110; non *Indigofera indica* Mill. 1768, nec. Lam. 1789.
Indigofera aspalathifolia Roxb., Fl. Ind. 3: 371. 1832, based on *Aspalathus indica* L.
Lespedeza juncea Wall. ex Wight & Arn., Prod. 199. 1854, nomen in synon.

Shrub, much branched; young stems appressed-pubescent with white hairs, later glabrescent; stipules minute, setaceous; leaves sessile, digitate, subfasciculate, 1–5-foliolate; leaflets linear to obovate, often involute, about 2–3(–7) mm long, 1 mm wide or less, obtuse at the apex, cuneate at the base, glabrous above, sparsely appressed-pubescent beneath; flowers about 5–6 mm long, solitary, axillary, on pedicels about 5 mm long; bracts minute, setaceous; calyx 1–1.5 mm long with deltoid lobes, sparsely pubescent; petals reddish; fruit linear, straight, 5–15 mm long, 1.5 mm wide, sparsely pubescent, (3–)5–8-seeded.

Distr. Southern India to northern Ceylon, in sandy areas.
Vern. Rat-kohamba (S); chivanarvempu, sivanarvum (T).
Specimens Examined. WITHOUT EXACT LOCALITY: *Macrae s.n.* (K);

Jaffna & *Gardner*, Batticaloa, Nov. 1858, *Thwaites C.P. 1455* (PDA); without data *C.P. 1455* (BM, K, P). JAFFNA DISTRICT: Jaffna, dyke, *s. coll. s.n.* (K); Kankesanturai, Feb. 1890. *Trimen? s.n.* (PDA), Mar. 1923, *de Alwis s.n.* (PDA), Jan. 1914, *s. coll. s.n.* (PDA), *Sumithraarachchi 792* (CAS, K, US); Kayts jetty, *Rudd 3278* (GH, K, PDA, SFV, US); near Velanai, *Bernardi 14273* (US); Point Pedro, *Simpson 9276* (BM); Jaffna Forest Reserve, *Worthington 4582* (K). MANNAR DISTRICT: Talaimannar, 16 July 1916, *"J. M. S." s.n.* (PDA). VAVUNIYA DISTRICT: Just S. of Mullaittivu, NE. coast, *Fosberg & Balakrishnan 53510* (K, SFV, US). PUTTALAM DISTRICT: Wilpattu Natl. Park, Kali Villu, *Fosberg et al. 50958* (K), *50959* (GH, SFV, UC, US), *Mueller-Dombois 68091105* (K, PDA, US), *Wirawan et al. 1011* (CAS, K, US). COLOMBO DISTRICT: Colombo, sea shore, *Macrae 383* (BM, K).

5. Indigofera oblongifolia Forssk., Fl. Aegypt.-Arab. 137. 1775; Gamble, Fl. Pres. Madras 311. 1918; Alston in Trimen, Handb. Fl. Ceylon 6: 73. 1931; Gillett, Kew Bull. Addit. Ser. 1: 116. 1958; Backer & Bakh. f., Fl. Java 1: 592. 1963; Ali, Fl. W. Pakistan, no. 100, 71, fig. 10, G–N. 1977. Type: *Forsskal s.n.* Yemen (C).

Indigofera lotoides Lam., Dict. 3: 247. 1789, exc. syn. Type: *Sonnerat s.n.*, India ?, as "Cape of Good Hope" (P).
Indigofera paucifolia Delile, Fl. Aegypt. 107, t. 37, fig. 22. 1813; Wight & Arn., Prod. 201. 1834; Wight, Ic. Pl. Ind. Or. 331. 1840–43; Thw., Enum. Fl. Zeyl. 83. 1859; Hook. f., Fl. Br. Ind. 2: 97. 1876; Trimen, Handb. Fl. Ceylon 2: 25. 1894. Type: *Delile s.n.*, Egypt (P).
Indigofera argentea Buch.-Ham. ex Roxb., Fl. Ind. 3: 374. 1832, non L. 1771. Type: *Buchanan-Hamilton s.n.*, India (BM ?), "Herb. Banks" fide Wight & Arn. (l.c.)
Indigofera desmodioides Baker, Kew Bull. 1894: 331. 1894, non Baker 1887. Type: *Bent 185*, Aden (K).

Erect shrub, to about 2 m tall, much branched; young stems densely white or grey appressed-pubescent; stipules lanceolate, acuminate, 3–3.5 mm long, caducous; leaves imparipinnate, (1–)3–5(–7)-foliolate; leaflets alternate or subopposite, oblong to lanceolate or obovate, obtuse or breviacuminate, about 1–3.5 cm long, 4–13 mm wide, pubescent on both surfaces; inflorescences racemose or subspicate, many-flowered; flowers about 5–10 mm long; calyx 2–2.5 mm long, pubescent, with deltoid teeth; petals reddish, the vexillum pubescent on the outer face; fruit terete, constricted between the seeds, straight or slightly curved, about 8–21 mm long, 1–1.5 mm in diameter, appressed-pubescent, sometimes glabrescent, 2–8-seeded.

Distr. In dry regions of Ceylon, Pakistan, India, Java, and tropical Africa.
Vern. Kuttukarasmatti, nante (T).

Specimens Examined. WITHOUT PRECISE LOCALITY: Matalan, Feb. 1889, *Nevill s.n.* (PDA); Jaffna, Aripo, Kalpitiya (as "Calputia"), *Gardner*, Trincomalee, *Glenie, Thwaites C.P. 1454* (PDA); without data; *s. coll. s.n.* (BM, CAS, K, P, PDA). JAFFNA DISTRICT: Jaffna, Feb. 1890, *Trimen* ? *s.n.* (PDA); about 6 mi. from Kilinochchi, *Simpson 9243* (BM, PDA). MANNAR DISTRICT: Between Thirukatheeswan and Mannar, *Kundu & Balakrishnan 617* (US). PUTTALAM DISTRICT: Island in Puttalam lagoon, *Alston 1228* (PDA); Karativu & Kalpitiya, Aug. 1883, *Trimen* ? *s.n.* (PDA).

6. Indigofera spicata Forssk., Fl. Aegypt.-Arab. 138. 1775; Gillett, Kew Bull., Addit. Ser. 1: 119, 138. 1958; Gillett, Fl. Trop. East Africa, Papil. 317. 1971; Backer & Bakh. f., Fl. Java 1: 591. 1963; Saldanha & Nicolson, Fl. Hassan 258. 1976; Verdcourt, Man. New Guinea Leg. 353. 1979. Type: *Forsskal.* Yemen, Bolgose (C).

Indigofera hendecaphylla Jacq., Collect. 2: 358. 1789; Jacq., Ic. Pl. Rar. t. 570. 1788–1789. Type: *Jacquin* ?, "Guinea," cultivated at Vienna.
Indigofera endecaphylla Auctt.; Alston in Trimen, Handb. Fl. Ceylon 6: 73. 1931; Macmillan, Trop. Pl. & Gard. ed. 5, 28. 1962; Baker in Hook. f., Fl. Br. Ind. 2: 98. 1876.
Indigofera kleinii Wight & Arn., Prod. 204. 1834. Type: *Klein,* Ceylon ?; Wight Cat. 855.

Gillett (l.c. p. 119) gives a more extensive synonymy for this species.

Herbaceous perennial from a thick rootstock; stems erect or prostrate, sparsely pubescent or subglabrous; stipules lanceolate, attenuate, to about 1 cm long, subglabrous; leaves subsessile, imparipinnate, 5–11-foliolate; leaflets alternate, or sometimes subopposite, predominantly obovate, about 0.5–3 cm long, 2–10 mm wide, obtuse at the apex, cuneate at the base, glabrous above, appressed-pubescent below; inflorescences axillary, subspicate or racemose, many-flowered; bracts lanceolate, attenuate, about 2 mm long; flowers about 5 mm long; calyx 2 mm long with attenuate lobes, sparingly pubescent; petals pink to purplish-red, the vexillum pubescent on the outer face; fruit linear, straight, reflexed, commonly 2–2.5 cm long, 2 mm in diameter, somewhat tetragonous with thickened margins, moderately appressed-pubescent, 5–10-seeded.

Distr. Ceylon, India, Southeast Asia, Yemen, Africa, Madagascar, Mascarene Isl., and Australia, often in weedy places, sometimes introduced in pastures.

Uses. Verdcourt (op. cit. p. 355) mentions that although this species is widely grown as a pasture legume it has proved to be poisonous and to cause abortion in cattle. Macmillan (l.c.) includes it in his list of plants suitable as cover crops.

Specimens Examined. WITHOUT EXACT LOCALITY: *Fairchild 1084*

(UC). KANDY DISTRICT: Peradeniya, *Cooray 68100409R* (SFV, US), *Comanor 485* (CAS, SFV, US); on A5, 3 mi. S. of. Gampola, *Maxwell & Jayasuriya 856* (US). NUWARA ELIYA DISTRICT: Between Dimbulla and Kotmale, culvert 10/12, 1500 m alt., *Maxwell et al. 924* (US); McDonald's Valley, below Hakgala, 1000 m alt., *Rudd & Balakrishnan 3176* (GH, PDA, SFV, US). KEGALLE DISTRICT: Kegalle, *Amaratunga 1653* (PDA). RATNAPURA DISTRICT: Depedene (Deep deen) on slopes of Mt. S. of Ratnapura, *Fosberg 56624* (US). BADULLA DISTRICT: Between Boralanda and Palugama, *Hepper 4592* (US).

7. Indigofera trifoliata L., Cent. Pl. 2: 29. 1756; L., Amoen. Acad. 4: 327. 1759; Wight, Ic. Pl. Ind. Or. 314. 1840; Baker in Hook. f., Fl. Br. Ind. 2: 96. 1876; Alston in Trimen, Handb. Fl. Ceylon 6: 73. 1931; Ali, Bot. Not. 111: 552. 1958; Saldanha & Nicolson, Fl. Hassan 258. 1976; Ali, Fl. W. Pakistan, no. 100, 78. 1977; Verdcourt, Man. New Guinea Leg. 357. 1979. Type: India, Herb. Linn. 923.3 (LINN).

Suffrutescent perennial; stems suberect or trailing, to about 60 cm long, appressed-pubescent, later glabrescent; stipules minute, setaceous; leaves digitately 3-foliolate; leaflets oblanceolate to oblong-elliptic, obtuse, mucronate, 9–25(–40) mm long, 5–12 mm wide, appressed-pubescent on both surfaces, glandular-dotted beneath; inflorescences axillary, racemose, subsessile, few-flowered; flowers 3–4.5 mm long; calyx 3–3.5 mm long, pubescent, with lanceolate lobes; petals reddish; fruit linear, straight, reflexed, 1–1.7 cm long, 1.5 mm wide, sparsely appressed-pubescent, tetragonous, with thickened margins, 6–10-seeded.

Distr. Ceylon, India, Pakistan, Indonesia, Philippines, Taiwan, New Guinea, and Australia.

Specimens Examined. POLONNARUWA AND BADULLA DISTRICTS: Minneriya on border of tank, May 1858, between "N. Ellia", Badulla, Dec. 1859, illegible, *Ferguson, Thwaites, C.P. 3592* (PDA); without data, *C.P. 3592* (GH, K, P).

Note. Alston (l.c.) has questioned the occurrence of this species in Ceylon. The rather mixed collection included in *Thwaites, C.P. 3592* should be reexamined to check for uniformity in the various specimens. The material at BM and K, recently studied, does appear to be referable to *I. trifoliata*.

8. Indigofera barberi Gamble, Fl. Pres. Madras 1: 306, 310. 1918; Gamble, Kew Bull. 1919: 222. 1919; Alston in Trimen, Handb. Fl. Ceylon 6: 73. 1931. Syntypes: *Beddome s.n.,* S. India, Cuddapah hills (K); *Barber 1076,* S. India, Melpat, S. Arcot (K); *Bourne 869,* S. India, Shevary Hills, Salem (K).

Suffrutescent herb to about 60 cm high; stems erect, appressed-pubescent; stipules setaceous, caducous; leaves pinnately 3-foliolate; terminal petiolule about

1 cm long, the lateral petiolules much shorter; leaflets obovate, mucronate, about 10 mm long, 3–5 mm wide, pubescent with minute lax hairs on both surfaces, glandular-dotted beneath; inflorescences axillary, racemose, sessile; flowers small; calyx villous, about 1 mm long with subulate teeth; petals dark pink; fruit linear, straight, about 8–15 mm long, somewhat torulose, 1 mm in diameter, 2–4-seeded, laxly-pubescent, the sharp apex curved upward.

Distr. Southern India and Ceylon ?

Note. It is possible that some of the material of *Thwaites C.P. 3572* assigned to *I. trifoliata* will on reexamination prove to be *I. barberi* as suggested by Alston. There has been confusion between *I. barberi* and *I. trifoliata*. Gamble, in his 1919 publication of *I. barberi,* stated, ''I have long hesitated about this species. I agree with Dr. Barber in considering it as coming between *I. trifoliata* and *I. trita* L. f. and I think it best to describe it as new, for I cannot consider it even as a variety of either''.

9. Indigofera trita L. f., Suppl. Pl. 335. 1781; Wight, Ic. Pl. Ind. Or. 315. 1840; 386. 1840–43; Thw., Enum. Pl. Zeyl. 83. 1859; Baker in Hook. f., Fl. Br. Ind. 2: 96. 1876; Trimen, Handb. Fl. Ceylon 2: 25. 1894; Ali, Bot. Not. 111: 553. 1958; Gillett, Fl. Trop. East Africa, Papil. 303. 1971; Saldanha & Nicolson, Fl. Hassan 258. 1976; Ali, Fl. West Pakistan, no. 100, Papil. 78. 1977; Verdcourt, Man. New Guinea Leg. 356. 1979. Type: India, Herb. Linn. 923. 9 (LINN).

Indigofera subulata Vahl ex Poir. in Lam., Enc. Suppl. 3: 150. 1813. Type: *Thonning, s.n.,* Ghana (P holotype; C isotype).
Indigofera scabra Roth, Nov. Pl. Sp. 359. 1821. Type: *Heyne s.n.,* India, Madras Prov. (K isotype).
Indigofera flaccida Koen. in Roxb., Fl. Ind. 3: 375. 1832.
Indigofera subulata Vahl ex Poir. var. *scabra* (Roth) Meikle, Kew Bull. 21: 352. 1950; Gillett, Kew Bull., Addit. Ser. 1: 100. 1958.
Indigofera trita L. f. var. *scabra* (Roth) Ali, Bot. Not. 3: 558. 1958.
Indigofera trita L. f. var. *subulata* (Poir.) Ali, Bot. Not. 3: 558. 1958.

Suffrutescent perennial, to about 1 m tall; stems divaricate, densely white-appressed-pubescent; stipules about 2 mm long, setaceous; leaves pinnately 3-foliolate; stipels minute, 0.3 mm long or less, setaceous, caducous; leaflets oblong to obovate, 5–10(–26) mm long, 1.5–10(–20) mm wide, obtuse or emarginate, glabrous to subglabrous above, appressed-pubescent beneath; inflorescences axillary, spicate-racemose, subsessile; bracts deltoid-cucullate, 1 mm long or less; flowers about 5 mm long; calyx 2–2.5 mm long, pubescent, with attenuate lobes; petals salmon or brick-red, the vexillum pubescent on the outer face; fruit linear, straight, spiniform at the apex, 2–3 cm long, 1.5–2 mm in diameter, tetragonous, divaricate or sometimes reflexed, 6–10-seeded.

Distr. Tropical Asia, northern Australia, and in Africa.

Vern. Wal-awari (S).

Specimens Examined. WITHOUT EXACT LOCALITY: *Mrs. & Col. Walker s.n.* (K). Batticaloa, 1846, *Gardner* & May 1858, Atakalan Korale, illegible, 1837 (?), Trincomalee, *Glenie, Thwaites C.P. 1463* (PDA); without data, *C.P. 1463* (GH, K, P, PDA). ANURADHAPURA DISTRICT: Near Kalpe, Aug. 1885, *Trimen?* (PDA); Isurumuni Vihare, *Hepper & Jayasuriya 4651* (US). AMPARAI DISTRICT: Kumana, Panama coast, *Balakrishnan 590* (US). HAMBANTOTA DISTRICT: Ruhuna Natl. Park, Yala, *Cooray 70032702R* (SFV, US); Wirawila, near lake, *Rudd 3094* (GH, PDA, SFV, US).

Notes: The distinctions between *I. trita* and *I. subulata* are somewhat unclear. The conclusion by Ali (l.c. 1958, p. 557) that the whole complex is best treated as a single species is being followed here.

Moon, Cat. 54. 1824, cited *I. cinerea* Willd., Alu-awari (S), from Kandy, which, according to Index Kewensis is a synonym of *I. trita* L. f., but I have seen no voucher for it.

10. Indigofera karnatakana Sanjappa, Taxon 32: 120. 1983.

Indigofera tenuifolia Rottl. ex Wight & Arn., Prod. 200. 1834; Thw., Enum. Pl. Zeyl. 83. 1859; Baker in Hook. f., Fl. Br. Ind. 2: 95. 1876; Trimen, Handb. Fl. Ceylon 2: 24. 1894; Alston in Trimen, Handb. Fl. Ceylon 6: 72. 1931; Ali, Fl. W. Pakistan, no. 100: 79. 1977. Type: *Wight 864*, India, Mysore (holotype E; isotype K), non Lam. 1789.

Herbaceous annual, to about 20 cm tall; stems diffuse, glabrous or subglabrous; stipules setaceous, 1 mm long or less; leaves pinnately 7–9-foliolate; leaflets opposite, obovate to oblanceolate, 5–10 mm long, 1–2(–4) mm wide, obtuse, apiculate, sparsely appressed-pubescent on both surfaces; stipels minute, setaceous, caducous; inflorescences axillary, racemose, 3–8-flowered, longer than the leaves; bracts minute, setaceous; flowers 4–5 mm long; calyx 1.5–2.5 mm long, pubescent, lobes setaceous; petals red; fruit 1.5–2.5 cm long, 2–3 mm in diameter, straight, divaricate, sparsely appressed-pubescent, somewhat torulose, 6–10-seeded; seeds cylindrical, about 2 mm long, pitted, hilum lateral.

Distr. Ceylon and southern India, in sandy areas.

Specimens Examined. POLONNARUWA DISTRICT: Mannampitiya, *Alston s.n.* (PDA); 4 mi. E. of Welikanda, *Jayasuriya et al. 710* (US); between Habarane and Polonnaruwa, mi. marker 62, *Rudd & Balakrishnan 3142* (GH, K, PDA, SFV, US). TRINCOMALEE/BADULLA DISTRICTS: Trincomalee and Bintenne, *Gardner, Thwaites C.P. 1462* (PDA); without data, *C.P. 1462* (GH, K, PDA). AMPARAI DISTRICT: Between Amparai and Maha Oya, mi 34–35, *Rudd & Balakrishnan 3233* (BM, PDA, SFV, US). MONERAGALA DISTRICT: Near Bibile, Uva, Jan. 1888, *Trimen? s.n.* (PDA).

11. Indigofera constricta (Thw.) Trimen, Cat. 23. 1885; Trimen, Handb. Fl. Ceylon 2: 27. 1894; Cooke, Fl. Pres. Bombay 1: 319. 1901; Gamble, Fl. Pres. Madras 308, 312. 1918.

Indigofera flaccida var. *β constricta* Thw., Enum. Pl. Zeyl. 411. 1864; Baker in Hook. f., Fl. Br. Ind. 2: 99. 1876, in note under *I. tinctoria* L. Type: *Thwaites C.P. 3811*, Matale East (as "Mettelle East"), June 1863 (PDA, holotype; BM, K, isotypes).

Shrub, erect, to about 1 m tall; stems spreading, appressed-pubescent; stipules minute, setaceous; leaves imparipinnate, 7–11-foliolate; leaflets opposite, deciduous, elliptic to elliptic-oblong, about 2–2.5 cm long, 1–1.5 cm wide, rounded or slightly emarginate, apiculate at the apex, rounded at the base, sparsely appressed-pubescent on both surfaces; inflorescences axillary, racemose, about as long as the leaves, many-flowered; calyx sericeous, campanulate with deltoid lobes; petals not seen; fruit linear, slightly curved, acute at the apex, about (3–)3.5–5 cm long, somewhat tetragonous, constricted between the seeds, sparsely appressed-pubescent, 3–12-seeded.

Distr. In forest areas of Ceylon and southern India, apparently rare.

Specimens Examined. MATALE DISTRICT: Etanwela (as "Et-tangwella") (on same sheet as holotype PDA); without locality, "by Bot. & Mycologist Peradeniya 14/3/18" (PDA); *Macrae 385* (BM).

12. Indigofera wightii Grah. ex Wight & Arn., Prod. 202. 1834; Baker in Hook. f., Fl. Br. Ind. 2: 99. 1876; Trimen, Handb. Fl. Ceylon 2: 27. 1894; Cooke, Fl. Pres. Bombay 1: 319. 1901; Gamble, Fl. Pres. Madras 308, 313. 1918; Saldanha & Nicolson, Fl. Hassan 258. 1976. Type: Wallich, Cat. 5458 (K).

Indigofera inamoena Thw., Enum. Pl. Zeyl. 83. 1859. Type: *Thwaites, C.P. 3513*, Ceylon, Kalupahane near Haldummulla, as "Caloopahane between Hapootelle and Balangodde", about 2000 ft elev., "Ap. 1856 & 1863" (PDA holotype; isotypes BM, GH, K, P).

Small erect shrub; stems canescent-appressed-pubescent, striate; stipules minute, subulate; leaves imparipinnate, 13–21-foliolate; leaflets elliptic to lanceolate-oblong or obovate, about 8–20 mm long, 2.5–5 mm wide, obtuse, apiculate at the apex, acute at the base, densely canescent-sericeous on both surfaces; inflorescences axillary, racemose, subsessile, shorter than the leaves; flowers about 7–8 mm long; calyx 3–4 mm long, sericeous, with acute, deltoid lobes; petals red or orange-red, the vexillum pubescent on the outer face; fruit finely canescent, linear, straight, cylindrical, about 2.5–4 cm long, 2.5–3 mm in diameter, 8–12-seeded; seeds truncate-cylindrical.

Distr. Ceylon and southern India.

Specimens Examined. Type material, cited above.

13. Indigofera tinctoria L., Sp. Pl. 751. 1753; Moon, Cat. 54. 1824; Wight & Arn., Prod. 202. 1834; Wight, Ic. Pl. Ind. Or. 365. 1840–43; Thw., Enum. Pl. Zeyl. 411. 1864; Baker in Hook. f., Fl. Br. Ind. 2: 99. 1876; Trimen, Handb. Fl. Ceylon 2: 26. 1894; Macmillan, Trop. Pl. & Gard. ed. 5, 416. 1962; Backer & Bakh. f., Fl. Java 1: 591. 1963; Gillett, Fl. Trop. East Africa, Papil. 308. 1971; Saldanha & Nicolson, Fl. Hassan 258. 1976; Ali, Fl. W. Pakistan, no. 100: 82. 1977; Allen & Allen, Legum. 342. 1981. Type: Ceylon, *P. Hermann*, Herm. Herb. 3: 20 (holotype BM; isotype Herm. Herb. Zeyl. 44, L).

Indigofera indica Lam., Enc. Meth. 3: 245. 1789, tab. 626, fig. 1. 1823. Apparently based, at least in part, on *I. tinctoria* L.
Indigofera sumatrana Gaertn., Fruct. 2: 307, t. 148. 1791. Type unknown.

Suffrutescent herb, to about 2 m tall; stems erect, appressed-pubescent, later glabrescent; stipules subulate, about 2 mm long; leaves pinnately 5–13(–19)-foliolate; stipels minute, setaceous, caducous; leaflets opposite, elliptic to obovate, about 5–30 mm long, 5–15 mm wide, rounded at the apex, acute at the base, glabrous above, appressed-pubescent beneath, darkening on drying; inflorescences axillary, spicate-racemose, sessile, many-flowered; flowers about 5–7 mm long; calyx 1–1.5 mm long, pubescent, lobes deltoid, about as long as the tube; petals salmon-pink; fruit linear, straight or slightly curved, 2–3(–3.5) cm long, 2 mm in diameter, moderately appressed-pubescent to glabrous, (3–)8–12-seeded.

Distr. Widespread in the Old World tropics and introduced in America.

Uses. This species was formerly a major source of blue dye. According to Macmillan (l.c.), *I. tinctoria* was superseded by *I. arrecta*, known as Natal indigo, and *I. sumatrana*, Java indigo. The latter is now considered to be taxonomically synonymous with *I. tinctoria* although it was thought to be a better source of the dye.

Vern. Nil-awari (S); indigo (E).

Specimens Examined. JAFFNA DISTRICT: Jaffna, Feb. 1890, *Trimen ? s.n.* (PDA); between Jaffna causeway and Kayts, *Rudd 3273* (BM, GH, PDA, SFV, US). ANURADHAPURA DISTRICT: Hinukkiriyawa, W. of Habarane, *Jayasuriya et al. 1392* (K, US). POLONNARUWA DISTRICT: Between Habarane and Kantalai near 123 mi. marker, *Rudd & Balakrishnan 3125* (PDA, SFV, US); Polonnaruwa, sacred area, *Ripley 322* (K, US), *393* (US); Polonnaruwa, 7 May 1927, *F.W. de Silva s.n.* (PDA, UC). KURUNEGALA DISTRICT: Wanduragala, quarry near Kurunegala, *Rudd 3332* (US). PUTTALAM DISTRICT: Wilpattu Natl. Park, near Eerige Ara confluence with Moderagama Ara, *Fosberg et al. 50801* (SFV, US). MATALE DISTRICT: Sigiriya, 7 Sept. 1885, *Trimen ? s.n.* (PDA). MANNAR DISTRICT: Between Murunkan and Madhu Road, *Simpson 9384* (BM). NUWARA ELIYA DISTRICT: McDonald's Valley, below Hakgala, *Rudd & Balakrishnan 3176* (K, PDA, US). BATTICALOA AND TRINCOMALEE DISTRICTS: Batticaloa, Nov. 1857,

"Trinco 1863, Glenie", *Thwaites C.P. 3591* (PDA). TRINCOMALEE DISTRICT: Near China Bay airport, *Rudd & Balakrishnan 3133* (PDA, SFV, US). KALUTARA DISTRICT: Kalutara, Dec. 1882, *Trimen? s.n.* (PDA). COLOMBO DISTRICT: Negombo, *Simpson 7936* (BM). BADULLA DISTRICT: Near Jangula, Badulla, *Simpson 8256* (BM). GALLE DISTRICT: Bentota, seashore, March 1887, *Trimen? s.n.* (PDA). HAMBANTOTA DISTRICT: Ruhuna Natl. Park, Patanagala, *Cooray 70032105* (US); Sithul Pahuwa, *Fosberg & Sachet 52915* (K, SFV, US); south of Situlpahuwa, *Mueller-Dombois 68102106* (PDA, US); Andunoruwa Wewa, *Mueller-Dombois & Cooray 67120606* (US); Bata-ata, *Alston 1227* (PDA).

Note. A collection previously identified as *I. tinctoria, Rudd 3295,* appears to be *I. arrecta.* The stems are woodier and the fruit somewhat shorter.

14. Indigofera parviflora Heyne ex Wight & Arn., Prod. 201. 1834; Baker in Hook. f., Fl. Br. Ind. 2: 97. 1879; Trimen, Handb. Fl. Ceylon 2: 26. 1894; Cooke, Fl. Pres. Bombay 1: 317. 1901; Gamble, Fl. Pres. Madras 311. 1918; Alston in Trimen, Handb. Fl. Ceylon 6: 73. 1931; Gillett, Kew Bull. Addit. Ser. 1: 126. 1958; Gillett, Fl. Trop. East Africa, Papil. 321. 1971. Type: India, *Heyne* in Wallich herb. *5457* (K).

Indigofera deflexa A. Rich., Tent. Fl. Abyss. 1: 178. 1847. Syntypes: Sudan, *Kotschy 14,* in 1839; Ethiopia, in 1840; *Schimper 1467* (K),
Indigastrum deflexum (A. Rich.) Jaub. & Spach, Ill. Pl. Or. t. 492. 1859.

Herbaceous annual, to about 6 dm tall; stems much-branched, finely canescent-strigillose; stipules linear, to about 2–3 mm long; leaves imparipinnate, 5–13-foliolate; leaflets opposite, oblong-lanceolate to oblanceolate, 1.5–2.5 cm long, 4–10 mm wide, obtuse to acute, apiculate, glabrous on the upper surface, appressed-pubescent beneath; inflorescences racemose, sessile or subsessile, 6–12-flowered; flowers about 7 mm long; calyx pubescent, 2–4 mm long with deltoid-attenuate lobes; petals lilac; fruit linear, straight, appressed-pubescent, glabrescent, 2.5–3.5 cm long, 2 mm in diameter, acute, recurved at the apex, 15–20-seeded.

Distr. India, Ceylon, tropical Africa, and northern Australia.

Specimens Examined. COLOMBO DISTRICT: Colombo, in 1865, *Ferguson s.n.* (PDA).

15. Indigofera suffruticosa Mill., Gard. Dict. ed. 8, *Indigofera* no. 2. 1768; Alston in Trimen, Handb. Fl. Ceylon 6: 73. 1931; Gillett, Kew Bull., Addit. Ser. 1: 105. 1958; Backer & Bakh. f., Fl. Java 1: 592. 1963; Verdcourt, Man. New Guinea Leg. 355. 1979; White in Fl. Panamá, Ann. Missouri Bot. Gard. 67: 713. 1980. Type: "*Unknown coll. s.n.* ex Miller herb. BM holotype" fide Gillett (l.c. p. 106); isotype BM.

Indigofera anil L., Mant. 2: 272. 1771; Trimen, Handb. Fl. Ceylon 2: 27. 1894. Type: "Habitat in India", Herb. Linn. 923.20 (LINN)?

Additional synonymy is given by Gillett (l.c.) and White (l.c.).

Shrub to about 3 m tall; stems appressed-pubescent, glabrescent with age; stipules filiform, to about 2–3 mm long; leaves imparipinnate, 9–17-foliolate; stipels minute, filiform; leaflets opposite, oblong, elliptic, or obovate, 0.5–3(–4) cm long, 3–15 mm wide, rounded to subacute, apiculate at the apex, acute at the base, lightly appressed-pubescent to glabrous on the upper surface, densely pubescent beneath; inflorescences axillary, racemose, sessile, many-flowered, shorter than the leaves; bracts filiform to lanceolate-subulate, to about 2 mm long, caducous; flowers 4–5 mm long; calyx pubescent, 1–1.5 mm long with deltoid, subulate lobes; petals salmon-pink to brick-red; fruit arcuate, about 1–1.5 cm long, 2–2.5 mm in diameter, subterete with thickened margins, appressed-pubescent, 3–8-seeded.

Distr. A native of tropical and subtropical America, widely introduced in the Old World tropics.

Vern. Añil; West Indian Indigo (E).

Specimens Examined. NUWARA ELIYA DISTRICT: Uda Pusselawe, Sept. 1885, *W.F. (Ferguson) s.n.* (PDA). KURUNEGALA DISTRICT: Mawatagama, *Amaratunga 1082* (PDA). KANDY DISTRICT: Kandy catchment, *Worthington 6940* (K); Peradeniya, Exp. Sta., in 1916, *s. coll. s.n.* (PDA), 21.10.1913, *s. coll. s.n.* (PDA). BADULLA DISTRICT: Ambewela Road, near Albion Estate Bungalow, "A.M.S. 9.10.06" (PDA).

Note. This species, like *I. tinctoria,* formerly was much used as a source of blue dye. It apparently was introduced to Ceylon and other parts of the Old World at an early period.

16. Indigofera galegoides DC., Prod. 2: 225. 1825; G. Don, Gen. Hist. 2: 209. 1832; Arn., Nov. Actorum Acad. Caes. Leop.-Carol. Nat. Cur. 18 (1): 329. 1840; Thw., Enum. Pl. Zeyl. 83. 1859; Baker in Hook. f., Fl. Br. Ind. 2: 100. 1876; Trimen, Handb. Fl. Ceylon 2: 28. 1894; Gamble, Fl. Pres. Madras 308, 313. 1918; Alston in Trimen, Handb. Fl. Ceylon 6: 73. 1931; Backer & Bakh. f., Fl. Java 1: 592. 1963. Type: Ceylon, *Leschenault s.n.* (P).

Indigofera uncinata Roxb., Fl. Ind. 3: 382. 1832, non G. Don 1832.

Shrub or small tree, to about 2.5 m tall; stems glabrous or minutely appressed-pubescent; stipules filiform, about 2 mm long; leaves imparipinnate, 15–23-foliolate; stipels minute, filiform; leaflets opposite or subopposite, elliptic to obovate, 1–5 cm long, 0.5–2 cm wide, obtuse, apiculate at the apex, obtuse at the base, appressed-pubescent with minute hairs on both surfaces; inflorescences axillary, racemose, subspicate, many-flowered, much shorter than the leaves;

bracts subulate, about 1 mm long; flowers about 1–1.3 cm long; calyx pubescent, about 2 mm long with short, deltoid lobes; petals pinkish, the vexillum sericeous on the outer face; fruit linear, straight, 5.5–7.5 cm long including an apical beak about 1–1.3 cm long, 2 mm wide, moderately appressed-pubescent to glabrous, 8–18-seeded.

Distr. Ceylon and southern India, Java, and south China, sometimes introduced.

Uses. Sometimes planted as an ornamental or as a cover crop.

Vern. Veliveriya (S).

Specimens Examined. WITHOUT EXACT LOCALITY: *Walker 17* (K), *1643* (K). MATALE DISTRICT: Matale. *Gardner, Thwaites C.P. 1461* (PDA); *s. coll. C.P. 1461* (BM). KANDY DISTRICT: Hantane, 3000 ft, *Gardner 474* (K); Royal Botanic Gardens, Peradeniya, 11-9-1907, *s. coll. s.n.* (PDA); Kadugannawa, "Aug. 1884, *W.F."* *Ferguson s.n.* (PDA); Aladeniya, *Amaratunga 681* (PDA). AMPARAI DISTRICT: Padagoda, S. of Inginiyagala, *Fosberg & Sachet 53088* (GH, K, SFV, UC, US). COLOMBO DISTRICT: Colombo, *Ferguson, Thwaites C.P. 1461* (PDA); without data, *Thwaites C.P. 1461* (BM, GH, K, P). KALUTARA DISTRICT: Welipenna, Pasdun Korale, March 1887, *Trimen ? s.n.* (PDA).

17. Indigofera zollingeriana Miq., Fl. Ind. Bat. 1: 310. 1855; Backer & Bakh. f., Fl. Java 1: 592. 1963; Verdcourt, Man. New Guinea Leg. 357. 1979; Paul, Trop. Agric. (Ceylon) 109: 27–35. 1953. Type: "In de kustreken van Zuidoost-Java, zeldzaam, bijv. bij. Siri Gontjo, door Zollinger ontdekt."

Indigofera teysmannii Miq., Fl. Ind. Bat. 1: 1083. 1858. Type: "Sumatra, in de kustreken van Siboga (Teysm.)".

Erect shrub or small tree, to about 5 m tall; stems subsericeous with minute brown or white appressed hairs, glabrescent with age; stipules attenuate, to about 8 mm long; leaves imparipinnate, 11–17-foliolate; stipels minute, acicular; leaflets lanceolate to ovate or elliptic-oblong, 2–7.5 cm long, 1.2–3.2 cm wide, acute to obtuse, mucronate at the apex, rounded to acute at the base, pubescent with minute appressed hairs on both surfaces; inflorescences axillary, racemose or subspicate, many-flowered; flowers about 10 mm long; calyx brown-sericeous, about 2 mm long, subtruncate, the carinal lobe longest, about 0.5 mm long or less; petals pink or whitish or dark purple; vexillum sericeous on the outer face, shorter than the keel; keel petals pointed, cucullate, pubescent, each with a lateral auricle about 1.5 mm long; fruit glabrous, linear, tetragonous-subcylindrical, 3.5–4.5 cm long including a beak about 2 mm long, 4–5 mm in diameter, about 12-seeded.

Distr. Malesia, South China, Taiwan; apparently introduced in Ceylon.

Specimens Examined. KANDY DISTRICT: Between Mahiyangana and

Kandy, culvert 10/6, *Maxwell 1027* (SFV, US); Kundesale, E. of Kandy, *Worthington 6807* (K); Kandy, *Worthington 6923* (K), *7170* (K); Madugoda, *Worthington 6966* (K).

18. Indigofera colutea (Burm. f.) Merr., Philipp. J. Sci. 19: 355. 1921; Alston in Trimen, Handb. Fl. Ceylon 6: 73. 1931; Gillett, Kew Bull. Addit. Ser. 1: 65. 1958; Ali, Bot. Not. 111: 548. 1958; Gillett, Kew Bull. 24: 483. 1970; Gillett, Fl. Trop. East Africa, Papil. 266, fig. 41. 1971; Verdcourt, Man. New Guinea Leg. 351. 1979.

Galega colutea Burm. f., Fl. Ind. 172. 1768, non sensu Willd. 1803. Type: "Habitat in India", Plukenet, Phytogr. 112, t. 166, fig. 3. 1692; based on specimen from India, Herb. Sloane, vol. 95, fol. 185 (BM), fide Ali (l.c. 1958).
Indigofera viscosa Lam., Enc. Meth. 3: 247. 1789; Wight, Ic. Pl. Ind. Or. 404. 1840–43; Thw., Enum. Pl. Zeyl. 83. 1859; Baker in Hook. f., Fl. Br. Ind. 2: 95. 1876; Trimen, Handb. Fl. Ceylon 2: 24. 1894. Type: "A plant of unknown origin cultivated in Paris (P, holo)" fide Gillett (l.c. 1971, p. 268).

Additional synonymy is given by Ali and Gillett in the references cited above.

Herbaceous annual or short-lived perennial, to almost 1 m tall; stems erect or spreading, densely pubescent with stalked glands; stipules setaceous, 2–4 mm long; leaves imparipinnate, (5–)7–11(–15) foliolate; stipels apparently lacking; leaflets opposite, elliptic to oblong or obovate, (4–)5–14 mm long, (2–)3–4 mm wide, obtuse at apex and base, appressed-pubescent on both surfaces; inflorescences glandular-pubescent, racemose, about 3–6(–20)-flowered; flowers 4–4.5 mm long; calyx appressed-pubescent, 2 mm long with setaceous lobes; petals pinkish to brick-red; fruit linear, terete, straight, spreading, densely pubescent with stalked, glandular and crispate hairs, (1–)1.5–2.5 cm long, 1.5 mm in diameter, 8–14-seeded.

Distr. Widespread from Pakistan, India, and Ceylon to Indonesia, Africa, Australia, and New Zealand.

Specimens Examined. WITHOUT EXACT LOCALITY: *Col. Walker s.n.* (P); *Koenig s.n.* (BM); "Vigitapoora" (?), *Gardner*, Oma Oya, July 1853, May 1856, *Thwaites, C.P. 1459* (PDA); without data, *C.P. 1459* (BM, K, P). POLONNARUWA DISTRICT: Minneriya (as "Mineri", Sept. 1885), *Trimen ? s.n.* (PDA). PUTTALAM DISTRICT: Karativu I. (as "Karativoe I."), Aug. 1883, *Trimen ? s.n.* (PDA). KURUNEGALA DISTRICT: Wetakeyapotha, *Alston 42* (PDA), *1226* (PDA). MATALE DISTRICT: Sigiriya, *Amaratunga 258* (PDA). BATTICALOA DISTRICT: Keeli-Kudah, *Waas 2134* (K, US). HAMBANTOTA DISTRICT: Ruhuna Natl. Park, Yala Bungalow, *Mueller-Dombois 69030801* (US); about 1.5 mi. beyond Tissamaharama on road to Kataragama, *Rudd 3102* (GH, K, PDA, SFV, US).

19. Indigofera hirsuta L., Sp. Pl. 751. 1753; Moon, Cat. 54. 1824; Thw., Enum. Fl. Zeyl. 83. 1859; Baker in Hook. f., Fl. Br. Ind. 2: 98. 1876; Trimen, Handb. Fl. Ceylon 2: 26. 1894; Gillett, Kew Bull. Addit. Ser. 1: 109. 1958; Ali, Bot. Not. 111: 559. 1958; Gillett, Kew Bull. 14: 290. 1960; Gillett, Fl. Trop. East Africa, Papil. 310, fig. 45. 1971. Type: Ceylon, *P. Hermann,* Hermann Herb. 1: 60 (BM). Probable isotype: *P. Hermann,* Paris. Biblio. Inst. France, Burman coll. Pl. Zeyl. 59.

Herbaceous annual, to about 1–1.5 m tall; stems erect or spreading, pubescent with spreading or crispate dark to light brownish hairs; stipules pubescent, setaceous, to about 1 cm long; leaves imparipinnate, 5–7(–9)-foliolate; stipels apparently lacking; leaflets opposite, elliptic to obovate, 1–2(–4) cm long, 5–10(–20–30) mm wide, obtuse at the apex, rounded to acute at the base, pubescent with subappressed hairs on the upper surface, lax hairs beneath; inflorescences racemose with peduncles 2–4.5 cm long, densely hirsute with brownish hairs, many-flowered; flowers about 5 mm long; calyx 4–4.5 mm long, brown-hirsute, with filiform lobes; petals pinkish; fruit linear, straight, subtetragonous, 1–2 cm long, about 2 mm wide, densely hirsute with dark and light brown hairs, 6–9-seeded.

Distr. Widespread as a weed in Ceylon and India to Indonesia, Africa, and northern Australia; adventive in tropical America.

Vern. Boo-awari (S).

Specimens Examined. WITHOUT EXACT LOCALITY: *Moon s.n.* (BM); Puttalam, *Gardner,* Badulla, Apr. 1859, Batticaloa, Mar. 1858, *Thwaites, C.P. 1456* (PDA); without data, *C.P. 1456* (BM, GH, K, P); *C.P. 1457* (K), *C.P. 1458* (K), *C.P.1459* (K), Naragam, June 1881,*Trimen ? s.n.* (PDA). POLONNARUWA DISTRICT: Minneriya (as "Mineri"), Aug. 1885, *Trimen ? s.n.* (PDA). MATALE DISTRICT: Dambulla Rock, *Amaratunga 525* (PDA); *Rudd & Balakrishnan 3109* (GH, K, PDA, SFV, US); Summit of Dambulla Hill, *Simpson 9777* (BM). KANDY DISTRICT: Exp. Sta. Peradeniya, in 1916, *s. coll. s.n.* (PDA). AMPARAI DISTRICT: Between Siyambalanduwa and Inginiyagala, *Rudd & Balakrishnan 3222* (K, PDA, SFV, US). MONERAGALA DISTRICT: Senanayake Samudra, *Maxwell & Jayasuriya 744* (CAS, US). HAMBANTOTA DISTRICT: Ruhuna Natl. Park, Padikema, *Cooray 70032601R* (CAS, PDA, US).

Note. Gillett (l.c. 1971) indicates that *I. astragalina* DC. also occurs in Ceylon and has so annotated a sheet of *Thwaites C.P. 1457* (K) as that species. Ali (l.c.) believes that it and *I. hirsuta* are synonymous. So far as the collections from Ceylon are concerned, I am citing them all as *I. hirsuta.* The collection *Rudd & Balakrishnan 3222* has generally lighter-coloured hairs on young parts than our *3109* but, otherwise, I can note no particular differences in number of leaflets, length of peduncle, or colour of mature fruit.

20. Indigofera glabra L., Sp. Pl. 751. 1753; Moon, Cat. 54. 1824; Trimen,

Handb. Fl. Ceylon 2: 23. 1894; Cooke, Fl. Pres. Bombay 1: 316. 1901; Gamble, Fl. Pres. Madras 306, 311. 1918; Ali, Bot. Not. 111: 572. 1958. Type: Ceylon, *P. Hermann*, Herm. Herb. 3: 27 (BM).

Indigofera pentaphylla Murr., Syst. Veg., ed. 13. 564. 1774, non Burch. ex Harvey & Sond., 1862; Wight, Ic. Pl. Ind. Or. 385. 1840–43; Thw., Enum. Pl. Zeyl. 311. 1864; Baker in Hook. f., Fl. Br. Ind. 2: 95. 1876. Type not cited.

Indigofera fragrans Retz., Obs. 4: 29. 1786. Type: *Koenig s.n.*, India ?, Ceylon ? (BM, isotype)?

Annual, to about 1m tall; stems erect, glabrous or sparsely villous; stipules ciliate, acicular, to about 5 mm long; leaves imparipinnate, (1–3)5(–7)-foliolate, sometimes variable on the same plant; stipels lacking; leaflets opposite, elliptic to suborbicular or obovate, obtuse at the apex, rounded to acute at the base, 5–15 mm long, 3–10 mm wide, subappressed-pubescent on both surfaces; inflorescences axillary, racemose, few-flowered; flowers about 5 mm long; calyx pubescent, 2 mm long with attenuate lobes; petals reddish-pink; fruit linear, straight, subtetragonous, glabrous, (1–)1.5–2 cm long, 1.5 mm in diameter, septate within; seeds cubiform, yellowish with darker mottling, about 1 mm in diameter.

Distr. Ceylon and southern India.

Specimens Examined. ANURADHAPURA DISTRICT: Between Medawachchiya and Vavuniya, mi. marker 98/1, *Rudd 3266* (K, PDA, SFV, US). POLONNARUWA DISTRICT: Between Habarane and Polonnaruwa, near mi. Marker 62, *Rudd & Balakrishnan 3141* (PDA, SFV, US). PUTTALAM DISTRICT: Wilpattu Natl. Park, Kumbu Wila, *Maxwell & Jayasuriya 800* (SFV, US); Kali Villu, *Cooray 70020202R* (US), *Wirawan et al. 1007* (US), *Fosberg et al. 50963* (CAS, US); Mannar-Puttalam Rd., *Cooray 70020119R* (US). MATALE DISTRICT: Dambulla Rock, *Rudd 3111* (K, PDA, US), 29 Sept. 1926, *de Silva s.n.* (PDA). AMPARAI DISTRICT: Between Amparai and Maha Oya, mile markers 19–20, *Rudd & Balakrishnan 3228* (CAS, K, PDA, US). COLOMBO DISTRICT: Colombo, Dec. 1859, *Ferguson, Thwaites C.P. 3524* (K, PDA).

*   *   *

In addition to the 20 preceding species of *Indigofera* a number of species have been introduced in the experimental gardens. A few have been collected elsewhere as represented by herbarium vouchers.

Indigofera arrecta A. Rich., Tent. Fl. Abyss. 1: 184. 1847; Macmillan, Trop. Pl. & Gard., ed. 5. 416. 1962. A native of Ethiopia, introduced as a source of "Natal" indigo. Jaffna District: Between Elephant Pass and Paranthan, *Rudd 3295* (GH, PDA, SFV, US). This species is distinguished with difficulty from

I. tinctoria. The fruits of this collection appear to be those of *I. arrecta.*

Indigofera decora Lindl., J. Hort. Soc. London 1: 68. 1846; Macmillan, Trop. Pl. & Gard., ed. 5. 65, 176. 1962. (*I. incarnata* (Willd.) Nakai). A native of China, introduced as an ornamental. Nuwara Eliya District: Hakgala, naturalized in the Royal Botanic Gardens, *s. coll. s.n.* (K), 26. 6.1919, *s. coll. s.n.* (PDA).

Indigofera dosua Buch.-Ham. ex D. Don, Prod. Fl. Nepal 244. 1825. (*I. stachyodes* Lindl.). A native of Nepal and northern India. Kandy District: Kelliewatte (Queensbury Gap), in tea, *Worthington 1776* (K).

Tribe ROBINIEAE

(Benth.) Hutchins., Gen. Fl. Pl. 1: 366. 1964; Polhill & Sousa in Polhill & Raven, Adv. Leg. Syst. 283. 1981. Type genus: *Robinia* L.

Trees, shrubs, or herbs, sometimes with gland-tipped or glandular-based hairs; leaves paripinnate, imparipinnate, 1-foliolate, or simple; stipules narrow, inserted close to the axil, sometimes intrapetiolar or spinous; stipels present or absent; leaflets usually opposite or subopposite; flowers singly inserted on the axillary or leaf-opposed rachides, or the inflorescences reduced to axillary fascicles; bracts small or sometimes showy; bracteoles present or absent; pedicels often jointed; hypanthium short to well developed; calyx bilabiate, subtruncate, or with vexillar lobes joined in part; petals glabrous or puberulent; vexillum sometimes with basal appendages, callouses, or inflexed auricles; stamens diadelphous 9:1 with the vexillar filament free or adnate above; anthers subuniform; ovary stipitate, usually with numerous ovules; style sometimes hardened, sometimes pubescent above; stigma small, terminal, introrse; fruits woody to chartaceous, dehiscent, sometimes septate between the seeds; seeds ovoid to oblong-reniform or oblong with a small lateral or subapical hilum.

A tribe of about 21 genera in tropical to warm temperate America, or pantropical (*Sesbania*).

KEY TO THE GENERA

1 Leaves paripinnate, 18–110-foliolate; stipels present or absent; bracteoles present; fruit transversely septate within..**1. Sesbania**
1 Leaves imparipinnate, (5–)7–17-foliolate; stipels absent; bracteoles absent; fruit not septate within ....**2. Gliricidia**

1. SESBANIA

Scop., Introd. 308. 1777, nom. cons.; Gillett, Kew Bull. 17; 19–159. 1963; Allen & Allen, Legum. 604. 1981. Type species; *S. sesban* (L.) Merr., based on *Aeschynomene sesban* L.

Agati Adans., Fam. Pl. 2: 326, 513. 1763. Type species: *A. grandiflora* (L.) Poir. based on *Robinia grandiflora* L.

Additional synonyms are given by Hutchinson, Gen. Fl. Pl. 1: 404. 1964.

Trees, shrubs, or herbs, unarmed or with prickles; leaves alternate, paripinnate; leaflets relatively small, usually numerous; stipules present, usually caducous; stipels present or absent; inflorescences axillary, racemose; bracts and bracteoles present, usually caducous; flowers small to large and showy; calyx campanulate, lobed, truncate, or subbilabiate; petals commonly yellow, often with red or purplish markings, sometimes red, or white; vexillum sometimes appendaged on the inner surface near the base; stamens diadelphous 9:1 with vexillar filament free; anthers uniform, dorsifixed; ovary many-ovuled, sessile or stipitate; style usually glabrous, sometimes pubescent near the apex; stigma small, capitate; fruit linear, 2-valved, dehiscent or subdehiscent, compressed or cylindric, sometimes with 4-angles or wings, transversely septate between the seeds; seeds reniform to oblong.

A genus of about 50 species widespread in tropical to warm temperate regions.

KEY TO THE SPECIES

1 Flowers white to red, about (5–)7.5–10 cm long; fruit 7–9 mm wide; trees...............
....**1. S. grandiflora**
1 Flowers yellowish, sometimes with brown or purplish markings on the petals, about 3 cm long or less; fruit 2–6 mm wide; herbs, shrubs, or small trees
  2 Leaflets glabrous or nearly so
    3 Flowers 10–17 mm long; calyx 4–5 mm long; fruit 2–5 mm wide with septa 4–8 mm apart
      4 Appendages at base of vexillum with free tips about (1.5–)2–5 mm long; stems, leaves, and inflorescences not aculeate; seeds olive-green, usually dark-mottled........**2. S. sesban**
        5 Vexillum yellow, often purplish-speckled.............................var. **sesban**
        5 Vexillum entirely purple...var. **bicolor**
      4 Appendages at base of vexillum wedge-shaped, truncate; stems, leaves, and inflorescences sparsely to densely aculeate; seeds brown, not mottled.....................**3. S. bispinosa**
    3 Flowers 20 mm long; calyx 7–8(1–10) mm long; fruit about 5 mm wide with septa about 10 mm apart...**4. S. macrantha** var. **levis**
  2 Leaflets pubescent on one or both surfaces
    6 Flowers 5–10 mm long; leaflets 2–4 mm wide, glabrous above, sericeous or subsericeous below; fruit 3–4 mm wide, glabrous; seeds brown, minutely dark-spotted...........**5. S. sericea**
    6 Flowers 25–30 mm long; leaflets 6–8 mm wide, tomentulose to subsericeous on both surfaces; fruit 5–6 mm wide, villous when young, puberulent to subglabrous at maturity; seeds yellowish-brown, not spotted...**6. S. speciosa**

1. Sesbania grandiflora (L.) Poir. in Lam., Enc. 7: 127. 1806 (as "*Sesban grandiflorus*"); Moon, Cat. 54. 1824; Baker in Hook. f., Fl. Br. Ind. 2: 115. 1876; Trimen, Hanb. Fl. Ceylon 2: 35. 1894; Alston in Trimen, Handb. Fl. Ceylon 6: 76. 1931; Macmillan, Trop. Pl. & Gard. ed. 5, 296, 302. 1962; Ali, Fl. W. Pakistan, no. 100, Papil. 87. 1977; Verdcourt, Man. New Guinea Leg. 360, fig. 83. 1979.

Robinia grandiflora L., Sp. Pl. 722. 1753, Type: India, Herb. Linn. 922.1 (LINN).
Aeschynomene grandiflora (L.) L., Sp. Pl. ed. 2. 1060. 1762.
Coronilla grandiflora (L.) Willd., Sp. Pl. 3: 1145. 1802.
Agati grandiflora (L.) Desv., J. Bot. 2, 1: 120. 1813; Wight & Arn., Prod. 215. 1834.
Agati grandiflora var. *albiflora* Wight & Arn., Prod. 215. 1834.
Agati grandiflora var. *coccinea* Wight & Arn., Prod. 216. 1834, non sensu type of *Aeschynomene coccinea* L. f.

Trees, to about 10 m tall; young stems pubescent, glabrate with age; leaves paripinnate with axis about 12–30(–60) cm long, 30–60-foliolate; leaflets oblong, 2–4 cm long, 5–15 mm wide, glabrous or appressed-pubescent on both surfaces, glaucous, obtuse, emarginate at the apex, rounded at the base; stipules obliquely ovate, 6–10 mm long; stipels minute, caducous; inflorescences 2–4-flowered, pedicels solitary or in pairs; bracts and bracteoles linear-oblong, obtuse, 3–4 mm long, 1 mm wide, caducous; flowers about 5–10 cm long; calyx 2–3 cm long, subtruncate to subbilabiate, glabrous; petals red, pink, or white, (5–)7–8.5(–10) cm long, glabrous; vexillum without basal appendages; stamen-sheath 3.5–6 cm long, curved; fruit 20–60 cm long, straight or somewhat curved, 7–9 mm wide, 5 mm thick, 15–50-seeded with septa 9–11 mm apart, the margins thickened; seeds reddish-brown, about 1 cm apart, 5 mm long, 4 mm wide, 1.5–2 mm thick.

Distr. Possibly native to Indonesia but widely planted as an ornamental in the tropics of Old and New Worlds.

Uses. The leaves and flowers are eaten as a vegetable and used for cattle fodder, Various parts of the plant are sometimes used medicinally. In the Philippines the clear gum is used as a substitute for gum arabic.

Vern. Katuru-murungá (S); akatti, agati-keerai (T).

Specimens Examined. MATALE DISTRICT: Nalanda Exp. Sta., *Worthington 3072* (K). TRINCOMALEE DISTRICT: Trincomalee, *Glenie s.n.* (PDA). KANDY DISTRICT: Royal Botanic Gardens, Feb. 1892, *Trimen ? s.n.* (PDA); Peradeniya, *Fairchild & Dorsett 311* (UC, US), *414* (UC, US); Kandy, Katukelle, Piachaud Gardens, *Rudd & C. Fernando 3320* (PDA, SFV, US), *3321* (PDA, US); Gampola, *Soejarto 4816* (US).

Notes. Both red-flowered and white-flowered trees are found in Ceylon, the white, sudu (S), the red, ratu (S), according to Moon, l.c.

Burbidge (Austral. J. Bot. 13: 115. 1965) recognizes another, similar species, *S. formosa* (F, Muell.) N. Burbidge, in Australia. She separates it from *S. grandiflora* on the basis of a conspicuously 5-lobed calyx, never subtruncate or bilabiate; the conspicuous basal lobe of the keel petals; cohesion of tips of wing petals; and more lanceolate and mucronate leaflets. From the specimens I have examined I can see no significant differences.

Many authors, including Wight and Arnott, have mistakenly identified the red-flowered form of *S. grandiflora* with *S. coccinea* (L.f.) Poir. However, examination of microfiches of the types of the two taxa, as well as additional collections of both, shows them to represent quite different species.

2. Sesbania sesban (L.) Merr., Philipp. J. Sci. 7: 235. 1912; Gillett, Kew Bull. 17: 112. 1963; Gillett, Fl. Trop. East Africa, Papil. 339. 1971; Ali, Fl. W. Pakistan, no. 100, Papil. 89. 1977.

Aeschynomene sesban L., Sp. Pl. 714. 1753. Type: Egypt, Hasselquist, Linn. Herb. 922.12 (LINN).

Coronilla sesban (L.) Willd., Sp. Pl. 3: 1147. 1802.

Sesbania aegyptiaca Poir., in Lam., Enc. 7: 128. 1806, as *"Sesban aegyptiacus"*, based on *Ae. sesban* L.; Pers., Syn. Pl. 2: 316. 1807; Moon, Cat. 53. 1824; Wight & Arn., Prod. 214. 1834; Wight, Ic. Pl. Ind. Or. 32. 1838; Thw., Enum. Pl. Zeyl. 84. 1859; Baker in Hook. f., Fl. Br. Ind. 2: 114. 1876; Trimen, Handb. Fl. Ceylon 2: 34. 1894; Alston in Trimen, Handb. Fl. Ceylon 6: 76. 1931; Macmillan, Trop. Pl. & Gard., ed. 5, 28, 118, 205. 1962.

Trees or shrubs, to about 7 m tall; young stems puberulent, glabrescent with age; leaves with axis 3–14 cm long, 14–56-foliolate; stipules 3–7 mm long, puberulent, caducous; leaflets linear, obtuse, 6–25 mm long, 2.5–3(–6) mm wide, glabrous or subglabrous above, puberulent below; inflorescences axillary, racemose, to about 15 cm long, 3–20-flowered; bracts and bracteoles linear-lanceolate, to about 3 mm long, pilose, caducous; flowers 12–15(–17) mm long; pedicels 5–12 mm long; calyx puberulent at the margin, otherwise glabrous, 5 mm long including deltoid teeth 1 mm long; petals yellow, the vexillum often with purplish speckles or entirely purple; appendages at base of vexillum with free tips (1.5–)2–5 mm long; ovary 30–50-ovulate, glabrous or pilose; style 5 mm long, usually glabrous; fruit glabrous, about 30 cm long, 2–5 mm wide, straight or slightly curved, 20–40-seeded with septa 4–8 mm apart; seeds olive-green, usually dark-mottled, 3–3.5 mm long, 2 mm wide, 1.6 mm thick.

var. **sesban** Gillett, Kew Bull. 17: 100, 102, 112. 1963.

The vexillum is yellow like the other petals or, often, is speckled with purple.

Distr. Probably native to India; introduced elsewhere in the tropics.

Vern. Chittakatti (T).

Specimens Examined. ANURADHAPURA AND TRINCOMALEE DISTRICTS: Habarane, in 1848, *Gardner s.n.*, Trincomalee, in 1869, *Glenie s.n.*, *Thwaites, C.P. 1512* (PDA). KANDY DISTRICT: Kandy, *Walker 141* (K), *s.n.* (K); Peradeniya, Exp. Sta., in 1916, *s. coll. 175* (K); Royal Botanic Gardens, herbarium ground, May 1887, *Trimen ? s.n.* (PDA); students' garden, in 1926, *Alston s.n.* (PDA).

var. **bicolor** (Wight & Arn.) F.W. Andr., Fl. Pl. Sudan 2: 232. 1952.

Sesbania aegyptiaca var. *bicolor* Wight & Arn., Prod. 214. 1834; Gillett, Kew
Bull. 17: 100, 112. 1963. Type: India, *Wight 906.*
Sesbania atropurpurea Taub., Bot. Jahrb. Syst. 23: 188. 1896. Type: Africa,
Sudan Republic, cultivated, *Schweinfurth 799, 796* (B syntypes, K isosyntypes).

The vexillum is entirely purple.

Distr. Probably native to India; introduced elsewhere in the tropics.
Vern. Senehe-kola (S).
Specimens Examined. KURUNEGALA DISTRICT: Bathalagoda,
cultivated in garden, *Waas 223* (SFV, UC, US). MATALE DISTRICT: Am-
bana, off Elahera-Naula Road, cultivated, *Jayasuriya 298* (US). KANDY
DISTRICT: Kadugannawa, Poilakande Estate, in 1927, *Alston s.n.* (PDA); Kandy,
Worthington 3041 (K).

3. Sesbania bispinosa (Jacq.) W.F. Wight, U.S.D.A. Bur. Pl. Industr. Bull.
137: 15. 1909; Gillett, Kew Bull. 17: 129. 1963; Gillett, Fl. Trop. East Africa,
Papil. 349. 1971; Ali, Fl. W. Pakistan, no. 100, Papil. 90. 1977.

Aeschynomene bispinosa Jacq., Ic. Pl. Rar. 3: 13, t. 564. 1793. Type: Specimen
and origin not known of plant cultivated in Vienna before 1788; represented
by Jacquin's tab. 564 (l.c.).
Aeschynomene aculeata Schreb., Nova Acta Phys.-Med. Acad. Caes. Leop.-
Carol. Nat. Cur. 4: 134. 1770. Type: A plant grown in Halle, Germany from
seed obtained in Malabar, India.
Coronilla aculeata Willd., Sp. Pl. 3: 1147. 1802, based on *Ae. bispinosa* Jacq.
Sesbania aculeata (Willd.) Poir. in Lam., Enc. 7: 128. 1806, as "*Sesban
aculeatus*"; Moon, Cat. 53. 1824; Wight & Arn., Prod. 214. 1834; Thw.,
Enum. Pl. Zeyl. 84. 1859; Baker in Hook. f., Fl. Br. Ind. 2: 114. 1876;
Trimen, Handb. Fl. Ceylon 2: 34. 1894; Alston in Trimen, Handb. Fl. Ceylon
6: 76. 1931.

Herbs, sometimes suffrutescent, to about 3 m tall; young stems glabrous or
nearly so, sparsely to rather densely aculeate; leaves with axis usually aculeate,
to about 20–35 cm long, 20–100-foliolate; stipules linear-lanceolate, 6–10 mm
long, adaxially pubescent; leaflets oblong, 5–10(–20) mm long, 2–4 mm wide,
obtuse, mucronulate, glabrous on both surfaces; stipels minute, caducous; inflorescences racemose, about (1–)3–12-flowered; peduncles and pedicels often
aculeate; bracts and bracteoles linear, caducous; flowers 10–12(–13) mm long;
pedicels 6–11 mm long; calyx about 4–5 mm long, glabrous except puberulent
along the margin and inside teeth; petals yellow with brownish markings; vexillum with wedge-shaped, truncate basal appendages within; fruit glabrous,
somewhat curved, about 15–25 cm long, 2–3 mm wide, with a beak about 1 cm

long, 35–40-seeded, with septa about 5 mm apart; seeds brown, about 3 mm long, 1.5 mm wide, 1.1 mm thick.

Distr. A weed, widespread in the Old World tropics and in Jamaica, in wet areas, cultivated land, especially in rice fields.

Specimens Examined. ANURADHAPURA DISTRICT: Dambulla-Anuradhapura Road, near culvert 47/6, *Cooray 70013108R* (US); between Anuradhapura and Galkulama, *Rudd 3305* (K, PDA, SFV, US); Borupangala, Hinukkiriyawa, W. of Habarane, *Jayasuriya et al. 1388* (K, US); Wilpattu Natl. Park, near Eerige Ara confluence with Moderagama Ara, *Fosberg et al. 50778* (US). KURUNEGALA DISTRICT: Kurunegala, "Koh"? (illegible), *Gardner,* "Hayurabutte" (?), July 1851, *Thwaites C.P. 1513* (BM, K, PDA). TRIN-COMALEE DISTRICT: China Bay, *Worthington 1244* (K); Mahaweli Ferry, Mahaweli Ganga, about 20 km from mouth, *Fosberg 56398* (SFV, UC, US). BATTICALOA DISTRICT: N. of Kalkudah on road to Elephant Point, *Rudd & Balakrishnan 3148* (K, PDA, SFV, US). AMPARAI DISTRICT: Gal Oya reservoir, near spillway, *Comanor 565* (K, PDA, SFV, US); Hot Springs area, 1 mi. N. of Maha Oya, *Davidse & Sumithraarachchi 9001* (US). MONERAGALA and AMPARAI DISTRICTS: Between Siyambalanduwa and Inginiyagala, *Rudd & Balakrishnan 3214* (GH, K, PDA, SFV, UC, US). HAMBANTOTA DISTRICT: Wirawila, near lake, *Rudd 3092* (PDA, US); Tissamaharama, *Simpson 9917* (BM); Ruhuna Natl. Park, Patanagala, *Mueller-Dombois & Cooray 68012828* (PDA, US), *Fosberg 50346* (US); Buttawa, on bank of Ara, *Cooray 69121301R* (GH, SFV, UC, US).

4. Sesbania macrantha Welw. ex Phil. & Hutch. var. **levis** Gillett, Kew Bull. 17: 118. 1963; Gillett, Fl. Trop. East Africa, Papil. 341. 1971. Type: Zimbabwe (Rhodesia), Salisbury and Marandellas Districts, *Eyles 1514* (BM, holotype; PRE, isotype).

Sesbania cinerascens sensu Baker in Oliver, Fl. Trop. Africa 2: 134. 1871, pro parte, non quoad typum.

Small tree, 3–4 m tall; young stems glabrous or nearly so, finely striate; stipules lanceolate, 5 mm long; leaves with axis to 25 cm long, glabrous, 80–86-foliolate; leaflets oblong, 5–22 mm long, 3–6 mm wide, obtuse at apex and base, mucronulate, glabrous on both surfaces; inflorescences racemose, axillary, 10–12-flowered, the axis unarmed or sparsely and minutely subaculeate, glabrous; bracts lanceolate, 3–4 mm long, caducous; bracteoles linear, 5 mm long, caducous; flowers about 2 cm long on pedicels 1–1.5 cm long; calyx glabrous outside, puberulent within, 7–8(–10) mm long, teeth 2–3 mm long; petals yellow; vexillum dark-speckled on back, the appendages without free tips; wings and keel with long, hook-like auricles at base of blades; fruit glabrous, 15–19 cm long, about 5 mm wide, subsessile, about 13–16-seeded, lightly margined, septa about

10 mm apart; seeds reddish-brown, oblong, 7 mm long, 3.5 mm wide, the hilum white, lateral.

Distr. Tropical and South Africa; introduced in Ceylon.

Specimens Examined. BADULLA DISTRICT: Along A5 from Passara to Bibile, culvert 101/7, planted in tea, *Maxwell 1023* (SFV, US); between Welimada and Hakgala, *Wirawan 746* (GK, K, PDA, US). There also are two specimens at PDA as *S. cinerascens,* collected in the "Exp. Sta.", one by *G.A. Ramanayake,* in 1952, the other without collector's name, in 1949. These, possibly, are the same as the *Maxwell* collection, but should be rechecked.

Note. Gillett, l.c. 1971, mentions that *S. macrantha* is occasionally grown as a shade plant for coffee.

5. Sesbania sericea (Willd.) Link, Enum. Hort. Berol. 2: 244. 1822; Alston in Trimen, Handb. Fl. Ceylon 6: 76. 1931; Gillett, Kew Bull. 17: 133. 1963; Backer & Bakh. f., Fl. Java 1: 597. 1963; Gillett, Fl. Trop. East Africa, Papil. 350. 1971.

Coronilla sericea Willd., Enum. Hort. Berol. 773. 1809. Type: a plant cultivated at Berlin from seeds supplied by Thonning from Ghana and apparently not preserved; Neotype: Ceylon, near Colombo, *Ferguson in Thwaites C.P. 3850* (K); isoneotypes (BM, P, PDA).
Sesbania pubescens DC., Prod. 2: 265. 1825. Type: Ghana, *Thonning* (G).
Agati sericea (Willd.) Hitchc., Annual Rep. Missouri Bot. Gard. 4: 75. 1893.

Suffrutescent herb, to about 3 m tall; stems sericeous, glabrescent, striate; leaves with axis 4–15 cm long, 20–50-foliolate; stipules linear-subulate, about 5 mm long, caducous; leaflets oblong, (8–)10–25 mm long, 2–4 mm wide, rounded at apex and base, mucronulate, sometimes slightly emarginate, glabrous above, sericeous or subsericeous beneath; inflorescences racemose, about 2–6 cm long, (1–)2–6-flowered; bracts and bracteoles linear-lanceolate, acuminate; flowers 5–10 mm long on pedicels to 0.5 mm long; calyx about 5 mm long, puberulent at base and margin, otherwise glabrous, the teeth deltoid, 1 mm long or less; petals pale yellow, sometimes flecked with purple; vexillum with narrow, wedge-shaped, truncate appendages at base within; fruit glabrous, sessile, 10–16 cm long, 3–4 mm wide, straight or slightly curved, 15–30-seeded with septa 5 mm apart; seeds brown, minutely darker-spotted, 2.5–3.5 mm long, 2 mm wide.

Distr. Ceylon, Africa, southern United States, Belize, West Indies, northern South America, in wet ground; place of origin uncertain, possibly Ceylon.

Specimens Examined. TRINCOMALEE DISTRICT: Trincomalee, in 1889, *Glenie s.n., Thwaites, C.P. 1512* in part (PDA).

6. Sesbania speciosa Taub., Pflanzenw. Ost.-Afr. C. 213. 1895; Gillett, Kew Bull. 17: 120. 1963; Gillett, Fl. Trop. East Africa, Papil. 342. 1971; Verdcourt, Man. New Guinea Leg. 362. 1979. Syntypes: Tanganyika, E. Usambara Mts.,

Mashewa, *Holst 3508* (B destroyed; isosyntype K); Bagamoyo District, Kingoni R., *Hildebrandt 960* (B destroyed; isosyntype BM); Uzaramo, *Stuhlmann* (B destroyed).

Sesbania pubescens DC. var. *grandiflora* Vatke, Oesterr. Bot. Z. 28: 215. 1878.
   Type: Tanganyika, Bagamoyo District, Kingoni R., *Hildebrandt 960* (B destroyed; isotype BM).

Suffrutescent herb, to 2–3 m tall; stems unarmed, puberulent to tomentulose; leaves with axes 18–32 cm long, subtomentulose, about 50-foliolate; stipules lanceolate-ovate, asymmetrical, falcate, about 1.5 cm long, 5 mm wide, attenuate, tomentulose; leaflets oblong, 2–3 cm long, 6–8 mm wide, obtuse, mucronate, sometimes emarginate at the apex, acute to rounded at the base, tomentulose to subsericeous on both surfaces; inflorescences racemose, axillary, 8–20-flowered; bracts and bracteoles caducous; flowers 2.5–3 cm long on pedicels 1–2 cm long; calyx glabrous, 1 cm long with teeth 2–3 mm long; petals bright yellow; vexillum with purplish-brown spots and basal appendages within with a short free tip 0.5–0.7 mm long; ovary densely pubescent; style puberulent, decreasingly toward the apex; fruit 23 cm long, 5–6 mm wide, somewhat villous when young, puberulent to subglabrous when mature, septa about 5 mm apart, 40–60-seeded; seeds yellowish-brown, 3–4 mm long, 2.5–3 mm wide, 1.5 mm thick.

Distr. A native of East Africa but introduced elsewhere as a green manure.
   Specimens Examined. POLONNARUWA DISTRICT: Between Habarane and Kantalai near mile marker 123, *Rudd & Balakrishnan 3123* (K, PDA, SFV, UC, US). KANDY DISTRICT: Peradeniya, Exp. Sta., in 1952, *Ramanayake s.n.* (PDA).

2. GLIRICIDIA

H.B.K., Nov. Gen. & Sp. fol. ed. 6: 309. 1824, quarto ed. 6: 393. 1824; Allen & Allen, Legum. 300. 1981. Type species: *G. maculata* H.B.K., a synonym of *G. sepium* (Jacq.) Kunth. ex Walpers.

Hybosema Harms, Feddes Repert. Spec. Nov. Regni Veg. 19: 66. 1923. Type species: *H. ehrenbergii* (Schlecht) Harms, based on *Robinia ehrenbergii* Schlecht.

Trees or shrubs, unarmed; leaves alternate, imparipinnate; stipules small; stipels lacking; flowers showy, in axillary racemes, often appearing before the leaves; bracts small, caducous; bracteoles absent; calyx campanulate with 5 teeth or subtruncate; petals pinkish to purple or, sometimes, white; stamens diadelphous 9:1 with the vexillar filament free; anthers uniform; ovary stipitate, 7–12-ovulate; style glabrous, inflexed; stigma terminal, capitate, usually papillose; fruit linear to linear-oblong, short-stipitate, compressed, dehiscent with the valves twisting spirally; seeds suborbicular, compressed.

A genus of 4 or 5 species native to tropical America, sometimes introduced into the Old World tropics.

Gliricidia sepium (Jacq.) Walp., Repert. 1: 679. 1842; Worthington, Ceylon Trees 176. 1959; Verdcourt, Man. New Guinea Leg. 348, fig. 78. 1979.

Robinia sepium Jacq., Enum. Pl. Carib. 28. 1760; Jacq., Select. 211, t. 179, fig. 101. 1763. Type: Cartagena, Colombia, *Jacquin s.n.*

Robinia hispida L., Mant. 101. 1767, in part as to Jacquin citation.

Robinia maculata H.B.K., Nov. Gen. Sp. Quarto ed. 393. 1824. Type: Mexico, *Humboldt & Bonpland* (P).

Lonchocarpus maculatus (H.B.K.) DC., Prod. 2: 260. 1825.

Lonchocarpus sepium (Jacq.) DC., Prod. 2: 260. 1825.

Robinia variegata Schlecht., Linnaea 12: 301. 1838. Type: *Schiede s.n.*, Actopan, Mexico.

Gliricidia maculata (H.B.K.) Walp., Repert. 1: 679. 1842; Macmillan, Trop. Pl. & Gard. ed. 5, 39, 85, 211, 487. 1962.

Gliricidia maculata var. *multijuga* Micheli in Donn. Sm., Bot. Gaz. 20: 284. 1895. Type: Santa Rosa, Guatemala, *Heyde & Lux 3296* (US).

Gliricidia lambii Fernald, Bot. Gaz. 20: 533. 1895. Type: Rosario, Sinaloa, Mexico, *Lamb 451* (GH).

Gliricidia sepium f. *maculata* (H.B.K.) Urban, Symb. Ant. 289. 1900.

Trees, to about 10(–15) m tall; young stems puberulent or subsericeous, glabrescent with age; bark grey to brownish with white lenticels; stipules about 2 mm long, lanceolate to ovate, pubescent; leaves imparipinnate, (5–)7–17-foliolate, deciduous; leaflets ovate to elliptic or lanceolate, about (1.5–)3–7(–8.5) cm long, (1–)2–3(–5) cm wide, acute at the apex, rounded or acute at the base, sericeous or subsericeous on both surfaces, glabrescent, with purplish blotches usually conspicuous when dry; inflorescences axillary, racemose, many-flowered, especially conspicuous when tree is leafless; bracts ovate, about 1 mm long; flowers 1.5–2 cm long; calyx subtruncate, glabrous or puberulent, 4–6 mm long; petals glabrous, 1.5–2 cm long, commonly pinkish or pink and white, sometimes lilac-pink or, less commonly, completely white; fruit dehiscent, 10–15 cm long, 1–1.5(–2) cm wide, short-stipitate, glabrous, compressed; seeds brown, lenticular, compressed, about 1 cm long, 9 mm wide.

Distr. Native to tropical America but widely introduced in the Old World tropics.

Uses. Commonly planted as shade for cacao, tea, and coffee. Because of quick rooting it is often planted as living fences. The leaves are poisonous to some animals, the roots especially so to rodents. According to Standley (Fl. Guatemala, Fieldiana, Bot. 24: 266. 1946) fresh crushed leaves are applied as poultices to sores caused by ulcers or gangrene, and fresh leaves are placed in hens' nests to remove parasites.

Vern. Kona (S and T); madera (fide Macmillan); madre de cacao, mata ratón, and many other local names (Latin America).

Specimens Examined. PUTTALAM DISTRICT: Wilpattu, along borders of estate beside Puttalam-Eluvamkulam Rd., *Cramer 4078* (US). KANDY DISTRICT: Botanical Gardens, Peradeniya, "*J.M.S. s.n.* 8-5-18" (PDA); Kandy, Hillcrest, *Worthington 1670* (K), *6970* (K), *7219* (K). COLOMBO DISTRICT: Colombo, School of Agriculture, *Drieberg s.n.* (PDA). MONERAGALA DISTRICT: Between Wellawaya and Moneragala, *Rudd & Balakrishnan 3213* (PDA, US). HAMBANTOTA DISTRICT: Ruhuna Natl. Park, Hambantota-Tissa Rd., *Comanor 838* (GH, US); road from Tissamaharama to Kataragama, *Mueller-Dombois & Cooray 68013112* (PDA, US).

Note. *Gliricidia sepium* and *G. maculata* have frequently been cited as combined by Steudel, Nom. Bot. ed. 2, 1: 688. 1840 (or 1841) but these combinations are now considered to be invalid because they were treated as synonyms under *Lonchocarpus*.

Tribe TEPHROSIEAE

(Benth.) Hutchins., Gen. Fl. Pl. 1: 394. 1964; Geesink in Polhill & Raven, Adv. Leg. Syst. 245. 1981. Type genus: *Tephrosia* Pers.

Galegeae subtribe Tephrosiinae Benth. In Benth. & Hook., Gen. Pl. 1: 444. 1865, as "*Tephrosieae*".
Millettieae Miq., Fl. Ind. Bat. 1: 137. 1855. Type genus: *Millettia* Wight. & Arn.

Trees, shrubs, or lianas; leaves usually imparipinnate, many-foliolate, rarely 1-foliolate, or digitately 3–7-foliolate, or pinnately 3-foliolate; stipules present, often caducous; stipels present or absent; leaflets usually opposite; inflorescences racemose, paniculate, terminal or leaf-opposed, often with the flowers in fascicles, in pseudoracemes, or pseudopanicles; bracts present; bracteoles present or absent; flowers small to medium-sized; calyx usually campanulate, truncate, or 4–5-lobed, or toothed; stamens 10 with the filaments united or the vexillar filament free; anthers uniform; ovary 2–many-ovulate; style glabrous or bearded; stigma capitate; fruit dehiscent or indehiscent; seeds oblong, orbicular, or reniform, usually compressed.

The tribe comprises about 50 genera, mostly tropical.

Included by Geesink (l.c.) in this tribe are the genera *Aganope* Miq., *Derris* Lour., and *Pongamia* Vent. which already have been included in my treatment of the tribe Dalbergieae.

KEY TO THE GENERA

1 Leaflets with numerous, straight, closely-parallel, lateral veins mostly extending to the margin; fruit elastically dehiscent with valves coiling; herbs or shrubs..............**1. Tephrosia**

1 Leaflets with relatively fewer lateral veins usually anastomosing, not extending to the margin; fruit indehiscent or tardily dehiscent; trees or shrubs, sometimes scandent
  2 Stamens with alternate filaments slightly dilated at the apex; stipels lacking; bracteoles absent; small trees or shrubs...**2. Mundulea**
  2 Stamens with filaments filiform, not dilated; stipels usually present; bracteoles present but caducous; trees or shrubs, sometimes scandent......................................**3. Millettia**

1. TEPHROSIA

Pers., Syn. Pl. 2: 328. 1807, nom. cons.; Wood, Contr. Gray Herb. 170: 193–384. 1949; Ali, Biologia (Lahore) 10: 23–37. 1964; Brummitt, Bol. Soc. Brot. 41, ser. 2: 219–393. 1967; White, Fl. Panama, Ann. Missouri Bot. Gard. 67: 777. 1980; Allen & Allen, Legum. 645. 1981; Bosman & de Haas, Blumea 28: 421–487. 1983. Type species: *T. villosa* (L.) Pers., non (Michx.) Pers., based on *Cracca villosa* L.

Cracca L., Sp. Pl. 752. 1753, non Medik. 1789, nec Benth. 1853.

Additional synonyms are given by Wood (l.c. p. 233), Ali (l.c. pp. 23, 24), White (l.c.), and Bosman & de Haas (l.c. p. 436).

Perennial herbs or shrubs; stems erect, decumbent, or prostrate; leaves commonly imparipinnate, 3–many-foliolate, sometimes digitately 3–7-foliolate, or sometimes simple or unifoliolate; stipules present, usually herbaceous, rarely spinescent; stipels lacking; leaflets with numerous, straight, closely parallel lateral veins mostly extending to the margin; inflorescences racemose or pseudoracemose, axillary, leaf-opposed or terminal, few–many-flowered; bracts persistent or caducous; bracteoles usually lacking; flowers small to medium-sized: calyx campanulate with 5 subequal teeth or lobes; petals white to purple or reddish with the vexillum pubescent on the outer face; stamens with the vexillar filament free at the base only or completely free from the others; anthers uniform; ovary sessile, usually 4–16-ovulate; style curved, glabrous or bearded; stigma terminal, minutely capitate or penicillate; fruit sessile, linear or linear-oblong, straight to falcate, compressed, dehiscent with the valves coiling, 1–many-seeded; seeds oblong-reniform to elliptic or quadrate.

A genus of about 300–400 species in warm-temperate and tropical areas.

KEY TO THE SPECIES

1 Flowers with bearded style
  2 Leaflets unequal with the terminal leaflet usually considerably longer than the laterals; leaves 1–13-foliolate..**1. T. tinctoria**
  2 Leaflets essentially equal in size; leaves (5–)7–29-foliolate
    3 Fruit lightly pubescent to subglabrous
      4 Flowers about 8–10 mm long; fruit 5 mm wide; leaves (5–)7–11-foliolate..............
      ....**2. T. senticosa**

4 Flowers about 12–15 mm long; fruit 3.5–5 mm wide; leaves 9–17-foliolate. . . . .
. . . .**3. T. maxima**
3 Fruit densely pubescent
   5 Flowers 17–30 mm long; petals white; calyx lobes glabrous or subglabrous at the tip; fruit
      6–10 mm long, 8–9 mm wide; leaves 13–27-foliolate. . . . .**4. T. candida**
   5 Flowers 25–35 mm long; petals white to violet; calyx lobes pubescent; fruit 10–14 cm long,
      12–16 mm wide; leaves 11–29-foliolate. . . . .**5. T. vogelii**
1 Flowers with glabrous style
6 Stipules spinous. . . . .**6. T. spinosa**
6 Stipules not spinous
   7 Fruit moderately pubescent to subglabrous
      8 Stems erect; inflorescences several- to many-flowered, about 5–15 cm long; flowers purplish
         or pink; fruit appressed-pubescent, glabrescent; flowers 7–10 mm long. . . . .
         . . . .**7. T. purpurea**
      8 Stems prostrate; inflorescences few-flowered, about 2 cm long or less; flowers white to salmon-
         pink, 6–8 mm long; fruit crisp-pubescent. . . . .**8. T. pumila**
   7 Fruit densely pubescent
      9 Flowers 8–10 mm long, white to bright pink or purplish; calyx lobes long, filiform; fruit 3–3.5
         cm long, 4–7 mm wide, oblong-falcate. . . . .**9. T. villosa**
      9 Flowers 10–13 mm long, white to pale lilac; calyx lobes short, deltoid; fruit 4.5–5.5 cm long,
         4–5 cm wide, essentially straight. . . . .**10. T. noctiflora**

Specimens cited as type material of *Cracca maxima* L. and *C. purpurea* L.
at the Biblio. Inst. France, Paris are in the Bibliothéque de l'Institute de France,
Paris, in the herbarium of Paul Hermann, basis of the Thesaurus Zeylanicus of
J. Burman (see A. Lourteig in Taxon 15: 23–33. 1966).

1. Tephrosia tinctoria (L.) Pers., Syn. Pl. 2: 329. 1807; Wight & Arn., Prod.
211. 1834; Wight, Ic. Pl. Ind. Or. 388. 1840–43; Thw., Enum. Pl. Zeyl. 84.
1859; Baker in Hook. f., Fl. Br. Ind. 2: 111. 1876; Trimen, Handb. Fl. Ceylon
2: 31. 1894; Alston in Trimen, Handb. Fl. Ceylon 6: 74. 1931.

Cracca tinctoria L., Sp. Pl. 752. 1753. Lectotype: Ceylon, *P. Hermann*, Herm.
   Herb. 3: 28 (BM); syntype: Herm. Herb. 4: 38 (BM).
Galega tinctoria (L.) L., Syst. Nat. ed. 10. 1172. 1759.
Galega heyneana Roxb., Fl. Ind. 3: 384. 1832. Type: "Reared in the Botanic
   Garden [Calcutta] from seed sent by Mr. B. Heyne, from Mysore, where the
   plant is indigenous."
Tephrosia heyneana (Roxb.) Wall. ex Wight & Arn., Prod. 211. 1834, in
   synonymy.
Tephrosia tinctoria (L.) Pers. var. *intermedia* Grah. ex Wight & Arn., Prod.
   211. 1834; Baker in Hook. f., Fl. Br. Ind. 2: 112. 1876. Type: *Wall. Cat. 5632.*
Tephrosia tinctoria (L.) Pers. var. *pulcherrima* Wight ex Baker in Hook. f., Fl.
   Br. Ind. 2: 112. 1876; Trimen, Handb. Fl. Ceylon 2: 31. 1894. Type: *Wight
   672*, India, Courtallum, Sept. 1835 and Feb. 1836 (K).
Tephrosia pulcherrima (Wight ex Baker) Gamble, Fl. Pres. Madras 2: 316, 319.
   1918; Alston in Trimen, Handb. Fl. Ceylon 6: 74. 1931.

Shrubby perennial, to about 1 m high; young stems velutinous to sericeous, glabrescent with age; stipules deltoid to ovate, subulate, 3–6 mm long, 1–1.5 mm wide at the base, striate, pubescent to subglabrous; leaves imparipinnate, 1–13-foliolate; leaflets subcoriaceous, usually variable in size on the same leaf with the basal leaflets smallest and terminal leaf largest, elliptic to oblong, obtuse, 0.5–5(–9.5) cm long, 0.3–1.5(–2.7) cm wide, glabrous, nitid above, sericeous or subsericeous beneath, sometimes rufous-pubescent along the midvein; inflorescences racemose, axillary, usually longer than the leaves, 3–12-flowered; bracts lanceolate to ovate-lanceolate, attenuate, pubescent, about 3 mm long; flowers 7–10 mm long, subsessile on pedicels about 3 mm long or less; calyx pubescent, 4–5 mm long with subulate teeth; petals bright pink to reddish-orange; style bearded; fruit linear-oblong, apiculate, slightly curved upward at the apex, velutinous to subsericeous, 4–7 cm long, 0.5 cm wide, 7–12-seeded; seeds olive green to dark brown, about 3 mm long, 2 mm wide.

Distr. Ceylon and southern India.

Uses. Trimen (l.c.) mentions that an "inferior Indigo is obtained from this and is or was in use by the Singalese".

Vern. Alu-pila (S).

Specimens Examined. WITHOUT EXACT LOCALITY: *Macrae 152* (BM), *295* (BM, K), *s.n.* (K); *Walker 1078* (P), *s.n.* (GH, K, P); "Central Prov.", *Thwaites C.P. 1449* (PDA); without data, *Thwaites C.P. 1449* (BM, GH, K, P, US); *s. coll.* "donné par M. Bonpland en 1833", "Willdenow dedit" (P). MATALE DISTRICT: Matale, *Gardner s.n., Thwaites C.P. 1449* (PDA). KANDY DISTRICT: Peradeniya Junction, *F.W. de Silva s.n.* in 1928 (PDA); Peradeniya, *Gardner 213* (BM, K), *Mueller-Dombois 67110911* (PDA); Kandy, *herb. de Poli 57* (P), *Thomson s.n.* (K), *Macrae 295* (K), *Worthington 6923* (K); Nilambe, Galaha, *Worthington 2484* (K). NUWARA ELIYA DISTRICT: Boralanda-Nuwara Eliya Road, *Sohmer & Sumithraarachchi 10070* (GH). AMPARAI DISTRICT: Between Amparai and Maha Oya, mi. markers 19–20, *Rudd & Balakrishnan 3234* (K, SFV, US). COLOMBO DISTRICT: Dewalapola, *Amaratunga 1442* (PDA); Elston Estate, Puwakpitiya, *Simpson (Foote) 8971* (BM, PDA). BADULLA DISTRICT: Between Welimada and Badulla, mi. marker 69, *Rudd & Balakrishnan 3192* (BM, GH, PDA, SFV, UC, US); Thotulagalla Ridge, Haputale, *Fosberg 51843* (US); 2 mi. S. of Welimada at road to Boralanda near mi. marker 2/2, *Mueller-Dombois 67112316* (PDA, US). MONERAGALA/ AMPARAI DISTRICTS: Between Siyambalanduwa and Inginiyagala, *Rudd & Balakrishnan 3224* (SFV, US).

Notes. As indicated in the synonymy above, Gamble elevated var. *pulcherrima* to specific status, characterized by leaves with few leaflets, sometimes only one, with the terminal leaflets much longer than the laterals. Alston (l.c.) referred the *Gardner* collections from Matale to this taxon. The *Walker* collections are examples of *T. tinctoria* var. *intermedia,* with more numerous leaflets, often

9–13, nearly equal in size, with looser pubescence. The remaining collections from Ceylon run the gamut, even within a given population. It seems preferable to treat *I. tinctoria* as one variable species as did Wight & Arnott, Thwaites, Trimen, and others.

Wight & Arnott (l.c.) cited varieties but referred to *T. tinctoria* as "a most beautiful but very variable species: our specimens exhibit all the different gradations from simple leaves to 6 pairs of leaflets; indeed one individual before us has one leaf simple, while all the others are pinnated." Thwaites (l.c.) commented, "Although the extremes of this species vary so exceedingly in habit, degree of pubescence, and size of leaflets, stipules, bracts, and calyx, it seems impossible, as Wight and Arnott remark, to separate them."

2. Tephrosia senticosa (L.) Pers., Syn. Pl. 2: 330. 1807; Baker in Hook. f., Fl. Br. Ind. 2: 112. 1876; Trimen, Handb. Fl. Ceylon 2: 30. 1894; Gamble, Fl. Pres. Madras 2: 317, 319. 1918; Alston in Trimen, Handb. Fl. Ceylon 6: 74. 1931; Bosman & de Haas, Blumea 28: 470. 1983.

Cracca senticosa L., Sp. Pl. 752. 1753. Lectotype: Ceylon, *P. Hermann*, Herb. Herm. 1: 72 (BM). Syntypes or isotypes, Herb. Herm. 2: 2 (BM); Herb. Herm. 142 (L).

Galega senticosa (L.) L., Amoen. Acad. 3: 19. 1756; L., Sp. Pl. ed. 2, 1063. 1762; Moon, Cat. 55. 1824.

Shrubby perennial, to about 1 m high; young stems densely appressed-pubescent; stipules deltoid, attenuate, pubescent to subglabrous, 4–5 mm long, 1 mm wide at the base; leaves imparipinnate, (3–5–)7–11-foliolate; leaflets essentially equal in size, 1–2.5 cm long, 5–12 mm wide, elliptic to oblong or obovate-oblong, obtuse or slightly emarginate at the apex, rounded to cuneate at the base, subcoriaceous, glabrous, nitid above, sericeous beneath; inflorescences pseudoracemose, axillary, or leaf-opposed, about as long as the leaves, with few flowers toward the apex of the peduncle; bracts deltoid, lanceolate, attenuate, pubescent; flowers about 8–10 mm long on short pedicels; calyx pubescent, 3.5–4 mm long with attenuate lobes; petals reddish-orange with vexillum sericeous on the outer face; style bearded; fruit linear-oblong, essentially straight, 4–6.5 cm long, 5–6 mm wide, sparsely pubescent with appressed or subappressed hairs, about 6–9-seeded.

Distr. Ceylon and southern India.

Specimens Examined. KURUNEGALA DISTRICT: "Hiyare", *J.M. Silva 222* (PDA); Dodangaslanda, *Amaratunga 1172* (PDA). MATALE DISTRICT: Madakumbura, past Rattota, *Jayasuriya et al. 996* (K, US). ANURADHAPURA DISTRICT: "Pangwella", in 1865, *Thwaites C.P. 1449* in part (PDA); without data (K). COLOMBO DISTRICT: Jaela, in 1913, *s. coll. s.n.* (PDA); Mirigama, *Amaratunga 1252* (PDA). KEGALLE DISTRICT: Arandara Estate, Kegalle, *Alston 1721* (K, PDA).

Note. This appears to be rather closely related to *T. tinctoria*.

3. Tephrosia maxima (L.) Pers., Syn. Pl. 2: 329. 1807; Wight & Arn., Prod. 213. 1834; Thw., Enum. Pl. Zeyl. 84. 1859; Trimen, Handb. Fl. Ceylon 2: 32. 1894; Pl. 27. 1894; Gamble, Fl. Pres. Madras 1: 316, 319. 1918; Macmillan, Trop. Pl. & Gard., ed. 5, 136. 1962.

Cracca maxima L., Sp. Pl. 752. 1753. Lectotype: Ceylon, *P. Hermann*, Herb. Herm. 3: 56 (BM). Syntype: Herb. Herm. 1: 31 (BM). Probably type material, Biblio. Inst. France, Paris, coll. Burman, Pl. Zeyl. p. 75, in part.
Galega maxima (L.) L., Syst. Nat. ed. 10, 1172. 1759; Moon, Cat. 55. 1824.
Tephrosia purpurea L. var. *maxima* (L.) Baker in Hook. f., Fl. Br. Ind. 2: 113. 1876.

Suffrutescent perennial, to about 1 m tall; stems erect to prostrate, pubescent or glabrous; stipules deltoid, attenuate, to about 5–7 mm long; leaves imparipinnate, 9–19-foliolate; leaflets oblong to obovate-oblong, approximately equal in size, 1–2 cm long, 3–7 mm wide, obtuse to acute or emarginate at the apex, acute to cuneate at the base, glabrous above, finely appressed-pubescent to glabrous beneath; inflorescences racemose, axillary or pseudoterminal, usually longer than the leaves, few-flowered; bracts ovate to lanceolate, acuminate, 3–5 mm long, caducous; flowers 12–15 mm long on pedicels 5–10 mm long; calyx subsericeous, 4–5 mm long with attenuate lobes; petals pinkish to violet; style lightly bearded; fruit linear, slightly curved upward toward the apex, 4.5–7 cm long, 3.5–4.5 mm wide, sparsely pubescent with short, appressed hairs, 10–14-seeded; seeds quadrate to reniform, olive to dark brown, slightly mottled, sometimes with some endocarp tissue adhering, about 3 mm long, 2.5 mm in diameter.

Distr. Ceylon and southern India.
Specimens Examined. WITHOUT EXACT LOCALITY: *Walker s.n.* (K). JAFFNA AND POLONNARUWA DISTRICTS: Jaffna, in 1846, *Gardner s.n.*, Minneriya (as "Minery"), *Thwaites C.P. 1444* (PDA); without data, *C.P. 1444* (BM, GH, K, P). ANURADHAPURA DISTRICT: Dambulla-Anuradhapura Rd., near mi. marker 52/5, *Sohmer 8079* (US); between Madawachchiya and Vavuniya, *Rudd 3263* (GH, K, PDA, SFV, UC, US). POLONNARUWA DISTRICT: 1 mi. N. of Dimbulagala, *Jayasuriya et al. 719* (K, US); Giritale, near shore of Giritale Wewa, *Townsend 73/222* (K, US). VAVUNIYA DISTRICT: S. of Mullaittivu, *Fosberg 53516* (SFV, US). PUTTALAM DISTRICT: Puttalam Lagoon Isthmus, *Maxwell & Jayasuriya 826* (SFV, US); Paramakanda, *Faden & Faden 77/71* (US). MATALE DISTRICT: 6.5 mi. N. of Dambulla, *Fosberg 53418* (SFV, US). TRINCOMALEE DISTRICT: Kantalai, Aug. 1885, *Trimen ?* (PDA). BATTICALOA DISTRICT: Gunner's Quoin, *Tirvengadum et al. 242* (US); Mannampitiya, at base of ascent to Gunner's Quoin, *Cramer et al. 3995* (K, US). HAMBANTOTA DISTRICT: Ruhuna Natl. Park, just E. of Elephant Rock, *Fosberg & Sachet 52873* (US).

Note. Superficially, *T. maxima* does suggest that it might be a variety of *T. purpurea* with larger flowers and longer pods, as reduced by Baker. However, the pubescent style indicates other relationships as well, and it seems to be best treated as a separate species.

4. Tephrosia candida Roxb. ex DC., Prod. 2: 249. 1825; Wight & Arn., Prod. 210. 1834; Baker in Hook. f., Fl., Br. Ind. 2: 111. 1876; Wood, Contr. Gray Herb. 170: 374. 1949; Macmillan, Trop. Pl. & Gard., ed. 5, 28, 30, 487. 1962; Backer & Bakh. f., Fl. Java 1: 595. 1963; Verdcourt, Man. New Guinea Leg. 340. 1979; Bosman & de Haas, Blumea 28: 445. 1983. Type: India, Bengal, reared in Calcutta from seed collected by Carey in the north of Bengal.

Robinia candida Roxb., Hort. Beng. 56. 1814, nom. nud.; Roxb., Fl. Ind. 3: 327. 1832.

Shrub, to about 3.5 m tall; stems velutinous; stipules linear-subulate or filiform, pubescent, about 5 mm long; leaves imparipinnate, 13–27-foliolate; leaflets oblong to lanceolate, 2–7.5(–9) cm long, 5–13 mm wide, acute or obtuse at the apex, rounded at the base, glabrous above, subsericeous beneath; inflorescences ax-illary, leaf-opposed, or terminal, pseudoracemose, with flowers in groups of 2–6 along the rachis; bracts linear-lanceolate, to about 5–6 mm long; flowers 1.7–3 cm long on pedicels about 1–1.5 cm long; calyx 5–7 mm long with subglabrous, rounded to acute lobes, otherwise sericeous; petals white; style bearded; fruit linear-oblong, 6–10 cm long, 7–10 mm wide, 5–13-seeded, densely pubescent with subappressed hairs; seeds brown, reniform, about 4 mm long, 3 mm wide.

Distr. A native of Southeast Asia, introduced in Ceylon and elsewhere in the tropics.

Uses. Cultivated as a fish-poison plant, as tea shade, and as a green-manure cover-crop.

Vern. Boga-medeloa.

Specimens Examined. WITHOUT EXACT LOCALITY: *Dorsett & Fair-child 366* (UC). AMPARAI DISTRICT: Gal Oya Reservoir, *Comanor 580* (BM), *581* (K). MATALE DISTRICT: Matale, Kent Estate, escape from cultivation, *Worthington 4812* (BM, K); Dombawela, *Sumithraarachchi 725* (K). MONERAGALA/AMPARAI DISTRICTS: between Siyambalanduwa and In-giniyagala, *Rudd & Balakrishnan 3220* (K, PDA, SFV, US).

5. Tephrosia vogelii Hook. f. in Hook., Niger Fl. 296. 1849; Baker in Oliver, Fl. Trop. Africa 2: 210. 1871; Wood, Contr. Gray Herb. 170: 376. 1949; Mac-millan, Trop. Pl. & Gard., ed. 5. 460. 1962; Backer & Bakh. f., Fl. Java 1: 595. 1963; Gillett, Fl. Trop. East Africa, Papil. 210. 1971; Verdcourt, Man. New Guinea Leg. 346. 1979; Bosman & de Haas, Blumea 28: 478. 1983. Syn-types: Nigeria, on the Niger [Quorra] and Fernando Po, *Vogel* (K).

Cracca vogelii (Hook. f.) Kuntze, Rev. Gen. Pl. 1: 175. 1891.

Suffrutescent herb, to about 4 m tall; stems erect, densely fulvo-velutinous or tomentulose; stipules lanceolate, 1–2 cm long, caducous; leaves imparipinnate, 11–29-foliolate; leaflets elliptic to oblong or obovate-oblong, 2–6(–9) cm long, 1–2(–2.5) cm wide, obtuse, apiculate at the apex, rounded to acute at the base, moderately subappressed-pubescent above, densely pubescent beneath, somewhat concolorous; inflorescences terminal or leaf-opposed, pseudoracemose, many-flowered; bracts ovate, to about 12 mm long, 9 mm wide, pubescent, caducous; flowers 2.7–3.2 cm long on pedicels 1–2.3 cm long; calyx 1–1.5 cm long, tomentulose, the carinal lobe longest, deltoid-attenuate, the other lobes obtuse; petals white or sometimes tinged with pink or violet; style bearded; fruit linear-oblong, densely velutinous or tomentulose, 6–14 cm long, 1–1.7 cm wide, 5–16-seeded; seeds dark brown, reniform, 7–8 mm long, 4–5 mm wide.

Distr. A native of Africa, introduced in Ceylon and elsewhere in the tropics.

Uses. Planted in tea and coffee as shade, as an ornamental and cultivated as a fish-poison plant.

Specimens Examined. KANDY DISTRICT: Roadside on private road through Raxawa Tea Estate, E. of Dolosbage, *Grupe 193* (US); Kandy, *Fairchild & Dorsett 374* (UC, US); on Kandy-Mahiyangana Rd., c. 6 mi. NE. of Hunnasgiriya, near mi. marker 27/16, *Davidse & Jayasuriya 8393* (US); Peradeniya, Exp. Sta. *s. coll. s.n.* (PDA). NUWARA ELIYA DISTRICT: Between Hakgala and Nuwara Eliya, *Mueller-Dombois 67090106* (US); between Kandy and Nuwara Eliya, mi. marker 30/7, *Rudd & Balakrishnan 3160* (GH, K, SFV, US); along Horton Plains to Hakgala Road, near Warwick Tea Factory, *Sumithraarachchi et al.* (CAS, SFV, US). KEGALLE DISTRICT: Kadugannawa, *Worthington 7022* (K). RATNAPURA DISTRICT: Above Pinnawala on Balangoda Rd., *Comanor 1095* (K, PDA, SFV, US); N. of Pinnawala, Adam's Peak Sanctuary, S. of divide, *Maxwell & Jayasuriya 912* (SFV, US). BADULLA DISTRICT: Along highway through tea plantations, 2 mi. SW. of Haputale, *Davidse & Sumithraarachchi 8921* (US); near Koslanda, culvert 116/9, *Maxwell & Jayasuriya 764* (SFV, US); Boragas, road to Fort MacDonald, *Cramer 3510* (K, US).

6. Tephrosia spinosa (L. f.) Pers., Syn. Pl. 2: 330. 1807; Roxb., Fl. Ind. 3: 383. 1832; Wight & Arn., Prod. 214. 1834; Wight, Ic. Pl. Ind. Or. 372. 1840–43; Thw., Enum. Pl. Zeyl. 411. 1864; Baker in Hook. f., Fl. Br. Ind. 2: 112. 1876; Trimen, Handb. Fl. Ceylon 2: 30. 1894; Gamble, Fl. Pres. Madras 2: 318, 320. 1918; Macmillan, Trop. Pl. & Gard., ed. 5, 366. 1962; Backer & Bakh. f., Fl. Java 1: 593. 1963; Bosman & de Haas, Blumea 28: 472. 1983.

Galega spinosa L. f., Suppl. Pl. 335. 1781. Type: India, Coromandel, *Koenig s.n.* in 1779, Herb. Linn. 924.6 (LINN). Isotype ?: "Ind. Or. Koenig," (BM).

Small shrub, to about 75 dm high; stems spreading, often decumbent, silvery-appressed-pubescent; stipules spiniform, divaricate, 3–12 mm long, persistent; leaves imparipinnate, 5–9(–13)-foliolate; leaflets obovate-oblong to cuneate, 5–13 mm long, 2–6 mm wide, truncate, emarginate, or rounded, mucronate at the apex, cuneate at the base, glabrous or nearly so above, sericeous beneath; inflorescences fasciculate, axillary, (1–)2–3-flowered; flowers about 6 mm long on pedicels 4–7 mm long; calyx subsericeous, 2–3 mm long, with attenuate teeth; petals red; style glabrous; fruit linear-falcate, appressed-pubescent, glabrescent, 1.5–2.5 cm long, 3–4 mm wide, 2–5-seeded.

Distr. Ceylon, southern India, and Java.
Uses. Both Trimen and Macmillan refer to this as a "Bazaar drug."
Vern. Mukavaliver (T).
Specimens Examined. MANNAR DISTRICT: Near Giant's Tank, *Sayaneris s.n.* (PDA).

7. Tephrosia purpurea (L.) Pers., Syn. Pl. 2: 329. 1807; Wight & Arn., Prod. 213. 1834; Thw., Enum. Pl. Zeyl. 84. 1859; Baker in Hook. f., Fl. Br. Ind. 2: 112. 1876, excl. vars.; Trimen, Handb. Fl. Ceylon 2: 31. 1894; Macmillan, Trop. Pl. & Gard. ed. 5, 28, 199, 366. 1962; Ali, Biologia (Lahore) 10: 28. 1964; Brummitt, Bol. Soc. Brot. ser. 2, 41: 240. 1967; Ali, Fl. W. Pakistan, no. 100, Papil. 63. 1977; Bosman & de Haas, Blumea 28: 464, 1983.

Cracca purpurea L., Sp. Pl. 752. 1753. Lectotype: Ceylon, *P. Hermann*, Herb. Herm. 1: 37 (BM). Syntype: Herb. Herm. 3: 22 (BM); probably type material, Biblio. Inst. France, Paris.
Galega purpurea (L.) L., Syst. Nat., ed. 10, 1172. 1759; Moon, Cat. 55. 1824.
Tephrosia hamiltonii Drumm. ex Gamble, Fl. Pres. Madras 1: 317, 320. 1918; Alston in Trimen, Handb. Fl. Ceylon 6: 75. 1931. Type: India, Deccan, *Wight 898* (K).

Additional synonymy is given by Ali (1964, l.c.) and Brummitt (l.c.).

Perennial or long-lived annual, sometimes suffrutescent, to about 1 m tall; stems erect or spreading, glabrous or puberulent; stipules linear to deltoid, subulate, to about 3 mm long, glabrous or puberulent; leaves imparipinnate, (7–)9–21-foliolate; leaflets elliptic to oblong or obovate, 1–2.5(–3) cm long, 3–13 mm wide, obtuse to emarginate or truncate, mucronate at the apex, acute or cuneate at the base, glabrous or appressed-pubescent above, sericeous, sometimes glabrescent beneath; inflorescences pseudoracemose, usually longer than the leaves, many-flowered; bracts filiform, pubescent, about 2 mm long; flowers 4–9 mm long on pedicels 3–4 mm long; calyx pubescent, about 3 mm long with attenuate teeth; petals pink to purple; style glabrous; fruit lightly to densely appressed-pubescent, linear-oblong, 3–4.5 cm long, 4 mm wide, 5–7-seeded; seeds ellipsoid, light yellowish to dark brown, sometimes mottled, 4 mm long, 2 mm in diameter.

Distr. Widespread in the Old World tropics and, apparently, introduced into the New World.

Uses. As a green-manure cover-crop, as a sand-binder, and as a "common village medicine for children."

Vern. Kavali, kolinchi (T); gam-pila, pila (S).

Specimens Examined. WITHOUT EXACT LOCALITY: *Gardner 214* (K); *Thwaites C.P. 1445* (BM); *Kelaart s.n.* (K, P); *Macrae 159* (BM); *Raynaud s.n.* (P); *Thunberg s.n.* (UPS). JAFFNA DISTRICT: Kankesanturai, *Sumithraarachchi 797* (US); Jaffna, *Gardner* in 1845, *Thwaites C.P. 1445* in part (K, PDA). MANNAR DISTRICT: Talaimannar, 16.7.1916, *J.M.S. [Silva] s.n.* (PDA). POLONNARUWA DISTRICT: Between Habarane and Minneriya, *Bremer & Bremer 913* (US). ANURADHAPURA DISTRICT: Anuradhapura, *Hepper & Jayasuriya 4655* (K, US); *Maxwell & Jayasuriya 796* (SFV, US). ANURADHAPURA/VAVUNIYA DISTRICTS: Issanbassawa, between Medawachchiya and Vavuniya, mi. marker 97/2, *Rudd 3255* (K, PDA, US). PUTTALAM DISTRICT: Karaitivu Isl., *Grupe 118* (US); Kandakuli, *Cramer 4672* (US); Daluwa village, W. of Puttalam Lagoon, *Maxwell & Jayasuriya 746* (US); Wilpattu Natl. Park, Marai Villu, *Cooray 70020120R* (BM, US); Kali Villu, *Wirawan et al. 1012* (GH, K, SFV, UC, US); *Mueller-Dombois 67051815* (PDA, US), *Fosberg et al. 50965* (GH, K, UC, US), *50266* (US). KURUNEGALA DISTRICT: Wanduragala, quarry near Kurunegala, *Rudd 3331* (K, PDA, SFV, US); Dekanduwela, 3 mi. W. of Rambe, *Maxwell & Jayasuriya 785* (US); Kelimune, near mi. marker 34 on Kurunegala-Puttalam Rd., *Faden & Faden 77/66* (US); "Kalpitiya", *Amaratunga 328* (PDA). MATALE DISTRICT: Kandy-Dambulla Rd., mi. marker 36/4, *Sohmer 8061* (GH); Dambulla Rock, *Rudd 3110* (K, PDA, SFV, US), *Kostermans 23527* (A, BM, K); Dik Patana, mi. marker 38/2, *Tirvengadum et al. 19* (US); Between Naula and Dambulla, mi. marker 36/4, *Nowicke et al. 329* (K, US), *330* (US); Dambulla-Trincomalee Road, mi. marker 99/3, *Comanor 737* (K, SFV, UC); Dambulla, *Amaratunga 1358* (PDA). KANDY DISTRICT: Hantane, *Gardner 214* (BM, K); Peradeniya, *Alston 2210* (PDA), in 1925, *J.M. de Silva s.n.* (PDA); Kandy-Kurunegala Road, *Soejarto & Balasubramaniam 4819* (K, US). NUWARA ELIYA DISTRICT: Horton Plains, *s. coll. s.n.* (PDA); Ramboda, *Gardner 212* (K), *s.n.* (K). TRINCOMALEE DISTRICT: Near China Bay airport, *Rudd 3131* (K, PDA, SFV, US); Trincomalee, *Comanor 780* (K, SFV, UC, US). BATTICALOA DISTRICT: Batticaloa, *Kundu & Balakrishnan 200* (US); 2 mi. NE. of Rukam, *Stone 11183* (US); Sandiveli, 1 mi. N. of Sittandikudi, *Fosberg & Jayasinghe 57162* (K, US). AMPARAI DISTRICT: Between Siyambalanduwa and Inginiyagala, *Rudd & Balakrishnan 3217* (GH, K, PDA, SFV, UC, US); between 7 and 8 mi. W. of Pottuvil, *Fosberg & Sachet 53014* (US). COLOMBO DISTRICT: Colombo,

Thwaites C.P. 1445 in part (PDA), *Macrae 379* (BM, K); Wattarama, *Amaratunga 1280* (PDA); without data, *C.P. 1445* (BM, P); Negombo, July 1854, *Thwaites C.P. 1445* (PDA), *Senaratne 3250* (PDA), *Rudd 3066* (GH, K, PDA, US), *3071* (GH, PDA, US). BADULLA DISTRICT: Mahiyangane-Bibile Rd., c. 5 mi. W. of Bibile, *Sohmer et al. 8322* (GH, US); N. of Welimada, near Ft. MacDonald, near culvert 8/8, *Maxwell & Jayasuriya 896* (SFV, US). MONERAGALA DISTRICT: S. of main dam on Senanayake Samudra, *Maxwell & Jayasuriya 746* (US); E. shore of Senanayake Samudra, S. of dam, *Mueller-Dombois & Comanor 67072619* (PDA, US). GALLE DISTRICT: Galle, Dec. 1853, *Thwaites C.P. 1445* in part (PDA). MATARA DISTRICT: Mawella, *Cramer 3440* (US). HAMBANTOTA DISTRICT: Bundala Sanctuary, *Bernardi 14174* (K, US); Ruhuna Natl. Park, Rakina Wewa, near Gonalabbe Lewaya, *Fosberg 50266* (US); Rakinawala, *Cooray 67120714R* (GH, K, PDA, US); Yala, *Comanor 667* (K, SFV, US); Patanagala, *Mueller-Dombois 67120826* (PDA, US); Buttawa, *Cooray 69111702R* (SFV, US); near Situlpahuwa ruins, *Wirawan 633* (GH, K, PDA, US); Uraniya, *Mueller-Dombois 68050311* (K, PDA, US), *Mueller-Dombois et al. 69010510* (GH, K, UC, US); 200 m W. of Karaugaswala junction, *Mueller-Dombois & Cooray 67121055* (PDA, US).

Note. As Ali (l.c., p. 31) has pointed out there are no satisfactory criteria for distinguishing *T. purpurea* from *T. hamiltonii*. In the field, however, some specimens from inland, with slightly larger, pink flowers, appear to be a bit different from those along the coast, with smaller, purplish flowers. Unfortunately, the correlation of size and colour is not completely consistent; for that reason, with some reluctance, I am treating the two species as synonyms.

8. Tephrosia pumila (Lam.) Pers., Syn. Pl. 2: 330. 1807; Alston in Trimen, Handb. Fl. Ceylon. 6: 75. 1931; Backer & Bakh. f., Fl. Java 1: 593. 1963; Ali, Biologia (Lahore) 10: 27. 1964; Brummitt, Bol. Soc. Brot. ser. 2, 41: 258. 1967; Gillett, Fl. Trop. East Africa, Papil. 184. 1971; Verdcourt, Man. New Guinea Leg. 344. 1979; Bosman & de Haas, Blumea 28: 461. 1983.

Galega pumila Lam., Enc. Meth. 2: 599. 1786. Type: Madagascar, *Sonnerat s.n.* (P).
Tephrosia purpurea (L.) Pers. var. *pumila* (Lam.) Baker in Hook. f., Fl. Br. Ind. 2: 113. 1876.

Additional synonymy is given by Ali, l.c., and by Bosman & de Haas, l.c.

Annual or perennial herb; stems pubescent with spreading hairs, procumbent, to about 1 m long; stipules linear, subulate, to 3–4 mm long; leaves imparipinnate, 7–13-foliolate; leaflets obovate to oblanceolate or obovate-cuneate, (4–)10–20 mm long, 3–8 mm wide, obtuse at the apex, cuneate at the base, glabrous or pubescent above, pubescent with lax hairs beneath; inflorescences pseudoracemose, terminal, axillary, or leaf-opposed, 1–3-flowered; bracts subulate, 1–3 mm long; flowers 6–8 mm long on pedicels 3–4 mm long; calyx villous, 2–3 mm long with subulate lobes; petals usually white, sometimes salmon-pink; style glabrous; fruit linear-oblong, 2.5–3(–4) cm long, 3.4–4 mm wide, crisp-pubescent, 8–14-seeded; seeds subreniform, about 3 mm long, 2 mm wide, tan and brown-mottled.

Distr. Widespread in tropical Asia and Africa.

Specimens Examined. RATNAPURA DISTRICT: Emilipitiya, crop rotation station, *Paul s.n.* in 1951 (PDA). ANURADHAPURA DISTRICT: Anuradhapura, *Rudd 3304* (PDA, SFV, US). POLONNARUWA DISTRICT: Between Habarane and Kantalai, *Rudd 3117* (PDA, SFV, US); Polonnaruwa, *Senaratne 3506* (PDA). MATALE DISTRICT: Between Dambulla and Nalanda, Jan. 1896, *Trimen*? *s.n.* (PDA). TRINCOMALEE DISTRICT: Trincomalee, *Worthington 1096* (K). HAMBANTOTA DISTRICT: About 1.5 mi. beyond Tissa rest house on road to Kataragama, *Rudd 3105* (PDA, SFV, US); Ruhuna Natl. Park, near turnoff to Andunoruwa Wewa, *Cooray 69121003R* (K, US); between Andunoruwa and Komawa Wewa, *Mueller-Dombois 69010705* (GH, SFV, US); Buttawa, *Fosberg et al. 51031* (PDA, US); near Buttawa Bungalow, *Cooray & Balakrishnan 69010910R* (GH, SFV, US), *Mueller-Dombois 68020205A* (US), *Mueller-Dombois & Cooray 68040604* (PDA, US); Patanagala, *Cooray 70032107R* (US).

9. **Tephrosia villosa** (L.) Pers., Syn. Pl. no. 23. 329. 1807, non (Michx.) Pers., no. 17. 1807; Thw., Enum. Pl. Zeyl. 84. 1859; Baker in Hook. f., Fl. Br. Ind. 2: 113. 1876; Trimen, Handb. Fl. Ceylon 2: 33. 1894; Backer & Bakh. f., Fl. Java 1: 594. 1963; Ali, Biologia (Lahore) 10: 25. 1964; Brummitt, Bol. Soc. Brot. ser. 2, 41: 220. 1967; Gillett, Fl. Trop. East Africa, Papil. 190. 1971; Bosman & de Haas, Blumea 28: 476. 1983.

Cracca villosa L., Sp. Pl. 752. 1753. Type: Ceylon, P. Hermann, Herb. Herm. 1: 31. (BM). Syntypes or isotypes: Herb. Herm. 3: 9 (BM); Herb. Herm. 6 (L).
Galega villosa (L.) L., Syst. Nat., ed. 10, 1172. 1759; Moon, Cat. 55. 1824.
Galega hirta Buch.-Ham., Trans. Linn. Soc. London (Bot.) 13: 546. 1822. Type: India, Mysore, *Buchanan-Hamilton s.n.* (BM).
Galega incana Roxb., Fl. Ind. 3: 385. 1832. Type: India, *Roxburgh s.n.* (BM).
Tephrosia incana (Roxb.) Wight. & Arn., Prod. 212. 1834; Wight, Ic. Pl. Ind. Or. 371. 1840–43.
Tephrosia villosa (L.) Pers. var. *argentea* Thw., Enum. Pl. Zeyl. 84. 1859.
Tephrosia hirta (Buch.-Ham.) Benth., Trans. Linn. Soc. London (Bot.), Index 101. 1866; Gamble, Fl. Pres. Madras 1: 316, 318. 1918.

Annual or perennial herbs, sometimes suffrutescent, to about 1 m tall; stems pubescent with white appressed-hairs; stipules deltoid-subulate, to about 8 mm long; leaves imparipinnate, 9–19-foliolate; leaflets obovate to elliptic or cuneate-oblong, 1–2 cm long, 5–10 mm wide, obtuse at the apex, acute or cuneate at the base, glabrous to subsericeous above, sericeous beneath; inflorescences pseudoracemose, terminal or leaf-opposed, longer than the leaves, many-flowered; bracts linear-deltoid, subulate, pubescent to subglabrous, 2–3 mm long; flowers 8–10 mm long, subsessile; calyx villous, 5 mm long with filiform lobes; petals white to bright pink or purplish; style glabrous; fruit oblong-falcate, tomentulose, 2.5–3.5 cm long, 4.5–5 mm wide, 6–8-seeded; seeds reniform, dark brown, about 3 mm long, 2 mm wide.

Distr. Widespread in tropical Asia and Africa.

Vern. Boo-pila, bu-pila (S).

Specimens Examined. WITHOUT EXACT LOCALITY: *Fraser 174* (BM, US); *Reynaud s.n.* (P); *Walker 1506* (GH), *s.n.* (K); *Macrae s.n.* (BM); *Thwaites, C.P. 1448* (K); Anuradhapura, *Gardner* in 1848, Negombo, July 1854, *Thwaites C.P. 1447* (PDA); without data, *C.P. 1447* (BM, GH, K, PDA). Weyakeyapotha, *Alston s.n.* in 1923 (PDA); Batá-atá, *Alston 1230* (PDA). JAFFNA DISTRICT: Jaffna, in 1846, *Thwaites, C.P. 1446* (PDA); without data, *C.P. 1446* (K); between Paranthan and Pooneryn, *Rudd 3298* (BM, CAS, K, PDA, US). ANURADHAPURA DISTRICT: North bank of Rajangana Reservoir, 18 mi. S. of Anuradhapura, *Maxwell & Jayasuriya 790* (SFV, US); Maradankadawala, *Jayasuriya et al. 630* (US); Maha Kanadarawa, *Sumithraarachchi 739* (US); Habarana, *Fosberg 57045* (US). POLONNARUWA DISTRICT: Polonnaruwa, Sacred Area, *Dittus 71102501* (US); Minneriya, *Comanor 1215* (GH, K, SFV, US). PUTTALAM DISTRICT: Wilpattu Natl. Park, W. end of Mail Villu, *Mueller-Dombois et al. 69043019* (K, SFV, US); Marai Villu, *Cooray 70020122R* (K, UC, US); Kali Villu, *Wirawan et al. 1013* (GH, K, SFV, UC, US); *Fosberg et al. 50691* (GH, K, UC, US), *50959* (BM, GH, K, SFV, UC, US). KURUNEGALA DISTRICT: Wanduragala, near Kurunegala, *Rudd 3330* (PDA, SFV, US); Kelimune, near mi. marker 34 on Kurunegala-Puttalam Road, *Faden & Faden 75/65* (US). MATALE DISTRICT: Kirinda, *Trimen ?*, Dec. 1882 (PDA); Sigiriya, *Simpson 9197* (BM). KANDY DISTRICT: Peradeniya, Exp. Sta., *Wright s.n.* (PDA). TRINCOMALEE DISTRICT: Kantalai, tank bed, *Worthington 2050* (K); Trincomalee, near China Bay airport, *Rudd 3132* (K, PDA, US), *3136* (PDA, SFV, US); Dambulla-Trincomalee Road between mi. markers 94–95, *Comanor 733* (CAS, GH, UC, US). AMPARAI DISTRICT: SE. of Senanayake Samudra, Inginiyagala, *Fosberg & Sachet 53095* (GH, SFV, UC, US); between Siyambalanduwa and Inginiyagala, *Rudd & Balakrishnan 3219* (PDA, US), *3221* (PDA, US). COLOMBO DISTRICT: Colombo, *Macrae 380* (K). HAMBANTOTA DISTRICT: Ruhuna Natl. Park, Buttawa Bungalow, *Fosberg & Sachet 52882* (GH, SFV, UC, US); *Cooray 70031908R* (US), *Worthington 5683* (K); About 1.5 mi. beyond Tissa on road to Kataragama, *Rudd 3100* (PDA, SFV, US).

10. Tephrosia noctiflora Bojer ex Baker in Oliver, Fl. Trop. Afr. 2: 112. 1871; Alston in Trimen, Handb. Fl. Ceylon 6: 75. 1931; Backer & Bakh. f., Fl. Java 1: 594. 1963; Brummitt, Bol. Soc. Brot., ser. 2, 41: 228. 1967; Gillett, Fl. Trop. East Africa, Papil. 182. 1971; Verdcourt, Handb. New Guinea Leg. 344. 1979; Bosman & de Haas, Blumea 28: 458. 1983. Type: Zanzibar, *Bojer s.n.* (K).

Tephrosia hirta sensu Thw., Enum. Pl. Zeyl. 84. 1859, non Buch.-Ham.
Tephrosia hookeriana sensu Baker in Hook. f., Fl. Br. Ind. 2: 113. 1876; Trimen,
    Handb. Fl. Ceylon 2: 32. 1894, non Wight & Arn.
Cracca noctiflora (Baker) Kuntze, Rev. Gen. Pl. 1: 175. 1891.
Tephrosia hookeriana Wight & Arn. var. *amoena* Prain in King, J. Asiat. Soc.
    Bengal 66, 2: 85. 1897. Lectotype: Malacca, Aug. 1889, *Derry 270.*
Tephrosia subamoena Prain in King, op. cit. p. 86. 1897, invalid name, merely
    suggested in a note.

Annuals or perennials, sometimes suffrutescent, to about 1.5 m tall; stems densely pubescent with whitish to fulvous subappressed hairs; stipules filiform, villous, 5–8 mm long; leaves imparipinnate, (5–)13–19(–25)-foliolate; leaflets oblong-obovate to oblanceolate, mucronate, rounded, emarginate or subtruncate at the apex, acute to cuneate at the base, 1.5–4 cm long, 5–10 mm wide, glabrous above, appressed-pubescent beneath; inflorescences racemose, terminal or sometimes axillary, many-flowered; bracts deltoid-subulate, 1–3 mm long, caducous; flowers 10–13 mm long on pedicels 2–4 mm long; calyx densely pubescent, about 5 mm long, with deltoid lobes; petals white to lilac; style glabrous; fruit linear-oblong, essentially straight or slightly curved upward at the apex, fulvo-villous, 4–5 cm long, 5–6 mm wide, (6–)8–14-seeded; seeds brown, reniform, about 3–4 mm long, 2.5 mm wide.

Distr. Native to Africa but introduced into Ceylon, India and Java.

Uses. Gillett, l.c., states that *T. noctiflora* has been cultivated as a fish-poison and as a cover crop.

Vern. Ela-pila (S).

Specimens Examined. KANDY DISTRICT: Hantane, *Gardner 215* (BM, K); Peradeniya, Exp. Sta. in 1916, *s. coll. s.n.* (PDA), in 1925, *de Silva s.n.* (PDA), *175* (PDA, UC); Peradeniya, above University, *Maxwell & Jayasuriya 844* (US). AMPARAI DISTRICT: E. shore of Senanayake Samudra, S. of dam, *Mueller-Dombois & Comanor 67072611* (PDA, US); S. of Dam on Senanayake Samudra, *Maxwell & Jayasuriya 749* (SFV, US); Hatpatha, S. shore of Senanayake Samudra, Gal-Oya Natl. Park, *Jayasuriya 2078* (K). COLOMBO DISTRICT: Colombo, *Macrae s.n., Thwaites C.P. 2776* (PDA). MONERA-GALA/AMPARAI DISTRICTS: Between Siyambalanduwa and Inginiyagala, *Rudd & Balakrishnan 3215* (K, PDA, US).

Note. Backer & Bakh. f, in Fl. Java, l.c., mention that the flowers are closed during the greater part of the day, expanding at about 4 p. m. and remaining open to about 8:30 p.m.

2. MUNDULEA

(DC.) Benth. in Miq., Pl. Jungh. 3: 248. 1852; Allen & Allen, Legum. 450. 1981; Geesink in Polhill & Raven, Ady. Leg. Syst. 258. 1981.

Tephrosia Pers. sect. *Mundulea* DC., Prod. 2: 249. 1825. Lectotype species: *M. suberosa* (DC.) Benth., based on *Tephrosia suberosa*, a synonym of *M. sericea* (Willd.) A. Chev.

Small trees or shrubs, to about 7 m tall; leaves alternate, imparipinnate; leaflets opposite or subopposite; stipules minute; stipels absent; inflorescences racemose, axillary, terminal, or pseudoterminal; bracts minute; bracteoles lacking; flowers of medium size; calyx campanulate with 5 short teeth, or lobes, the upper (vexillar) often connate; petals bluish to pink or purple, usually pubescent, the wings slightly adherent to the keel, the keel petals connate; stamens 10 with the vexillar filament sharply bent, free at the base but above united with the others, the filaments alternately dilated toward the apex, pubescent above; anthers uniform, elliptic; ovary sessile, many-ovuled; style bent above, hardened; glabrous; stigma terminal, minutely capitate; fruit indehiscent or tardily dehiscent, linear-oblong, compressed, many-seeded; seeds oblong to reniform, brown, non-arillate, the hilum lateral.

A genus of about 15 species, chiefly in Madagascar.

Mundulea sericea (Willd.) A. Chevalier, Compt. Rend. Hebd. Séances Acad. Sci. 180: 1521. 1925; Gillett, Fl. Trop. East Africa, Papil. 155, fig. 28. 1971; Verdcourt, Man. New Guinea Leg. 335, fig. 75. 1979.

Cytisus sericeus Willd., Sp. Pl. 3: 1121. 1802; Willd., Nov. Actorum Acad. Caes. Leop.-Carol. Nat. Cur. 4: 204. 1803; Moon, Cat. 53. 1824. Type: Southern India, Tranquebar, *s. coll.* (holotype B-Willd.), fide Gillett (l.c.).
Tephrosia ? sericea (Willd.) DC., Prod. 2: 249. 1825.
Robinia suberosa Roxb., Hort. Beng. 56. 1814, nomen; Roxb., Fl. Ind. 3: 327. 1832. Type: Cultivated at Calcutta. Type material ?, BM, K.
Robinia sennoides Roxb., Hort. Beng. 56. 1814, nomen; Roxb., Fl. Ind. 3: 328. 1832, nom. illeg., based on *Cytisus sericeus* Willd.
Tephrosia suberosa DC., Prod. 2: 249. 1825; Hook., Ic. Pl. t. 120. 1837; Thw., Enum. Pl. Zeyl. 84. 1859. Type: Cultivated at Calcutta, *Wallich cat. 5628* (isosyntype K).
Mundulea suberosa (DC.) Benth. in Miq., Pl. Jungh. 248. 1852; Baker in Hook. f., Fl. Br. Ind. 2: 110. 1876; Trimen, Handb. Fl. Ceylon 2: 29. 1894; Alston in Trimen, Handb. Fl. Ceylon 6: 74. 1931; Macmillan, Trop. Pl. & Gard., ed. 5, 135. 1962.

Shrub or small tree, to about 7 m tall; young stems velutinous, glabrescent; leaves 9–17-foliolate, the axis about 4–10 cm long, pubescent; stipules deltoid,

1–3 mm long; leaflets opposite to subopposite, ovate to oblong or lanceolate, 1–5 cm long, 4–15(–17) mm wide, commonly acute or acuminate, sometimes attenuate, sometimes rounded at the apex, cuneate or rounded at the base, glabrous or subsericeous above, sericeous or subsericeous beneath, the secondary veins moderately or scarcely conspicuous; inflorescences racemose, terminal or axillary, about 5–10 cm long, few- to many-flowered with two flowers at each node; bracts deltoid, 0.5–2 mm long; flowers 17–25 mm long; calyx pubescent, about 2 mm long; petals pinkish or reddish-violet to purple, the vexillum sericeous on the outer face; fruit linear-oblong, about 6–8(–10) cm long, 6–8 mm wide, sericeous to velutinous, commonly 6–8-seeded, thickened along both sutures; seeds reniform or subreniform, dark green to brown, sublustrous, 4–5 mm long, 3 mm wide, the hilum elliptic, lateral.

Distr. Dry rocky hills of southern India and Ceylon, tropical and South Africa, and Madagascar.

Uses. Macmillan (l.c.) includes this species, as *M. suberosa,* in his list of ornamental plants indigenous to Ceylon, as suitable for dry regions. Allen and Allen (l.c.) cite it as one of the fish-poison plants, which also is toxic to reptiles and insects.

Vern. Kang-bendi-gas, wal-buruta, gal-buruta (S); pilavaiam (T).

Specimens Examined. WITHOUT EXACT LOCALITY: *Walker 1423* (K), *s.n.* (K). ANURADHAPURA DISTRICT: Habarane, Sept. 1885, *Trimen ?* (PDA), *Alston 513* (PDA), *Worthington 4875* (BM, K), *6968* (K), *Fosberg 57016* (CAS, US); Borupangala, Hinukkiriyawa, W. of Habarane, *Jayasuriya et al. 1391* (K, US); forest W. of Ritigala Hill, *Balakrishnan & Jayasuriya 1112* (US); Ritigala Hills, E. slopes, *Sohmer & Sumithraarachchi 10774* (GH, US); Ritigala Strict Natural Reserve, E. slope towards Unakanda, *Jayasuriya 1094* (K, US); Gonawalpola–Kekirawa Road, mi. marker 4/4, *Sumithraarachchi & Sumithraarachchi 745* (US). POLONNARUWA DISTRICT: W. slope of Gunner's Quoin, *Huber 437* (US). MATALE DISTRICT: Dambulla, *Thwaites C.P. 1486* (K, PDA), without data, *C.P. 1486* (BM, GH, K, P, US), *Gardner, Thwaites C.P. 1486* (PDA), *Amaratunga 254* (PDA), *1359* (PDA), *2797* (PDA), *Simpson 9774* (BM), *Rudd 3112* (K, PDA, US); Dambulla Rock Hill, *Worthington 2797* (BM, K); Erawalagala Mtn., E. of Kandalama Tank, 6 mi. E. of Dambulla, *Davidse & Sumithraarachchi 8086* (US). AMPARAI DISTRICT: 7–8 mi. W. of Pottuvil, *Fosberg & Sachet 53009* (GH, K, SFV, UC, US). HAMBANTOTA DISTRICT: Ruhuna Natl. Park, near Patanagala Beach, *Mueller-Dombois 67083111* (PDA, US); near Padikema Rocks, *Wirawan 793* (K, PDA, SFV, US).

3. MILLETTIA

Wight & Arn., Prod. 263. 1834; Dunn, J. Linn. Soc. Bot. 41: 123–243. 1912; Allen & Allen, Legum. 435. 1981. Type species: *M. rubiginosa* Wight & Arn.

Trees or shrubs, sometimes scandent; leaves alternate, imparipinnate; leaflets alternate or opposite; stipules small, caducous; stipels usually present; inflorescences paniculate, lateral or terminal, rarely racemose; bracts and bracteoles caducous; flowers of medium size; calyx broadly campanulate, truncate or 5-toothed; petals purple to rose or white; standard pubescent or glabrous, sometimes calloused; wings free from the keel; stamens 10, usually monadelphous or, sometimes, diadelphous with the vexillar filament free, anthers ovate, uniform; ovary usually sessile, linear, 3–many-ovuled; style filiform, curved, glabrous; stigma small, terminal; fruit tardily dehiscent, glabrous or pubescent, linear-oblong, compressed or turgid, coriaceous or slightly woody, 1- or few-seeded; seeds orbicular or reniform.

A genus of about 100 species in tropical Africa and Asia. Only one species, *M. dura*, is known from Ceylon.

Millettia dura Dunn, J. Bot. 49: 221. 1911; Dunn, J. Linn. Soc. Bot. 41: 223. 1912; Dunn, Bot. Mag. t. 8959. 1923; Gillett in Fl. Trop. East Africa, Papil. 144. 1971.

Trimen stated that "Ceylon has no species of *Millettia*" (Handb. Fl. Ceylon 2: 29. 1894), and Macmillan (Trop. Pl. & Gard. ed. 5. 1962) lists none. However. Worthington (Ceylon Trees 177. 1959) includes *Millettia* sp. as "possibly *M. dura* or *oblata*". From the photograph, supported by voucher specimens, it appears to represent *M. dura* Dunn, "a form with unusually glabrous leaflets (as also in some African material, particularly specimens from Uganda)", (G. Lewis, in litt.). According to Worthington, he introduced this ornamental tree with blue-mauve or violet flowers, in 1934, from Kikiyu, East Africa.

Specimens Examined. KANDY DISTRICT: Poilakanda, Kadugannawa, *Worthington 560* (K), *804* (K), *834* (K).

Tribe AESCHYNOMENEAE

(Benth.) Hutch., Gen. Fl. Pl. 1: 470. 1964. Type: *Aeschynomene* L.

Hedysareae subtribe Aeschynomeninae Benth. in Benth. & Hook. f., Gen. Pl. 1: 448. 1865, as "Aeschynomeneae".
Coronilleae subtribe Aeschynomeninae (Benth.) Schulze-Menz in Engler's Syllabus 2: 236. 1964.

Shrubs or herbs, sometimes scandent, rarely small trees; leaves paripinnate or imparipinnate, 1–many-foliolate; stipules present, sometimes appendiculate below the point of attachment, sometimes spinescent; stipels usually absent; inflorescences commonly racemose or paniculate, sometimes fasciculate or subcymose, or the flowers solitary; flowers papilionoid; calyx campanulate with subequal lobes or teeth, or bilabiate with the vexillar lip entire to bifid, the carinal lip entire to trifid; petals commonly yellowish, sometimes almost white, or orange, sometimes purplish; stamens 10, usually monadelphous, or diadelphous 5:5, but sometimes with 1 or more free filaments; anthers uniform and versatile or, in *Arachis, Stylosanthes,* and *Zornia,* dimorphic; ovary 1–many-ovulate, the style filiform, glabrous, with a minute terminal stigma; fruit lomentaceous, usually breaking into 1-seeded articles or, in *Arachis,* torulose, unjointed, and geomorphic; seeds mostly reniform with a small lateral hilum, sometimes ovoid or subovoid with the hilum near one end.

In the preliminary key to the tribes of the faboid legumes (Revised Handbook Flora of Ceylon 1: 429. 1980) the genera with lomentaceous fruit were all assigned to the Hedysareae. On the basis of more recent studies it is now believed that separation into four tribes is a more natural arrangement. (Advances in Legume Systematics, ed. Polhill & Raven. 1981.) Of those tribes only the Aeschynomeneae and the Desmodieae are known in Ceylon.

KEY TO THE GENERA

1 Leaves 5–many-foliolate
  2 Leaflets alternate; fruit exserted at maturity; calyx not scarious
    3 Calyx bilabiate; stipules peltate, appendiculate below the point of attachment............
    ....**1. Aeschynomene**
    3 Calyx campanulate with 5 subequal lobes; stipules attached at the base...**2. Ormocarpum**
  2 Leaflets opposite; fruit plicate, enclosed by a somewhat scarious calyx.........**3. Smithia**
1 Leaves 2–4-foliolate
  4 Leaflets 2-foliolate; calyx-tube not elongated at the base; stipules and bracts relatively large, conspicuous, usually with dark puncta....................................**4. Zornia**
  4 Leaflets 3- or 4-foliolate; calyx-tube elongate-filiform at the base; stipules and bracts relatively inconspicuous, not punctate
    5 Fruit small, compressed, the articles to about 2.5 mm wide; leaves 3-foliolate...........
    ....**5. Stylosanthes**
    5 Fruit terete, to about 15 mm wide; leaves 4-foliolate....................**6. Arachis**

1. AESCHYNOMENE

L., Sp. Pl. 713. 1753; L., Gen. Pl. ed. 5. 319. 1754. Type species: *A. aspera* L.

Herminiera Guill. & Perr. in Guill., Perr. & A. Rich., Fl. Sénég. 1: 201. 1832.

Herbs, shrubs, or small trees; leaves alternate, pinnate or subimparipinnate, 5–many-foliolate; leaflets small in most species; stipules attached at the base or peltate, appendiculate below the point of attachment; stipels lacking; inflorescences racemose, sometimes paniculate, terminal or axillary with few to many flowers; flowers with calyx campanulate with 5 subequal lobes or bilabiate with carinal lip entire to trifid, the vexillar lip entire to bifid; petals yellowish, sometimes with red or purplish markings; stamens 10, monadelphous or diadelphous 5:5; anthers uniform and versatile; fruit (1–)2- about 18-articulate, compressed, straight

or contorted; seeds reniform with a lateral hilum; chromosome numbers
2n = (18) 20, 40.

A genus of about 150 species, tropical to warm temperate, in America, Africa,
and Asia, introduced in Australia.

KEY TO THE SPECIES

1 Leaflets 1-costate, elliptic, oblong, or obovate
  2 Plants herbaceous, unarmed; flowers 7–20 mm long; fruit straight or but slightly curved
    3 Flowers 15–20 mm long; vexillum and keel petals usually pubescent; fruit with articles about
      10 mm long, 7–8 mm wide...**1. A. aspera**
    3 Flowers 7–10 mm long; petals glabrous; fruit with articles (3–)5–6 mm long, 3–6 mm wide
      ....**2. A. indica**
  2 Plants shrubby, thorny; flowers 25–45 mm long; fruit strongly curved or spirally contorted
    ....**3. A. elaphroxylon**
1 Leaflets 2–several-costate, somewhat falcate
  4 Fruit glabrous to puberulent or glandular hispidulous, definitely articulated, reticulate-veiny near
    the margin, usually muricate; flowers 5–10, commonly 6–8 mm long.....**4. A. americana**
  4 Fruit villous or hispid, articulations weak or lacking, the surface without conspicuous venation
    or murication; flowers 3–9 mm, commonly 3–5 mm long..................**5. A. villosa**

1. Aeschynomene aspera L., Sp. Pl. 713. 1753; Burm. f., Fl. Ind. 169. 1768;
Moon, Cat. 54. 1824; Wight, Ic. Pl. Ind. Or. t. 299. 1840; Thw., Enum. Pl.
Zeyl. 85. 1859; Baker in Hook. f., Fl. Br. Ind. 2: 152. 1876; Trimen, Handb.
Fl. Ceylon 2: 39. 1894; Rudd, Reinwardtia 5: 29. 1959; Macmillan, Trop. Pl.
& Gard. ed. 5, 412. 1962; Ali, Fl. W. Pakistan, no. 100: 339. 1977. Type:
Ceylon, *P. Hermann*, Herb. vol. 2, fol. 52 (lectotype BM; probable isotypes L,
Herm. Herb. fol. 33; Biblio. Inst. France, Paris, Coll. Pl. Zeyl. p. 128). Paratype:
Hort. Ups. LINN 922.3).

Aeschynomene lagenaria Lour., Fl. Cochinch. 446. 1790. Type: Cochinchina
    (Viet Nam), *Loureiro s.n.* (BM).
Hedysarum lagenarium (Lour.) Roxb., Fl. Ind. ed. 2, 3: 365. 1832.

Erect herb, sometimes suffrutescent, to about 1–2 m tall; stems glabrous to
moderately hispid; stipules peltate-appendiculate, about 20 mm long; leaves 8–17
cm long, 60–100-foliolate, the petiole and rachis glabrous to sparsely hispidulous;
leaflets linear-oblong, 5–10 mm long, about 1.5 mm wide, glabrous, 1-costate,
the apex subacute, mucronulate, the base asymmetrical, the margin entire to
ciliolate; inflorescences axillary, racemose, 1–few-flowered; peduncles and
pedicels hispid; bracts cordate, about 3–5 mm long, 1.5–3 mm wide, hispid;
bracteoles ovate, rounded to acute, hispid, about 4–5 mm long, 2 mm wide;
flowers 15–20 mm long; calyx hispid, bilabiate, 6–8 mm long; petals orange-
yellow, pubescent in part; fruit about 6–10 cm long, 4–7-articulate, compressed,
glabrous to moderately hispid, dark brown at maturity, straight or but slightly
curved, one margin entire, the other crenate, the stipe 10–15 mm long, the basal

article usually aborted, tapering to the stipe, the apical article also aborted, tapering to the usually persistent style, the other articles subquadrate,10 mm long, 7–8 mm wide, papillose to verrucose at the margin and at the centre over the seeds; seeds reniform, black, about 7 mm long, 4–5 mm wide, 1.5 mm thick.

Distr. In water or wet places, Ceylon, eastern India, Southeast Asia, and Malaysia.

Uses. The pith of the stems is sometimes used to make articles such as fishing floats, toys, and sun hats.

Vern. Maha-diya-siyambala (S); attuneddi (T); pith plant, shola (E).

Specimens Examined. HAMBANTOTA AND MATARA DISTRICTS: "Tanks nr. Mootetisse, Gardner", "Paddy fields, Matara, Dec. 1853", *Thwaites C.P. 1514* (BM, K, NY, P, PDA). ANURADHAPURA DISTRICT: Between Galkulama and Tirappane, *Rudd & Jayasinghe 3311* (PDA, SFV, US). PUT-TALAM DISTRICT: Wilpattu Natl. Park, Mail Villu, *Mueller-Dombois 68082107* (PDA, US), *Cooray 70020101R* (PDA, US). MATALE DISTRICT: Dambulla, Pahala Veva, *Simpson 8092* (BM). COLOMBO DISTRICT: Gampaha, *Simpson 8873* (NY); Muturajawela, *Amaratunga 143* (PDA). HAMBANTOTA DISTRICT: Tissamaharama Tank, Dec. 1882, *Trimen*? *s.n.* (PDA); Ruhuna Natl. Park, Patanagala, *Mueller-Dombois 67120823* (PDA, US); Nabadagas Wewa, *Fosberg et al. 51080* (PDA); near Meynet Wewa, *Cooray 70032304R* (PDA, US); Yala, Katagamuwa tank, *S. Ripley 120* (US).

Note. The specimen cited as a probable isotype of *A. aspera* is at the Biblio-théque de l'Institut de France, Paris, in the herbarium of Paul Hermann, basis of the Thesaurus Zeylanicus of J. Burman (see A. Lourteig in Taxon 15: 23–33. 1966).

2. Aeschynomene indica L., Sp. Pl. 713. 1753; Burm. f., Fl. Ind. 169. 1768; Moon, Cat. 54. 1824; Wight, Ic. Pl. Ind. Or. 405. 1840–43; Thw., Enum. Pl. Zeyl. 85. 1859; Baker in Hook. f., Fl. Br. Ind. 2: 151. 1876; Trimen, Handb. Fl. Ceylon 2: 38. 1894; Rudd, Reinwardtia 5: 30. 1959; Ali, Fl. W. Pakistan, no. 100: 338. 1977. Lectotype: India, Malabar. Neli tali. Rheede, Hort. Mal. 9: 31, t. 18. 1689. Syntype: *Van Royen* herb. *384* (L).

Aeschynomene pumila L., Sp. Pl. ed. 2, 1061. 1763; Burm. f., Fl. Ind. 170. 1768; Moon, Cat. 54. 1824; Trimen, Handb. Fl. Ceylon 2: 39. 1894. Lec-totype: Ceylon, *P. Hermann s.n.* (BM; isotypes S-Linnaean herb. *s.n.* 62: 12. Isotype ? : *Van Royen* herb. *385* (L). Syntype: India, Malabar. Malam-todda-vali, Rheede, Hort. Mal. 9: 37, t. 21 (cited by Linnaeus as Niti-todda-vali, t, 20). 1689.

Smithia aspera Roxb., Hort. Beng. 56. 1814; Roxb., Fl. Ind. ed. 2, 3: 343. 1832. Type ?: India, Bengal, *Roxburgh s.n.* (BM).

Hedysarum neli-tali Roxb., Fl. Ind. ed. 2, 3: 365. 1832; Wight, Ic. Pl. Ind. Or. 405. 1840. 1840–1843, in synon. under *Aeschynomene indica* L.

Additional synonymy for this species is given in Rudd, 1959, cited above.

Erect herb, sometimes suffrutescent, about 1–2.5 m tall; stems glabrous to moderately hispid; stipules peltate-appendiculate, 10–15 mm long; leaves about 5–10 cm long, 50–70-foliolate, the petiole and rachis glabrous or sparsely hispidulous; leaflets elliptic-oblong, 2–10 mm long, 1–2.5 mm wide, entire or, rarely, ciliate-denticulate, glabrous, 1-costate, rounded at apex and base; inflorescences axillary, racemose, few-flowered, the peduncles and pedicels glabrous to hispidulous; bracts ovate, acuminate, about 5 mm long, 1–2 mm wide; bracteoles lanceolate-ovate, acute, 2–4 mm long, 1 mm wide; flowers 7–10 mm long; calyx glabrous, bilabiate, 4–6 mm long; petals yellow, sometimes almost white, sometimes purplish, glabrous; fruit 3–4 cm long, 5–10(–12)-articulate, compressed, glabrous to moderately hispid, sometimes muricate, usually dark brown at maturity, the upper edge essentially straight, the lower crenate, the stipe somewhat recurved, 4–10 mm long, the articles subquadrate, (3–)5–6 mm long, 3–6 mm wide; seeds reniform, dark brown, 3–4 mm long, 2–3 mm wide, 1 mm thick.

Distr. A common plant of wet places such as near tanks and in paddy fields of Old World tropics and subtropics, Japan, and in the southeastern United States.

Uses. The plants are sometimes used as green manure. The pith is used as a substitute for that of *Aeschynomene aspera*.

Vern. Heen-diya-siyambala, diya-siyambala (S).

Specimens Examined. WITHOUT EXACT LOCALITY: *Col. Walker 5* (PDA), *s. coll. s.n.* (P); "Mootetisse, illegible, Habarana, *Gardner*", "Ooma oya, July 1853", *Thwaites C.P. 1515* (BM, K, NY, P, PDA); JAFFNA DISTRICT: Near Kayts, *Rudd 3279* (PDA, SFV, US); between Paranthan and Pooneryn, *Rudd 3297* (US). MANNAR DISTRICT: Near Giant's tank, "*Sayaneris coll.*", Feb. 1890 (PDA); Mannar, along banks of Thaladikulam, *Cramer 2893* (US). ANURADHAPURA DISTRICT: s. loc., *Jayasuriya et al. 566* (US); between Anuradhapura and Galkulama, *Rudd 3307* (PDA, SFV, US). TRINCOMALEE DISTRICT: Trincomalee, near China Bay, *Rudd & Balakrishnan 3135* (PDA, US). PUTTALAM DISTRICT: Puttalam, Nov. 1881, *Nevill* (PDA); near lagoon, *Jayasuriya & Moldenke 1451* (US); Wilpattu Natl. Park, Atha Villu, *Mueller-Dombois et al. 69042706* (PDA, SFV, US). KURUNEGALA DISTRICT: Kurunegala, *Amaratunga 50* (PDA); Andiambalama, *Simpson 8589* (BM). NUWARA ELIYA DISTRICT: Madanwela, Hanguranketa area, mile marker 19/11, *Rudd 3249* (PDA, SFV, UC, US). AMPARAI DISTRICT: Inginiyagala, *Rudd & Balakrishnan 3225* (PDA, SFV, US). BADULLA DISTRICT: Mahiyangana-Bibile Rd., 5 mi. West of Bibile, *Sohmer et al. 8323* (US). MONERAGALA DISTRICT: Okkampitiya, 25 June 1953, *M.B.H. 23* (PDA); near Wellawaya, *Rudd & Balakrishnan 3208* (PDA, SFV, US). HAMBANTOTA DISTRICT: Wirawila, *Rudd 3093* (PDA, SFV, US); Ruhuna Natl. Park, 1 mi. NW. of Yala Bungalow, near bridge, *Mueller-Dombois*

& *Cooray 68040701* (PDA, US); south of bridge, *Mueller-Dombois 68050320* (PDA, US); near Meynet Wewa, *Cooray 70032323R* (PDA, US); Patangala, *Fosberg et al. 51168* (SFV, UC, US).

Note. *Aeschynomene pumila* L. appears to be a somewhat depauperate form of *A. indica* from drier areas.

3. Aeschynomene elaphroxylon (Guill. & Perr.) Taub. in Pflanzenfam. 3, 3: 319, f. 124 A–C. 1894; Leonard, Fl. Congo Belge 5: 261. 1954; Hepper in Hutch. & Dalz., Fl. West Trop. Africa, ed. 2, 578, f. 168. 1958; Rudd, Reinwardtia 5: 28. 1959; Verdcourt in Fl. Trop. East Africa, Papil. 375. 1971.

Herminiera elaphroxylon Guill. & Perr. in Guill., Perr. & A. Rich., Fl. Sénég. 1: 201, t. 21. 1832; Macmillan, Trop. Pl. & Gard. ed. 5, 412. 1962. Type: Senegal, N'Gher (Panié-Foul), island at mouth of R. Marigot de Taoué, *Perrottet s.n.* (P).
Smithia elaphroxylon (Guill. & Perr.) Baillon, Bull. Mens. Soc. Linn. Paris 1: 404. 1883.

Shrub or small tree, (1–)2–8(–12) m tall, the trunk pithy; stems hispid and also armed with thorns about 2–15 mm long; stipules densely pubescent on the outer face, ovate-lanceolate, appendiculate below the point of attachment, the upper portion acute, 7–13 mm long, 3–5 mm wide, the appendage erose, 2–4 mm long, 2–4 mm wide; leaves 4–16 cm long, 20–40-foliolate, the petiole and rachis pubescent and spiny; leaflets oblong-elliptic, 5–30 mm long, 4–10 mm wide, 1-costate, entire, retuse or emarginate, mucronulate, the base rounded, the upper surface glabrous, the lower coarsely pubescent and the secondary veins usually dark; inflorescences axillary, racemose, usually 2–4-flowered, the axes hispid like the stems; bracts and bracteoles ovate, acute, about 5–15 mm long, 3–10 mm wide, pubescent; flowers 3–4.5 cm long; calyx 2–2.5 cm long, hispid, bilabiate, the lips essentially entire or briefly 2- and 3-dentate; petals orange-yellow, pubescent in part; fruit about 5–17 cm long, 6–17-articulate, glandular hispid, spirally contorted, short-stipitate, the articles trapezoidal, about 5–10 mm long, 7–9 mm wide; seeds black, about 5 mm long, 3 mm wide.

Distr. In water or wet places of tropical Africa and Madagascar. Cultivated as an ornamental in other tropical areas of the Old World and South America.

Uses. In Africa the light weight wood is used for floats and rafts.

Vern. Nile pith-tree, ambach, ambash.

Specimens Examined. KANDY DISTRICT: Peradeniya, Botanic Garden, in lake, 18 Dec. 1907, *s. coll. s.n.* (PDA), 8 Dec. 1926, *J.M. Silva s.n.* (PDA).

Note. This shrub was observed growing in the lake in early 1970 but in April of that year it was cut. I do not know if it still exists.

4. Aeschynomene americana L., Sp. Pl. 713. 1753; Rudd, Reinwardtia 5: 25.

1959; Verdcourt, Man. New Guinea Leg. 367. 1979. Lectotype: Jamaica, Sloane, Hist. 1: 186, t. 118, f. 3. 1696; typotype (BM). Syntype ? or isotypotype ?, *van Royen 384* (L).

Additional synonymy is given in Rudd, 1959, cited above.

Erect or decumbent herb, to about 2 m tall; stems glandular-hispid to subglabrous; stipules peltate-appendiculate, glabrous or somewhat hispid at the point of attachment, striate, usually ciliate, (5–)10–25 mm long, 1–4 mm wide, the upper portion attenuate, 2–3 times longer than the lower, acute, or erose portion; leaves 2–7 cm long, about 20–60-foliolate, the petiole and rachis hispidulous; leaflets glabrous, subfalcate, 2–several-costate, 4–15 mm long, 1–2 mm wide, ciliate, apiculate, the base asymmetrically rounded; inflorescences axillary, racemose, few-flowered, the axes hispidulous; bracts cordate, acuminate or, sometimes, truncate-flabelliform, about 2–4 mm long, 2–3 mm wide, glabrous, ciliate; bracteoles linear to linear-ovate, 2–4 mm long, 1–1.5 mm wide, acute to acuminate, serrate-ciliate; flowers about 5–10 mm long; calyx glabrous to hispidulous, bilabiate, 3–6 mm long; petals yellowish to tan, usually with red or purplish lines, glabrous; fruit 2–3 cm long, 3–9-articulate, the stipe about 2 mm long, the articles semicircular, the upper margin essentially straight, the lower curved, 3–6 mm long, 2.5–5 mm wide, glandular-hispidulous, usually muricate, the margins thickened, reticulate-veiny; seeds dark brown, 2–3 mm long, 1.5–2 mm wide.

Distr. Widespread in tropical and subtropical America, especially in wet places; inroduced in Ceylon, Southeast Asia, Philippines, and New Guinea.

Uses. Grown as a green-manure cover crop and as hay but often becomes a weed.

Vern. Thornless mimosa (W.R.C. Paul, Trop. Agric. (Ceylon) 107: 15–20. 1951).

Specimens Examined. KURUNEGALA DISTRICT: Narammala, in paddy field, *Amaratunga 1214* (PDA). KANDY DISTRICT: Peradeniya, Botanic Garden, Nov. 1893, *Trimen ? s. n.* (PDA), *de Silva 714* (K, PDA), *Fosberg 50667* (PDA, SFV, US).

Note. There is so much variability in the size and indument of the fruit that it seems preferable to omit any reference to subspecific taxa in this treatment.

5. Aeschynomene villosa Poir. in Lam., Enc. Suppl. 4: 76. 1816; Rudd, Contr. U.S. Natl. Herb. 32: 32. 1955; Rudd, Reinwardtia 5: 27. 1959; Verdcourt, Man. New Guinea Leg. 368. 1979. Type: Puerto Rico, savannas, *Ledru s.n.* (P).

Additional synonymy is given in Rudd, 1959.

Herbs, erect to prostrate, up to about 1 m tall; stems hispid; stipules peltate-appendiculate, usually hispid, especially at the point of attachment, striate, ciliate,

(5–)10–15 mm long, 1–1.5 mm wide, the upper portion attenuate, slightly longer than the lower attenuate or erose portion; leaves about 2–7 cm long, 20–50-foliolate, the petiole and rachis glandular-hispidulous; leaflets 3–15 mm long, 1–3(–4) mm wide, glabrous, subfalcate, 2–several-costate, ciliate or entire, apiculate, the base asymmetrically rounded; inflorescences axillary, racemose, 3–10(–15)-flowered, the axes hispidulous; bracts cordate, acuminate, 1.5–6 mm long, 1–2 mm wide, ciliate; bracteoles ovate-lanceolate, acute to acuminate, 1–4 mm long, 0.5–1 mm wide, ciliate; flowers 3–7 mm long; calyx hispidulous, bilabiate, 2–4 mm long; petals pale yellow to purplish, glabrous; fruit 1–2 cm long, 3–7(–9)-seeded, the stipe 1.5–2 mm long, ˙the articulations distinct or sometimes lacking, the articles suborbicular, (2–)2.5–3(–4) mm in diameter, glandular-villous or hispid, the venation inconspicuous, the margins often breaking away from the body of the articles; seeds blackish, 2–2.5 mm long, 1.5–2 mm wide.

Distr. A weedy plant of tropical and subtropical America, in wet or dry places. Introduced in Ceylon, Malaysia, Australia, and New Guinea.

Specimens Examined. Plants grown at the Agricultural Research Center, Fort Pierce, Florida, United States, from seed collected in Ceylon, Maha Illuppallama, Dry Zone Station, Oct. 23, 1978, *IRFL 2299, P.I. 420271* (SFV, US), sent by Dr. A.E. Kretschmer, Jr.

Note. This is a potential weedy escape often confused with *A. americana*. The material introduced in the Old World appears to be referable to the typical variety *villosa*.

2. ORMOCARPUM

Beauv., Fl. d'Oware 1: 95, t. 58. 1806, nom. cons. Type species: *O. verrucosum* Beauv.

Diphaca Lour., Fl. Cochinch. 453. 1790. Type species: *D. cochinchinensis* Lour.
Rathkea Schum. & Thonn., Beskr. Guin. Pl. 355. 1827. Type species: *R. glabra* Schum. & Thonn.
Hormocarpus Spreng., Gen. 594. 1831, based on *Ormocarpum* Beauv.
Russelia Koenig. ex Roxb., Fl. Ind. ed. 2, 3: 364. 1832, nomen in synon., non Jacq. 1760 nec L. f. 1781.
Acrotaphros Steud. ex A. Rich., Tent. Fl. Abyss. 1: 207, t. 38. 1847. Type species: *A. bibracteata* Steud. ex A. Rich.
Saldania Sim, Forests Fl. Port. E. Afr. 41: 1909. Type species: *S. acanthocarpa* Sim.

Shrubs or small trees; leaves alternate, usually imparipinnate, 1- or ∞-foliolate; leaflets alternate or subalternate; stipules striate, attached at the base; stipels lacking; inflorescences usually racemose or the flowers solitary; calyx campanulate with 5 subequal lobes, acute or attenuate; petals yellowish, sometimes with reddish or purplish lines; stamens 10, monadelphous or diadelphous 5:5, 4:1:5, or essentially free; anthers uniform; fruit straight, 1–9-articulate; seeds subreniform or asymmetrically ellipsoid with the hilum near one end; chromosome numbers 2n = 24, 26.

A genus of about 20 species in Africa, Madagascar, southern Asia to the Philippines, northern Australia, and Fiji.

KEY TO THE SPECIES

1 Straggling shrub or small tree to about 5 m tall; young stems glabrous or glandular-hispid; flowers about 10–15 mm long, the calyx 5–6 mm long, glabrous or glandular-hispid; fruit to about 8 cm long, the articles 10–14 mm long, 5–6 mm wide, glabrous to glandular hispid.............
....1. O. sennoides
1 Shrub or small tree to about 7.5 m tall; young stems glabrous; flowers about 10–24 mm long, the calyx 8–12 mm long, glabrous; fruit to about 17 cm long, the articles glabrous, 15–25 mm long, 7–9 mm wide...2. O. cochinchinense

1. Ormocarpum sennoides (Willd.) DC., Prod. 2: 315. 1825; Wight & Arn., Prod. 216. 1834; Wight, Ic. Pl. Ind. Or. 297. 1840; Thw., Enum. Pl. Zeyl. 85. 1864; Baker in Hook. f., Fl. Br. Ind. 2: 152. 1876; Trimen, Handb. Fl. Ceylon 2: 39. 1894; Gillett, Kew Bull. 20: 332. 1966; Gillett, Fl. Trop. East Africa, Papil. 335. 1971; Verdcourt, Kirkia 9: 362. 1974.

Hedysarum sennoides Willd., Sp. Pl. 3: 1207. 1803; Moon, Cat. 54. 1824; Roxb., Fl. Ind. ed. 2, 3: 364. 1832. Syntypes: India, *Klein 214* (B, K-photo); Nandaradsh, *s. coll. 740* (B, K-photo).
Ormocarpum cochinchinense (Lour.) Merr., Philipp. J. Sci. 5: 76. 1910, pro parte, non quoad typum, fide Gillett, l.c. 1966.

Straggling shrub or small tree, to about 5 m tall; young stems glabrous or viscid with swollen-based hairs; stipules lanceolate to ovate, acuminate, glandular-hispid, 3–4(–7) mm long, 1.5 mm wide; leaves about 6–16 cm long, 6–17-foliolate, sometimes to 24-foliolate, the petiole and rachis glabrous to glandular-setose; leaflets glabrous, about 0.5–1.5(–2) cm long, 2–10(–12) mm wide, elliptic to elliptic-oblong, obtuse or slightly emarginate, apiculate, rounded to subacute at the base; inflorescences axillary, racemose, 2–13-flowered, the axes glandular-setose; bracts somewhat scarious, ovate-acuminate, setose along the margin, often trifid, 1.5–2 mm long; bracteoles ovate-acuminate, 2–3 mm long; flowers about 10–15 mm long; calyx 5–7 mm long, glabrous or glandular-hispid; petals yellow, sometimes with reddish lines, glabrous; ovary 5–8(–11)-ovulate, glabrous to glandular-hispid; fruit to about 8 cm long, longitudinally striate, glabrous to densely glandular-hispid, commonly 1–4(–6)-articulate, the articles about 10–14 mm long, 5–6 mm wide, the stipe about 2 mm long.

Distr. Africa, India, and Ceylon, usually in shady woodland.

KEY TO THE SUBSPECIES

1 Ovary and fruit densely glandular-hispid..........................**1.** subsp. **sennoides**
1 Ovary and fruit glabrous or glandular-hispid only along margins........**2.** subsp. **hispidum**

1. subsp. **sennoides**

Characterized by densely glandular-hispid fruit.

Distr. India and Ceylon.
Specimens Examined. WITHOUT EXACT LOCALITY: Gonagama, *Gardner s.n.* in 1848; *Thwaites* ? in Apr. 1854; Bibile, Mar. 1858, *Thwaites C.P. 1438* (BM, K, P, PDA). ANURADHAPURA DISTRICT: Vijitapura, near Kalawewa, Feb. 1888, Trimen ? (PDA). HAMBANTOTA DISTRICT: Tissamaharama, 20 Dec. 1882, Trimen ? (PDA); Ruhuna Natl. Park, Andunoruwa Wewa, *Mueller-Dombois & Cooray 67120605* (PDA); Situlpahuwa- Palatupana Rd., *Comanor 901* (K, PDA, SFV, US); S. of Situlpahuwa, *Mueller-Dombois 68102104* (K, PDA).

2. subsp. **hispidum** (Willd.) Brenan & Léonard, Bull. Jard. Bot. Etat. 24: 103, f. 14, 2. 1954; Léonard, Fl. Congo Belge 5: 244. 1954; Gillett, Kew Bull. 20: 334. 1966.

Cytisus hispidus Willd., Sp. Pl. 3: 1121. 1803. Type: "Guinea", probably Ghana, *s. coll. s.n.* (B, K-photo).

Characterized by fruits with the articles glabrous or hispid only along the margins.

Distr. Africa and Ceylon.
Specimens Examined. MATALE DISTRICT: Matale, in 1819, *Moon s.n.* (BM).
Note. Gillett (1966, p. 334) refers to a citation by Léonard of *Thwaites 1423*. I have not seen that specimen but suspect that it was a misreading of *C.P. 1438*, in many cases a mistake easily made. Since Thwaites' C.P. numbers are a medley of collections, one sheet could easily be of subsp. *hispidum*.
Alston (in Trimen, Handb. Fl. Ceylon 6: 77. 1931) substituted the name *O. cochinchinense* (Lour.) Merr. as an earlier synonym for *O. sennoides* as used by Trimen in 1894. However, the two species are now believed to be distinct.

2. Ormocarpum cochinchinense (Lour.) Merr., Philipp. Jour. Sci. 5: 76. 1910; Gillett, Kew Bull. 20: 335. 1966.

Diphaca cochinchinensis Lour., Fl. Cochinch. 454. 1790. Type: Cochinchina (Viet Nam), *Loureiro s.n.* (BM).
Dalbergia diphaca Pers., Syn. Pl. 2: 276. 1807, based on *Diphaca cochinchinensis* Lour.

Parkinsonia orientalis Spreng., Syst. 4, Cur. Post. 170. 1827, based on *Diphaca cochinchinensis* Lour.

Ormocarpum orientale (Spreng.) Merr., Interp. Rumph. Herb. Amboin. 266. 1917; Backer & Bakh. f., Fl. Java 1: 598. 1963; Verdcourt, Man. New Guinea Leg. 364. 1979.

Ormocarpum glabrum Teijsm. & Binn., Tijdschr. Ned.-Indie 27: 56. 1864; Backer & Bakh. f., Fl. Java 1: 598. 1963 as synonym of *O. orientale*. Type: Seram (Ceram), Indonesia, *Teijsmann s.n.*

Shrub or small tree, to about 7.5 m tall; young stems yellowish-brown, glabrous; leaves 9–20-foliolate; leaflets glabrous, about 1.5–5 cm long, 0.7–2 cm wide, oblong-obovate, usually obtuse, emarginate at the apex, rarely acute, the base rounded to acute; inflorescences axillary, short, fasciculate; flowers 10–24 mm long; calyx about 8–12 mm long; petals whitish, sometimes with red or purple lines; ovary glabrous or sparsely pubescent, 1–8-ovulate; fruit to about 20 cm long, longitudinally striate, glabrous, 1–7-articulate, the articles about 15–25 mm long, 7–9 mm wide.

Distr. In moist areas of Southeast Asia, the Celebes and Moluccas; introduced elsewhere.

Uses. According to Verdcourt (p. 365) the leaves are used as a vegetable.

Specimens Examined. KANDY DISTRICT: Peradeniya, Botanic Gardens, 11 Aug. 1899, *s. coll. s.n.* (PDA), *Alston 359* (K), *813* (K).

Note. My thanks to Dr. Robert Geesink of the Rijksherbarium, Leiden, for comments and information concerning this species and its synonyms.

3. SMITHIA

Ait., Hort. Kew ed. 1, 3: 496, t. 13. 1789, nom. cons. Type species: *S. sensitiva* Ait.

Damapana Adans., Fam. 2: 323. 1763, nom. rej. Type: "H.M.9. t. 38" [*Smithia sensitiva* Ait.].

Petagnana Gmel., Syst. 1078. 1791. Type species: *P. sensitiva* (Ait.) Gmel.

Herbs or small shrubs; leaves alternate, paripinnate, about 10–28-foliolate; leaflets small; stipules appendiculate below the point of attachment; stipels lacking; inflorescences axillary, subumbellate, scorpioid-cymose, or terminal, paniculate; flowers with calyx sometimes scarious, accrescent, enclosing the fruit at maturity, bilabiate, the lips entire or slightly toothed or lobed; petals yellow, sometimes with brownish or reddish markings; stamens 10, diadelphous 5:5 with filaments alternately long and short, the anthers uniform; fruit (1–)2–9-articulate, commonly about 5 or 6-articulate, plicate, usually not exserted; chromosome number 2n = 38.

A genus of about 30 species in the Old World tropics, chiefly in Asia and Madagascar.

KEY TO THE SPECIES

1 Calyx scarious, rigid, with close parallel veins, the lips entire, acute; flowers in axillary inflorescences; leaflets oblong or linear-oblong
  2 Flowers (1–)2–6(–10) in short, simple, axillary racemes; leaflets oblong or linear-oblong, glabrous above, bristly on midrib and along margins; calyx lips with a few scattered bristles........
  ....**1. S. sensitiva**
  2 Flowers 1–4, axillary; leaflets oblong, glabrous above, very bristly-ciliate along the margins and midrib on lower surface, with additional, smaller bristles; calyx with lower lip glabrous or with a few bristles, especially at apex....................................**2. S. conferta**
1 Calyx membranous with anastomosing veins, the lips lobed; flowers in terminal panicles; leaflets oval-oblong to obovate..**3. S. blanda**

1. Smithia sensitiva Ait., Hort. Kew 3: 496, tab. 13. 1789; Roxb., Fl. Ind. ed. 2, 3: 343. 1832; Baker in Hook. f., Fl. Br. Ind. 2: 148. 1876; Trimen, Handb. Fl. Ceylon 2: 37. 1894; van Meeuwen in van Steenis, Reinwardtia 5: 444. 1961. Type: India, *Koenig s.n.* (BM).

Petagnana sensitiva (Ait.) Gmel., Syst. 1119. 1791.

Herbaceous annual; stems erect or spreading, to about 1.5 m long, glabrous; leaves 6–24-foliolate; leaflets linear-oblong, 4–15 mm long, 2–3 mm wide, bristly-ciliate along the margins and midrib on lower surface, otherwise glabrous; flowers (1–)2–6(–10) in short, simple, axillary racemes at the ends of peduncles usually longer than the leaves; calyx scarious, rigid, about 6–9 mm long with close, parallel veins, the lips entire, acute, with a few scattered bristles; petals yellow, sometimes marked with red, 8–9 mm long; fruit about 4–6(–7)-articulate, the articles minutely papillose, about 1.7 mm long and wide.

Distr. In moist areas, pastures, etc., a native of "East Indies", apparently introduced in Ceylon, India, and Australia.

Uses. According to Roxburgh (l.c.) it "makes excellent hay".

Specimens Examined. MATALE DISTRICT: "Kaluganga, Lagalla-Beckett", *Thwaites C.P. 3946* (PDA); "Lenadore, near Nalande", Feb. 1893, *Trimen ? s.n.* (PDA).

Note. Aiton states that this species was introduced to India in 1785 by Koenig. Trimen cites *S. sensitiva* sensu Moon as *S. conferta.*

2. Smithia conferta Smith in Rees, Cyclop. 33, no. 2. 1816; Moon, Cat. 54. 1824, as *S. sensitiva;* Wight & Arn., Prod. 220. 1834; Thw., Enum. Pl. Zeyl. 85. 1859; Baker in Hook. f., Fl. Br. Ind. 2: 149. 1876; Trimen, Handb. Fl. Ceylon 2: 27. 1894; Alston, Kandy Fl. 32, f. 168. 1938; van Meeuwen in van Steenis, Reinwardtia 5: 445. 1961. Type: Australia ("New Holland"), *Banks* (BM).

Smithia geminiflora Roth, Nov. Pl. Sp. 352. 1821; Gamble, Fl. Pres. Madras
2: 328, 329. 1918. Type: India, *Heyne,* "India orientali".
Smithia sensitiva var. β Wight & Arn., Prod. 220. 1834, based in part on *S.
geminiflora* Roth.
Smithia geminiflora Roth var. *conferta* (Rees) Baker in Hook. f., Fl. Br. Ind.
2: 149. 1876.
Smithia conferta Rees var. *geminiflora* (Roth) Cooke, Fl. Pres. Bombay 1: 336.
1901.

Herbaceous annual; stems spreading, to about 1 m long, glabrous; leaves
(4–)8–16-foliolate; leaflets oblong, obtuse or subacute, mucronate, about 6–12
mm long, 2–4 mm wide, glabrous above, bristly-ciliate along the margins and
along the midrib on lower surface; flowers 1–4 in the axils of upper leaves;
bracteoles scarious, to about 5 mm long, acute, with a long bristle at the apex
and a few on the back; calyx scarious, rigid, about 4–6 mm long, with close,
parallel veins, the lips entire, acute, glabrous except for a tuft of bristles at apex
of lower lip or with a few scattered bristles on back; petals yellow, sometimes
with reddish markings, about 8–12 mm long; fruit 3–6(–7)-articulate, the articles
usually papilose, about 2 mm long and wide.

Distr. Ceylon and southern India, in moist open areas at elevations up to
about 1300 metres.

Specimens Examined. WITHOUT PRECISE LOCALITY: *"Mrs. & Col.
Walker" s.n.* (K); "Hantane, July 1847, Gardner", "Maturata, Aug. 1853",
Thwaites C.P. 2777 (BM, K, P, PDA); "Hantane, 2300 ft", *Gardner 216* (BM,
K). KALUTARA DISTRICT: "Caltura", *Macrae 233* (BM, K); Delgoda, 23
Mar. 1919, *F. Lewis & J.M.S. (de Silva ?)* (PDA). BADULLA DISTRICT:
"Dikwelle & Ella, Uva", Sept. 1890, *Trimen? s.n.* (PDA); "between Welimada
& Atampitiya", Jan.1888, *Trimen ? s.n.* (PDA).

3. Smithia blanda Wall. ex Wight & Arn., Prod. 221. 1834; Wight, Ic. Pl. Ind.
Or. 986. 1845; Thw., Enum. Pl. Zeyl. 85. 1859; Trimen, Handb. Fl. Ceylon
2: 37. 1894. Type: India, *Wallich no. 5669* (K holotype; BM isotype).

Smithia racemosa Heyne in Wall., Cat. ex Wight & Arn., Prod. 221. 1834. Type:
Wallich no. 5670
Smithia paniculata Arn., Nov. Actorum Acad. Caes. Leop.-Carol. Nat. Cur. 18:
330. 1836. Type: Ceylon, Nuwara Eliya ("monte Neueri-Ellia ad scaturigines
alt. 1000 hexap."), *Walker, s.n.* (E-GL).
Smithia blanda var. *paniculata* (Arn.) Baker in Hook. f., Fl. Br. Ind. 2: 151. 1876.
Smithia blanda var. *racemosa* (Heyne ex Wight & Arn.) Baker in Hook. f., Fl.
Br. Ind. 2: 151. 1876.

Perennial or, sometimes, annual herb; stems erect or spreading, hispid or glabrous, to about 6 dm tall; leaves 4–12-foliolate; leaflets oval-oblong to obovate or cuneate, about (3–)5–15(–20) mm long, (1–)2–3(–6) mm wide, obtuse, the upper surface glabrous, the lower glabrous or with a few setae, the margins entire or bristly-ciliate; flowers about 10–14 mm long in terminal panicles; calyx membranous with anastomosing veins, 4–7 mm long, the lips lobed, hispidulous; petals bright yellow marked with red; fruit 1–4(–7)-articulate, the articles suborbiculate, 2–2.5 mm in diameter, reticulate, glabrous.

Distr. In moist areas of forest and patana, up to about 2000 metres or more, in Ceylon and India.

Specimens Examined. WITHOUT LOCALITY: *"Col. Walker" s.n.* (K, P); *Col. Walker 171* (PDA); *G. Thomson s.n.* (K); *"Dr. Maxwell" s.n.* (K); *Gardner 12* (K). NUWARA ELIYA DISTRICT: *"Rambodda & N. Ellia, Gardner"*, *Thwaites C.P. 58* (BM, PDA); "Rambodda, 2300 ft", *Gardner 217* (K), "3500 ft", *s. coll. s.n.* (BM); "Neuer Ellia, 6000 ft", *Gardner 218* (BM, BR, K); Pattipola, Horton Plains, Sept. 1890, *Trimen ? s.n.* (PDA); Patana, Sita Eliya, 3/10/06, *A.M.S. (Silva ?)* (PDA); World's End, Bogowanthalawa footpath, *Waas 157* (SFV, US); Patanas, opposite Hakgala Gardens, *Balakrishnan 1061* (K, US); between Ramboda and Nuwara Eliya, roadmarker 36/A5, 1180 m elev. *Mueller-Dombois 67112309* (US); between Nuwara Eliya and Kande Ela Reservoir, *Mueller-Dombois 68051803* (US); Nuwara Eliya, *Carrick 1262* (K); high patana above Deltota Estate, Pussellawa, *Simpson 8761* (BM, PDA). BADULLA DISTRICT: Ambawela, 1940 m elev., *Mueller-Dombois & Comanor 67091318* (PDA, US); Haputale, patana on hill above Monamaya, *Rudd & Balakrishnan 3204* (PDA, SFV, US).

Note. As mentioned by Trimen (p. 38) this is a variable plant. Among the few specimens known from Ceylon there is considerable variation in size of plants and in indument.

4. ZORNIA

Gmel., Syst. Nat. 2: 1076, 1096. 1791. Type species: *Z. bracteata* [Walt.] Gmel.

Myriadenus Desv., J. Bot. 3: 121. 1813. Type species: *M. tetraphyllus* Desv.

Herbs, annual or perennial; leaves alternate, digitately paripinnate with 1 or 2 pairs of leaflets, rarely 3-foliolate; leaflets sometimes glandular-punctate; stipules peltate, appendiculate below the point of attachment; stipels lacking; inflorescences terminal, spicate or rarely racemose, 1–many-flowered; bracts and bracteoles conspicuous, peltate, usually persistent, often dark-glandular-punctate; flowers usually enclosed by bracteoles; calyx essentially campanulate, 5-lobed, the vexillar lobes somewhat joined; petals yellowish, glabrous; stamens 10, monadelphous with anthers alternately long, sub-basifixed and shorter, dorsifixed; ovary sessile, 5–8-ovulate; style filiform; stigma minute, capitate; fruit about 2–15-articulate, compressed, constricted between the seeds, the articles usually suborbicular, reticulate, often with barbed hairs; seeds subreniform, usually black or dark brown; chromosome number 2n = 20.

A pantropical and warm temperate genus of about 80 species.

KEY TO THE SPECIES

1 Fruit spiny with barbed setae
  2 Mature articles of fruit about 4–4.5 mm long, 3–4 mm wide; the spines barbed at the tip only; leaflets ovate, 10–20 mm long, 7–12 mm wide, glabrous except along the midrib on lower sur-face; bracts somewhat acute at base.................................. **1. Z. diphylla**
  2 Articles of fruit about 2–2.5 mm long, 1.5–2 mm wide; the spines retrorsely barbed; leaflets lanceolate, (0.5–)10–30 mm long, 2–5(–7) mm wide; bracts acute at base ... **2. Z. gibbosa**
1 Fruit mostly not spiny, appressed-pubescent, sometimes with a few short bristles; leaflets ovate, (2–)5–15 mm long, 2–6 mm wide.......................................**3. Z. walkeri**

1. Zornia diphylla (L.) Pers., Syn. 2: 318. 1807; Mohlenbrock, Webbia 16: 67. 1961; Dandy & Milne-Redhead, Kew Bull. 17: 73. 1963.

Hedysarum diphyllum L., Sp. Pl. 747. 1753. Lectotype: Ceylon, *P. Hermann*, Herm. Herb. vol. 2, fol. 14 (BM).

Hedysarum diphyllum var. *β* L., Sp. Pl. 747. 1753, based on Burm., Thes. Zeyl. 114, t. 50, fig. 1. 1737. Typotype: Ceylon, *P. Hermann* (Biblio. Inst. France, Paris, coll. Pl. Zeyl. p. 77).

Hedysarum conjugatum Willd., Sp. Pl. 3: 1178. 1803, based on *H. diphyllum* var. *β* L. Moon, Cat. 54. 1824.

Zornia conjugata (Willd.) Smith in Rees, Cyclop. 39, no. 3. 1820; Thw., Enum. Pl. Zeyl. 85. 1859; Alston in Trimen, Handb. Fl. Ceylon 6: 79. 1931.

Zornia zeylonensis Pers., Syn. Pl. 2: 318. 1807, based on *Hedysarum conjugatum* Willd.

Zornia diphylla var. *zeylonensis* (Pers.) Benth. in Mart., Fl. Bras. 15 (1): 82. 1859; Baker in Hook. f., Fl. Br. Ind. 2: 148. 1876.

Zornia diphylla var. *β conjugata* (Willd.) Trimen, Handb. Fl. Ceylon 2: 35. 1894.

Perennial herb; stems erect or prostrate, glabrous or puberulent, to about 60 cm long; leaves 2-foliolate; stipules about 5–10 mm long, glabrous, epunctate or sparsely punctate; leaflets ovate, (5–)10–20 mm long, (2–)5–13 mm wide, acute, ciliate, otherwise glabrous, apparently epunctate; inflorescences 5–10 cm long; bracts ovate-lanceolate, acute, 8–10 mm long, 3–5 mm wide, the auricle about 2 mm long, epunctate; flowers about 10–12 mm long; calyx 3–4 mm long, glabrous; fruit exserted at maturity, 3–4-articulate, the articles 4–4.5 mm long, 3–4 mm wide, glabrous except for numerous spines with only a few reflexed barbs at the apex.

Distr. Apparently restricted to Ceylon and India.
Vern. Maha-aswenna ("Mahaswaenna" fide Hermann) (S).

Specimens Examined. WITHOUT LOCALITY: *Sir J.S. Mackenzie s.n.* (K); *Col. Walker s.n.* (K). PUTTALAM DISTRICT: Wilpattu Natl. Park, Kali Villu, *Cooray 70020201R* (PDA, SFV, US), *Cooray & Balakrishnan 69050104* (K, PDA, SFV, US). COLOMBO DISTRICT: Colombo, *Thwaites C.P. 3600* (BM, K, P), Aug. 1858, *Ferguson s.n.* (PDA); Negombo, 2 July 1930, *de Silva s.n.* (PDA); Kochchikade, sandy coconut estate, *Simpson 7967* (BM).

2. Zornia gibbosa Spanoghe, Linnaea 15: 192. 1841; Mohlenbrock, Webbia 16: 112. 1961; Ali, Fl. West Pakistan, no. 100. 1977. Type: Timor, Kupang (Koepang), *Spanoghe*? (untraceable fide Ali). Neotype: Vietnam (Cochinchina), Cai-Cong, *Thorel 1426* (US), selected by Mohlenbrock, Webbia 16: 113. 1961.

Zornia angustifolia Smith in Rees, Cyclop. 39, no. 1. 1820. pro parte, nom. il-leg. based on *Hedysarum diphyllum* L.; Thw., Enum. Pl. Zeyl. 84. 1859. *Zornia graminea* Spanoghe, Linnaea 15: 192. 1841. Type: Timor.

Annual; stems prostrate to suberect, 5–50 cm long, glabrous to puberulent or strigillose; leaves 2-foliolate; stipules lanceolate, acuminate above and below the point of attachment, 5–9 mm long, glabrous, punctate; leaflets lanceolate, about 8–20 mm long, 2–6 mm wide, acute, apiculate at the apex, rounded at the base, glabrous or nearly so, sometimes dark-punctate; inflorescences about (2–)5–11 cm long; bracts 6–13 mm long, 3–4 mm wide, acute, glabrous or ciliate, punctate, the auricle 1–2.5 mm long; flowers (5–)10–12 mm long; calyx about 3 mm long, glabrous, sometimes ciliolate; petals yellow, sometimes with red stripes; fruit usually exserted at maturity, 4–6(–7)-articulate, the articles 2–2.5 mm in diameter, puberulent and with numerous retrorsely-barbed bristles.

Distr. Widespread in southern Asia from West Pakistan to China, Malesia, and Australia.

Specimens Examined. JAFFNA DISTRICT: 38 mi. from Elephant Pass, bed of canal, *Simpson 9308* (BM, PDA), *9309* (BM, PDA); between Paranthan and Pooneryn, *Rudd 3296* (PDA, US). ANURADHAPURA DISTRICT: Ritigala, *Waas 329* (US); between Medawachchiya and Vavuniya, mile marker 98/1, *Rudd 3260* (US); N. of Akirikanda, mile 98, *Fosberg & Balakrishnan 53468* (K, SFV, US), *53474* (SFV, US). PUTTALAM DISTRICT: W. of Kala Oya, *Jayasuriya & Moldenke 1437* (K, US); Wilpattu Natl. Park, 2.5 mi. S. of Maduru Odai, 5 mi. S. of Marai Villu, *Fosberg et al. 50848* (CAS, SFV, US). BATTICALOA DISTRICT: N. of Kalkudah on road to Elephant Point, *Rudd & Balakrishnan 3146* (PDA, SFV, US). MATALE DISTRICT: "Cent. Prov. 1837, Dambulla, 1868", *Thwaites C.P. 3598* (BM, PDA); Dambulla Rock, *Amaratunga 259* (PDA). HAMBANTOTA DISTRICT: Yala, main park road, *Comanor 660* (PDA, SFV, US); Ruhuna Natl. Park, Karaugaswala, *Mueller-Dombois & Cooray 67121041* (PDA, US), near Buttawa Bungalow, *Cooray 70031607R* (PDA, SFV, US), *Mueller-Dombois & Cooray 68020204R* (PDA, US), *Mueller-Dombois et al. 69010534* (PDA, SFV, US), *Cooray & Balakrishnan 69010920R* (PDA, US), *69011017R* (CAS, PDA, SFV, US); Kohambagaswala, *Cooray 69112703R* (PDA, US), *Cooray & Balakrishnan 69012113R* (PDA, SFV, US); Padikema, near Patanagala, *Fosberg et al. 51177* (PDA, SFV, US). AMPARAI DISTRICT: a few mi. S. of Panama, N. of Yala, *Koyama et al. 14019* (US).

Note. This species was cited by J.G. Baker in Hook. f., Fl. Br. Ind. 2: 148. 1876, Trimen, Handb. Fl. Ceylon 2: 35. 1894, and Alston in Trimen, Handb. Fl. Ceylon 6: 79. 1931 as *Zornia diphylla* L. The true *Z. diphylla* was given as *Z. conjugata* (Willd.) Smith.

3. Zornia walkeri Arn., Nov. Actorum Acad. Caes. Leop.-Carol. Nat. Cur. 18: 330. 1836; Thw., Enum. Pl. Zeyl. 85. 1859; Alston in Trimen, Handb. Fl. Ceylon 6: 77. 1931; Mohlenbrock, Webbia 16: 89. 1961. Type: Ceylon, *Col. Walker, s.n.* (E-GL).

Zornia diphylla L. var. *walkeri* (Arn.) Baker in Hook. f., Fl. Br. Ind. 2: 149. 1879; Trimen, Handb. Fl. Ceylon 2: 35. 1894.

Perennial herb; stems erect or spreading, to about 30 cm long, glabrous or nearly so; leaves 2-foliolate; stipules glabrous, ciliolate, epunctate or sometimes punctate, lanceolate above the point of attachment, attenuate, about 2–7 mm long, 1–1.5 mm wide, the auricle 1–1.5 mm long; leaflets ovate to lanceolate, (3–)5–15 mm long, 2–7 mm wide, acute or subacute, glabrous or nearly so, sometimes ciliolate, punctate or epunctate; inflorescences about 3–10 cm long; bracts about 6–10 mm long, 3–5 mm wide, acute, ciliate, otherwise glabrous, usually punctate, the auricle about 1 mm long; flowers about 9 mm long; calyx 3 mm long, glabrous; petals yellow; fruit 2–4-articulate, sometimes exserted at maturity, the articles 2 mm in diameter, usually exhibiting reddish reticulation, pubescent with minute appressed hairs, usually lacking bristles.

Distr. Apparently endemic to Ceylon at relatively high elevations.

Specimens Examined. WITHOUT LOCALITY *Thwaites C.P. 524* in part (BM, K, P); *s. coll. C.P. 3599* (BM, K, NY); "*Mrs. & Col. Walker*" *15* (P); "*Col. Walker*" *47* (K). KANDY DISTRICT: Hunnasgiriya, 2 Dec. 1926, *Alston s.n.* (PDA); between Madugoda and Hunnasgiriya, mile marker 24/19, *Rudd & Balakrishnan 3241* (K, PDA, SFV, US); Peradeniya, Hantane patana above University Circuit Bungalow, *Mueller-Dombois & Cooray 67111603* (PDA); Nugagalla, Hunnasgiriya, alt. 1100 m, *Jayasuriya 943* (K, PDA, US). NUWARA ELIYA DISTRICT: Hewaheta, Oct. 1852, *Thwaites C.P. 3599* (PDA); Galagama, Feb. 1846, *Gardner ?, Thwaites C.P. 524* (K); between Ramboda and Nuwara Eliya, above miles marker 36 A5, patana, 1180 m elev., *Mueller-Dombois 67112304* (PDA, US). BADULLA DISTRICT: Haputale, 4800 ft, 23 Mar. 1906, *Willis s.n.* (PDA); Haputale, patana on hill above Monamaya, *Rudd & Balakrishnan 3202* (CAS, K, PDA, SFV, US); Craig, Bandarawela; *Willis s.n.* (PDA); Badulla, Jan. 1888, *Trimen ? s.n.* (PDA).

Note. *Thwaites C.P. 524* is a mixture, later numbered as *C.P. 3598* and *3599.* The collections from lower elevations are referable to *Zornia gibbosa*, i.e. *3598*; those from higher to *Zornia walkeri, C.P. 3599.*

The original description of *Z. walkeri* does not give a collector's name but in a preface to the paper Arnott indicates that the species included are based on "a rich collection of plants made in Ceylon within these few years by the exertions of Colonel Walker, secretary to the Governor of that island, and his lady". Mr. A.J.C. Grierson, at the Royal Botanic Garden, Edinburgh, has kindly written to me that "we do have a specimen that came from the Glasgow herbarium of *Zornia walkeri* which we have tentatively marked as holotype because the name is in Arnott's handwriting but there is no mention on the sheet of Colonel Walker or his lady, nor any number." Presumably, that specimen is correctly indicated as the holotype. The Walker collections at K and P may or may not be isotypes.

5. STYLOSANTHES

Sw., Prod. Veg. Ind. Occ. 108. 1788; Sw., Fl. Ind. Occ. 3: 1280, t. 25. 1806. Lectotype species: *S. procumbens* Sw., nom. illeg. based on *Hedysarum hamata* L. = *Stylosanthes hamata* (L.) Taub.

Perennial herbs or subshrubs; stems erect or spreading, glabrous or hispid, sometimes glandular-hispid; leaves alternate, pinnate, subdigitate, (1–)3-foliolate; leaflets with resin deposits; stipules bilobed, adnate to the petiole, persistent; stipels lacking; inflorescences terminal or axillary, capitate or spicate, 1–many-flowered, some inflorescences with a filiform axis rudiment; bracts 1-3-"foliolate"; bracteoles hyaline; flowers with a long, filiform receptacle; calyx campanulate with 5 subequal lobes, filiform at the base; petals glabrous, yellow to orange, sometimes with purplish lines; stamens 10, monadelphous with the anthers alternately long, subbasifixed and shorter, dorsifixed; ovary sessile, 2–3-ovulate; style filiform with the lower portion persistent, usually recurved or revolute; stigma minute, terminal; fruit 1–2-articulate, compressed, usually reticulate, the terminal article beaked; seeds brown, compressed, subovoid or lenticular, somewhat beaked toward the hilum; chromosome number 2n = 20.

A pantropical and warm temperate genus of about 25 species.

KEY TO THE SPECIES

1 Fruit pubescent with a well developed beak.
  2 Beak of fruit curved but not uncinate, usually shorter than the terminal article; fruit usually with both articles fertile; flowers subtended by a rudimentary or vestigial axis. .. **1. S. fruticosa**
  2 Beak of fruit uncinate, usually longer than the terminal article; fruit usually with only the terminal article fertile; rudimentary axis lacking .... **3. S. humilis**
1 Fruit glabrous or nearly so with a short beak; flowers lacking a rudimentary axis.........
  .... **2. S. guianensis**

1. Stylosanthes fruticosa (Retz.) Alston in Trimen, Handb. Fl. Ceylon 6: 77. 1931; Mohlenbrock, Ann. Missouri Bot. Gard. 44: 318. 1957; Verdcourt, Kew Bull. 24: 59. 1970.

Hedysarum hamatum L., Syst. Veg. ed. 10. 1170. 1757, pro parte quoad Burm., Thes. Zeyl. 226, t. 106, f. 2. 1737.

Arachis fruticosa Retz., Obs. Fasc. 5: 26. 1788. Type: "Zeylonae, tranquebariae aridis", *Koenig*.

Stylosanthes mucronata Willd., Sp. Pl. 3: 1166. 1803 based on *Arachis fruticosa* Retz.; Burm., Fl. Ind. 167. 1768; Baker in Hook. f., Fl. Br. Ind. 2: 148. 1876.

Stylosanthes bojeri Bog., Linnaea 12: 68 1838. Type: Zanzibar, *Bojer*.

Stylosanthes flavicans Baker in Oliver, Fl. Trop. Africa 2: 156. 1871. Type: Nile Land, Kordofan, *Kotschy 425* (K).

Suffrutescent herb with a woody rootstock; stems erect or prostrate, to about 1 m tall, puberulent and hispid with thick-based glandular setae; stipules about 5–10 mm long; leaves 3-foliolate, the axis about 3–10 mm long; leaflets elliptic to lanceolate, 5–20(–33) mm long, 1–5(–9) mm wide, acute to obtuse at the apex and base, apiculate, glabrous or subglabrous above, glabrous to puberulent below, setose along the margins and veins; inflorescences axillary, spicate, 1–1.5(–4) cm long, about 4–10-flowered, subtended by stipule-like primary bracts 8–20 mm long and secondary bracts 3–6 mm long; bracteoles 2, 3–5 mm long, hyaline; flowers subsessile, subtended by a plumose, vestigial axis-rudiment; receptacle 4.5–7 mm long; calyx lobes 2–4 mm long, 0.7–1 mm wide; petals yellow, 5–7 mm long, sometimes marked with red, sometimes with a white standard; fruit 4–9 mm long, 1- or 2-articulate, the articles pilose, reticulate, 1.5–4 mm long, 1.5–2.5 mm wide, the beak curved, 1.5–2.5 mm long; seeds chestnut brown, 2.5–3 mm long, 2–2.5 mm wide, 1–1.2 mm thick.

Distr. Ceylon, India, tropical Africa, Madagascar, Malaysia, and the Lesser Sunda Islands.

Specimens Examined. WITHOUT PRECISE LOCALITY: "Mineri, Habarana, Jaffna, *Gardner &*" *Thwaites*, Mar. 1858, *C.P. 1451* (PDA), s. loc., s. coll. *C.P. 1451* (BM, K, PDA). JAFFNA DISTRICT: Between Mulliyan and Chempiyanpattu, c. 32 mi. ESE of Jaffna, *Townsend 73/111* (K, US); near Karativu, *Bernardi 15323* (US). VAVUNIYA DISTRICT: 4 mi. N. of Mankulam, *Fosberg & Balakrishnan 53554* (K, SFV, US); Mankulam, *Alston 1433* (PDA). TRINCOMALEE DISTRICT: Between Trincomalee and Kanniyai, 24 July 1921, *de Silva s.n.* (PDA). ANURADHAPURA DISTRICT: Mihintale, 19 Aug. 1925, *de Alwis s.n.* (PDA); Tantirimale, *Sumithraarachchi 753* (SFV, US); Ritigala, *Waas 330* (K, US); between Medawachchiya and Vavuniya, mi. marker 98/1, *Rudd 3259* (K, PDA, SFV, US). POLONNARUWA DISTRICT: Polonnaruwa, *Townsend 73/227* (K, US). PUTTALAM DISTRICT: Karativu I. Aug. 1883, *Trimen? s.n.* (PDA); Kalpitiya, *Simpson 9177* (BM); W. of Kala Oya, *Jayasuriya*

& *Moldenke 1436* (K, US); Hill overlooking ocean, W. of Pomparippu Sanctuary, *Maxwell & Jayasuriya 807* (SFV, US); Wilpattu Natl. Park, Kokmotae bungalow on Moderagam Aru, *Maxwell & Jayasuriya 813* (US). MATALE DISTRICT: Dambulla, *Kostermans 23530D* (K, US). c. 8 mi. ESE of Dambulla, hills above the Mahaweli Project Anicut, *Davidse 7432* (US). BATTICALOA DISTRICT: N. of Kalkudah on road to Elephant Point, *Rudd & Balakrishnan 3147* (CAS, PDA, SFV, US). HAMBANTOTA DISTRICT: Ruhuna Natl. Park, Walaskema Rocks, *Mueller-Dombois, Comanor & Cooray 67100112* (PDA, US); between Buttawa Plain and Buttawa Bungalow, *Wirawan 672* (K, PDA, US); near Gonnelabbe Lewaya, *Fosberg & Sachet 52901* (K, SFV, US); near Buttawa Bungalow, *Mueller-Dombois & Cooray 68040603* (US), *69010539* (K, SFV, US), *Cooray & Balakrishnan 69011025R* (K, SFV, US), *69012011R* (K, SFV, US), *Cooray 69120206R* (K, UC, US); Uraniya, *Mueller-Dombois & Balakrishnan 69010519* (US).

2. Stylosanthes guianensis (Aubl.) Sw., Kongl. Svenska Vetenskapsakad. Handl. 10: 301. 1789; Mohlenbrock, Ann. Missouri Bot. Gard. 44: 330. 1957; Verdcourt in Fl. Trop. East Africa, Papil. 438. 1971; t'Mannetje, Austral. J. Bot. 25: 347–362. 1977; Verdcourt, Man. New Guinea Leg. 373. 1979.

Trifolium guianense Aubl., Hist. Pl. Guian. Fr. 2: 776; 4: pl. 309. 1775. Type: French Guiana, Macouria, *Aublet s.n.* (BM).

Extensive synonymy including six varieties is given by t'Mannetje, 1977, cited above.

Herb, usually suffrutescent, erect, to about 1–1.5 m tall, sometimes prostrate; stems glabrous to hispid; stipules 4–25 mm long; leaves 3-foliolate, the axis 2–15 mm long; leaflets elliptic-oblong to lanceolate, commonly 1–4 cm long, 2–10 mm wide, acute, apiculate at the apex, cuneate at the base, glabrous to puberulent, sometimes hispidulous; inflorescences terminal or axillary, spicate and few-flowered to subglobose, 1.5 cm in diameter, 2–40-flowered; primary bracts 10–22 mm long; secondary bracts 2.5–5.5 mm long; bracteole 1, 2–4.5 mm long; axis rudiment lacking; flowers with receptacle 4–8 mm long; calyx lobes 3–5 mm long, 1–1.5 mm wide; petals 4–8 mm long, yellow to orange with reddish streaks; fruit with only the terminal article fertile, 2–3 mm long, 1.5–2.5 mm wide, glabrous or nearly so, sometimes minutely pubescent near the apex, the beak 0.1–0.5 mm long, strongly inflexed; seeds pale brown, about 2.2 mm long, 1.5 mm wide, 0.8 mm thick.

Distr. Native to tropical America; introduced elsewhere in the tropics.

Uses. Planted as a cover crop and nitrogen source. More recently this and other species of *Stylosanthes* are being studied for use as pasture and forage legumes. (t'Mannetje et al. in Advances in Legume Science, ed. Summerfield and Bunting, 537–551. 1980; Burt et al. l.c. 553–558.)

Specimens Examined. MONERAGALA DISTRICT: Namaloya-Mullegama, *Jayasuriya 2044* (K). BADULLA DISTRICT: Rahangala, between Ohiya and Boralanda, *Rudd & Balakrishnan 3186* (K, SFV, PDA, US). RATNAPURA DISTRICT: Vicinity of Rakwana to Bulutota Pass, *Maxwell et al. 983* (K, US). MATARA DISTRICT: Deniyaya, *Tirvengadum & Balasubramaniam 328* (K, US).

3. Stylosanthes humilis H.B.K., Nov. Gen. & Sp. 6: 506, t. 594. 1824; Mohlenbrock, Ann. Missouri Bot. Gard. 44: 345. 1957; Nooteboom in van Steenis, Reinwardtia 5: 449. 1961; Verdcourt, Man. New Guinea Leg. 373. 1979. Type: Venezuela, Amazonas, Carichana, Orinoco R., *Humboldt & Bonpland*.

Stylosanthes sundaica Taub., Verh. Bot. Vereins Prov. Brandenburg 32: 21. 1890.
   Type: Malaysia.

Herb, usually perennial, prostrate or erect, about 1–5 dm high; stems glabrous to puberulent or hispid; stipules 6–10 mm long, often reddish; leaves with axis 5–10 mm long; leaflets lanceolate, acute at apex and base, 5–30(–45) mm long, 1–5 mm wide, glabrous above, puberulent to setulose below; inflorescences axillary, spicate, about 1 cm long, dense, about 3–10-flowered; bracts 1–3-"foliolate"; bracteoles (1)2, 2–3 mm long, flowers with receptacle 5 mm long, an axis rudiment usually lacking; calyx lobes acute, about 1.5 mm long; petals 3–4 mm long, yellow to orange, sometimes with reddish streaks; fruit 1.5–2.5(–4) mm long excluding uncinate beak (1.5–)5–7 mm long, usually only the terminal article fertile, 1.5–2.5 mm wide, puberulent.

Distr. Native to tropical America, Mexico to northern South America; introduced in Malaysia, New Guinea, and Australia; in Ceylon apparently only in experimental plantings.
   Specimens Examined. MATALE DISTRICT: Pelwehera, Agricultural Station, 1 Nov. 1949, *Senaratna s.n.* (PDA). KANDY DISTRICT: Peradeniya, Royal Botanic Gardens, 5 Aug. 1953, *Appuhamy s.n.* (PDA).

6. ARACHIS

L., Sp. Pl. 741. 1753; L., Gen. Pl. ed. 5. 329. 1754. Type species: *A. hypogaea* L.

Herbs, prostrate to erect, annual or perennial; leaves alternate, paripinnate with 2 pairs of leaflets or, in 1 or 2 species, 3-foliolate; stipules with lower part adnate to the petiole; stipels lacking; flowers axillary, solitary or in short spikes or racemes, the receptacle long, filiform; calyx long, filiform at base, somewhat bilabiate at the apex with the vexillar lobes connate, the carinal lobes separate; petals yellowish, sometimes with reddish lines; stamens 8–10, monadelphous with anthers alternately long and basifixed and short, versatile; ovary subsessile, at the base of the receptacle tube, to 14-ovulate; style long, filiform, the stigma minute, terminal; fruit essentially indehiscent, terete, submoniliform, not articulate, commonly 1–6-seeded, developing underground, the ovary pushed into the soil by the reflexed, lengthening, gynophore; seeds ovoid to oblong, the cotyledons fleshy, the hilum near one end; chromosome numbers 2n = 20, 40.

A genus of about 20 known species with, possibly, 40 as yet undescribed; native to South America, introduced in tropics and warm temperate regions elsewhere.

For additional information concerning *Arachis* see Advances in Legume Science, eds. Summerfield and Bunting 469–535. 1980.

Arachis hypogaea L., Sp. Pl. 741. 1753; Moon, Cat. 52. 1824; Baker in Hook. f., Fl. Br. Ind. 2: 161. 1876; Verdcourt in Fl. Trop. East Africa, Papil. 442. 1971; Ali, Fl. W. Pakistan, no. 100, 357. 1977; Verdcourt, Man. New Guinea Leg. 381. 1979. Type: "Brazil and Peru", grown at Hort. Ups. LINN 909.1.

Herbs or subshrubs, usually annual in cultivation, perennial when abandoned, the stems mostly decumbent, to about 30 dm long, angular, sparsely pilose, glabrescent; stipules lanceolate, slightly falcate, acuminate; leaves 4-foliolate, the petiole longer than the leaf-bearing rachis; leaflets obovate to elliptic, about (1–)2–7 cm long, 1–4 cm wide, rounded to subacute at the apex, obliquely obtuse to acute at the base, the surfaces essentially glabrous, the petiolules pilose; flowers usually solitary, with numerous, narrow bracts at the base; calyx pilose, to about 6 cm long, the tube 5 cm long, the lobes 1 cm long; petals about 7–13 mm long, yellow, usually with reddish lines; stamens 8 or 9; fruit torulose, developing underground, 2–6 cm long, 1.5 cm in diameter, 2–4(–6)-seeded; seeds with a reddish, papery coat, ovoid or subovoid, 1–2 cm long, 8–12 mm in diameter, the hilum near one end.

Distr. Native to Brazil but cultivated in the tropics and warm temperate areas, sometimes occurring as an escape. It apparently was introduced to Ceylon at an early date, having been collected by J.G. König in the 18th century.

Vern. Rata-kaju (S); nella-kadalai (T); peanut, ground-nut, earthnut, monkeynut, pinder, goober (E).

Specimens Examined. WITHOUT EXACT LOCALITY: "Zeylona", *König s.n.* (BM). KANDY DISTRICT: Peradeniya, Royal Botanic Gardens, June 1895, *s. coll. s.n.* (PDA), 10 July 1913, *s. coll. s.n.* (PDA). AMPARAI DISTRICT: Dune just N. of Arugam Bay, in garden, *Fosberg & Sachet 53044* (US).

Note. According to Macmillan (Trop. Pl. & Gard. ed. 5: 294. 1962), groundnuts in Ceylon have not proved a commercial success but are sometimes grown in gardens.

Tribe GENISTEAE

(Adans.) Benth. in Benth. & Hook. f., Gen. Pl. 1: 439. 1865; Polhill, Bot. Syst. 1: 328. 1976; Bisby in Polhill & Raven, Adv. Leg. Syst. 409. 1981. Type genus: *Genista* L.

Leguminosae sect. Genisteae Adans., Fam. Pl. 2: 320. 1763.

Shrubs, herbs, or small trees, branches sometimes spine-tipped, or winged; leaves alternate, opposite, or crowded, digitately 3–many-foliolate, 1-foliolate, or simple, sometimes as spinous phyllodes; stipules free, fused to the branch and persistent, or sometimes absent; stipels absent; flowers in terminal racemes or heads, sometimes solitary, sometimes axillary; bracts variable, sometimes 3-lobed; bracteoles usually present; calyx usually bilabiate, rarely spathaceous, sometimes with 5 subequal lobes; petals yellow, white, bluish to purple, sometimes variegated; stamens monadelphous in a closed tube; anthers dimorphic, alternately long-basifixed and short-dorsifixed; ovary sessile or stipitate, 2–many-ovuled; style glabrous, usually slender, bent or curved; fruit usually linear to oblong, dehiscent, sometimes impressed between the seeds, rarely inflated, or compressed and somewhat winged, or globose and indehiscent, 2–many-seeded; seeds reniform or orbicular to ovate-oblong, the hilum in a sinus.

KEY TO THE GENERA

1 Stems virgate, not spiny, with few, simple, oblanceolate leaves; calyx spathaceous . . . . **3. Spartium**
1 Stems not virgate, often spiny, with 3-foliolate leaves; calyx bilabiate
  2 Leaves at maturity mostly reduced to spinose phyllodes; inflorescences axillary . . . . **2. Ulex**
  2 Leaves not becoming spinose; leaflets small, obovate to elliptic; inflorescences terminal . . . . **1. Genista**

Only three species of this tribe have been reported from Ceylon. None is native but they have been introduced and are cited by Macmillan in his "Tropical Planting and Gardening", ed. 5, 1962, and are represented by herbarium vouchers.

1. Genista canariensis L., Sp. Pl. 709. 1753; Macmillan, l.c. p. 176. *Cytisus canariensis* (L.) Kuntze; *Teline canariensis* (L.) Webb & Berth. Cape broom (E). A native of the Canary Islands, introduced elsewhere as an ornamental. Two collections seen: Royal Botanic Gardens, Hakgala, Sept. 1894, *s. coll. s.n.* (PDA); 4 Oct. 1899, *s. coll. s.n.* (PDA).

2. Ulex europaeus L., Sp. Pl. 741. 1753; Trimen, Handb. Fl. Ceylon 2: 7. 1894; Alston in Trimen, Handb. Fl. Ceylon 6: 68. 1931; Macmillan, l.c. p. 71. Gorse, furze, whin (E). A native of western Europe, introduced elsewhere as an ornamental and barrier hedge, and as a soil binder. In Ceylon it has become naturalized and in some areas a noxious weed. Specimens Examined: Nuwara Eliya District: Nuwara Eliya, in 1914, *s. coll. s.n.* (PDA), *Simpson 9912* (BM, PDA), *Cramer 2865* (US), *Amaratunga 218* (PDA), *1546* (PDA), *Maxwell & Jayasuriya 778* (US), *Rudd & Balakrishnan 3162* (K, PDA, US). Horton Plains, *Comanor 446* (PDA, GH, K, US).

3. Spartium junceum L., Sp. Pl. 708. 1753; Macmillan, l.c. p. 192. Spanish broom, weaver's broom (E). A native of the Mediterranean region, introduced elsewhere. The only Ceylon collection seen is: Royal Botanic Gardens, Hakgala, 28 Sept. 1899, *s. coll. s.n.* (PDA). In addition to use as an ornamental the plant fibres are sometimes used in making rope and cords.

Tribe CROTALARIEAE

(Benth.) Hutchins., Gen. Fl. Pl. 1: 364. 1964; Polhill, Bot. Syst. 1: 317. 1976; Polhill in Polhill & Raven, Adv. Leg. Syst. 399. 1981. Type genus: *Crotalaria* L.

Genisteae (Bronn) Benth. subtribe Crotalariineae Benth., in Benth. & Hook. f., Gen. Pl. 1: 440. 1865, as "Crotalarieae".

Herbs or shrubs; leaves alternate, generally digitately 3(–7)-foliolate, or sometimes 1-foliolate or simple; stipules present, sometimes decurrent; flowers small to medium-sized, in terminal, leaf-opposed, few–many-flowered racemes or heads, or sometimes axillary, or solitary; bracts and bracteoles usually present; calyx with 5 subequal lobes or, sometimes bilabiate; petals yellow to white, pink, bluish or purple; stamens monadelphous in a sheath open on the vexillar side or, rarely, with the vexillar filament free; anthers dimorphic or subequal; style straight, curved, or geniculate, glabrous or pubescent; stigma terminal, minute, capitate; fruit dehiscent or indehiscent, turgid or compressed, sometimes impressed between the seeds, 1–many-seeded; seeds obliquely-cordiform to reniform or discoid; hilum small; aril sometimes conspicuous.

Sixteen genera, chiefly African but with some ranging into Asia, southern Europe, Australia, and America.

KEY TO THE GENERA

1 Fruit linear-oblong, somewhat compressed; petals white to pale pink or lilac; stamens with anthers subequal, subspherical, dorsifixed; style straight; leaves 3-foliolate .... **1. Rothia**

1 Fruit oblong to globose or lenticular, usually inflated; petals yellow, often with reddish or purplish markings, white, or bluish; stamens with anthers dimorphic, alternately long, basifixed and short, dorsifixed; style curved or geniculate; leaves 3-, or 5-foliolate, or simple .... **2. Crotalaria**

1. ROTHIA

Pers., Syn. Pl. 2: 302, 638. 1807, nom. cons., non Schreb. 1791, nec Borkh. 1792, nec Lam. 1792; Polhill, Bot. Syst. 1: 326. 1976. Type species: *R. trifoliata* (Roth) Pers., based on *Dillwynia trifoliata* Roth, a synonym of *R. indica* (L.) Druce.

Dillwynia Roth, Catolecta 3: 71. 1806, non J.E. Smith 1805. Type species: *D. trifoliata* Roth.
Westonia Spreng., Syst. 3: 152, 230. 1826. Type species: *W. humifusa* (Willd.) Spreng.
Goetzea Reichb., Consp. 150. 1828, nom. rejec., non *Goetzea* Wydler 1830.
Xerocarpus Guill., Perr. & A. Rich., Fl. Seneg. Tent. 169. 1832. Type species: *X. hirsutus* Guill., Perr. & A. Rich.
Harpelema Jacq. f., Ecloge Pl. Rar. 2: 6, t. 129. 1844.

Annual herbs; stems pròstrate or diffuse; leaves digitately 3-foliolate; stipules linear to oblong or lanceolate; stipels lacking; flowers small, solitary or in few-flowered terminal or leaf-opposed racemes; bracts small, linear; bracteoles minute or lacking; calyx campanulate with 5 subequal lobes; petals white to pinkish or lilac; vexillum glabrous or nearly so; keel petals slightly coherent; stamens monadelphous with the filament sheath open on the vexillar side; anthers small, subequal, subspherical, dorsifixed; ovary sessile, many-ovuled; style glabrous, essentially straight; stigma terminal, minutely capitate; fruit sessile, tardily dehiscent, linear-oblong, somewhat compressed, many-seeded; seeds obliquely cordiform, somewhat compressed; hilum orbicular.

A genus of two widespread species, one in Asia from western Pakistan eastward and in Australia, the other in Africa. Only one, *R. indica*, occurs in Ceylon.

Rothia indica (L.) Druce, Bot. Soc. Exch. Club Brit. Isles Rep. 1913, 3: 423. 1914.

Trigonella indica L., Sp. Pl. 778. 1753. Lectotype: Ceylon, *P. Hermann*, Herm. Herb. 3: 24 (BM). Isotypes or syntypes: Herm. Herb. 1: 74 (BM); fol. 109 (L); Paris, Biblio. Inst. France, Burman coll. Pl. Zeyl. 71 in part.
Dillwynia trifoliata Roth, Catolecta 3: 71. 1805.
Rothia trifoliata (Roth) Pers., Syn. Pl. 2: 638. 1807; Wight, Ic. Pl. Ind. Or. 199. 1839; Baker in Hook. f., Fl. Br. Ind. 2: 63. 1876; Trimen, Handb. Fl. Ceylon 2: 7. 1894; Gamble, Fl. Pres. Madras 1: 279. 1918; Saldanha & Nicolson, Fl. Hassan 263. 1976.

Annual herb, to about 20 cm high; stems moderately pubescent; stipules linear to lanceolate, 1.5–2.5 mm long; leaves with petiole about 5 mm long or less and blades obovate or oblanceolate, 5–10 mm long, 2–4 mm wide, obtuse or subacute, cuneate at the base, the upper surface sparsely pubescent to glabrous, the lower surface moderately to densely pubescent with subappressed hairs; venation inconspicuous; bracts lanceolate, 2–3 mm long; bracteoles linear, about 1.5 mm long; flowers about 7–8 mm long; calyx puberulent, 4.5 mm long with acute lobes about 2.5 mm long; petals pale pink, 7–8 mm long; fruit sessile, linear, straight to falcate, moderately pubescent, about 2.5–3.5(–4.5) mm long, 1.5–2 mm wide, 1 mm thick, many-seeded; seeds cordiform, smooth, light tan, about 1.5 mm long, 1 mm wide.

Distr. In dry, sandy places in Pakistan, India, Ceylon, Southeast Asia, and northern Australia.

Note. The specimen cited as an isotype or syntype at Paris, Biblio. Inst. France is in the Bibliothéque de l'Institut de France, Paris, in the herbarium of Paul Hermann, basis of the Thesaurus Zeylanicus of J. Burman (see A. Lourteig in Taxon 15: 23–33. 1966).

Specimens Examined. WITHOUT EXACT LOCALITY: *Walker s.n.* (K), *1049* (GH), "Oma Oya", "Naware", Jaffna, *Gardner*, Batticaloa, *Thwaites C.P. 1452* (PDA); without data, *Thwaites, C.P. 1452* (BM, CAS, GH, K, US). ANURADHAPURA DISTRICT: Between Medawachchiya and Vavuniya, mi. marker 98/1, *Rudd 3262* (PDA, SFV, US). POLONNARUWA DISTRICT: Polonnaruwa, Govt. Farm, *Senaratna 3503* (PDA). PUTTALAM DISTRICT: Chilaw, dunes, *Simpson 8146* (BM); Marawila, 2 Aug. 1914, *Bamlees s.n.* (PDA); Wilpattu National Park, Mannar-Puttalam Road, mi post 36, *Cooray 70020117R* (K, PDA, SFV, US). MATALE DISTRICT: Dambulla Rock, 16 Jan. 1896, *Trimen ? s.n.* (PDA). TRINCOMALEE DISTRICT: Near China Bay Airport, *Rudd 3134* (K, PDA, SFV, US). BATTICALOA DISTRICT: Manresa, *Jayasuriya et al. 703* (US). COLOMBO DISTRICT: Negombo, July 1930, *de Silva s.n.* (PDA); Uswetakeiyawa, *Amaratunga 99* (PDA); Muturajawela, *Amaratunga 1107* (PDA). HAMBANTOTA DISTRICT: Ruhuna National Park, near Buttawa Bungalow, *Mueller-Dombois et al. 69010543* (PDA, SFV, US), *Cooray & Balakrishnan 69010909* (GH, PDA, SFV, US), *Mueller-Dombois & Cooray 68020205* (PDA, US); Rakina Wewa, near Gonalabbe Lewaya, *Fosberg 50265* (PDA, US); Yala, on road to Hinwewe tank, *Nowicke & Jayasuriya 406* (US); Rakinawala water hole, *Mueller-Dombois 68102207* (US), *Cooray 69120717* (PDA, US); Hambantota, *Alston 1910* (PDA).

2. CROTALARIA

L., Sp. Pl. 714. 1753; L., Gen. Pl. ed. 5, 320. 1754; Polhill, Kew Bull. 22: 169–348. 1968; Polhill, Bot. Syst. 1: 323. 1976; Allen & Allen, Legum. 192. 1981. Lectotype species: *C. lotifolia* L.

Goniogyna DC., Ann. Sci. Nat. Bot., sér. 1, 4: 91. Jan. 1825. Lectotype species: *G. hirta* (Willd.) Ali, based on *Hallia hirta* Willd.
Heylandia DC., Prod. 2: 123. Nov. 1825. Lectotype species: *H. hebecarpa* DC.

Priotropis Wight & Arn., Prod. 1: 180. 1834. Type: *P. cytisoides* (Roxb.) Wight & Arn., based on *Crotalaria cytisoides* Roxb.

Herbs or shrubs; leaves simple, 1-foliolate, or digitately 3–(–7) foliolate; stipules filiform to foliaceous, often decurrent, or lacking; stipels lacking; inflorescences usually terminal or leaf-opposed, sometimes axillary racemes, sometimes heads or fascicles, or the flowers solitary; bracts and bracteoles commonly present; flowers small to medium sized; calyx campanulate with 5 subequal lobes or subbilabiate; petals usually yellow, sometimes white or bluish; vexillum usually with 2 calluses inside near the base, glabrous or pubescent on the outer face; keel often beaked; stamens monadelphous with filaments joined in a sheath open at least at the base; anthers alternately long, basifixed and short, dorsifixed; style curved or geniculate, thicker at the base, usually with 1 or 2 lines of hairs on the upper part; stigma usually small, rarely bilobed; fruit subsessile to long-stipitate, usually inflated, dehiscent, 1–many-seeded; seeds obliquely cordiform to oblong-reniform, with a definite hilar sinus, sometimes with a conspicuous aril.

About 550 species, worldwide in the tropics to warm temperate areas.

Specimens cited as at Paris, Biblio. Inst. France are in the Bibliothéque de l'Institut de France, Paris, in the herbarium of Paul Hermann, basis of the Thesaurus Zeylanicus of J. Burman (see A. Lourteig in Taxon 15: 23–33. 1966).

KEY TO THE SPECIES

1 Leaves simple
  2 Flowers blue, purplish, or white; leaves lanceolate, ovate, rhomboid, or elliptic; stems striate, 4-angled .... **1. C. verrucosa**
  2 Flowers yellow, sometimes with red, purplish, or brownish markings; leaves elliptic, oblong, ovate, obovate, or linear; stems terete or angled, sometimes striate
    3 Stipules relatively large, decurrent as persistent wings along the stem
      4 Flowers 2–2.5 cm long; leaves uniformly ovate to elliptic-ovate, or obovate, with secondary veins evident; apex of stipules 5–15 mm wide
        5 Plants shrubby, to about 1 m high; leaves elliptic-ovate, usually obtuse, about 5–6 cm long, 4–5 cm wide; flowers about 2.5 cm long .... **2. C. wightiana**
        5 Plants erect or decumbent, to about 5–7 dm high; leaves ovate to obovate, acute or subacute, 1.5–4 cm long, 0.5–2.5 cm wide; flowers about 2 cm long .... **3. C. scabrella**
      4 Flowers about 1–1.5 cm long; leaves dimorphic, elliptic, ovate, oblong, or lanceolate, with secondary veins inconspicuous; apex of stipules 3–5 mm wide .... **4. C. bidiei**
    3 Stipules usually small, not decurrent, or sometimes lacking
      6 Stems relatively robust, usually erect, often suffrutescent or shrubby, about 0.5–5 m tall
        7 Fruit glabrous
          8 Calyx villous, longer than the petals; flowers 2–3 cm long........... **5. C. calycina**
          8 Calyx glabrous to variously pubescent, shorter than the petals; flowers 1.5–2.5 cm long
            9 Leaves oblanceolate to obovate-oblong, minutely appressed-pubescent to sericeous beneath; stipules filiform to subulate, 1–1.5 mm long; flowers about 1.5–2.5 cm long ...**6. C. retusa**
            9 Leaves elliptic to suborbicular or elliptic-obovate, villous to lunate beneath; stipules transversely lunate, apiculate, about 10–15 mm long, 7–10 mm wide; flowers about 2.5 cm long
            .... **7. C. lunata**

7 Fruit pubescent
  10 Inflorescences racemose; bracts and bracteoles small, linear to lanceolate or falcate, to about 7 mm long
    11 Flowers 1–2 cm long; leaves appressed-pubescent beneath
      12 Leaves oblong to oblanceolate; flowers on pedicels about 5 mm long; calyx finely fulvo-velutinous; fruit fulvo- to brown-subvelutinous............ **8. C. juncea**
      12 Leaves elliptic to ovate; flowers on pedicels 10–15 mm long; calyx sericeous or subsericeous; fruit puberulent, sometimes glabrescent with age... **9. C. walkeri**
    11 Flowers about 3 cm long; leaves patent-pubescent beneath. .... **10. C. multiflora**
  10 Inflorescences paniculate; bracts and bracteoles ovate, elliptic, or cordate, 2–15 mm long
    13 Bracts and bracteoles about 5–15 mm long, ovate to elliptic, obtuse or subacute; panicle dense, compact.. **11. C. berteroana**
    13 Bracts and bracteoles about 2–4(–8) mm long, cordate, acute; panicle loose and spreading .... **12. C. lunulata**
6 Stems relatively slender, usually procumbent or decumbent, sometimes suffrutescent, less than 1 m long
  14 Stipules present
    15 Fruit glabrous
      16 Racemes leaf-opposed, 1–4-flowered; flowers 1–1.5 cm long on pedicels 3–6 mm long; leaves narrowly oblong or obovate to broadly elliptic, 1–6.5 cm long, 5–20 mm wide .... **13. C. ferruginea**
      16 Racemes mostly terminal, about 5–13-flowered; flowers 1.2–2 cm long on pedicels 7–10 mm long; leaves linear-oblong, 1.5–1.8 cm long, 2–13 mm wide .... **14. C. mysorensis**
    15 Fruit pubescent
      17 Flowers 7–9 mm long; calyx 4–6 mm long; fruit 1.2–2 cm long, 6–8 mm wide; stems terete .... **15. C. evolvuloides**
      17 Flowers 10–12 mm long; calyx 8–10 mm long; fruit about 1.5–2.5 cm long, 4–5 mm wide; stems 3-angled ....**16. C. triquetra**
  14 Stipules lacking
    18 Fruit lenticular, 1-seeded, villous to glabrous, (3–)5–6 mm long, 2.5–3 mm wide; flowers solitary .... **17. C. hebecarpa**
    18 Fruit inflated, 6–20-seeded
      19 Inflorescences racemose, 1–15(–20)-flowered
        20 Flowers 3–7 mm long
          21 Fruit subglobose, (9–)13–15 mm long, (8–)10–12 mm wide, villous; inflorescences 1- or 2-flowered; flowers 5–7 mm long................ **18. C. angulata**
          21 Fruit oblong, 10–16 mm long, (3–)4–5 mm wide, glabrous; inflorescences 1–5-flowered; flowers 3–5 mm long................... **19. C. prostrata**
        20 Flowers 7–13 mm long
          22 Fruit oblong, 10–15 mm long, 4–5 mm wide; flowers 10–13 mm long ....**20. C. albida**
          22 Fruit ellipsoid or subglobose, 4–7 mm long, 4–5 mm wide; flowers 7–10 mm long
            23 Flowers about 10 mm long; fruit subglobose, about 7 mm long, 5 mm wide; leaves linear .... **21. C. linifolia**
            23 Flowers about 7 mm long; fruit ellipsoid, finely punctate, 4–7 mm long, 4 mm wide; leaves linear-oblong to elliptic................. **22. C. montana**
      19 Inflorescences capitate-racemose or sometimes umbellate, 1–8-flowered........ ....**23. C. nana**
1 Leaves compound, commonly 3-foliolate, sometimes 4- or 5-foliolate (in *C. quinquefolia*)

24 Fruit small, 4–6 mm long, subglobose, 1- or 2-seeded
  25 Leaflets usually glabrous above, appressed-pubescent beneath; petals 4–5(–10) mm long, about twice as long as the calyx . . . . **24. C. medicaginea**
  25 Leaflets crisp-pubescent on both surfaces; petals (5–)6–8 mm long, about 3 times as long as the calyx . . . .**25. C. willdenowiana**
24 Fruit larger, (10–)20–55 mm long excluding stipe, many-seeded
  26 Surface of fruit pubescent, sometimes glabrescent
    27 Fruit and stems pilose with hairs mostly spreading or crispate . . . . **26. C. incana**
    27 Fruit and stems subsericeous with hairs appressed or lax
      28 Calyx appressed-pubescent; stipules present but sometimes caducous
        29 Leaflets predominantly obovate, obtuse; fruit clavate or oblong-cylindrical, less than 1 cm in diameter
          30 Flowers about 1 cm long on pedicels 2–3 mm long; fruit clavate, 2–2.5 cm long, 5–6 mm in diameter . . . . **27. C. clavata**
          30 Flowers about 1.3–1.5 cm long on pedicels 3–4 mm long; fruit oblong-cylindrical, 3–4.5 cm long, 6–8 mm in diameter . . . . **28. C. pallida**
        29 Leaflets elliptic, acute or subacute; fruit ellipsoid, 3–4 cm long, 1–1.5 cm in diameter; bracts linear-filiform, crispate . . . . **29. C. micans**
      28 Calyx glabrous or nearly so; stipules lacking . . . . **30. C. zanzibarica**
  26 Surface of fruit glabrous
    31 Leaves 3-foliolate; flowers about 3 cm long; keel with prominent beak, not twisted; stipe of fruit 2–3 cm long . . . . **31. C. laburnifolia**
    31 Leaves 3–5-foliolate; flowers about 1.5–1.8 cm long; keel twisted; stipe of fruit 0.5–1(–1.5) cm long . . . . **32. C. quinquefolia**

1. Crotalaria verrucosa L., Sp. Pl. 715. 1753; Moon, Cat. 52. 1824; Wight, Ic. Pl. Ind. Or. 200. 1839; Thw., Enum. Pl. Zeyl. 81. 1859; Baker in Hook. f., Fl. Br. Ind. 2: 77. 1876; Trimen, Handb. Fl. Ceylon 2: 15. 1894; Backer & Bakh. f., Fl. Java 1: 582. 1963; Polhill, Fl. Trop. East Africa, Papil. 959. 1971; Verdcourt, Man. New Guinea Leg. 584. 1979; Windler & McLaughlin, Ann. Missouri Bot. Gard. 67: 613. 1980. Type: Ceylon, *P. Hermann.* Herb. Herm. 3: 4 (BM). Isotypes: Biblio. Inst. France, Paris, pp. 14, 53.

Numerous synonyms are given by Windler and McLaughlin, l.c.

Annual, to about 1 m tall; stems much branched, erect or spreading, striate, 4-angled, pubescent with subappressed hairs, sometimes glabrescent; stipules foliaceous, ovate to falcate, acuminate, 5–20 mm long, 4–13 mm wide, persistent or deciduous; leaves simple, on short petioles 2–3 mm long, ovate to lanceolate, rhomboid, or elliptic, about 1–12 cm long, 0.7–8 cm wide, obtuse or acute at the apex, acute at the base, appressed-pubescent on both surfaces; inflorescences racemose, leaf-opposed, pseudoterminal, longer than the leaves, many-flowered; bracts lanceolate, caudate, about 1.5 mm long; bracteoles at base of pedicels or also near midpoint, filiform to lanceolate, 1–4 mm long; flowers 11–16 mm long on pedicels 3–5 mm long; calyx appressed-pubescent, 7–13 mm long with deltoid, acute lobes longer than the tube; petals blue, purplish, or white, glabrous; fruit sessile, inflated, oblong-cylindrical, pubescent with subappressed hairs, about 2.5–3(–5) cm long, 10–12 mm in diameter, 10–16-seeded; seeds yellowish to brown, obliquely cordiform, 3–4.5 mm long, 3–4 mm wide.

Distr. A native of Asia now widespread in tropics of the Old and New Worlds.

Vern. Nil-andana-hiriya, "jac beerie gha," yak bairiye (S); kilukiluppai (T); blue andana (E).

Specimens Examined. WITHOUT EXACT LOCALITY: *Stadhouder 62* (P); *Fraser 34* (US); *Gardner 203* (K). JAFFNA DISTRICT: *Gardner, Thwaites, C.P. 1273* (PDA); between Jaffna and Elephant Pass, *Rudd 3283* (K, PDA, US). ANURADHAPURA DISTRICT: Anuradhapura, W. shore of Nuwara Wewa, *Sohmer 8092* (US); Galapitagala, *Waas 363* (K, US); W. of main Dambulla-Anuradhapura Rd., mile marker 49, *Townsend 73/179* (K, US); N. bank of Rajangana Reservoir, *Maxwell & Jayasuriya 792* (SFV, US). POLONNARUWA DISTRICT: Between Habarane and Kantalai, *Rudd & Balakrishnan 3118* (K, PDA, US). PUTTALAM DISTRICT: Wilpattu Natl. Park, near Eerige Ara confluence with Moderagama Ara, *Fosberg et al. 50788* (K, US); Periya Naga Villu, *Mueller-Dombois et al. 69043017* (PDA, SFV, US); Mail Villu, W. end, *Mueller-Dombois et al. 69043020* (GH, K, PDA, SFV, UC, US); Kali Villu, *Cooray 70020220R* (GH, K, PDA, SFV, UC, US), *Maxwell & Jayasuriya 803* (US); Maradanmaduwa Tank, *Ripley 32* (US). KURUNEGALA/PUTTALAM DISTRICTS: Kurunegala-Puttalam Rd., Mile marker 57, *Amaratunga 120* (PDA). MATALE DISTRICT: *Kostermans 24433* (GH, K, US). KANDY DISTRICT: Kandy, *Thomson s.n.* (K), *Champion s.n.* (K), *Worthington 7239* (K); Peradeniya, April 1889, *Trimen ?* (PDA); in 1845, *Thomson s.n.* (K); Doluwa, *Fosberg 51804* (US); Urugala, *Amaratunga 1479* (PDA); Marassana, *Maxwell 991* (US). TRIN-COMALEE DISTRICT: Vicinity of Kantalai Reservoir, *Jayasuriya et al. 634* (US); Kantalai tank bed, *Worthington 2048* (K). AMPARAI DISTRICT: Between Siyambalanduwa and Inginiyagala, *Rudd & Balakrishnan 3216* (K, PDA, US). COLOMBO DISTRICT: Near coconut plantation, S. of Negombo, *Simpson 8885* (BM); near Mt. Lavinia, *Rudd 3073* (K, PDA, US). GALLE DISTRICT: Koggala, *D. B. & D. Sumithraarachchi 1036* (US); Induruwa, S. of Kalutara, *Rudd 3081* (GH, K, PDA, US); Galle-Matara Rd., *Sohmer et al. 8877* (GH, US). HAMBANTOTA DISTRICT: Main Yala Rd., *Comanor 820* (GH, K, US); Ruhuna Natl. Park, Katagamuwa, *Cooray 69121110R* (K, PDA, SFV, US).

2. Crotalaria wightiana Graham ex Wight & Arn., Prod. 181. 1834; Thw., Enum. Pl. Zeyl. 81. 1859; Trimen, Handb. Fl. Ceylon 2: 12. 1894; Gamble, Fl. Pres. Madras 1: 281, 291. 1918; Alston in Trimen, Handb. Fl. Ceylon 6: 70. 1931. Type: Wall. Cat. 5358a. India, Dindygul hills, at an elevation of 3500 ft.

Crotalaria rubiginosa sensu auctt. in part, non Willd., Sp. Pl. 3: 973. 1802; Moon, Cat. 52. 1824; Trimen, Handb. Fl. Ceylon 2: 11. 1894 in part; Back. & Bakh. f., Fl. Java 1: 579. 1963.

Crotalaria rubiginosa var. *wightiana* (Grah. ex Wight & Arn.) Baker in Hook.
f., Fl. Br. Ind. 2: 69. 1876.

Shrubby herb to about 1 m tall; stems densely pubescent with shining rusty
or golden subappressed hairs; stipules decurrent as a persistent narrow wing along
the entire length of the internode, triangular-ovate, mucronate, about 1.5 cm wide
at the apex; leaves simple, elliptic-ovate, usually obtuse, mucronate at the apex,
rounded at the base, to about 5–6 cm long, 4–5 cm wide, shining, rusty-pubescent
with subappressed hairs on both surfaces; venation evident; inflorescences lateral,
racemose, 2–5-flowered; bracts and bracteoles lanceolate; flowers about 2.5 cm
long; calyx 1.5–2 cm long, pubescent, with lanceolate lobes; petals pale yellow;
fruit glabrous, ellipsoid, about 5 cm long, 1.5 cm in diameter, 30–40-seeded.

Distr. Ceylon and southern India at high elevations.
Specimens Examined. WITHOUT EXACT LOCALITY: In 1819, *Moon
s.n.* (BM); *Gardner s.n.* (GH); *Walker 364* (K), *s.n.* (K); Elephant Plains, 6000
ft, *Gardner 198* (BM, K, PDA).

3. Crotalaria scabrella Wight & Arn., Prod. 181. 1834; Prain, J. Roy. Asiat.
Soc. Bengal 66: 350. 1897; Gamble, Fl. Pres. Madras 1: 281, 291. 1918; Alston
in Trimen, Handb. Fl. Ceylon 6: 69. 1931. Type: *Wight, Cat. no. 692* (K).

Crotalaria rubiginosa sensu auctt. in part, non Willd. Sp. Pl. 3: 973. 1802; Wight,
Ic. Pl. Ind. Or. 885. 1843–45.
Crotalaria rubiginosa var. *scabrella* (Wight & Arn.) Baker in Hook. f., Fl. Br.
Ind. 2: 69. 1876.

Erect or procumbent herb, to about 5–7 dm high; stems densely pubescent
with greyish to yellowish subappressed hairs; stipules decurrent, extending as
a narrow wing almost the length of the internode, triangular at the apex, 5–15
mm wide, acuminate; leaves simple, ovate to obovate, acute or subacute,
mucronate at the apex, 1.5–4 cm long, 0.5–2.5 cm wide, moderately pubescent
with lax hairs above, densely pubescent beneath; venation evident; inflorescences
racemose, commonly 2–3-flowered; bracts and bracteoles lanceolate or ovate-
lanceolate; flowers about 2 cm long; calyx 1.5–2 cm long, sericeous, with
lanceolate lobes; petals yellow, scarcely exserted; fruit glabrous, ellipsoid, about
4 cm long, 1.5 cm in diameter, 20–30-seeded.

Distr. Ceylon and southern India.
Specimens Examined. WITHOUT EXACT LOCALITY: *Thwaites C.P.
2772* (BM in part, P). NUWARA ELIYA DISTRICT: Hakgala, *Trimen ? s.n.*
(PDA); between Nuwara Eliya and Badulla, mi. marker 54 on A5, *Maxwell 1015*
(US); Hakgala, *Maxwell & Jayasuriya 904* (SFV, US). BADULLA DISTRICT:
Karandagolla Bridge, Wellawaya to Ella, *Hepper & de Silva 4777* (K, US);
Bandarawela-Haputale Rd. near 10/12 mi. marker, *Sumithraarachchi 931* (US);

Bandarawela, Uva, Sept. 1890, *Trimen ? s.n.* (PDA); between Welimada and Badulla, mi. marker 69, *Rudd & Balakrishnan 3189* (K, PDA, US); between Haputale and Boralanda, *Amaratunga 468* (PDA).

4. Crotalaria bidiei Gamble, Kew Bull. 1917: 27. 1917; Gamble, Fl. Pres. Madras 1: 281, 291. 1918; Alston in Trimen, Handb. Fl. Ceylon 6: 70. 1931. Syntypes: South India, *Bidie*; *Beddome*; *Barber 5627* (K).

Suffrutescent herb to about 45 cm high; stems erect or suberect, moderately to densely pubescent with yellowish, rusty, or grey patent hairs; stipules decurrent, forming a narrow wing the length of the internode, deltoid, acuminate, recurved, about 3–5 mm wide at the apex; leaves dimorphic, elliptic, ovate, oblong, or lanceolate, 1–7 cm long, 1–4 cm wide, obtuse at apex and base, sparsely to moderately pubescent with lax hairs above, moderately pubescent beneath; venation inconspicuous; inflorescences axillary, racemose, 1–few-flowered; peduncles about 2–5 cm long; bracts and bracteoles ovate, acuminate; flowers about 1–1.5 cm long; calyx patent-pubescent, 1–1.5 cm long with lanceolate, acuminate lobes; petals yellow, scarcely exserted; fruit glabrous, ellipsoid, 3.5–4 cm long, 1–1.5 cm in diameter, many-seeded.

Distr. Ceylon and southern India.

Specimens Examined. WITHOUT EXACT LOCALITY: In 1845, *Thomson s.n.* (K); *Walker s.n.* (K); *Moon 607* (BM); in 1819, *s.coll. s.n.* (BM). KANDY DISTRICT: Peradeniya Gardens, *Alston 816* (K, UC). NUWARA ELIYA DISTRICT: Road from Kandy to Maturata, culvert 20/5, *Maxwell 1001* (US in part). RATNAPURA DISTRICT: Galagama, *Gardner, Thwaites,* May 1856, *C.P. 2772* (PDA), without data, *C.P. 2772* (BM in part). BADULLA DISTRICT: Near Lunugala, Jan. 1888, *Trimen ? s.n.* (PDA). MONERAGALA DISTRICT: 7 mi. E. of Bibile, *Fosberg & Sachet 53153* (K, SFV, US); Wellawaya, *Hepper & de Silva 4772* (K, US).

5. Crotalaria calycina Schrank, Pl. Rar. Hort. Monac. t. 12. 1817; Thw., Enum. Pl. Zeyl. 82. 1859; Baker in Hook. f., Fl. Br. Ind. 2: 72. 1876; Trimen, Handb. Fl. Ceylon 2: 14. 1894; Gamble Fl. Pres. Madras 1: 286. 1918; Backer & Bakh. f., Fl. Java 1: 581. 1963; Polhill in Fl. Trop. East Africa, Papil. 949. 1971; Ali, Fl. W. Pakistan, no. 100, Papil. 45. 1977; Verdcourt, Man. New Guinea Leg. 577. 1979. Type: Plate 12, Pl. Rar. Hort. Monac. 1817.

Crotalaria anthylloides sensu D. Don, Prod. Fl. Nep. 241. 1825; Wight & Arn., Prod. 181. 1834, non Lam. 1786; Moon, Cat. 52. 1824.
Crotalaria stricta Roxb., Hort. Beng. 54. 1814 ("1813"); Roxb., Fl. Ind. ed. 2, 3: 265. 1832, non Roth, 1821.
Crotalaria roxburghiana DC., Prod. 2: 129. 1825, nom. nov. for *C. stricta* Roxb.

Annual, to about 1 m tall; stems erect, densely pubescent with fulvous, appressed or subappressed hairs; stipules minute, inconspicuous; leaves simple, linear-oblong or linear-lanceolate, 6.5–14.5 cm long, 6–10 mm wide, acute at apex and base, glabrous above, pubescent beneath with appressed or subappressed hairs; inflorescences racemose, terminal, about 4–12-flowered or, sometimes axillary, 1–2-flowered; bracts and bracteoles lanceolate, acute or caudate, 6–10(–18) mm long; bracteoles inserted at base of calyx; flowers 2–3 cm long, on pedicels about 1 cm long; calyx 2–3 cm long, exceeding the petals and fruit, 5–lobed, subbilabiate, densely villous with fulvous or brownish spreading hairs; petals yellow, sometimes with maroon streaks; fruit glabrous, sessile, oblong-cylindrical, 2–3 cm long, 8 mm wide, about 20–35-seeded; seeds light brown, about 3 mm long.

Distr. Widespread in tropical Asia, Africa, and Australia.

Vern. Gorandiya (S).

Specimens Examined. WITHOUT LOCALITY: *Col. Walker 37* (K), *s. coll. s.n.* (GH, K, PDA). KANDY DISTRICT: Madugoda patana, *Worthington 1576* (K); Mottawa patana, *Worthington 6561* (K); Loolewatte Road, *Simpson 9438* (BM); Peradeniya, *Mueller-Dombois & Cooray 67103111* (PDA, US). NUWARA ELIYA DISTRICT: Hakgala, Sept. 1881, *Trimen? s.n.* (PDA); Roadbank Govindahela & patana, Hakgala, 27 Feb. 1906, *A.W.S. (Silva) s.n.* (PDA); between Kandy and Maturata, culvert 20/5, *Maxwell 1001* (US in part). KANDY/NUWARA ELIYA/RATNAPURA DISTRICTS: Galagama, May 1856, "Hantain", in 1847, Nuwara Eliya, Elephant Plains, in 1859, *Thwaites, C.P. 526* (PDA); without data, *C.P. 526* (BM, GH, K, P, PDA). BADULLA DISTRICT: Near Palugama, mi. marker 59 on A5, *Maxwell & Jayasuriya 902* (SFV, US); between Welimada and Badulla, *Rudd & Balakrishnan 3196* (PDA, US); 2 mi. S. of Welimada, near mi. marker 2/2 on way to Boralanda, *Mueller-Dombois & Cooray 68011421* (US); near Nilgala, Uva, Jan. 1888, *Trimen? s.n.* (PDA); near Bandarawela, *Simpson 8655* (BM, PDA). MONERAGALA DISTRICT: Koslanda, *Simpson 8351* (BM).

6. Crotalaria retusa L., Sp. Pl. 715. 1753; Moon, Cat. 52. 1824; Wight & Arn., Prod. 187. 1834; Wight, Ic. Pl. Ind. Or. 200. 1839; Thw., Enum. Pl. Zeyl. 81. 1859; Baker in Hook. f., Fl. Br. Ind. 2: 75. 1876; Trimen, Handb. Fl. Ceylon 2: 15. 1894; Backer & Bakh. f., Fl. Java 1: 580. 1963; Polhill in Fl. Trop. East Africa, Papil. 958. 1971; Saldanha & Nicolson, Fl. Hassan 244. 1976; Ali, Fl. W. Pakistan, no. 100, Papil 49. 1977; Verdcourt, Man. New Guinea Leg. 583. 1979. Lectotype: Ceylon, P. Hermann, Herb. Herm. 2: 21 (BM). Syntypes: Herb. Herm. 2: 84; 4: 51; 4: 78 (BM). Type material: P. Hermann, Herb. Zeyl. fol. 6 (L).

Crotalaria retusa L. var. *maritima* Trimen, Handb. Fl. Ceylon 2: 15. 1894. Type: Ceylon, Puttalam-Kalpitiya, "on the seashore", Aug. 1883, *Trimen s.n.* (PDA).

Annual or perennial herb, to about 1.3 m tall; stems erect, or sometimes pro-strate, striate, pubescent with short, appressed hairs; stipules filiform to subulate, 1–5 mm long, caducous; leaves simple, oblanceolate to obovate-oblong, 2–10.5 cm long, 0.8–4 cm wide, obtuse or emarginate at the apex, acute to cuneate at the base, glabrous above, minutely appressed-pubescent to sericeous beneath; in-florescences racemose, terminal, many-flowered; bracts and bracteoles filiform to lanceolate, attenuate, to about 6 mm long, caducous; bracteoles inserted at base of calyx or at midpoint of pedicel; flowers about 1.5–2.5 cm long on pedicels 5–8 mm long; calyx 1–1.5 cm long, glabrous to moderately pubescent with short appressed hairs; lobes about twice as long as the tube; petals yellow, sometimes with reddish veins; fruit glabrous, oblong-cylindrical, 3–4.5 cm long, about 1(–1.8) cm wide, 15–20-seeded; seeds light brown to black, 3–5 mm long.

Distr. A pantropical weed, sometimes cultivated.

Uses. According to Irvine (Woody plants of Ghana 371. 1961) this species is cultivated as a fibre and as a cover crop; the roots and leaves are sometimes used medicinally although the plant is also reported as being poisonous. Verd-court, l.c., states that it is dangerous for livestock, being one of the causes of "Kimberley horse disease."

Vern. Kaha-andana-hiriya, "andenehirya gha" (S); kilukiluppai (T).

Specimens Examined. WITHOUT LOCALITY: *Stadhouter 30* (P); *Leschenault s.n.* (P); *Walker s.n.* (P). Batticaloa, Jaffna, B (illegible), *Gardner, Thwaites C.P. 1247* (PDA); without data, *C.P. 1274* (BM, K). ANURADHAPURA DISTRICT: Mihintale, near Dagoba, *Cramer 4150* (US). PUTTALAM DISTRICT: Puttalam Lagoon Isthmus, cart track from Daluwa to highway, *Maxwell & Jayasuriya 827* (US). MATALE DISTRICT: Sigiri, *Bernardi 14347* (US). KANDY DISTRICT: Galagedara, *Amaratunga 951* (PDA); Wattegama, *Worthington 7234* (K). TRINCOMALEE DISTRICT: Kantalai, 3 mi. W. of Mahaweli Ganga ferry road, *Sohmer & Sumithraarachchi 10796* (GH); Kantalai, tank bed, *Worthington 2049* (K). BATTICALOA DISTRICT: Panichchankerni, *Jayasuriya et al. 701* (US); Keeli-Kudah, *Waas 2132* (SFV, US). COLOMBO DISTRICT: Mt. Lavinia, along railroad tracks just above beach, *Rudd 3074* (K, PDA, US); Jaela, 8–12–13, *s. coll. s.n.* (PDA). BADULLA DISTRICT: Badulla, *Simpson 8228* (BM); Bandarawela, Uva, Sept. 1890, *Trimen ? s.n.* (PDA); Uraniya, *Mueller-Dombois 68050308* (K, US). HAMBAN-TOTA DISTRICT: About 1 mi. E. of Bundala, *Davidse 7763* (US); Amaduwa, *Jayasuriya 2251* (K); Ruhuna Natl. Park, Patanagala, *Comanor 884* (GH, K, US), *885* (K, US), *Fosberg 50343* (K, SFV, US), *Mueller-Dombois 67082507* (PDA, US); beach behind Yala camp site, *Comanor 681* (GH, K, UC, US); Butawa, *Worthington 5218* (K); Andonoruwa, W. of Yala Bungalow, *Fosberg et al. 51149* (US); East of Kirinda in dunes behind *Spinifex* zone, *Simpson 9937* (BM, PDA). GALLE DISTRICT: Koggala, *Sumithraarachchi 1034* (US); Induruwa, S. of Kalutara, *Rudd 3080* (GH, K, PDA, US).

Note. Some of the plants growing in coastal sand dunes are prostrate, more densely pubescent, and with shorter inflorescences. These have been designated by Trimen as var. *maritima*. Otherwise they do not appear to differ from typical *C. retusa*.

7. Crotalaria lunata Beddome ex Polhill, *Crotalaria* in Africa and Madagascar 373. 1982. Type: India, Madras, Anamalai and Pulney Hills, *Beddome* (K holotype).

Crotalaria lanata Beddome, Madras J. Lit. Sci., ser. 2, 19: 178. 1858; Beddome, Ic. Ind. Or. 1: 22, tab. 105. 1870; Baker in Hook. f., Fl. Br. Ind. 2: 77. 1876; Gamble, Fl. Pres. Madras 1: 287, 297. 1918; Verdcourt, Man. New Guinea Leg. 580. 1979; non Thunb. 1800. Type: as for *C. lunata*.
Crotalaria lunata Beddome, Madras J. Lit. Sci., ser. 3, 1: 42. 1864, nomen.

Shrub, to 5 m tall; stems angled, villous, or tomentulose; stipules transversely lunate, about 1−1.5 cm long, 7−10 mm wide, apiculate; leaves simple, elliptic to suborbicular, or elliptic-obovate, 8–15 cm long, 3–9 cm wide, obtuse, apiculate at the apex, rounded at the base, glabrous above, villous to lanate beneath; inflorescences racemose, terminal, 16–20-flowered; bracts and bracteoles lanceolate or ovate-lanceolate, 5–10 mm long; bracteoles below base of calyx; flowers about 2.5 cm long; calyx villous or tomentulose, 1.5–2 cm long with acuminate lobes twice as long as the tube; petals yellow; standard with greenish or orange stripes, glabrous on the outer face; keel with twisted beak; fruit glabrous, ellipsoid or oblong-clavate, 6–6.5 cm long including stipe about 1 cm long, 1.5–2.5 cm wide, many-seeded; seeds reniform, black or dark brown, 5–7 mm long, 4–5 mm wide.

Distr. Native to southern India but introduced elsewhere, including Ceylon, Africa, New Guinea, and Australia, as an ornamental and as shade for tea and coffee; sometimes occurring as an escape.

Specimens Examined. NUWARA ELIYA DISTRICT: Liddesdale Estate, among tea, *Worthington 5632* (K); Hakgala, Botanic Garden, cultivated, *Simpson 9026* (BM); near Hakgala, growing wild, *Ash 508* (PDA).

8. Crotalaria juncea L., Sp. Pl. 714. 1753; Moon, Cat. 52. 1824; Thw., Enum. Pl. Zeyl. 81. 1859; Baker in Hook. f., Fl. Br. Ind. 2: 79. 1876; Trimen, Handb. Fl. Ceylon 2: 16. 1894; Macmillan, Trop. Pl. & Gard. ed. 5, 28, 404. 1962; Backer & Bakh. f., Fl. Java 1: 582. 1963; Polhill in Fl. Trop. East Africa, Papil. 950. 1971; Ali, Fl. W. Pakistan, no. 100, Papil. 43. 1977; Verdcourt, Man. New Guinea Leg. 579. 1979. Type: India, *s. coll.* (LINN 895.11).

Annual herb, to about 2.5 m tall; stems erect, ribbed, subappressed-pubescent; stipules filiform, about 2 mm long, caducous; leaves simple, oblong to oblanceolate, (2.5–)3–10(–13) cm long, 0.5–1.5(–2.2) cm wide, obtuse or subacute, glabrous or sparsely pubescent above, subsericeous beneath; inflorescences terminal, racemose, many-flowered; bracts lanceolate-oblong, acute, 3–5 mm long; bracteoles linear, 1.5–5 mm long, inserted at base of calyx; flowers 1–2 cm long on pedicels about 5 mm long; calyx 1.5–2 cm long, subbilabiate, lobes lanceolate, 3–4 times the length of the tube, finely fulvo-velutinous; petals 1.5–2 cm long, bright yellow with dark reddish or brown streaks; wings shorter than the keel and vexillum; fruit sessile, oblong-cylindrical, 2.5–3.5(–5.5) cm long, 1–1.7 cm in diameter, fulvo- to brown-subvelutinous, 6–15-seeded; seeds obliquely-cordiform, 4–6 mm long, light brown to black.

Distr. A native of India but widely distributed elsewhere in the tropics.

Uses. Widely cultivated as a cover crop and for fibre.

Vern. Hana (S); San hemp, Sunn hemp (E).

Specimens Examined. WITHOUT EXACT LOCALITY: "Kalawewa & Kurunegala, 1856, *Gardner,* Trincomalie, Glenie", *Thwaites C.P. 1264* (PDA), without data *C.P. 1264* (K, PDA). POLONNARUWA DISTRICT: Between Habarane and Kantalai, *Rudd & Balakrishnan 3130* (K, PDA, US). MATALE DISTRICT: Matale, *Simpson 9891* (BM, K). BADULLA DISTRICT: Badulla, Jan, 1888, *Trimen ? s.n.* (PDA); Nilgala, Uva, Jan, 1888, *Trimen ? s.n.* (PDA). KEGALLE DISTRICT: "Four Korales" Oct. 1882, *Trimen ? s.n.* (PDA).

9. Crotalaria walkeri Arn., Nov. Actorum Acad. Caes. Leop.-Carol. Nat. Cur. 18: 328. 1836; Trimen, Handb. Fl. Ceylon 2: 16. 1894, pl. 26. 1894; Gamble, Fl. Pres. Madras 1: 287, 297. 1918; Alston in Trimen, Handb. Fl. Ceylon 6: 69, 70. 1931; Macmillan, Trop. Pl. & Gard. ed. 5, 134. 1962. Saldanha & Nicolson, Fl. Hassan 244. 1976. Type: Ceylon, *Walker 115* (isotype K).

Crotalaria semperflorens Vent. var. *walkeri* (Arn.) Baker in Hook. f., Fl. Br. Ind. 2: 78. 1879.

Suffrutescent herb, to about 2.5–3 m tall; stems erect to semiscandent, striate, pubescent with minute appressed-hairs; stipules linear to falcate, attenuate, 1.5–3 mm long, 1 mm wide or less; leaves simple, elliptic to ovate, acute to obtuse, (1.5–)2–6(–9) cm long, (0.5–)1–4 cm wide, glabrous above, pubescent with minute, appressed hairs beneath; inflorescences terminal, racemose, few-flowered; bracts and bracteoles linear-attenuate; flowers 1.5–2 cm long on pedicels 1–1.5 cm long; calyx 10–12 mm long with attenuate lobes slightly longer than the tube, sericeous or subsericeous; petals bright yellow, sometimes with red venation, the standard glabrous on the outer face; fruit essentially sessile, the stipe about 5–7 mm long, puberulent, sometimes glabrescent with age, about 4–5 cm long, 2 cm wide, clavate-ellipsoid, 7–10-seeded; seeds dark brown, subreniform, 5–7 mm long, 3 mm wide.

Distr. Ceylon and southern India.

Uses. Trimen and Macmillan suggest using this species as an ornamental.
Specimens Examined. WITHOUT EXACT LOCALITY: *Walker s.n.*
(GH); *Leschenault s.n.* (P); *Macrae s.n.* (BM); *s. coll. 802* (BM); *Thomson s.n.*
(K); *Dr. Maxwell s.n.* (K); *Gardner 206* (P). KANDY DISTRICT: Adams Peak,
de Silva 47 (GH, PDA, UC); Hantane, *Gardner 205* (BM, BR, K, PDA); Kan-
dy, *Macrae 447* (BM, K); Katugastota, *Worthington 6880* (K); foot of Laksapana
Falls, *Jayasuriya et al. 1122* (US); slopes of Knuckles Peak, about 2 mi. E. of
Madulkele, *Davidse 8303* (US); summit of Sentry Box East Mt. behind Raxawa
Tea Estate E. of Dolosbage, *Theobald & Grupe 2355* (US). NUWARA ELIYA
DISTRICT: Horton Plains, 3–5–06, *"J.C.W.", s.n.* (PDA), *Gardner*? Feb. 1846,
Thwaites C.P. 280 (K), *Rudd & Balakrishnan 3178* (K, PDA, US), *3183* (K,
PDA, US); Horton Plains, *Gardner,* and Maturata, Sept. 18?8 (unclear), *Thwaites,*
C.P. 12 (280) (PDA); without data, *C.P. 12* (BM, GH, K, P); Totupola, Horton
Plains, Sept. 1890, *Trimen ? s.n.* (PDA); Horton Plains, Diyagama Road, *Com-*
anor 970 (GH, K, UC, US); at main road to Horton Plains Rest House, *Mueller-*
Dombois & Comanor 67070949 (PDA, US); Horton Plains; *Gould 13562* (UC),
Hepper 4444 (K, US), *Tirvengadum & Cramer 279* (K, US); Little World's End,
Grey-Wilson & Silva 3075 (K, US); between Horton Plains and Pattipola, *van*
Beusekom 1508 (US); Horton Plains near Farr Inn, *Meijer 639* (GH, K, US);
Hakgala, 30.9.06, *A.M.S. s.n.* (PDA), *Wilson s.n.* (A); forest below Hakgala,
Grey-Wilson & Silva 3056 (K, US); Nuwara Eliya, *Gardner 206* (BM, BR, K),
Mackenzie s.n. (K), *Worthington 2913* (BM, K); near Pundaloya, *Simpson 8907*
(BM); Ambewela Rd., Blackpool, *Tirvengadum & Cramer 86* (K, US), *Cramer*
& Tirvengadum 3937 (US); World's End, *Waas 151* (US); Pattipola, *Cramer*
4254 (K, US); Pidurutalagala, ½ mi. from beginning of trail to top, *Sohmer et*
al. 8381 (US); Moon Plains, 24.8.26, *de Silva s.n.* (PDA). RATNAPURA
DISTRICT: Karavita kanda, *Waas 386* (SFV, US). BADULLA DISTRICT:
Ambewela Junction, *Fosberg & Sachet 53242 a* (SFV); Boralanda-Ohiya Rd.,
Worthington 5582 (BM, K); Ambewela, 20.3.06, *A.M.S. s.n.* (PDA).

Note. The relationship between *C. walkeri* and *C. semperflorens* seems to
be very close as indicated by J.C. Baker's reduction of *C. walkeri* to a variety
of *C. semperflorens.* Typical *C. semperflorens* does not appear to occur in Ceylon.
It is an erect shrub with larger leaflets, stipules and flowers. The fruit generally
has a stipe a little longer than in *C. walkeri.* According to Gamble, l.c., *C.*
semperflorens is found at higher elevations than *C. walkeri.*

10. Crotalaria multiflora (Arn.) Benth., London J. Bot. 2: 478. 1843; Benth.
in Miq., Fl. Ind. Bat. 328. 1855; Thw., Enum. Pl. Zeyl. 81. 1859; Baker in
Hook. f., Fl. Br. Ind. 2: 69. 1876; Trimen, Handb. Fl. Ceylon 2: 11. 1894;
Alston in Trimen, Handb. Fl. Ceylon 6: 69. 1931.

Crotalaria bifaria L. f. var. ? *multiflora* Arn., Nov. Actorum Acad. Caes. Leop.-
    Carol. Nat. Cur. 18: 329. 1836. Type: Walker, "In Ceylono alt. 670 hexa-
    pod." holotype E-GL ?, not seen; isotypes ?, *Walker 16* (K), *s.n.* (K).

Robust perennial herb with stems to about 0.75 m long, somewhat prostrate, densely patent-pubescent; stipules linear-deltoid, attenuate, 5–7 mm long; leaves simple, variable in shape and size, ovate to narrowly lanceolate or rotundate, (2–)3–10 cm long, (0.4)1–3.5 cm wide, rounded to attenuate at the apex, apiculate, rounded at the base, glabrous to puberulent above, patent-pubescent beneath, especially along the midvein; petiole about 3 mm long; inflorescence terminal, racemose, many-flowered, on a long peduncle; bracts linear-deltoid like the stipules; bracteoles minute, linear-filiform, inserted about 5 mm below the calyx; flowers about 3 cm long on pedicels 1.2–1.5 cm long; calyx 2–2.5 cm long with lobes lanceolate, acute, puberulent; petals yellow with red or purple streaks on vexillum; fruit oblong-ovoid, sessile, about 3.5–5 cm long, 1.7–2 cm in diameter, 10–12-seeded, fulvo- or brown-silky-villous-pubescent.

Distr. Apparently endemic to Ceylon.

Specimens Examined. WITHOUT EXACT LOCALITY. *Macrae 49* (BM), *178* (BM); in 1847, *Gardner s.n.* (K); Hewahette, Oct. 1853, Habarana & Dols... illegible, *Gardner*, Elk Plains, Sept. 1857, *Thwaites C.P. 1268* (PDA); without data, *C.P. 1268* (BM, GH, K, P, PDA). KANDY DISTRICT: Udahentenne, *Amaratunga 911* (PDA); Hantane, 2500 ft, *Gardner 1207* (BM, K); Hewaheta, *Simpson 8760* (BM); Nawalapitiya, *Worthington 380* (K), *3156* (K). BADULLA DISTRICT: Bandarawela, Uva, Sept. 1890, *Trimen ? s.n.* (PDA); Haputale to Boralanda, 4700 ft, *Hepper 4590* (K, US); Haputale, Sept. 1890, *Trimen ? s.n.* (PDA); about ¼ mi. below Haputale, *"A.M.S."* [*Silva ?*] (PDA). RATNAPURA DISTRICT: Ratnapura, *de Alwis s.n.* (PDA).

Note. Trimen, l.c., refers to this as "the most ornamental of our species, with very large showy flowers."

11. Crotalaria berteroana DC., Prod. 2: 127. 1825 ("Berteriana"); Adams, Fl. Pl. Jamaica 347. 1972; Saldanha & Nicolson, Fl. Hassan 241. 1976; Verdcourt, Man. New Guinea Leg. 577. 1979. Type: A plant cultivated in Guadeloupe, W.I., *Bertero* (G-DC).

Crotalaria fulva Roxb., Fl. Ind. ed. 2, 3: 266. 1832; Wight & Arn., Prod. 183. 1834; Baker in Hook. f., Fl. Br. Ind. 2: 80. 1876; Trimen, Handb. Fl. Ceylon 2: 17. 1894; Gamble, Fl. Pres. Madras 1: 289, 298. 1918. Type: India, Calcutta Botanic Garden, grown from seed sent by Dr. Buchanan.

Shrub, to about 2.5 m high; stems fulvo-pubescent; stipules about 1 cm long, setaceous, caducous; leaves simple, subsessile, oblanceolate or obovate to elliptic, 4–9.5 cm long, 1.5–5 cm wide, obtuse, subsericeous above and beneath; inflorescences paniculate, dense, compact, terminal and axillary; bracts and bracteoles conspicuous, ovate to elliptic, sericeous, about 5–15 mm long, 5–12 mm wide, obtuse or subacute; flowers 15–20 mm long; calyx 10–15 mm long, sericeous, with lobes obtuse, longer than the tube; petals yellow or the keel

sometimes greenish; fruit about 1–1.3 cm long, ovoid, sessile, villous or tomentulose, 2-seeded, scarcely exserted from the calyx and bracteoles; seeds reniform,
light brown.

Distr. A native of the Old World tropics, cultivated and naturalized elsewhere
in both Old and New World tropics.

Specimens Examined. WITHOUT LOCALITY: *Macrae 801* (BM). KAN-
DY DISTRICT: Peradeniya, *Gardner 199* (BM, K); Botanic Garden, "weed",
Oct. 1894, *s. coll. s.n.* (PDA); Kandy, *Thomson s.n.* (K).

12. Crotalaria lunulata Heyne ex Wight & Arn., Prod. 183. 1834; Wight, Ic.
Pl. Ind. Or. 480. 1840–43; Thw., Enum. Pl. Zeyl. 81. 1859; Baker in Hook.
f., Fl. Br. Ind. 2: 80. 1876; Trimen, Handb. Fl. Ceylon 2: 17. 1894; Gamble,
Fl. Pres. Madras 1: 289. 1918. Type: India, Tanjore, *Heyne, Wall. Cat. 5378*,
Wight Cat. 663 (BM, K).

Crotalaria valetonii Backer, Bull. Jard. Bot. Buitenzorg 3, 2: 324. 1920; de Munk,
Reinwardtia 6: 217. 1962; Backer & Bakh. f., Fl. Java 1: 583. 1963. Type:
Malaysia, Madura, Tamberu, *Backer 21176* (UC syntype).

Herb, to about 30 cm tall; stems tomentulose; leaves simple, variable in size,
shape, and pubescences, elliptic, oblong, or obovate, (1–)2–6(–9) cm long,
(0.5–)1–2.5(–4) cm wide, obtuse to acute at the apex, cuneate at the base, the
surfaces densely appressed-pubescent to tomentulose; stipules glabrous, amplexicaul, minute, caducous; inflorescences paniculate, terminal, several-flowered;
bracts and bracteoles cordate, amplexicaul, acute, reflexed, about 2–4(–8) mm
long, glabrous above, pubescent beneath, usually turning black on drying; flowers
15–25 mm long; calyx puberulent, 10–12 mm long with attenuate lobes 8–10
mm long; petals yellow, the vexillum pubescent on the outer face; fruit sessile,
oval-oblong, about 15 mm long, 6 mm wide, 1-seeded, villous; seeds reniform-
cordiform, tan, about 5 mm long, 3 mm wide.

Distr. Ceylon, southern India, and Malaysia.

Specimens Examined. WITHOUT LOCALITY: *C.P. 1270* (K, US).
Milleniya Estate, Paragastota, cultivated (*J. Silva s.n.*)? *Alston 1720* (K). JAFF-
NA DISTRICT: *Gardner* in 1846, *Thwaites C.P. 1270* in part (PDA), as *Thwaites
1854, C.P. 1270* (BM, P); about 13 km S. of Pooneryn on road to Mannar, *Ber-
nardi 15344* (K, US); Chunnavil, Feb. 1890, *Trimen ? s.n.* (PDA). POLON-
NARUWA & BATTICALOA DISTRICTS: Polonnaruwa and Kalkudah, *Rudd
& Balakrishnan 3140* (PDA, US), *Worthington 5853* (K). PUTTALAM
DISTRICT: Wilpattu Natl. Park, Maduru Odai, 2.5 mi. S. of Marai Villu, *Fosberg
et al. 50863* (GH, K, SFV, UC, US); Marai Villu, *Wirawan et al. 930* (K, SFV,
US), *Cooray 70020123R* (K, SFV, US); Manikapola Uttu, *Cooray 70020244R*
(K, SFV, US), *Koyama et al. 13455* (K, US). KURUNEGALA DISTRICT:
Kurunegala, Oct. 1882, *Trimen ? s.n.* (PDA). KANDY DISTRICT: Kandy,

Nunna s.n. (K); Peradeniya, in 1845, *Thomson s.n.* (K). TRINCOMALEE DISTRICT: Kantalai, Aug. 1885, *Trimen ? s.n.* (PDA). BATTICALOA DISTRICT: Batticaloa, in 1858, *Thwaites C.P. 1270* in part (PDA). Panichchankerni, *Jayasuriya et al. 699* (US).

Note. As indicated above, there is considerable variation in pubescence in this species. Specimens that have been designated as *C. valetonii* are more densely pubescent than average, otherwise appear to be referable to *C. lunulata*. Other syntypes, not seen, are: Java, Res. Kediri, *Grutterink 3246*; Madura, *Backer 20509, 20738, 20792, 20875, 20900, 21187, 21255*.

13. Crotalaria ferruginea Grah. ex Benth., Lond. J. Bot. 2: 476, 570. 1843; Thw., Enum. Pl. Zeyl. 81. 1859; Baker in Hook. F., Fl. Br. Ind. 2: 68. 1876; Trimen, Handb. Fl. Ceylon 2: 10. 1894; Gamble, Fl. Pres. Madras 1: 282, 292. 1918; de Munk, Reinwardtia 6: 203. 1962; Backer & Bakh. f., Fl. Java 1: 581. 1963; Verdcourt, Man. New Guinea Leg. 578. 1979. Lectotype: Assam, *Jenkins, Wall. Cat. 5398* (K).

? *Crotalaria leioloba* Bartl. (as "*lejoloba*"), Ind. Sem. Hort. Gott. 2. 1837; Bartl., Linnaea 12: Littbl. Ber. 80. 1838; Walp., Rep. Bot. Syst. 1: 595. 1842; Benth., London J. Bot. 2: 570. 1843; Hochr., Candollea 2: 73. 1925; Alston in Trimen, Handb. Fl. Ceylon 6: 68. 1931; de Munk, Reinwardtia 6: 203. 1962. Type: Not known.
Crotalaria pilosissima Miq., Fl. Ind. Bat. 1, 1: 327. 1855. Type: Java; *Zollinger 2203* (U).
Crotalaria ferruginea var. *pilosissima* (Miq.) Baker in Hook. f., Fl. Br. Ind. 2: 68. 1876.

Suffrutescent herb, to about 75 cm tall; stems fulvo- or ferrugino-puberulent, erect to procumbent; stipules lanceolate, attenuate, spreading, about 3–10 mm long, 2–2.5 mm wide; leaves simple, narrowly oblong or obovate to broadly elliptic, obtuse, 1–5(–6.5) cm long, 0.5–2 cm wide, laxly fulvopubescent on both surfaces; inflorescences racemose, lax, leaf-opposed, long-peduncled, 1–4-flowered; bracts lanceolate, 6–10 mm long, spreading like the stipules; bracteoles lanceolate, 5–10 mm long, at base of calyx; flowers 1–1.5 cm long on peduncles 3–6 mm long; calyx villous, 1–1.3 cm long with attenuate lobes much longer than the tube; petals yellow, about as long as the calyx; fruit sessile, glabrous, oblong, inflated, 1.7–4 cm long, 7–15 mm wide, about 30–40 seeded; seeds subreniform, dark brown.

Distr. Ceylon, southern India and Southeast Asia.
Specimens Examined. WITHOUT EXACT LOCALITY: *Macrae 180* (BM); *Moon s.n.* (BM); *Walker s.n.* (K); *Leschenault s.n.* (P). KANDY DISTRICT: Kandy, *Moon 407* (BM); Pussellawa, Sangilipalama, close to Panaloya, *Cramer 3860* (K, US); Peradeniya, *Appuhamy s.n.* (PDA); Madugoda-

Urugala Road, *Simpson 8826* (BM); Corbett's Gap, Rangala, *Simpson 9442* (BM, PDA). NUWARA ELIYA DISTRICT: Maturata, *Gardner s.n.*, June 1848, *Thwaites*?, in 1853, "Rambodda", in 1854, *Thwaites C.P. 1265* (PDA); without locality, in 1854, *Thwaites C.P. 1265* (BM, K); St. Coombes, Talawakele, *Simpson 8900* (BM). Hakgala, patana, 1.10.06, *A.M.S. (Smith?) s.n.* (PDA); between Ramboda and Nuwara Eliya, above road marker 36/A5, *Mueller-Dombois 67112303* (PDA, US), *Mueller-Dombois & Cooray 67091311* (PDA, US). RATNAPURA DISTRICT: Ratnapura, 11.4.30, *de Alwis s.n.* (PDA); Haldummulla, *Simpson 8442* (BM). BADULLA DISTRICT: Way to Ambewela Patana, 31.10.1928, *de Silva s.n.* (PDA); near Lunugala, Uva, Jan. 1888, *Trimen? s.n.* (PDA); Ella Pass, Uva, Sept. 1890, *Trimen? s.n.* (PDA); Haputale, patana on hill above Monamaya, *Rudd & Balakrishnan 3203* (PDA, US); Fort MacDonald Valley, 31.5.06, *A.M.S. (Silva?)* (PDA); road Haputale to Bandarawela, *Maxwell & Jayasuriya 772* (US); Along A5 to Nuwara Eliya, culvert 33/12, *Maxwell & Jayasuriya 861* (US).

Note. Alston, l.c., in a revised key to *Crotalaria*, replaced without further explanation, *C. ferruginea*, as cited by Trimen, with *C. leioloba*. However, as pointed out by de Munk, p. 204, no type of *C. "lejoloba"* has been found and the identity remains uncertain.

14. Crotalaria mysorensis Roth, Nov. Pl. Sp. 338. 1821; Wight & Arn., Prod. 182. 1834; Thw., Enum. Pl. Zeyl. 82. 1859; Baker in Hook. f., Fl. Br. Ind. 2: 70. 1876; Trimen, Handb. Fl. Ceylon 2: 12. 1894; Gamble, Fl. Pres. Madras 1: 285, 295. 1918; Backer & Bakh. f., Fl. Java 1: 581. 1963; Ali, Fl. W. Pakistan, no. 100. Papil. 47. 1977; Verdcourt, Man. New Guinea Leg. 581. 1979. Type: India, Mysore, *Heyne.*

Crotalaria stipulacea Roxb., Fl. Ind. 3: 264. 1832. Type: India, Mysore, "from thence the seeds were sent by Mr. Heyne to the Botanic garden at Calcutta...".

Suffrutescent herb, to about 1 m tall; stem erect, densely patent-pubescent with silvery to fulvous or ferruginous hairs; stipules linear-acicular to lanceolate, about 3–25 mm long; leaves simple, linear-oblong, obtuse, apiculate at apex, rounded at base, 1.5–8 cm long, 2–13 mm wide, laxly pubescent on both surfaces; inflorescences racemose, terminal, or some lateral, about 5–13-flowered; bracts linear to lanceolate, 1–2 cm long; bracteoles linear-lanceolate, about 6–16 mm long, at base of calyx; flowers 1.2–2 cm long on pedicels 7–10 mm long; calyx 1.2–1.6 cm long, spreading-pilose; petals pale yellow, sometimes with red venation, scarcely exserted; fruit subsessile, glabrous, inflated, oblong, 1.8–3.5 cm long, 1–1.7 cm wide, 20–60-seeded.

Distr. Ceylon, India, Pakistan, Southeast Asia.

Specimens Examined. BATTICALOA DISTRICT: Batticaloa, Mar. 1858,

Thwaites C.P. 3594 (PDA). BADULLA DISTRICT: Uma-Oya, Apr. 1883, *Trimen ? s.n.* (PDA); Nilgala, Uva, Jan. 1888, *Trimen ? s.n.* (PDA).

15. Crotalaria evolvuloides Wight ex Wight & Arn., Prod. 188. 1834; Thw., Enum. Pl. Zeyl. 81, 410. 1859; Baker in Hook. f., Fl. Br. Ind. 2: 68. 1876; Trimen, Handb. Fl. Ceylon 2: 10. 1894; Gamble, Fl. Pres. Madras 1: 282, 292. 1918. Type: India, *Wall. Cat. 5410* (K holotype, BM isotype).

Perennial herb, sometimes suffrutescent at base; stems trailing, terete, puberulent with spreading hairs; stipules linear, about 1–3 mm long; leaves simple, oblong to elliptic or suborbicular, 5–20(–35) mm long, 5–15(–25) mm wide, obtuse, sometimes retuse at apex, glabrous above, sparsely to moderately puberulent beneath; inflorescences racemose, leaf-opposed, few-flowered; bracts cordate-lanceolate, acuminate, 1–2 mm long; bracteoles linear-lanceolate, about 1 mm long, at base of calyx, caducous; flowers about 7–9 mm long on pedicels 4–5 mm long; calyx puberulent, 4–6 mm long, with attenuate lobes 2–2.5 times the length of the tube; petals light yellow, sometimes with darker venation; vexillum puberulent along the midvein, otherwise glabrous; fruit ellipsoid, sessile, 1.2–2 cm long, 6–8 mm wide, shortly villous-pubescent, about 12-seeded.

Distr. Southern India and Ceylon.

Specimens Examined. WITHOUT LOCALITY: *Walker s.n.* (K), *299* (K); *Thwaites, C.P. 1266* in part (K); *Thwaites, C.P. 3593* (BM, K, P, PDA). ANURADHAPURA DISTRICT: 6.5 mi. NE. of Anuradhapura, *Fosberg & Balakrishnan 53460 a* (K, SFV, US). BADULLA DISTRICT: Ekiriyankumbura, Uva, Jan. 1888, *Trimen ? s.n.* (PDA); Uma Oya, Apr. 1883, *Trimen ? s.n.* (PDA); between Welimada and Badulla, mi. marker 69, *Rudd & Balakrishnan 3199* (K, PDA, US).

Note. Several of the small, trailing species are superficially similar in appearance and in some cases the collections are mixed. Caution is advised in matching new material against old herbarium specimens that may be incorrectly annotated.

16. Crotalaria triquetra Dalz., Hooker's J. Bot. Kew Gard. Misc. 2: 34. 1850; Thw., Enum. Pl. Zeyl. 410. 1864; Baker in Hook. f., Fl. Br. Ind. 2: 71. 1876; Trimen, Handb. Fl. Ceylon 2: 12. 1894; Cooke, Fl. Pres. Bombay 295. 1901; Gamble, Fl. Pres. Madras 1: 286, 296. 1918; Alston in Trimen, Handb. Fl. Ceylon 6: 68 1931; Saldanha & Nicolson, Fl. Hassan 244. 1976. Type: "provincia Malawan", *Dalzell & Gibson s.n.* (K).

Suffrutescent herb or subshrub, to about 50 cm high; stems strongly 3-angled, glabrous or with a few spreading hairs; stipules ovate to lanceolate, to 5 mm long, 2 mm wide; leaves simple, subsessile, oblong-ovate to oblong-elliptic, (2–)2.5–4(–5) cm long, (0.5–)1.5–2.3 cm wide, obtuse at the apex, rounded or cordate at the base, glabrous above, glabrous or subglabrous beneath; inflorescences racemose, long, lax, terminal and lateral, (1–)2–3-flowered; bracts and bracteoles minute; flowers about 10–12 mm long; calyx 8–10 mm long, sericeous, with linear-lanceolate, acute lobes; petals pale yellow, slightly exserted; vexillum glabrous or subglabrous on the outer face; fruit oblong, cylindrical, appressed-pubescent, 1.5–2.5 cm long, 4–5 mm wide, 15–20-seeded.

Distr. Southern India, probably introduced in Ceylon.

Specimens Examined. KANDY DISTRICT: Peradeniya, *Thwaites C.P. 3832* (PDA), cited as "rare".

17. Crotalaria hebecarpa (DC.) Rudd, Phytologia 54: 26. 1983.

Hallia hirta Willd., Sp. Pl. 3: 1169. 1802. Type: [*Koenig* ?], India, Tranquebar. (holotype B-Willd., microfiche 13750), non *Crotalaria hirta* Willd., 1803, nec Lag., 1816, nec Roth, 1821.

Goniogyna hebecarpa DC., Ann. Sci. Nat. Bot. 4: 92. Jan. 1825. Type: Ceylon, *Leschenault,* in 1823 (holotype G–DC; isotype P).

Goniogyna leiocarpa DC., Ann. Sci. Nat. Bot. 4: 92. Jan. 1825, based on *Hallia hirta* Willd., non *Crotalaria leiocarpa* Vog. 1843.

Heylandia hebecarpa DC., Prod. 2: 123. Nov. 1825; DC., Mém. Leg. 200. Feb. 1826, tab. 34. Jan. 1827, based on the same collection as *Goniogyna hebecarpa* DC. without reference to the earlier name.

Heylandia leiocarpa DC., Prod. 2: 123. Nov. 1825; DC., Mém. Leg. 200. Feb. 1826, based on *Hallia hirta* Willd., non *Crotalaria leiocarpa* Vog. 1843.

Crotalaria uniflora Koenig ex Roxb., Fl. Ind. ed. 2, 3: 271. 1832, based on *Hallia hirta* Willd., given as synonym but, possibly, intended as a new name for a duplicate of the same *Koenig* collection; non Baker in Oliver, 1871

Heylandia latebrosa sensu auctt., non (L.) DC.; Wight & Arn., Prod. 180. 1834; Thw., Enum. Pl. Zeyl. 81. 1859; Baker in Hook. f., Fl. Br. Ind. 2: 65. 1876; Trimen, Handb. Fl. Ceylon 2: 8. 1894; Gamble, Fl. Pres. Madras 1: 280. 1918.

Goniogyna hirta (Willd.) Ali, Taxon 16: 463. 1967, based on *Hallia hirta* Willd.; Polhill, Kew Bull. 22: 301. 1968; Ali, Fl. West Pakistan, no. 100, Papil. 39. 1977.

Prostrate annual; stems villous, to about 40 cm long; stipules lacking; leaves simple, subsessile, the petioles 1 mm long or less, blades obliquely ovate, 5–15(–20) mm long, 3–9(–13) mm wide, obtuse at the apex, cordate at the base, sparsely villous to glabrous on both surfaces, the margin usually ciliate, the secondary veins inconspicuous; bracts and bracteoles lanceolate, 1 mm long or less; flowers solitary, 4–6.5 mm long on pedicels about 2 mm long; calyx 2.5–4 mm long, villous, with attenuate lobes; petals yellow; vexillum lightly pubescent on the outer face; fruit ovate-lenticular, villous to glabrous, (3–)5–6 mm long, 2.5–3 mm wide, 1- or 2-seeded; seeds reniform-cordiform, light brown, mottled, 2 mm long, 1–1.5 mm wide.

Distr. Ceylon. India, and Pakistan.

Vern. "Boogota kola in C." (Marriott).

Specimens Examined. WITHOUT LOCALITY: In 1836, *Marriott s.n.* (G); *Walker s.n.* (K); in 1819, *Moon s.n.* (BM). JAFFNA DISTRICT: Jaffna, Feb. 1890, *Trimen ? s.n.* (PDA); near Kankesanturai, *Rudd 3281* (K, PDA, SFV, US); between Jaffna and Elephant Pass, *Rudd 3291* (K, PDA, US). ANURADHAPURA DISTRICT: Anuradhapura, *Amaratunga 1385* (PDA); Issan-bassawa, between Medawachchiya and Vavuniya, mi. marker 97/2, *Rudd 3256* (K, PDA, SFV, US); Mihintale Hill, *Balakrishnan & Jayasuriya 1122* (US). ANURADHAPURA, BADULLA, AND PUTTALAM DISTRICTS: Mihintale, "Calapt" . . . (illegible, Kalpitiya ?), & in 1848, *Gardner,* Ettampitiya ("At-tampittya", & Badulla), Apr. 1859, *Thwaites C.P. 1275* (PDA); without data, *Thwaites, C.P. 1275* (BM, G, GH, K, P, PDA, US). ANURADHAPURA OR PUTTALAM DISTRICT: Wilpattu Natl. Park, near Sadpuda Kallu (Occapu junction), *Fosberg et al. 50817* (SFV, US). POLONNARUWA DISTRICT: Between Habarane and Kantalai, *Rudd 3119* (K, PDA, SFV, US); Polonnaruwa, Govt. Farm, *Senaratna 3507* (PDA); Sacred area, *Dittus WD 71090602* (US); vicinity of Alutoya Bridge, *Jayasuriya et al. 676* (US). KURUNEGALA DISTRICT: Bingiriya, Marandavila Estate, in 1950, *"J.E.S." s.n.* (PDA). MATALE DISTRICT: Near Nalanda, in Nov. 1882, *Trimen ? s.n.* (PDA); Sigiriya, *Amaratunga 257* (PDA), *Bernardi 14343* (US). TRINCOMALEE DISTRICT: China Bay, *Amaratunga 575* (PDA). RATNAPURA DISTRICT: Belihuloya, *Kostermans 23612* (GH). BADULLA DISTRICT: Ella, in 1924, *"J.M.S." (Silva ?) s.n.* (PDA); Jangula, near Badulla, *Simpson 8254* (BM). HAMBANTOTA DISTRICT: Kirinde, sea shore, Dec. 1882, *Trimen ? s.n.* (PDA); about 1.5 mi. NE. of Tissamaharana, on road to Kataragama, *Rudd 3099* (PDA, SFV, US); Ruhuna Natl. Park, *Cooray 69121022R* (K, US); Yala, Buttawa Bungalow area, *Nowicke & Jayasuriya 398* (K, US), *Mueller-Dombois & Cooray 68020201* (PDA, US), *Cooray & Balakrishnan 69011018R* (K, SFV, US), *69010905R* (SFV, US); Kohambagaswala, *Mueller-Dombois & Cooray 68040620* (K, PDA, US); near Karaugaswala, *Mueller-Dombois & Cooray 68012806* (PDA, US), *Cooray 70031802R* (GH, K, SFV, US), *Cooray & Balakrishnan 6901211R* (SFV, US); road near Gonagala, *Comanor 851* (GH, K, SFV, US).

Note. Moon collected this species in 1819 but did not cite it in his catalogue. As Trimen noted, l.c. p. 8, it may be what was included, without description, Moon, Cat. 52. 1824, apparently as a new species, as *Crotalaria humifusa,* from Maturata.

18. Crotalaria angulata Mill., Gard. Dict. ed. 8, *Crotalaria* no. 9. 1768; Britten & E. Baker, J. Bot. 35: 228. 1897; Windler, Rhodora 76: 188. 1974; Saldanha

& Nicolson, Fl. Hassan 241. 1976. Type: India ?, as "Campeachy" [Campeche, Mexico], *Houstoun s.n.* (BM, holotype).

Astragalus biflorus L., Mant. 273. 1771. Type: "Hab. in Insula S. Johannae, Koenig" (LINN 895.20 holotype).

Crotalaria biflora (L.) L., Mant. 570. 1771, excluding synon. *C. nana* Burm., Wight & Arn., Prod. 190. 1834; Thw., Enum. Pl. Zeyl. 81. 1859; Baker in Hook. f., Fl. Br. Ind. 2: 66. 1876; Trimen, Handb. Fl. Ceylon 2: 9. 1894; Gamble, Fl. Pres. Madras 1: 282, 292. 1918; Alston in Trimen, Handb. Fl. Ceylon 6: 68. 1931; de Munk, Reinwardtia 6: 201. 1962.

Crotalaria nummularia Willd., Sp. Pl. 3: 979. 1802; Moon, Cat. 52. 1824. Type: "Habitat in India", collector not cited.

Prostrate annual; stems villous; stipules lacking; leaves simple, (5–)10–25 mm long, 4–13 mm wide, ovate to suborbicular, acute to obtuse, mucronate at the apex, cordate to rounded at the base, glabrous or villous above, villous beneath; flowers 5–7 mm long, solitary or binate, lateral or terminal, on villous peduncles 10–20 mm long and pedicels 2–15 mm long; bracts filiform, 1–2 mm long; bracteoles filiform, about 1 mm long on pedicel 2 mm below the calyx; calyx 5 mm long, villous with attenuate lobes; fruit sessile, inflated, subglobose, villous, (9–)10–15 mm long, (8–)10–12 mm wide, 15–20-seeded; seeds brown.

Distr. Ceylon and southern India.

Specimens Examined. WITHOUT EXACT LOCALITY: *Mrs. Walker s.n.* (K); *Gardner s.n., Thwaites, C.P. 3325* (PDA). JAFFNA DISTRICT: Jaffna, Feb. 1890, *Trimen ? s.n.* (PDA). TRINCOMALEE DISTRICT: Trincomalee, *Glenie s.n. Thwaites, C.P. 3736* (K, PDA); without data, *Thwaites C.P. 3736* (BM, P, PDA). HAMBANTOTA DISTRICT: Gonalabbe, *Fosberg et al. 51202* (SFV, US), *Cooray & Balakrishnan 69012102R* (US); near Buttawa Bungalow, *Cooray & Balakrishnan 69010913R* (K, SFV, US), *Cooray 69120220R* (US).

Note. Because of Miller's error in citing the type as from Campeche, Senn (Rhodora 41: 341. 1939), followed by various authors, accepted *C. angulata* as a North American species, synonymous with *C. rotundifolia* (Walt.) Gmelin, which is common in the southeastern United States and in Mexico.

19. Crotalaria prostrata Rottl. ex Willd., Enum. Hort. Berol. 747. 1809; de Munk, Reinwardtia 6: 211. 1962; Backer & Bakh. f., Fl. Java 1: 580. 1963; Ali, Fl. West Pakistan, no. 100, Papil. 49. 1977. Type: India, *Rottler*.

Crotalaria prostrata Roxb., Hort. Beng. 54. 1814; Roxb., Fl. Ind. 3: 270. 1832; Roxb. ex D. Don, Prod. Fl. Nepal 241. 1825; Wight & Arn., Prod. 189. 1834; Thw., Enum. Pl. Zeyl. 81. 1859; Baker in Hook. f., Fl. Br. Ind. 2: 67. 1876; Trimen, Handb. Fl. Ceylon 2: 9. 1894; Gamble, Fl. Pres. Madras 1: 282, 292. 1918; Alston in Trimen, Handb. Fl. Ceylon 6: 68. 1931. Type not cited by Roxburgh.

Slender annual; stems prostrate or decumbent to suberect, pubescent; stipules lacking; leaves simple, subsessile, elliptic-lanceolate to ovate or obovate-oblong, about (5–)8–40 mm long, 3–20 mm wide, obtuse or subacute at the apex, obliquely rounded to subcordate at the base, subappressed-villous on both surfaces; inflorescences racemose, leaf-opposed, 1–5-flowered; bracts subulate, 1–2 mm long; bracteoles lanceolate-filiform, 1–1.5 mm long; flowers 3–5 mm long on pedicels 1–2 mm long; peduncles about 2–2.5 cm long; calyx 3–5 mm long, densely pubescent, subvillous, with linear, attenuate lobes; petals yellow, slightly exserted; fruit glabrous, subsessile, linear-oblong, 10–16 mm long, (3–)4–5 mm wide, 12–20-seeded.

Distr. Ceylon, Pakistan, and Indonesia.

Specimens Examined. ANURADHAPURA DISTRICT: Mihintale Hill, *Balakrishnan & Jayasuriya 1124* (US). BADULLA AND BATTICALOA DISTRICTS: Uma-Oya, in 1848, Ettampitiya, Apr. 1854, Batticaloa, Mar. 1858, *Thwaites, C.P. 1266* (PDA); without data, *Thwaites, C.P. 1266* (BM, K, P, PDA). BADULLA DISTRICT: Ekiriyankumbura, Uva, Jan. 1888, *Trimen ? s.n.* (PDA).

20. Crotalaria albida Heyne ex Roth, Nov. Pl. Sp. 333. 1821; Thw., Enum. Pl. Zeyl. 82. 1859; Baker in Hook. f., Fl. Br. Ind. 2: 71. 1876; Trimen, Handb. Fl. Ceylon 2: 12. 1894; Alston in Trimen, Handb. Fl. Ceylon 6: 68. 1931; de Munk, Reinwardtia 6: 200. 1962; Backer & Bakh. f., Fl. Java 1: 580. 1963; Saldanha & Nicolson, Fl. Hassan 241. 1976; Ali, Fl. W. Pakistan, no. 100, Papil. 47. 1977; Verdcourt, Man. New Guinea Leg. 573. 1979. Type: India orientali, *Heyne.*

Several synonyms are given by de Munk, l.c. and Backer & Bakh. f., l.c.

Perennial herb, to about 60 cm tall, suffrutescent from a short woody base; stems usually numerous, slender, erect or somewhat decumbent, minutely appressed-pubescent; stipules lacking; leaves simple, oblanceolate to linear-spathulate, 1–2 cm long, 2–15 mm wide, obtuse, mucronulate, or subtruncate at the apex, cuneate at the base, glabrous or subglabrous above, appressed-pubescent with minute hairs beneath; inflorescences terminal, racemose, 6–20-flowered; bracts and bracteoles linear, setaceous, 1–2.5 mm long; bracteoles inserted at base of calyx; flowers 10–13 mm long, on pedicels 3 mm long, calyx 8–10(–16) mm long, with lobes much longer than the tube, subsericeous; petals about as long as the calyx, pale or bright yellow, sometimes with reddish streaks; fruit sessile, oblong, 10–15 mm long, 4–5 mm wide, glabrous, 6–12-seeded.

Distr. Southern Asia from India, Pakistan, and Ceylon to China and the Philippines.

Specimens Examined. WITHOUT LOCALITY: *Walker 17* (K), *81* (PDA), *s.n.* (GH); *Fraser 137* (CAS, US); *Thwaites C.P. 269* in part (BM, P, PDA), *2482* (K); *Townsend 73/128* (US); *Worthington 5646* (K). KANDY

DISTRICT: E. of Madugoda, culvert 30/4, *Maxwell & Jayasuriya 738* (US). NUWARA ELIYA DISTRICT: Deltota, *Appuhamy RA 9* (PDA); Maturata, Aug. 1837, *Thwaites C.P. 269* in part (PDA); Hakgala, *Maxwell & Jayasuriya 884* (SFV, US); near Hakgala junction of highway A5 at road to Ambawela, mi. marker 5/3, *Mueller-Dombois 68051822* (K, PDA, US); below Hakgala, *Rhind R3* (PDA); MacDonald's valley, below Hakgala, *Rudd & Balakrishnan 3172* (K, PDA, US); Horton Plains, *Dassanayake 214A* (K); between Ramboda and Nuwara Eliya, mi. marker 36/A5, *Mueller-Dombois 67112308* (US), *67112305* (US). RATNAPURA DISTRICT: Galagama, *s. coll.*, Feb. 1846, *Thwaites, C.P. 269* in part (K, PDA). BADULLA DISTRICT: Badulla, Jan. 1888, *Trimen ? s.n.* (PDA); 2 mi. S. of Welimada at road to Boralanda, mi. marker 2/2, *Mueller-Dombois 67091519* (PDA, US); Ettampitiya (as "Atampitiya Uva"), Jan. 1888, *Trimen ? s.n.* (PDA); Welimada, *Alston A13* (PDA); between Welimada and Badulla, mi. marker 69, *Rudd & Balakrishnan 3197* (GH, K, PDA, US); near Welimada, *Simpson 8328* (BM); Welimada-Dikwella, *Simpson 9625* (K); Haputale, Thotulagalla Ridge, *Fosberg 51846* (K, US); Haputale to Boralanda, *Hepper 4591* (K, US); Medowita, Weerakumbura, Keppetipola, *Jayasuriya & Cramer 774* (K, US); 2 mi. from Boralanda on road to Haputale, *Townsend 73/129* (K); Ohiya-Boralanda Road near mi. marker 9/5, *Sohmer & Sumithraarachchi 10061* (GH); Haputale to Bandarawela, *Maxwell & Jayasuriya 770* (SFV, US).

21. Crotalaria linifolia L. f., Suppl. 322. 1781; Trimen, Handb. Fl. Ceylon 2: 13. 1894. Presumed type: "ad margines agrorum prope Tanschuhr copiosissime", *Koenig,* Herb. Linn. 895.26 (LINN).

Crotalaria tecta Heyne in Roth, Nov. Sp. 334. 1821, non sensu auctt.
Crotalaria cespitosa Roxb., Fl. Ind. ed. 2, 3: 269. 1832. No type given.
Crotalaria diffusa Roxb. ex Wight & Arn., Prod. 190. 1834. Type: "E.I.C. Mus. tab. 367".

Herb with stems prostrate (or erect ?), densely appressed-pubescent; stipules lacking; leaves simple, linear, obtuse, to about 13 mm long, glabrous above, sericeous beneath; inflorescences racemose, terminal; flowers about 10 mm long; calyx about 10 mm long or slightly less; fruit glabrous, subglobose, about 7 mm long, 5 mm wide.

Distr. Ceylon, India and, possibly elsewhere.

Specimens Examined. GALLE DISTRICT: Galle, *Gardner s.n., Thwaites C.P. 1277* (BM, K, P, PDA).

Note. There has been confusion between *C. linifolia* L. f. and *C. tecta* Heyne in Roth, treated by various authors as two separate species. According to Polhill, in litt. 1976, "LINN 895.26, the presumed type of *Crotalaria linifolia* L. f., seems without doubt to be the species generally called *C. tecta*". Comparison of the *Gardner* collection, *Thwaites C.P. 1227,* at BM with a microfiche photograph of LINN 895.26 indicates a reasonable match. A *Koenig* collection, LINN 895.27, appears to be even better.

A discrepancy to be resolved is Trimen's description, "stems numerous, prostrate", whereas the original description by Linnaeus filius cites "caules erecti, filiformes, iuncei, simplices". Further examination is needed.

22. Crotalaria montana Roth, Nov. Pl. Sp. 335. 1821; Verdcourt, Man. New Guinea Leg. 581. 1979. Type: "India orient".

Crotalaria linifolia sensu auctt., non L. f.
Crotalaria tecta sensu Trimen, J. Bot. 27. 162. 1889; Trimen, Handb. Fl. Ceylon
    2: 14. 1894, non Heyne in Roth, 1821.

Erect herb to 30 cm tall or more; stems appressed-pubescent; stipules lacking; leaves simple, linear-oblong to elliptic, 5–15 mm long, 2–4 mm wide, obtuse at apex and base, glabrous above, appressed-pubescent beneath; inflorescences terminal, racemose, lax; flowers about 7 mm long; calyx pubescent; petals pale yellow; fruit glabrous, ellipsoid, finely punctate, 4–7 mm long, 4 mm wide, scarcely exceeding the calyx, about 10-seeded.

Distr. Ceylon, India, and southeast Asia.
Specimens Examined. POLONNARUWA DISTRICT: Minneriya ("Mineri"), dry paddy-fields, Sept. 1885, *Trimen ? s.n.* (PDA).
Note. Without an opportunity to reexamine the above cited collection, referred to *C. tecta* Heyne in Roth by Trimen, I am identifying it rather tentatively as *C. montana* Roth. However, this, like the collection cited under *C. linifolia* L. f., should be examined further.

23. Crotalaria nana Burm. f., Fl. Ind. 156. tab. 48, f. 2. 1768; Wight & Arn., Prod. 191. 1834; Moon, Cat. 52. 1824; Thw., Enum. Pl. Zeyl. 82. 1857; Baker in Hook. f., Fl. Br. Ind. 2: 71. 1876; Trimen, Handb. Fl. Ceylon 2: 13. 1894; Gamble, Fl. Pres. Madras 1: 284, 294. 1918; Alston in Trimen, Handb. Fl. Ceylon 6: 68. 1931; Backer & Bakh. f., Fl. Java 1: 580. 1963. Type: India, Malabar, Garcin. Herb.

Crotalaria umbellata Wight & Arn., Prod. 191. 1834; Thw., Enum. Pl. Zeyl.
    82. 1859; Gamble, Fl. Pres. Madras 1: 285, 294. 1918; Saldanha & Nicolson,
    Fl. Hassan 244. 1976. Type: India, Dindygul Hills, Wight Cat. 700; Wall.
    Cat. 5383 in part (K).
Crotalaria nana Burm. f. var. *umbellata* (Wight & Arn.) Trimen, Handb. Fl.
    Ceylon 2: 13. 1894.

Herbs about 10–60 cm tall; stems prostrate to erect, appressed-pubescent to subvillous; stipules lacking; leaves simple, oblong, 5–15 mm long, 2–4 mm wide, obtuse to subacute at apex and base, glabrous to sparsely villous above, villous to subvillous beneath; inflorescences capitate racemes, sometimes umbellate, 1–8-flowered; bracts minute, lanceolate; bracteoles minute; filiform, inserted below the calyx; flowers about 5 mm long on short pedicels; calyx about 4 mm long, villous with vexillar lobes joined almost to the apex, carinal lobes attenuate; petals yellow; fruit subglobose, slightly beaked, glabrous, often turning black on drying, about 5 mm long, 4 mm wide, 6–12-seeded.

Distr. Ceylon and India.

Specimens Examined. WITHOUT LOCALITY: *Thomson s.n.* (K, P); *Walker s.n.* (K). POLONNARUWA DISTRICT: 1 mi. N. of Dimbulagala, *Jayasuriya et al. 717* (US). PUTTALAM DISTRICT: Wilpattu Natl. Park, Mannar-Puttalam Rd., 36 mi. marker, *Cooray 70020118R* (SFV, US). MATALE DISTRICT: Dambulla Rock, 16 Nov. 1926, *de Silva s.n.* (PDA). NUWARA ELIYA DISTRICT: Nuwara Eliya, 6000 ft, *Gardner 201* (BM, K); "Rambod-da", *Gardner*, Maturata, Sept. 1853, *Thwaites C.P. 1276* (BM, P, PDA), without data, *C.P. 1276* (GH, PDA); St. Coombes Estate, Talawakele, *Simpson 8901* (BM). COLOMBO DISTRICT: Pamunugama, *Simpson 8930* (BM); Kochchikade, *Simpson 7965* (BM); Katukanda, Negombo, July 1854, *Thwaites C.P. 3301* (BM, PDA), *Thwaites C.P. 3301* (GH, K, P); Negombo, *Senaratne 3248* (PDA); Without exact locality, *Maxwell 1031* (US); Talahena, *Amaratunga 502* (PDA). BATTICALOA DISTRICT: N. of Kalkudah on road to Elephant Point, *Rudd & Balakrishnan 3143* (K, SFV, US). BADULLA DISTRICT: Passara, Uva, Feb. 1888, *Trimen ? s.n.* (PDA); Haputale, patana on hill above Monamaya, *Rudd & Balakrishnan 3201* (PDA, US). HAMBANTOTA DISTRICT: Ruhuna Natl. Park, Buttawa Plain, *Cooray 69121209R* (US); Uraniya, *Mueller-Dombois et al. 69010520* (US).

Note. In most cases the specimens from low elevations indicate plants of low stature with few-flowered inflorescences that compare very well with Burman's tab. 48, f. 2, cited above. Those from higher elevations generally are of more robust, taller plants with larger, many-flowered inflorescences. However, there are intermediates that cannot be exactly placed. Some gatherings include both large and small plants. For that reason, *C. umbellata* is here treated in synonymy under *C. nana*. The correct status can best be clarified by field studies to determine if the variations might be due to age of the plants, elevation, or true genetic differences.

24. Crotalaria medicaginea Lam., Enc. 2: 201. 1786; Wight & Arn., Prod. 192. 1834; Thw., Enum. Pl. Zeyl. 82. 1859; Baker in Hook. f., Fl. Br. Ind. 2: 81. 1876; Trimen, Handb. Fl. Ceylon 2: 18. 1894; Gamble, Fl. Pres. Madras 1: 287, 299. 1918; Alston in Trimen, Handb. Fl. Ceylon 6: 69, 70. 1931; de Munk, Reinwardtia 6: 208. 1962; Backer & Bakh. f., Fl. Java 1: 584. 1963; Ali, Fl. W. Pakistan, no. 100, Papil. 41. 1977; Verdcourt, Man. New Guinea Leg. 580. 1979. Type: "Indies orientalis", *Sonnerat* (P).

Crotalaria neglecta Wight & Arn., Prod. 192. 1834. Type: Wight Cat. 718.
Crotalaria luxurians Benth., Lond. J. Bot. 2: 578. 1843. Type: Ham. in Wall.
    Cat. 5434.
Crotalaria medicaginea var. *neglecta* (Wight & Arn.) Baker in Hook. f., Fl. Br.
    Ind. 2: 81. 1876.
Crotalaria medicaginea var. *luxurians* (Benth.) Baker in Hook. f., Fl. Br. Ind.
    2: 81. 1876; Trimen, Handb. Fl. Ceylon 2: 18. 1894; Ali, Biologia (Lahore)
    12: 25. 1966.

Annual or perennial herb, to about 1 m tall; stems appressed-pubescent; stipules
minute, filiform; leaves 3-foliolate; leaflets obovate to oblanceolate, about 3–12
mm long, 1–6 mm wide, obtuse, subtruncate, or emarginate at the apex, cuneate
at the base, glabrous above, appressed-pubescent beneath; inflorescences
racemose, terminal or leaf-opposed, few-flowered; bracts and bracteoles linear,
minute, the bracteoles at base of calyx or slightly below; flowers about (4–)5–10
mm long; calyx appressed-pubescent, 3–4 mm long with attenuate lobes about
as long as the tube; petals yellow; vexillum appressed-pubescent on the outer face;
keel prominently beaked, as long as or exceeding the vexillum; fruit 4–6 mm
long, 3–4 mm wide, subglobose, beaked, 2-seeded, appressed-pubescent, glabrescent with age; seeds black or dark brown.

Distr. Ceylon, India, Pakistan, Southeast Asia to Australia.
Specimens Examined. BATTICALOA AND JAFFNA DISTRICTS: Bat-
ticaloa, Jaffna, *Gardner* (illegible), *Thwaites C.P. 1278* (PDA); without data,
C.P. 1278 (BM, GH, K, P, PDA). AMPARAI DISTRICT: "From dry sandy
plains, Panama E.P. Aug. 1914" *s. coll. s.n.* (PDA). JAFFNA DISTRICT:
Kankesanturai, sand flats, Feb. 1890, *Trimen ? s.n.* (PDA). NUWARA ELIYA
DISTRICT: Fort MacDonald valley, patana, 11.3.06, *A W.S. s.n.* (PDA). TRIN-
COMALEE DISTRICT: Trincomalee, near the sea, *Alston 539* (PDA); Pirates'
Cove, Palwakka, *Worthington 731* (K). KANDY DISTRICT: Poilakande,
Kadugannawa, *Worthington 1155* (BM, K); Haragama, *Trimen ? s.n.* (PDA).
BADULLA DISTRICT: Ella Pass, Uva, Sept. 1890, *Trimen ? s.n.* (PDA). PUT-
TALAM DISTRICT: Wilpattu Natl. Park, Manikapola Uttu, *Cooray 70020247R*
(US).
Note. On the basis of herbarium specimens alone it is difficult to separate
varieties. According to Trimen, l.c., the above-cited collections from Haragama
and Ella Pass, Uva, represent var. *luxurians*.

25. Crotalaria willdenowiana DC., Prod. 2: 134. 1825; Wight & Arn., Prod.
191. 1834; Thw., Enum. Pl. Zeyl. 441. 1864; Baker in Hook. f., Fl. Br. Ind.
2: 81. 1876; Trimen, Handb. Fl. Ceylon 2: 18. 1894; Gamble, Fl. Pres. Madras
1: 289, 300. 1918; Alston in Trimen, Handb. Fl. Ceylon 6: 69. 1931. Type:
Based on *C. genistoides* Willd. non Lam.

Crotalaria genistoides Willd., Sp. Pl. 3: 987. 1802, non Lam. 1786. Type: "Habitat in aridis Indiae orientalis".

Perennial suffrutescent herb, to about 1 m tall; stems gray-puberulent; stipules minute, setaceous, persistent; leaves trifoliolate; leaflets oblanceolate, obtuse, emarginate at the apex, cuneate at the base, about 10–15 mm long, 2–4 mm wide, gray crisp-pubescent on both surfaces; inflorescences racemose, terminal or lateral, 2–8-flowered; flowers (5–)6–8 mm long; calyx 4–6 mm long; petals yellow; fruit subglobose, about 4 mm in diameter.

Distr. Southern India; apparently introduced in Ceylon.

Specimens Examined. COLOMBO DISTRICT: Near Colombo, "scarcely indigenous", *Ferguson*, Apr. 1864, *Thwaites C.P. 3853* (PDA).

26. Crotalaria incana L., Sp. Pl. 716. 1753; Thw., Enum. Pl. Zeyl. 82. 1859; Baker in Hook. f., Fl. Br. Ind. 2: 82. 1876; Trimen, Handb. Fl. Ceylon 2: 18. 1894; Alston in Trimen, Handb. Fl. Ceylon 6: 71. 1931; Senn, Rhodora 41: 350 1939; de Munk, Reinwardtia 6: 205. 1962; Backer & Bakh. f., Fl. Java 1: 584. 1963; Polhill in Fl. Trop. East Africa 869. 1971; Windler & McLaughlin, Ann. Missouri Bot. Gard. 67: 603. 1980. Syntypes: Jamaica, Herb. Sloane 6: 6 (BM-SL) and Barham & Lane, Herb. Sloane 67: 76 (BM-SL); Hort. Cliff. 358: 6 (BM-CLIFF).

Various synonyms are given by Senn, l.c., de Munk, l.c., and by Windler & McLaughlin, l.c.

Annual or perennial herb, sometimes suffrutescent, 1–2(–3) m tall; stems terete, pilose or puberulous; stipules filiform, 2.5–6(–20) mm long; leaves 3-foliolate with petiole 1–6(–8) cm long; leaflets elliptic to obovate or suborbicular, 1–6 cm long, 0.7–4 cm wide, obtuse to acute at the apex, acute to cuneate at the base, glabrous or nearly so above, pubescent with lax or appressed hairs beneath; inflorescences terminal or subterminal, racemose, many-flowered; bracts filiform, 1–10 mm long, caducous; bracteoles deltoid-subulate, about 2 mm long, inserted at base of calyx, caducous; flowers 10–15 mm long; pedicels about 2 mm long, extending to 5 mm long in fruit; calyx pubescent, 7–10 mm long, campanulate with deltoid-attenuate lobes 2–3 times the length of the tube; petals yellow, often with dark reddish or purplish venation; fruit brown-pilose, oblong-cylindrical, 2.5–3.5(–4) cm long, about 1 cm in diameter, 25–50-seeded; seeds brown, sometimes mottled, about 3 mm long.

Distr. A native of tropical America; adventive or introduced in the Old World, possibly as a cover crop.

Specimens Examined. WITHOUT EXACT LOCALITY: *Fraser 127* (US); *Walker 256* (P). KANDY DISTRICT: Peradeniya, *Gardner 207* (BM, GH, K, PDA), *Gardner s.n.*, *Thwaites C.P. 1269* (PDA), June 1890, *Trimen ? s.n.*

(PDA), *Simpson 8136* (BM); Kandy, *Thomson s.n.* (K); Ambale, *Senaratne 2872* (PDA); Haragama, *Alston s.n.*, 28.4.1926 (PDA). BADULLA DISTRICT: Ella Pass, Uva, Sept. 1890, *Trimen ? s.n.* (PDA).

27. Crotalaria clavata Wight & Arn., Prod. 194. 1834; Baker in Hook. f., Fl. Br. Ind. 2: 83. 1876; Trimen, Handb. Fl. Ceylon 2: 19. 1894; Alston in Trimen, Handb. Fl. Ceylon 6: 69. 1931; Gamble, Fl. Pres. Madras 1: 290, 301. 1918. Syntypes: Cunnawady near Dindygul, India, Wight Cat. 711 and Wight in Wall. Cat. 5424d (K).

Suffrutescent herb; stems finely appressed-pubescent; stipules setaceous, to about 1.5 mm long, spreading or recurved; leaves 3-foliolate; leaflets cuneate-obovate, retuse or obtuse, mucronulate at apex, 1–3 cm long, 0.5–2 cm wide, glabrous above, subsericeous beneath, sometimes glabrescent; inflorescences short, racemose, few-flowered, axillary or terminal; bracts setaceous; flowers about 1 cm long or less on pedicels 2–3 mm long; calyx minutely appressed-pubescent, 4–5 mm long with deltoid lobes about as long as the tube; petals yellow; fruit subsericeous, oblong-clavate, about 2.5 cm long, 5–6 mm in diameter, 10–12-seeded; seeds subreniform, 3 mm long, light brown.

Distr. Southern India and northern Ceylon.
Specimens Examined. MANNAR DISTRICT: Near Giant's tank, Feb. 1890, *"Sayaneris coll." s.n.* (K, PDA); Madhu Rd., semi-open dry evergreen forest, *Jayasuriya et al. 602* (US).
Note. This species superficially resembles *Crotalaria pallida* but can be distinguished by smaller leaflets and fruit, which is subsericeous, rather than puberulous with lax hairs or glabrescent as in *C. pallida*.

28. Crotalaria pallida Ait., Hort. Kew 3: 20. 1789; Polhill, Kew Bull. 22: 262. 1968; Polhill in Fl. Trop. East Africa, Papil. 905. 1971; Saldanha & Nicolson, Fl. Hassan 243. 1976; Verdcourt, Man. New Guinea Leg. 582. 1979. Type: Plant cultivated at Kew from seed collected by Bruce in Ethiopia (BM).

Crotalaria mucronata Desv., J. Bot. Agric. 3: 76. 1814. Type: Jamaica, *s. coll.* (P).
Crotalaria brownei Bert. ex DC., Prod. 2: 130. 1825.
Crotalaria striata DC., Prod. 2: 131. 1825; DC., Bot. Mag. 3200. 1832; Thw., Enum. Pl. Zeyl. 82. 1859, 410. 1864; Baker in Hook. f., Fl. Br. Ind. 2: 84. 1876; Trimen, Handb. Fl. Ceylon 2: 18. 1894; Alston in Trimen, Handb. Fl. Ceylon 6: 69. 1931; Gamble, Fl. Pres. Madras 1: 290, 301. 1918; Macmillan, Trop. Pl. & Gard., ed. 5, 28, 486. 1962. Type: India, *Leschenault* (G).
Crotalaria striata DC. var. β Thw., Enum. Pl. Zeyl. 410. 1864. Type: Ceylon, Peradeniya, *Gardner, Thwaites C.P. 3608* (holotype PDA); probable isotypes, without data, *C.P. 3608* (BM, K, P).

Crotalaria striata DC. var. *β acutifolia* Trimen, Handb. Fl. Ceylon 2: 19. 1894, based on Thw. var. *β*.

Annual or perennial herb, usually to about 1 m tall; stems erect, finely appressed-pubescent, glabrescent; stipules filiform, 1–3 mm long, caducous; leaves 3–foliolate with petiole 2–8 cm long; leaflets elliptic to obovate, 2–9(–13) cm long, 1.3–4.5 cm wide, acute or obtuse, sometimes emarginate at the apex, acute at the base, glabrous above, pubescent with short-appressed hairs beneath; inflorescences terminal or axillary, racemose, many-flowered; bracts filiform, 1–3 mm long, caducous; bracteoles filiform, 1–2 mm long, inserted at the base of the calyx, caducous; flowers about 1.3 cm long on pedicels 3–4 mm long; calyx pubescent, 5–6 mm long with acute to attenuate lobes about as long as the tube or slightly longer; petals yellow, often with reddish or brown striations, especially on the keel; fruit oblong-cylindrical, 3–4.5 cm long, 6–8 mm in diameter, puberulous with lax hairs, often glabrescent, 25–30-seeded; seeds olive green to brown.

Distr. Pantropical.

Uses. Macmillan, l.c. p. 28, suggested this species (as *C. striata*) for planting as a green manure, cover crop.

Specimens Examined. ANURADHAPURA DISTRICT: W. shores of Nuwara Wewa, *Sohmer 8083* (GH); between Anuradhapura and Galkulama, *Rudd 3310* (K, PDA, US). POLONNARUWA DISTRICT: Polonnaruwa, *Fosberg 51900* (US), *Kundu & Balakrishnan 217* (US), *Ripley 274* (US), *294* (US), *323* (US), *400* (K, US); between Habarane and Kantalai, near mi. marker 123, *Rudd 3129* (K, PDA, US); Gal Oya Reservoir near spillway, *Comanor 563* (K, PDA, US). PUTTALAM DISTRICT: Kali Villu, *Cooray 70020204R* (K, SFV, US). KURUNEGALA DISTRICT: Dekanduwela, 3 mi. W. of Rambe, *Maxwell & Jayasuriya 786* (US). KANDY DISTRICT: Peradeniya, *Alston 2119* (PDA), *Comanor 708* (GH, K, SFV, US), *Gardner 208* (BM, K), *Jayasuriya 2220* (K, US); Peradeniya-Gampola Road, *Cooray 70013001R* (GH, K, SFV, UC, US); Gampola, *Maxwell 925* (SFV, US); Galagedera, *Rudd 3335* (K, PDA, US); Wattegama, Hunnasgiriya, *Alston 1740* (PDA, UC). TRINCOMALEE DISTRICT: Trincomalee, *Gardner s.n.*, *Thwaites C.P. 3810* in part (PDA); without data, *Thwaites* in 1868, *C.P. 3810* (BM, GH, P); Trincomalee, *Simpson 9771* (BM), *Bernardi 15287* (US); Kantalai-Trincomalee Road, mile marker 136.2, *Comanor 760* (US). COLOMBO DISTRICT: N. of Colombo, *Ferguson s.n.*, *Thwaites, C.P. 3810* in part (K, PDA); Mirigama, *Amaratunga 1755* (PDA); near Negombo, *Rudd 3069* (K, PDA, US); Miriswatte, *Amaratunga 1694* (PDA). KEGALLE DISTRICT: Bulathkohupitya, *Amaratunga 1220* (PDA). MONERAGALA DISTRICT: Between Koslanda and Wellawaya, mi. marker 136, *Rudd 3210* (K, PDA, US). MONERAGALA/AMPARAI DISTRICTS: Between Siyambalanduwa and Inginiyagala, *Rudd & Balakrishnan 3218* (K, PDA, US). HAMBANTOTA DISTRICT: Ruhuna Natl. Park, Buttawa, *Cooray 69120307R* (K, US).

29. Crotalaria micans Link, Enum. Pl. Hort. Reg. Bot. Berol., part 2: 228. 1822; Windler & McLaughlin, Ann. Missouri Bot. Gard. 67: 606. 1980; Windler, Phytologia 50: 86. 1982. Type: "America meridionalis", *Humboldt 2172* (B-Willd. no. 13272).

Crotalaria anagyroides H.B.K., Nov. Gen. & Sp. 6: 404. 1824; Alston in Trimen,
   Handb. Fl. Ceylon 6: 69, 71. 1931; Backer & Bakh. f., Fl. Java 1: 584. 1963;
   Verdcourt, Man. New Guinea Leg. 573. 1979. Type: Venezuela, "prope
   Caracas", *Humboldt 598* (P).

Suffrutescent herb or shrub, to about 3.5 m tall; stems erect, terete, striate, fulvo-subsericeous; stipules linear-deltoid, about 4 mm long, caducous; leaves 3-foliolate with petiole 2.5–5 cm long; leaflets elliptic, (2–)4–7 cm long, 1–3 cm wide, acute or subacute at apex and base, glabrous or sometimes puberulous along the midvein above, pubescent with appressed or subappressed hairs beneath; inflorescences terminal or axillary, racemose, many-flowered; bracts linear-filiform, 7–10 mm long, crispate, caducous; bracteoles filiform, 6–10 mm long, inserted at about midpoint of pedicel; flowers about 12–18 mm long on pedicels 5–10 mm long; calyx 10–13 mm long, sericeous or subsericeous, campanulate with acute to attenuate lobes 1.5–2 times the length of the tube; petals yellow, sometimes with purple streaks; fruit ellipsoid, 3–4 cm long including stipe 5 mm long, 1–1.3 cm in diameter, puberulous, glabrescent, 10–20-seeded; seeds 3–3.5 mm long, light brown.

Distr. A native of tropical America, introduced in Old World tropics as green manure and as an ornamental.

Specimens Examined. KANDY DISTRICT: Kandy, *Fairchild & Dorsett 368* (UC, US); Primrose Hill, between Haloluwa and Mulgampola, *Fosberg 50662* (GH, K, SFV, US); Culvert 68/12 towards Kitulgala, *Maxwell 940* (SFV, US); Peradeniya, *Maxwell & Jayasuriya 842* (SFV, US), *849* (SFV, US), *Rudd & Balakrishnan 3324* (K, PDA, US); Hantane, above University, *Rudd 3052* (K, PDA, US); Kobbekaduwa, near Kandy, *Rudd 3336* (K, PDA, US); at base of sentry box E. of Mt. behind Raxawa tea estate E. of Dolosbage, *Grupe 186* (US); Laxapana Estate, *Simpson 9736* (BM). NUWARA ELIYA DISTRICT: Along A5 to Nuwara Eliya, near culvert 33/12, *Maxwell & Jayasuriya 863* (US). BADULLA DISTRICT: N. of Welimada near Ancient Site, Ft. MacDonald, near culvert 8/8, *Maxwell & Jayasuriya 897* (US); Rahangala, between Ohiya and Boralanda, *Rudd & Balakrishnan 3185* (K, PDA, US); between Welimada and Badulla, *Rudd & Balakrishnan 3188* (K, PDA, US), *3191* (K, PDA, UC, US). MONERAGALA DISTRICT: About 2 mi. W. of Diyaluma Falls on A4, culvert 129/4, *Maxwell & Jayasuriya 761* (SFV, US).

30. Crotalaria zanzibarica Benth., Lond. J. Bot. 2: 584. 1843; Baker in Oliver, Fl. Trop. Africa 2: 35. 1871; Polhill in Fl. Trop. East Africa, Papil. 911. 1971; Verdcourt, Man. New Guinea Leg. 584. 1979. Type: *Bojer,* Zanzibar (K).

Crotalaria usaramoensis E.G. Baker, J. Linn. Soc. Bot. 42: 346. 1914; Backer
& Bakh. f., Fl. Java 1: 584. 1963. Syntypes: Tanganyika, Uzaramo, *Stuhlmann*
(B destroyed, BM fragment); "Maogoro" [Morogoro ?], *Stuhlmann 8216* (B
destroyed).

Annual or perennial herb, to about 3 m tall; stems erect, ribbed, pubescent
with short appressed hairs; stipules lacking; leaves 3-foliolate; leaflets lanceolate
to elliptic-oblong, 3–10.5 cm long, 1–4 cm wide, acute, glabrous or sparsely
puberulous above, pubescent with short, appressed hairs beneath; petioles 2.5–6
cm long; inflorescences terminal, racemose, many-flowered; bracts linear, attenuate, 1–4 mm long; bracteoles linear, about 1–2.5 mm long, inserted at base
of calyx or slightly below; flowers about 1.5 cm long on pedicels 4–6 mm long;
calyx about 5 mm long, glabrous or nearly so, campanulate with short, deltoid-
subulate teeth about as long as the tube or shorter; petals orange or bright yellow
with purple stripes or dark spots near the base; fruit oblong-cylindrical, 3–4.5
cm long, with stipe 2–3 mm long, (0.7–)1–1.2 cm in diameter, about
50–70-seeded, pubescent with subappressed hairs; seeds dark red or reddish-
brown, oblique-cordiform, 2–2.8 mm long, 2 mm wide.

Distr. A native of tropical East Africa but introduced elsewhere in the tropics
and subtropics as green manure or cover crop.

Specimens Examined. WITHOUT LOCALITY: *Fairchild & Dorsett, Ar-
mour Exped., 365* (UC). KANDY DISTRICT: Kandy, Primrose Hill, between
Haloluwa and Mulgampola, *Fosberg 50661* (CAS, GH, K, SFV, US); Kandy,
Worthington 7201 (K); Peradeniya, *Amaratunga 779* (PDA), *Jayasuriya 2224*
(K, US), *Maxwell & Jayasuriya 843* (SFV, US); Gannoruwa, Central Agr. Exp.
Sta. *Burtt and Townsend 28* (K, US); Anniewatte, *Fosberg 52693* (US), *52694*
(SFV, US); Corbets Gap, *Worthington 3595* (K); near Pussellawa, near mi. marker
19 on A5, *Mueller-Dombois 67082401* (US); Hantane, above University Circuit
Bungalow, *Rudd 3052a* (US); near Arambekade, *Rudd 3317* (K, PDA, US);
Maskeliya, *van Beusekom 1533* (US); outskirts of Gampola, *Maxwell 926* (US).
NUWARA ELIYA DISTRICT: Between Ramboda and Nuwara Eliya, *Comanor
906* (K, US); Ramboda, *Tirvengadum & Jayasuriya 119* (K, US); along A5 to
Nuwara Eliya, near culvert 33/12, *Maxwell & Jayasuriya 862* (SFV, US). RAT-
NAPURA DISTRICT: Southern escarpment above Pinnawala on Balangoda Road,
Comanor 1096 (K, UC, US); east of Kalawana, *Maxwell et al. 973* (US), *974*
(K, US); vicinity of Rakwana, culvert 82/9 on A12, *Maxwell et al. 982* (K, US).

31. Crotalaria laburnifolia L., Sp. Pl. 715. 1753; Moon, Cat. 52. 1824; Thw.,
Enum. Pl. Zeyl. 82. 1859; Baker in Hook. f., Fl. Br. Ind. 2: 84. 1876; Trimen,
Handb. Fl. Ceylon 2: 19. 1894; Alston in Trimen, Handb. Fl. Ceylon 6: 69.
1931; Macmillan, Trop. Pl. & Gard., ed. 5, 104. 1962; de Munk, Reinwardtia
6: 206. 1962; Backer & Bakh. f., Fl. Java 1: 583. 1963; Polhill in Fl. Trop.
East Africa, Papil. 856. 1971. Lectotype: Ceylon, *P. Hermann*. Herb. Herm.

4: 1 (BM). Syntype: Herb. Herm. 2: 48 (BM). Type material: Biblio. Inst. France, Paris, p. 20 verso; Herb. Herm. fol. 73 (L).

Perennial herb, to about 2 m tall; stems erect, terete, glabrous or nearly so; stipules lacking; leaves 3-foliolate with petiole 3–7(–12) cm long; leaflets elliptic to oblong or obovate, 1–5(–12.5) cm long, 0.5–2.5(–5.5) cm wide, acute to obtuse at the apex, acute at the base, glabrous or nearly so above, glabrous or moderately pubescent with appressed or lax hairs beneath; inflorescences terminal, racemose, many-flowered; bracts subulate, 2–4 mm long; bracteoles filiform, 1 mm long, inserted about 3–5 mm below the base of the calyx; flowers about 3 cm long on pedicels 1–2 cm long; calyx 12–17 mm long with acute lobes slightly longer than the tube, glabrous; petals yellow sometimes marked with red; keel dark reddish with a prominent beak; fruit glabrous with stipe 2–3 cm long and seminiferous oblong-cylindric or subclavate body 3–5 cm long, 1–1.3 cm in diameter, about 10–30-seeded.

Distr. Ceylon, southern India, tropical Africa, and Malaysia to northeast Australia.

Uses. Macmillan, l.c., includes this species in his list of ornamental flowering shrubs.

Vern. Yak-bériya (S).

Note. Only the typical variety is known from Ceylon but Polhill, l.c., cites two additional varieties as occurring in Kenya, East Africa.

Specimens Examined. WITHOUT LOCALITY: *Thwaites, C.P. 367* (BM, PDA); *Leschenault s.n.* (P); *Thunberg s.n.* (UPS). KANDY DISTRICT: "Mr. Wright from Gangaruwa Village" (Gannoruwa), 18.11.13 (PDA); Gampola, Lantern Hill Estate, *Alston s.n.* (PDA); Hakkinda, Dec. 1930, *de Silva s.n.* (PDA); Kandy, *Thomson s.n.* (K); Peradeniya, *Senaratne 10102* (PDA); between Madugoda and Hunnasgiriya, mi. marker 24/19, *Rudd & Balakrishnan 3247* (K, PDA, US), mi. marker 26/4, *Rudd & Balakrishnan 3238* (K, PDA, US). JAFFNA DISTRICT: Between Paranthan and Pooneryn, *Rudd 3301* (K, PDA, US). ANURADHAPURA DISTRICT: Kalawewa ("Caluwawe"), May 1856, *s. coll. s.n.* (PDA). NUWARA ELIYA DISTRICT: From Uda Pussellawa to Nuwara Eliya, near culvert 12/1, *Maxwell 1011* (SFV, US). TRINCOMALEE DISTRICT: Trincomalee ("Trinquemalay", *s. coll. s.n.* (P); Foul Point, *Simpson 9656* (BM). AMPARAI DISTRICT: Pottuvil, Arugam Bay, *Comanor 603* (K, US). COLOMBO DISTRICT: "Env. de Colombo, *Sergent Moubet,* Dec. 1907" (P); Labugama, *Cramer 3344* (US). BADULLA DISTRICT: N. of Welimada, *Maxwell & Jayasuriya 900* (SFV, US); Between Ettampitiya and Badulla, culvert 77/3, *Maxwell 1021* (US). HAMBANTOTA DISTRICT: Ruhuna Natl. Park, Karaugaswala, *Comanor 882* (K, PDA, SFV, US), *Mueller-Dombois & Cooray 68012810* (US), *Cooray 69090801R* (US); Palatupana, *Cooray 69112317R* (US); near Plot R 28, *Cooray 69121207R* (US).

32. Crotalaria quinquefolia L., Sp. Pl. 716. 1753; Moon, Cat. 52. 1824; Thw., Enum. Pl. Zeyl. 82. 1859; Baker in Hook. f., Fl. Br. Ind. 2: 84. 1876; Trimen, Handb. Fl. Ceylon 2: 19. 1894; Alston in Trimen, Handb. Fl. Ceylon 6: 69. 1931; Senn, Rhodora 41: 348. 1939; de Munk, Reinwardtia 6: 212. 1962; Backer & Bakh. f., Fl. Java 1: 583. 1963; Verdcourt, Man. New Guinea Leg. 582. 1979. Type: India, Rheede, Hort. Mal. 9, tab. 28. 1678–1703.

Crotalaria heterophylla L. f., Suppl. 323. 1781. Type: Hortus Upsaliensis, probably grown from seed collected by Koenig in India. 895.36 (LINN).

Annual herb, to about 1.2 m tall; stems erect, ribbed; glabrous or nearly so; stipules lanceolate-subulate, 2–4 mm long; leaves 3–5-foliolate with petioles 1–3 cm long; leaflets lanceolate or oblanceolate to elliptic, 1.5–12.5 cm long, 0.5–2 cm wide, obtuse to subacute, sometimes shallowly emarginate at the apex, cuneate at the base, glabrous above, appressed-pubescent beneath; inflorescences racemose, terminal, many-flowered; bracts lanceolate, acuminate, 4–5 mm long; bracteoles filiform, 1–2 mm long, inserted at about mid-point on the pedicel, or slightly lower; flowers about 1.5–1.8 cm long on pedicels 5–10 mm long; calyx glabrous, campanulate with deltoid, acute, subequal lobes about as long as the tube; petals bright yellow, usually with red or purplish streaks; fruit glabrous, subellipsoid, 4–6.5 cm long including stipe 5–10(–15) mm long, about 1.2–2 cm wide, 25–40-seeded; seeds grayish to light brown, about 5 mm long, subreniform.

Distr. A native of tropical Asia; introduced in Jamaica, and, possibly, in other islands of the West Indies.

Specimens Examined. WITHOUT LOCALITY: *Thunberg s.n.* (UPS). JAFFNA DISTRICT: Chunavil, Feb. 1890, *Trimen ? s.n.* (PDA). MANNAR DISTRICT: Madhu Road, *Jayasuriya et al. 600* (US). ANURADHAPURA DISTRICT: Medawachchiya-Horowpotana Road, *Simpson 9396* (BM). POLON-NARUWA DISTIRCT: Between Polonnaruwa and Kalkudah, *Rudd 3139* (GH, K, PDA, US). KURUNEGALA DISTRICT: Polgahawela, near culvert 5/1 on Polgahawela-Narammala Road, *Cramer 3277* (US). KANDY DISTRICT: Nursery garden, seed from Polonnaruwa, *Alston 1224* (PDA). TRINCOMALEE DISTRICT: Mahaweli Ganga North Forest Reserve, *Jayasuriya et al. 641* (US). BATTICALOA DISTRICT: Panichchankerni, *Jayasuriya et al. 700* (US); between Kalkudah and Elephant Point, *Rudd 3149* (K, PDA, UC, US). BADULLA DISTRICT: Bintenne, *Gardner in 1848, Thwaites, C.P. 1272* (PDA); without data, *C.P. 1272* (BM, K). HAMBANTOTA DISTRICT: Tissamaharama, *Simpson 9924* (BM, PDA); Ruhuna Natl. Park, Andonoruwa Wewa, just W. of Yala Bungalow, *Fosberg et al. 51149* (US).

Note. Thwaites, l.c., parenthetically mentions "var. *trifolia*" in referring to the Gardner collection from Bintenne, *C.P. 1272*, but neither Trimen nor Alston take up the varietal epithet. Trifoliolate leaves are not uncommon in this species.

Tribe BOSSIAEEAE

(Benth.) Hutch., Gen. Fl. Pl. 1: 347. 1964; Polhill, Bot. Syst. 1: 305. 1976; Polhill in Polhill & Raven, Adv. Leg. Syst. 393. 1981. Type genus: *Bossiaea* Vent.

Genisteae Benth. subtribe Bossiaineae Benth. in Benth. & Hook. f., Gen. Pl. 1: 440. 1865, as "Bossiaeae".

Shrubs or herbs, sometimes suffrutescent; leaves simple, 1-foliolate, digitately 3-foliolate, imparipinnately 3–13-foliolate, or reduced to scales; stipules present or absent, sometimes spinescent; stipels usually absent; flowers of medium size, solitary or in axillary or leaf-opposed racemes; bracts and bracteoles present; calyx with 2 vexillar lobes broad and united into a 2-toothed lip; petals variously coloured; stamens monadelphous in a sheath open on the vexillar side, or with the vexillar and carinal filaments sometimes free in part; anthers essentially uniform, dorsifixed; style slender, curved; stigma small, terminal; fruit dehiscent, 2–many-seeded; seeds arillate.

The tribe includes ten genera endemic to Australia. Only one species, *Goodia lotifolia* Salisb., Parad. London, t. 41. 1806, has been collected in Ceylon, an introduction to the Royal Botanic Gardens, Hakgala, Jan. 1888, *s. coll.* (PDA). This is an erect, bushy shrub to about 3 m high with flowers about 1.3 cm long, the petals orange-yellow with reddish-brown markings.

Tribe DALBERGIEAE

Brogn., Diss Legum. 134. 1822. Type: *Dalbergia* L.f.

Trees, shrubs, or lianas; leaves imparipinnate, (3–)5–many-foliolate or sometimes 1-foliolate, alternate or (in *Platymiscium*) opposite or verticillate; leaflets opposite or alternate; stipules present, sometimes spinescent; stipels present or absent; inflorescences terminal or axillary, paniculate, racemose, or cymose; bracts and bracteoles present, sometimes caducous; flowers papilionaceous; calyx campanulate or cyathiform, truncate or with 5 lobes or teeth or, rarely (in *Fissicalyx*) spathaceous, 1 or 2-lobed; stamens 9 or 10, monadelphous or diadelphous, 9, 9:1, or 5:5, the anthers uniform or nearly so, dorsifixed or (in *Dalbergia*) basifixed; ovary 1–many-ovulate; style glabrous; stigma terminal; fruit indehiscent, chartaceous, coriaceous, or lignous, often samaroid or drupaceous; seeds 1–many, usually reniform or suborbicular, the hilum lateral or subapical.

The tribe Dalbergieae is here treated in the broad sense of Bentham (J. Linn. Soc. Bot. 4, suppl.: (1)–134. 1860). More recently, Hutchinson (Gen. Fl. Pl. 1: 380–393. 1964) has divided it into four tribes, the Dalbergieae, Pterocarpeae, Lonchocarpeae, and Geoffroeeae.

KEY TO THE GENERA
(Based on species in Ceylon)

1 Leaves alternate
  2 Leaflets alternate
    3 Flowers white, pink, or purplish; anthers basifixed, dehiscing apically; fruit oblong or orbicular
      ....**1. Dalbergia**
    3 Flowers yellow; anthers dorsifixed, dehiscing laterally; fruit compressed, orbicular or oblique-
      ly orbicular, usually with a marginal wing .... **2. Pterocarpus**
  2 Leaflets opposite or subopposite
    4 Fruit compressed, about 1 cm thick or less; trees, scandent shrubs, or lianas
      5 Valves of fruit sometimes thickened along the margins but not winged; trees or scandent shrubs
        6 Trees; fruit essentially sessile, coriaceous to ligneous; flowers white to pinkish........
          ....**3. Pongamia**
        6 Trees or scandent shrubs; fruit somewhat stipitate, membranous to subligneous; flowers violet
          to purple ....**Lonchocarpus**
      5 Valves of fruit winged along one or both margins; lianas
        7 Flowers with stamens monadelphous or subdiadelphous with the vexillar filament free only
          at the base; fruit membranous to subcoriaceous, winged along one or both margins. ....
          .... **4. Derris**
        7 Flowers with stamens diadelphous, the vexillar filament free; fruit coriaceous, narrowly winged
          along one margin .... **5. Aganope**
    4 Fruit drupaceous, more than 1 cm thick; trees........................... **Andira**
1 Leaves opposite or verticillate; flowers yellow; fruit oblong, compressed ..... **Platymiscium**

The technical characters separating *Aganope, Derris, Lonchocarpus*, and *Pongamia* are tenuous and controversial, as has been pointed out by various authors, notably Bentham (J. Linn. Soc. Bot. 4, suppl.: 18–24. 1860) and, more recently, by Polhill (Kew Bull. 25: 259–273. 1971).

1. DALBERGIA

L.f., Suppl. 52, 316. 1781, nom. cons. Type. *D. lanceolaria* L.f.

Trees, shrubs, or lianas; leaves 1 or (3–)5–many-foliolate, alternate; stipules present; stipels absent; leaflets alternate; inflorescences terminal or axillary, racemose or paniculate; bracts and bracteoles usually minute; flowers with calyx campanulate, 5-lobed, the carinal (lower) lobe usually longer than the others; petals white, pink, or purplish; stamens 9 or 10, monadelphous 9 or 10, or diadelphous 9 : 1 or 5 : 5, the anthers small, basifixed, dehiscing apically; ovary stipitate, 1–several-ovuled; fruit orbicular, oblong, or lunate, 1–several-seeded; seeds reniform, the hilum lateral; chromosome number 2n = 20.

A pantropical genus of about 100 species.

KEY TO THE SPECIES

1 Stamens 10, diadelphous 5:5 or, sometimes monadelphous

2 Tree; leaves (7–)10–15-foliolate; leaflets 1.5–4.5 cm long, 1–2.5(–3) cm wide, the petiolules about 5 mm long; flowers 7–10 mm long; fruit glabrous, oblong, straight, long-stipitate, 1-seeded and 3.5–5 cm long or 2–4-seeded and 5–10 cm long, 1.2–2 cm wide . . . . **1. D. lanceolaria**

2 Liana; leaves 4–6-foliolate; leaflets short-petiolulate, oval or obovate-oval, 3–3–5 cm long, 1.5 cm wide; flowers 6–10 mm long; fruit glabrous, short-stipitate, falcate-lunate, 2–5 cm long, 1.5 cm wide, 1-seeded . . . . **2. D. candenatensis**

1 Stamens 9, monadelphous

    3 Liana; leaves 1–5-foliolate; leaflets ovate, acuminate, glabrous or puberulent when young, 4–10 cm long, 2.5–7 cm wide; flowers 6–7 mm long; fruit finely pubescent, short-stipitate, 7.5–10 cm long, 1.8 cm wide, straight . . . . **3. D. pseudo-sissoo**

    3 Tree; leaves 3–7, commonly 5–6, -foliolate, leaflets orbicular or broadly obovate, long-petiolulate, 4–10 cm long, 2.5–7 cm wide; flowers 5–6 mm long; fruit glabrous, 1–3-seeded, straight, 4–7.5 cm long, 1.2 cm wide, the stipe about 5–10 mm long (introduced) . . . . **D. latifolia**

1. Dalbergia lanceolaria L.f., Suppl. 316. 1781. Benth., J. Linn. Soc. Bot. 4, suppl.: 45.·1860; Baker in Hook. f., Fl. Br. Ind. 2: 235. 1876; Trimen, Handb. Fl. Ceylon 2: 88. 1894; Gamble, Man. Ind. Timbers 253. 1922; Alston in Trimen, Handb. Fl. Ceylon 6: 86. 1931; Worthington, Ceylon Trees 170. 1959. Type: *J.G. König s.n.*, Ceylon in 1777 (lectotype LINN 886.3; isotypes LINN 886.1, 886.2 and ? LINN-Smith 1179.2).

Dalbergia frondosa Roxb., Hort. Beng. 53. 1814; Roxb., Fl. Ind. 3: 226. 1832; Wight & Arn., Prod. 266. 1834; Wight, Ic. Pl. Ind. Or. t. 266. 1840; Thw., Enum. Pl. Zeyl. 94. 1859. Type: *W. Roxburgh s.n.*, Ceylon or India ? (Type collection ? LINN-Smith 1179.8, "East India, Roxburgh" ?).

Dalbergia zeylanica Roxb., Hort. Beng. 53. 1814.

Dalbergia arborea Heyne in Roth, Nov. Pl. Sp. 330. 1821, non Willd. 1802.

Tree, to about 20 m tall; young stems glabrous; leaves (7–)10–15-foliolate; leaflets elliptic, broadly oblong, or somewhat obovate, obtuse or retuse, sericeous when young, glabrescent, (1.5–)2.5–4.5 cm long, 1–2.5(–3) cm wide, the petiolules about 5 mm long; inflorescences axillary, short-paniculate; flowers 7–10 mm long (May–August); calyx cyathiform, puberulent, 5 mm long; petals violet, glabrous; stamens usually 5 : 5; ovary glabrous, or villous at the base, 3-ovulate; fruit glabrous with stipe 1 cm long, straight, 1-seeded, 3.5–5 cm long or 2–4-seeded, to 10 cm long, 1.2–2 cm wide, lightly reticulate; seeds suborbicular, about 6–10 mm long.

Distr. In Ceylon and peninsular India in dry areas at elevations up to about 900 m.

Vern. Kala, bol-mara (S); Velaruvai (T).

Note. According to Worthington, the wood of this tree is white, soft, and not durable.

Specimens Examined. WITHOUT PRECISE LOCALITY: "Jaffna, El. Pass-*Gardner*, Deltotta, May 1851, Trincomalee, Aug. 1862", *Thwaites, C.P. 1496* (PDA); without locality, *s. coll. C.P. 1496* (BM, BR, K, P, US).

JAFFNA DISTRICT: Vaddakkachchi, *Worthington s.n.* (BM). VAVUNIYA DISTRICT: Vavuniya, July 1889, *Nevill s.n.* (PDA). ANURADHAPURA DISTRICT: Anuradhapura, Tissawewa Tank spillway, *Worthington 4242* (BM, Worthington Herb.); Maradankadawala, *Worthington s.n.* (BM). TRINCOMALEE DISTRICT: Trincomalee, *Worthington 720* (BM, Worthington Herb.), *1124* (Worthington Herb.). MATALE DISTRICT: Apuwatte, 2 mi. S. of Dambulla, *Fosberg & Balakrishnan 53378* (PDA, SFV, US). KANDY DISTRICT: "Forest above Peradeniya, *W. Wright 1855, C.P. 1496* var." (PDA); Andiatenne Bungalow, *Worthington 6565* (Worthington Herb.); Peradeniya, 26 Sept. 1927, *J.M. de Silva s.n.* (PDA), Aug. 1881, *s. coll. s.n.* (PDA), *Worthington 925* (Worthington Herb.). MONERAGALA DISTRICT: Bibile, *Kostermans 25295* (PDA, US); between Bibile & Badulla, *Kostermans 25325* (PDA, US).

2. Dalbergia candenatensis (Dennst.) Prain, J. Asiat. Soc. Bengal 70: 49. 1901; Alston in Trimen, Handb. Fl. Ceylon 6: 86. 1931.

Cassia candenatensis Dennst., Schl. Hort. Mal. 12. 1818.

Dalbergia monosperma Dalz., Hooker's J. Bot. & Kew Gard. Misc. 2: 36. 1850. Thw., Enum. Pl. Zeyl. 94. 1859; 413. 1864; Benth., J. Linn. Soc. Bot. 4, suppl.: 48. 1860; Baker in Hook. f., Fl. Br. Ind. 2: 337. 1876; Trimen, Handb. Fl. Ceylon 2: 89. 1894.

Dalbergia torta Graham in Wall., Cat. no. 5873. 1832, nomen; Graham ex Gray in Bot. U.S. Expl. Exped. 1: 458. 1854; Prain, J. Asiat. Soc. Bengal 66(2): 120. 1897; Prain, Ann. Roy. Bot. Gard. (Calcutta)10: 64, tab. 42. 1904.

Drepanocarpus monosperma (Dalz.) Kurz, For. Fl. Burm. 1: 337. 1877.

Scandent shrub or liana climbing by means of divaricate lateral twigs often hooked at the ends; young stems glabrous; leaves 4–7-foliolate; leaflets elliptic to obovate-elliptic, obtuse, or emarginate, the base acute, 2–5 cm long, about 1.5 cm wide, glabrous above, puberulent on the lower surface; petiolules short; inflorescences short-paniculate or racemose; flowers about 6–10 mm long (August–November); calyx subglabrous, about 5 mm long; petals white to pinkish; stamens diadelphous, 5 : 5; ovary glabrous, 2-ovulate; fruit glabrous, falcate-lunate, acute, 1-seeded, about 2.5–5 cm long, 1–1.5 cm wide.

Distr. Coastal areas, especially in mangrove swamps, of Ceylon, India, Malaysia, Philippines, Polynesia, China, and northern Australia.

Specimens Examined. WITHOUT EXACT LOCALITY: *Walker s.n.* (K); *Thwaites s.n.* (P); "O.C. & Pantura Nov. 1855, Trinco, *Glenie*", *Thwaites C.P. 243* (BM, BR, PDA). TRINCOMALEE DISTRICT: Koddiyar, *s. coll. s.n.* Aug. 1885, *Trimen* ? (PDA). COLOMBO DISTRICT: Muturajawela, *Amaratunga 350* (PDA).

3. Dalbergia pseudo-sissoo Miq., Fl. Ind. Bat. 1: 128. 1855; Alston in Trimen, Handb. Fl. Ceylon 6: 86. 1931. Type: *Hasskarl Cat. 283* partim.

Dalbergia championii Thw., Enum. Pl. Zeyl. 94. 1859; Benth., J. Linn. Soc.
Bot. 4, suppl.: 39. 1860; Baker in Hook. f., Fl. Br. Ind. 2: 231. 1876; Trimen,
Handb. Fl. Ceylon 2: 88. 1894. Type: *Thwaites C.P. 761*, "Hantane, *Gardner*. Deltotte, May 1851", Ceylon (holotype PDA; isotypes BM, BR, P).
Endospermum zeylanicum Champion ex Thw., Enum. Pl. Zeyl. 94. 1859, nomen
in synon.
Dalbergia radiata Graham in Wall., Cat. no. 5867. 1832, nomen; Graham ex
Prain, J. Asiat. Soc. Bengal 70: 45. 1901; Graham, Ann. Roy. Bot. Gard.
(Calcutta) 10: 60, tab. 36. 1904.
Dalbergia diversifolia Blume ex Miq., Fl. Ind. Bat. 1: 128. 1855.

Scandent shrub or liana, climbing by means of hooked lateral twigs; young
stems pubescent; leaves 1–5-foliolate; leaflets subcoriaceous, ovate, acuminate,
4–10 cm long, 2.5–7 cm wide, glabrous, nitid above, the lower surface glabrous
to densely pubescent with minute, appressed hairs; inflorescences axillary,
paniculate; flowers 6–7 mm long (April, May); calyx campanulate, pubescent,
3.5–4 mm long; petals white, glabrous; stamens 9, monadelphous; ovary
puberulent; fruit oblong, straight, short-stipitate, 7.5–10 cm long, 1.8 cm wide,
finely pubescent; seed oblong, compressed, about 3 cm long.

Distr. In moist areas at about 300–1200 m elevation, in Ceylon, southern
India, Borneo, and Java.

Uses. According to Worthington, note in Herbarium, "Elephants were using this creeper for towing logs to the saw. The creeper is passed through an
eye carved in the nose of the log".

Vern. Bambara wel (S); Hornet creeper (E).

Specimens Examined. WITHOUT EXACT LOCALITY: *Beddome 2402,
2403* (BM); *Champion s.n.* (K); *Walker 43* (PDA). ANURADHAPURA
DISTRICT: Summit of Ritigala, 24 Mar. 1905, *s. coll. s.n.* (PDA). KANDY
DISTRICT: Dolosbage, Apr. 1882, *s. coll. s.n.* (PDA); Imboolpitiya,
Nawalapitiya, *Worthington 906* (Worthington Herb.); Pussellawa, *Thwaites ? 772*
(K); Hantane, 2000–3000 ft, *Gardner 245* (K); East of Madugoda, Culvert 30/4,
Maxwell & Jayasuriya 737 (SFV, US); Maskeliya valley, *Jayasuriya et al. 748*
(US); Hunasgiriya, *Tirvengadum et al. 75* (US); road to Mahiyangane, *Kostermans 25146* (US); Laxapana-Maskeliya Double-cutting Road, *Kostermans 24098*
(US). RATNAPURA DISTRICT: Wewalwatte, *Tirvengadum et al. 172* (US);
"Kudawa side of Sinharaja Forest", *Mueller-Dombois 72042308* (US). GALLE
DISTRICT: Udalamatta, *Cramer et al. 3784* (US); Homodola, *Worthington 6022*
(Worthington Herb.).

Several species of *Dalbergia* have been introduced into Ceylon, chiefly in
the Royal Botanic Gardens, Peradeniya. *Dalbergia latifolia* Roxb. has been more
widely planted and for that reason has been included in the key to species. The
following cultivated species are known in Ceylon as represented by voucher herbarium specimens.

Dalbergia armata E. Mey., Comm. Pl. Afr. Austr. 152. 1836. Peradeniya, R.B.G., 13 March 1928, *A. Jayasinghe s.n.* (PDA).

Dalbergia assamica (?) Benth. in Miq., Pl. Jungh. 255. 1852. Wanarajah Estate, Dikoya, 21 March 1928, *Alston s.n.* (PDA).

Dalbergia latifolia Roxb., Pl. Corom. 2: tab. 113. 1799; Roxb., Fl. Ind. 3: 221. 1832; Wight, Ic. Pl. Ind. Or. 1156. 1846; Trimen, Handb. Fl. Ceylon 2: 88. 1894; Worthington, Ceylon Trees 171. 1959; Macmillan, Trop. Pl. & Gard. ed. 5, 216. 1962. Peradeniya R.B.G., Feb. 1888, *s. coll. s.n.* (PDA). Ooma Oya, Oct. 1880, "native collector", (K, PDA). Trincomalee, *Glenie s.n.* (PDA). Kadugannawa, *Worthington s.n.* (BM). Anuradhapura, *Worthington 26* (K), *4241* (BM, K), *4245* (BM, K), *6718* (K) between Trincomalee and Kandy, forest plantation, *Worthington 22* (K). Vern. Itti (T); Bombay rosewood (E). This species, a native of India, produces valuable wood for furniture, but Worthington states, l.c., "planted in Ceylon but growth poor; e.g. Dambulla."

Dalbergia melanoxylon Guill. & Perr., Fl. Seneg. Tent. 227, tab. 53. 1832. Peradeniya, R.B.G., *s. coll. s.n.* (PDA), 4 March 1926, *s. coll. s.n.* (PDA).

Dalbergia volubilis Roxb., Pl. Corom. 2: 48, tab. 191. 1805; Roxb., Fl. Ind. 3: 231. 1832, non (L.) Urban 1919. Peradeniya, R.B.G., *A. Jayasinghe 16* (PDA).

2. PTEROCARPUS

Jacq., Select. 283. Jan.–Jul. 1763, nom. cons. Type: *P. officinalis* Jacq., type cons.

Trees, sometimes shrubby; leaves (1–3)5–20-foliolate, alternate; stipules present, often small; stipels absent; leaflets alternate or subopposite; inflorescences axillary or terminal, racemose or paniculate; bracts and bracteoles small, often caducous; flowers with calyx turbinate or campanulate, somewhat asymmetrical at the base, subequally 5-lobed; petals yellow to orange, sometimes with a purplish or pink target spot on the vexillum; stamens 10, monadelphous or diadelphous 5 : 5, or 9 : 1 with the vexillar filament free or partially so, the anthers versatile, dorsifixed; ovary sessile or stipitate, 2–8-ovulate; fruit indehiscent, samaroid with a marginal wing, orbicular to subfalcate, 1–3(–4)-seeded; seeds reniform to elliptic, the hilum small, lateral to subapical; chromosome numbers $2n = 22, 24, 44$.

A pantropical genus of about 20 species.

KEY TO THE SPECIES

1 Leaflets ovate to obcordate, deeply emarginate (commonly in Ceylon) to acute; inflorescences predominantly terminal; flowers 12–19 mm long; calyx 6.5–8.5 mm long; young stems usually drying greyish-brown .... **P. marsupium**
1 Leaflets ovate, acuminate or acute, rarely obtuse; inflorescences predominantly axillary; flowers 16–20 mm long; calyx 5–6(–7) mm long; young stems usually drying dark brown or blackish ....**P. indicus**

Pterocarpus marsupium Roxb., Pl. Corom. 2: 9, tab. 116. 1799; Roxb., Fl. Ind. 3: 234. 1832; Wight & Arn., Prod. 266. 1834; Thw., Enum. Pl. Zeyl. 92. 1859; Benth., J. Linn. Soc. Bot. 4, suppl. 76. 1860; Beddome, Fl. Sylv. tab. 21. 1869; Baker in Hook., f., Fl. Br. Ind. 2: 239. 1876; Trimen, Handb. Fl. Ceylon 2: 90. 1894; Alston in Trimen. Handb. Fl. Ceylon 6: 86. 1931; Worthington, Ceylon Trees 173. 1959; Rojo, Pterocarpus, Phan. Monogr. 5: 58. 1972. Type: *Roxburgh* in Herb Wallich 5842A, ''Yeanganshaw of the Telings, Circar Mts.'' (lectotype K, fide Rojo).

Pterocarpus bilobus Roxb. ex G. Don, Gen. Hist. 2: 376. 1832; Moon, Cat. 51. 1824. Type: "Roxb. in herb. Lamb.'', ''Native of the East Indies'' (holotype possibly at G ?).
Lingoum marsupium (Roxb.) Kuntze, Rev. Gen. 1: 193. 1891.
Pterocarpus marsupium Roxb. forma *biloba* (G. Don) Prain, Ind. For. 26. 1900.

Tree, to about 30 m tall; young stems essentially glabrous, lenticellate, drying greyish-brown; leaves 4–8-foliolate; leaflets coriaceous, ovate to obcordate, 5.5–12(–15) cm long, 2.5–7 cm wide, deeply emarginate to acute, the base rounded to acute, the upper surface glabrous, nitid, the lower surface pubescent with minute, appressed hairs, glabrescent; inflorescences predominantly terminal, paniculate, sometimes axillary; flowers 12–19, commonly 15–17 mm long (June–September); calyx fulvo-sericeous, 6.5–8.5 mm long; petals bright yellow, glabrous; ovary brown-sericeous, 1 or 2-ovulate; fruit sericeous when young, essentially glabrous at maturity, orbicular, 1- or 2-seeded, about 4–5.5 cm in diameter, the marginal wing 1–2 cm wide, the stipe 7–10 mm long; seeds brown, nitid, compressed, subelliptic, 10–12 mm long, 5–6 mm wide, the hilum orbicular, about 1 mm in diameter, subapical; chromosome number 2n = 44.

Distr. Ceylon and India, in forest and patanas at elevations up to about 900 m.
Uses. The blood-red, resinous exudate from the trunk, known as Kino, is used as a medicine and for outward application, according to Trimen (l.c.). Macmillan (Trop. Pl. & Gard. ed. 5: 214. 1962) cites *P. marsupium* as a ''widespreading, handsome tree, fine, dark, hard timber.'' Worthington (l.c.) states that the timber is useful ''for the best furniture, carving, buildings.'' In India (Sudhir Kumar Das, Medic. Econ. & Useful Pl. of India 75, undated) leaf juice or paste is applied on wounds, erysipelas, ulcers and skin diseases. The gum is

a tonic for invalids and is taken in acute diarrhoea, acidity, digestive disorders, tapeworm, urethritis and applied on aching teeth.

Vern. Gammalu, gan-malu (S); utera-venkai, venkai (T).

Specimens Examined. WITHOUT LOCALITY: *Koenig s.n.* (BM). KANDY DISTRICT: Kundasale, *Gardner, Thwaites C.P. 1495* (BM, P, PDA); Nawalapitiya, *Worthington 379* (Worthington Herb); Imbulpitiya, Nawalapitiya, *Worthington 587* (BM, Worthington Herb.) BADULLA DISTRICT: Lunugala, *Worthington 2970* (BM); Petiyagoda, between Uraniya and Ekiriyankumbura, *Jayasuriya 378* (US); near Passara, Jan. 1888, *s. coll. s.n.* (PDA). MONERAGALA DISTRICT: Bibile, *Worthington 4954* (BM, Worthington Herb.), *5123, 5134* (Worthington Herb.); Wellawaya, *Rudd & Balakrishnan 3207* (PDA, SFV, US); Galorya National Park, Nilgala, *Jayasuriya 1962* (US).

In addition to *P. marsupium,* the only species of *Pterocarpus* native to Ceylon, two additional species have been introduced, *P. indicus* Willd., with two forms, and *P. santalinus* L.f. The more widely planted is *P. indicus,* which has been included in the key. Following are citations of voucher specimens.

Pterocarpus indicus Willd., Sp. Pl. 3: 904. 1800 forma **indicus**. Colombo, museum grounds, *s. coll. s.n.* 31 July 1918 (PDA); "By the great circle", 18 July 1918, *s. coll. s.n.* (PDA). Kandy, *Worthington 6646* (Worthington herb.), *6827* (Worthington herb.). Kegalle, *Worthington & Holme 7042* (Worthington herb.). Peradeniya, R.B.G., *Worthington 924* (BM). Poilakanda arboretum, Kadugannawa, *Worthington 648* (Worthington herb.). Trincomalee, Golf Club, *Worthington 6632* (Worthington herb.).

According to Worthington (Ceylon Trees 172. 1959) this species is "much planted in Colombo for shade; also in Kandy."

Pterocarpus indicus Willd. forma **echinatus** (Pers.) Rojo, Phan. Monogr. 5: 46. 1972. (*P. echinatus* Pers., Syn. Pl. 2: 277. 1807). Peradeniya, R.B.G., *s. coll. s.n.* Aug. 1890 (PDA); June 1896 (PDA); 10 May 1918 (PDA).

This form, like the typical form, is used in Ceylon chiefly as an ornamental tree.

Pterocarpus santalinus L.f., Suppl. 318. 1781. Peradeniya R.B.G., 16 March 1900, *s. coll. s.n.* (PDA).

Known as red sandal-wood, the red heartwood of this species is fragrant and resembles sandal-wood. "It affords a reddish-brown dye used for colouring woollen fabrics" (Macmillan, op. cit. 418). S.K. Das (l.c.) states that "the powdered or pasted wood is applied on boils and other skin eruptions and infections and on the forehead for headaches, the wood and bark brew taken orally relieves chronic dysentery, worms, haematemesis, weak vision, and hallucination."

3. PONGAMIA

Vent., Jard. Malm. 28, tab. 28. Dec. 1803, nom. cons. Type: *Pongamia glabra* Vent., nom. illeg., a synonym of *Pongamia pinnata* (L.) Pierre.

Trees, sometimes scandent; leaves imparipinnate, 5–9-foliolate, alternate; stipules present; stipels absent; leaflets opposite with one terminal; inflorescences axillary, racemose; bracts and bracteoles present; flowers with calyx campanulate, truncate or subtruncate; petals white to pinkish, usually with a yellow target spot on the vexillum; stamens 10, submonadelphous, the filaments united above but the vexillar filament free at the base; anthers uniform, elliptic, versatile, dorsifixed; ovary subsessile, 1- or 2-ovuled; fruit oblong, somewhat compressed, coriaceous or lignous, indehiscent, 1- or 2-seeded; seed reniform with a small lateral hilum; chromosome numbers 2n = 20, 22.

A monotypic genus of the Old World tropics.

Pongamia pinnata (L.) Pierre, Fl. For. Cochinch. tab. 385, in obs., 15 April 1899; Alston in Trimen, Handb. Fl. Ceylon 6: 87. 1931 *(as P. pinnata* (L.) Merr.); Worthington, Ceylon Trees 114. 1959.

Cytisus pinnatus L., Sp. Pl. 741. 1753. Type: India, Plukenet, Phytographia 104, fig. 3. 1691 (typotype BM-Sloane ?).

Robinia mitis L., Sp. Pl. ed. 2, 1044. 1763, based on *Cytisus pinnatus* L.

Galedupa indica Lam., Enc. 2: 594. 1786, excl. syn. *Caju galedupa* Rumph.; Roxb., Fl. Ind. 3: 239. 1832.

Pterocarpus flavus Lour., Fl. Cochinch. 431. 1790.

Galedupa pungum Gmel., Syst. 1086. 1792.

Dalbergia arborea Willd., Sp. Pl. 3: 901. 1802, non Heyne in Roth, 1821. Type: India (B-Willd. 13074).

Pongamia glabra Vent., Jard. Malm. 28, tab. 28. Dec. 1803. Type: *Ventenat* ? (now at G ?). Wight & Arn., Prod. 262. 1834; Wight, Ic. Pl. Ind. Or. 59. 1858; Thw., Enum. Pl. Zeyl. 92. 1859; Benth., J. Linn. Soc. Bot. 4, suppl.: 115. 1860; Baker in Hook. f., Fl. Br. Ind. 2: 239. 1876; Trimen, Handb. Fl. Ceylon 2: 91. 1894.

Pongamia xerocarpa Hassk., Retzia, ed. nov. 208. 1856.

Malaparius flavus Miq., Fl. Ind. Bat. 1: 1082. 1858.

Malaparius flavus Miq. var. *obtusata* Miq., Fl. Ind. Bat. 1: 1082. 1858.

Pongamia mitis (L.) Kurz, J. Asiat. Soc. Bengal 45 (2): 128. 1876.

Cajum pinnatum (L.) Kuntze, Rev. Gen. 1: 167. Nov. 1891.

Galedupa pinnata (L.) Taub. in Pflanzenfam. 3(3): 344. 1894.

Pongamia glabra Vent. var. *xerocarpa* (Hassk.) Prain, J. Asiat. Soc. Bengal 66: 456. 1897.

Pongamia mitis var. *xerocarpa* (Hassk.) Merr., Philipp. J. Sci. 5: 101. 1910.

Pongamia pinnata (L.) Pierre var. *xerocarpa* (Hassk.) Alston in Trimen, Handb. Fl. Ceylon 6: 87. 1931 (as *P. pinnata* Merr. var. *xerocarpa* Prain).

Tree, to about 30 m tall, sometimes scandent; stems glabrous; stipules (bud scales) cucullate, 3.5–4 mm long, 3 mm wide, striate, puberulent, caducous; leaves 5–9-foliolate, the axis 6–17 cm long, glabrous or nearly so; leaflets ovate to elliptic-ovate, (2–)3–10 cm long, 2–5 cm wide, acuminate, rounded at the base, the sur-faces glabrous or sparingly puberulent; inflorescences axillary, racemose, the axes glabrous or subglabrous, the flower buds nutant on pedicels about 3–5 mm long; bracts linear-deltoid, 3–4 mm long, 0.5–1 mm wide at the base, pubescent, caducous; bracteoles, at base of calyx, minute, linear-deltoid, about 0.5 mm long, pubescent; flowers 12–14 mm long; calyx campanulate, sericeous, 3–4 mm long, 4–5 mm in diameter, subtruncate or truncate, the teeth, or lobes, deltoid, about 0.25–0.3 mm long or less; petals white to pinkish, the vexillum sericeous or subsericeous, essentially straight, the keel petals cohering at the tip; anthers elliptic, 0.75–1 mm long; ovary subsessile, usually 2-ovuled, puberulent; fruit indehis-cent, glabrous, somewhat compressed, obliquely oblong or elliptic, 5–8 cm long, 2.5–3 cm wide, about 0.8–1 cm thick, essentially sessile, 1- or 2-seeded; seeds brown, lustrous, reniform, about 2 cm long, 1 cm wide, compressed, the hilum lateral, elliptic, about 1 mm long.

Distr. In Ceylon, India, Malaysia, northern Australia, Polynesia, Seychelles. Introduced as shade trees in tropical east Africa.

Vern. Gal-karanda, Karanda, Magul-karanda (S); Poona, Punku (T); Mullikulam tree, Indian Beech (E).

Specimens Examined. WITHOUT LOCALITY: *Fraser 41* (BM, US); *Macrae 227* (BM); *Walker 147, s.n.* (K). ANURADHAPURA DISTRICT: Culvert 39/7 on A 12, *Maxwell & Jayasuriya 797* (PDA, SFV, US); Thoru Wewa, 2 mi. E. of Maradankadawala, *Sumithraarachchi & Jayasuriya 213* (US); Tammannawa-Galebindunuwewa Road near 3/12 culvert, *Sumithraarachchi & D. Sumithraarachchi 740* (US). PUTTALAM DISTRICT: Sand Road towards Mundel Lake, parallel to A3 but across lagoon, *Maxwell & Jayasuriya 829* (PDA, US); Wilpattu Natl. Park, *Mueller-Dombois et al. 69042613R* (PDA, SFV, US), *Cooray 69092608R* (PDA, SFV, US), *704040R* (PDA, US), *van Beusekom 1625* (US); Puttalam, *Kundu & Balakrishnan 397* (PDA, US). MATALE DISTRICT: Dambulla, *Worthington 89* (Worthington Herb.); Madulkele, Raxawa Garden, *Worthington 145* (Worthington Herb.); Wahakotte, *Amaratunga 1580* (PDA). KANDY DISTRICT: Guru Oya, Teldeniya, *Balakrishnan & Jayasuriya 1212* (US, PDA); Peradeniya, Kosinna, *Alston s.n.* (PDA); Nitre Cave Valley, *Simpson 9450* (BM); Hakkinda, 9 May 1922, *s. coll. s.n.* (PDA); Doluwa Kande, Nov. 1922, *Livera s.n.* (PDA); Pallekelle Estate, Haragama, *Alston 859* (PDA). KURUNEGALA DISTRICT: Delwita, *Senaratne 2807* (PDA); Ibbagamuwa, *Amaratunga 1709* (PDA). BATTICALOA DISTRICT: Batticaloa, *Amaratunga 1580* (PDA); Kalkudah, in Rest House Compound, *Waas 638* (US). KALUTARA

DISTRICT: Kalutara (as "Caltura"), *Gardner, C.P. 1489* (BM, PDA); between Matugama & Nagoda, Mar. 1887, *s. coll. s.n.* (PDA); between Wadduwa & Panadura, 21 Feb. 1922, *s. coll. s.n.* (PDA). COLOMBO DISTRICT: Talahena, near lagoon edge, *Waas 730* (US). RATNAPURA DISTRICT: Belihul Oya, *Comanor 1108* (PDA, SFV, US); near Embilipitiya, *Simpson 9967* (BM, PDA). BADULLA DISTRICT: Meda Oya, R. Mahaweli, *Jayasuriya 394* (PDA, US). MONERAGALA DISTRICT: About 15 mi. S. of Wellawaya, crossing of highway A2 and Kirindi Oya at 183/2, *Davidse 7755* (US). GALLE DISTRICT: Udugama, *Worthington 2337* (BM, Worthington Herb.). HAMBANTOTA DISTRICT: Ruhuna National Park, *Fosberg et al. 51082* (PDA, US).

Pongamia occurring in Ceylon appears to be referable to *P. pinnata* var. *xerocarpa,* if one wishes to recognize the two varieties, characterized by Prain (J. Asiat Soc. Bengal 66: 456. 1897) as follows:

1 Leaflets usually 5, occasionally 7, oblong or ovate, 6.5–9 cm (2.5–3.5 in) wide, quite glabrous beneath; racemes always solitary, simple; pedicels 8 mm (0.35 in) long, their bracteoles subopposed and situated slightly above the middle (sea coasts, banks of tidal rivers and mangrove swamps on all the coasts; only occurs inland as a planted species)..................................
....var. **pinnata** (*P. glabra* var. *typica*)
1 Leaflets 7–9, very rarely 5, lanceolate, 2.5–3 cm (1–1.35 in) wide, usually sparingly puberulous on the midrib and main nerves beneath; racemes occasionally 5–7.5 cm (2–3 in), sometimes sparingly branched; pedicels 6 mm (0.25 in) long, the bracteoles opposed and placed close under the calyx (Ceylon, Pahang, Redak, Perak, Malacca, Java). ....var. **xerocarpa**

According to Worthington, these trees are planted for tea shade and as ornamentals. The timber is used for posts, wheels, and fuel. Trimen states that the seeds "afford an oil which is used in skin diseases" and Macmillan (Trop. Pl. & Gard. ed. 5, 210. 1962) that the "foliage relished by cattle in time of drought" and "juice of roots used for sores; also for cleaning the teeth and strengthening the gums." S.K. Das (Medic., Econ., Useful Pl. of India, 73. undated) gives numerous uses for *Pongamia,* the bark, root, seed, and leaf as vermifuge, insecticide, and anthelmintic; seed oil and leaf juice applied on skin diseases, leprosy, and rheumatism; leaf juice for gastric acidity, dropsy, dyspepsia, elephantiasis, piles, croup, whooping cough, and bronchitis, etc.

4. DERRIS

Lour., Fl. Cochinch. 432. 1790, nom. cons. Type: *D. trifoliata* Lour., type. cons.

Deguelia Aubl., Pl. Guian. 750, tab. 300. 1775. Type: *D. scandens* Aubl., non *Derris scandens* (Roxb.) Benth.
Dalbergia subgen. *Brachypterum* Wight & Arn., Prod. 264. 1834. Type: *Dalbergia scandens* Roxb.

Brachypterum (Wight & Arn.) Benth., Comm. Leg. Gen. 37. 1837 (preprint);
Benth., Ann. Wiener Mus. Naturgesch. 2: 101. 1838.

Derris sect. *Brachypterum* (Wight and Arn.) Benth., J. Linn. Soc. Bot. 4, suppl.:
101. 1860.

Lianas or scandent shrubs or, sometimes, trees; leaves about 3–23-foliolate,
imparipinnate, alternate; stipules and stipels present or absent; leaflets opposite;
inflorescences racemose or paniculate, axillary or terminal; bracts and bracteoles
usually small, caducous; flowers with calyx cyathiform, truncate or subtruncate;
petals white to violet or reddish; stamens 10, monadelphous or subdiadelphous
with the vexillar filaments free at the base but all filaments connate above, the
anthers uniform, dorsifixed; ovary sessile or short-stipitate, (1–)2–several ovul-
ed; style incurved; stigma small, terminal; fruit indehiscent, membranous to subcoriaceous, suborbicular to elliptic-oblong, compressed, one or both margins narrowly winged, 1–several-seeded; seeds reniform or suborbicular; chromosome
numbers 2n = 20, 22, 24, 26, 36.

A pantropical genus of about 70 species.

KEY TO THE SPECIES

1 Flowers 5–8 mm long; fruit with one or two marginal wings
  2 Fruit glabrous, elliptic, about 3–6 cm long, 1–2 cm wide, narrowly winged along both margins;
    flowers 5–7 mm long; calyx ciliolate, otherwise glabrous; leaves 5–9-foliolate...**1. D. parviflora**
  2 Fruit finely velutinous, elliptic to elliptic-oblong, 1.5–8 cm long, 1.5–2 cm wide, narrowly winged
    along one margin; flowers 6–8 mm long; leaves 5–7-foliolate............**2. D. benthamii**
1 Flowers 9–12 mm long; fruit with one marginal wing
  3 Leaves 9–19-foliolate; leaflets pubescent below, sometimes glabrescent with age; calyx pubescent
    4 Fruit sericeous, glabrescent, 1–5-seeded, oblong, 2.5–8 cm long, 0.9–1.3 cm wide, acute at
      apex and base; flowers 9–10 mm long................................**3. D. scandens**
    4 Fruit glabrous, 1-seeded, elliptic, 3.5–5 cm long, 1.8–2.2 cm wide, decurved-mucronate at apex,
      narrowed at base; flowers 12–15 mm long........................**4. D. canarensis**
  3 Leaves 3–7-foliolate; leaflets glabrous or nearly so; fruit elliptic to subreniform, glabrous, 3–5.5
    cm long, 2–3.5 cm wide; flowers 9–12 mm long; calyx ciliolate, otherwise glabrous or nearly
    so..**5. D. trifoliata**

1. Derris parviflora Benth., J. Linn. Soc. Bot. 4, suppl.: 105. 1860, nom. nov.
based on *Brachypterum elegans* Thw., non *Derris elegans* Benth. 1852; Thw.,
Enum. Pl. Zeyl. 413. 1864; Baker in Hook. f., Fl. Br. Ind. 2: 240. 1876; Trimen,
Handb. Fl. Ceylon 2: 92. 1894; Alston in Trimen, Handb. Fl. Ceylon 6: 87. 1931.

Brachypterum elegans Thw., Enum. Pl. Zeyl. 93. 1859. Type: Ceylon, Central
    Province, Deltota, July 1851, *Thwaites C.P. 2508* (holotype PDA; isotypes
    BM, K, P, PDA).

Deguelia parviflora (Benth.) Taub., Bot. Centralbl. 47: 388. 1891.

Liana or scandent shrub; young stems glabrous; leaves (5–)7–9-foliolate;
leaflets coriaceous or subcoriaceous, ovate to oblong or obovate-oblong,

(2–)2.5–4(–9) cm long, (1–)1.5–2.5(–4.5) cm wide, subacute or breviacuminate, blunt at apex or, sometimes, emarginate, rounded at the base, glabrous on both surfaces; inflorescences axillary, racemose or paniculate, the peduncles fasciculate, the pedicels verticillate; bracts deltoid, minute, mostly caducous; bracteoles minute, usually 1–1.5 mm below base of calyx; flowers about 5–7 mm long; sweet-scented; calyx reddish, ciliolate, otherwise glabrous or nearly so, 2–2.5 mm long, 2.5 mm in diameter, subtruncate, the teeth deltoid, about 0.5 mm long; petals white or pale purplish, the vexillum and wings glabrous, the keel puberulent near the apex; fruit sessile, 1 or 2-seeded, glabrous, elliptic, rounded at apex and base, about 3–6 cm long, 1–2 cm wide, narrowly winged along both margins, the wings 1–2 mm wide; seeds reniform, compressed, light brown, about 15 mm long, 5–6 mm wide.

Distr. Apparently endemic to Ceylon, in dry areas.

Vern. Kala-wel, sudu kala-wel (S).

Specimens Examined. JAFFNA DISTRICT: 4.5 mi. S. of Elephant Pass, *Simpson 8030* (BM). ANURADHAPURA DISTRICT: E. of main Dambulla-Anuradhapura Road near milestone 53, growing from edge of dry granite slab 250 yards from road side, *Townsend 73/183* (US); Wilpattu, Maradanmaduwa, *Ripley 36* (US); Wilpattu National Park, across Kalli Villu from Park Bungalow, *Maxwell & Jayasuriya 802* (PDA, SFV, US). POLONNARUWA DISTRICT: Polannaruwa-Batticaloa Road, mile 66, *Meijer & Balakrishnan 126* (US); Welikanda-Kandakaduwa Road, *Waas 609* (US). PUTTALAM DISTRICT: Wilpattu National Park, from Tala Wila on the Puttalam-Mannar Road, *Mueller-Dombois et al. 69042620* (PDA, SFV, US). KANDY DISTRICT: New Peradeniya Estate, 21 Nov. 1940, *s. coll. s.n.* (PDA). TRINCOMALEE DISTRICT: Great Sober Island, near beach, *Wheeler & Balakrishnan 12579* (US). BATTICALOA DISTRICT: Maha Oya area, 2 mi. NE. of Rukam, 30 m. elev., *Stone 11177* (US). AMPARAI DISTRICT: On road to Wadinagala, about 3 mi. S. of main dam on Senanayake Samudra, *Maxwell & Jayasuriya 750* (PDA, SFV, US). BADULLA DISTRICT: Between Bibile and Mahiyangana, culvert 4/1, *Maxwell 1024* (PDA, SFV, US); Uma Oya, June 1881, *s. coll. s.n.* (PDA). HAMBANTOTA DISTRICT: Ruhuna National Park, N. of Kumbukkan Oya ford to Kumana, *Cooray 68052706R* (PDA, US); Ruhuna National Park, Block 1, Rugamtota, Plot 31, on Menik Ganga, 10 m elev., *Fosberg 50179* (PDA, SFV, US), Block 2, Plot 11, past Uda Potana, *Comanor 04290867* (PDA, US), Block 2, 400 m. W. of Kumbukkan Oya, 2 mi. N. of jeep road, *Mueller-Dombois et al. 67100118* (PDA, US).

2. Derris benthamii (Thw.) Thw., Enum. Pl. Zeyl. 413. 1864; Gamble, Fl. Pres. Madras 387. 1918; Alston in Trimen, Handb. Fl. Ceylon 6: 87. 1931.

Brachypterum benthamii Thw., Enum. Pl. Zeyl. 93. 1859. Type: Ceylon: Uma-oya (''Ooma Oya''), July 1853, *Thwaites C.P. 2925* (holotype PDA; isotypes BM, BR, K, P, PDA).

Derris paniculata Benth., J. Linn. Soc. Bot. 4, suppl. 105: 1860; Baker in Hook. f., Fl. Br. Ind. 2: 242. 1878; Trimen, Handb. Fl. Ceylon 2: 93. 1894. Type: Ceylon: *Col. Walker 188* (holotype K).
Deguelia paniculata (Benth). Taub., Bot. Centralbl. 47: 388. 1891, as "*panniculata*".

Liana; young stems pubescent with short, reddish-brown hairs, glabrescent; leaves 5–7-foliolate; leaflets subcoriaceous, elliptic to oblong or obovate-oblong, 3–10 cm long, 1–4 cm wide, glabrous above, puberulent below, especially along the midvein, usually glabrous at maturity, the apex acuminate, blunt, the base subcuneate to rounded; inflorescences paniculate, terminal, to about 30 cm long; bracts and bracteoles minute, deltoid, caducous; flowers 6–8 mm long; calyx reddish, puberulent, 2–3 mm long, 2.5 mm in diameter, subtruncate; petals pink, glabrous; fruit sessile, 1 or 2-seeded, elliptic to elliptic-oblong, 1.5–8 cm long, 1.5–2 cm wide, finely brown-velutinous, winged along one margin, the wing 1–4 mm wide; seeds reniform, reddish-brown, nitid (mature seeds not seen).

Distr. Ceylon and southern India, in dry areas.
Vern. Han-kalawel (S).
Specimens Examined. WITHOUT EXACT LOCALITY: *Walker s.n.* (K). ANURADHAPURA DISTRICT: Habarane, *F. W. de Silva 37* (PDA), 30 June 1927, *s.n.* (PDA); Wilpattu National Park, Kokmote Bungalow on Moderagam Aru, along river, *Maxwell & Jayasuriya 814* (PDA, SFV, US), between Magul Illaima and Malimaduwa, *Wirawan et al. 1130* (PDA, SFV, US).

3. Derris scandens (Roxb.) Benth., J. Linn. Soc. Bot. 4, suppl.: 103. 1860; Thw., Enum. Pl. Zeyl. 413. 1864; Baker in Hook. f., Fl. Br. Ind. 2: 240. 1876; Trimen, Handb. Fl. Ceylon 2: 91. 1894; Gamble, Man. Ind. Timbers 263. 1922; Alston in Trimen, Handb. Fl. Ceylon 6: 87. 1931; Macmillan, Trop. Pl. & Gard. ed. 5, 127. 1962, non (Aublet) Pittier 1917, 1944.

Dalbergia scandens Roxb., Pl. Corom. 2: 49, tab. 192, May 1805; Moon, Cat. 51. 1824; Roxb., Fl. Ind. 3: 232. 1832; DC., Prod. 2: 417. 1825; Wight & Arn., Prod. 264. 1834; Wight, Ic. Pl. Ind. Or. 275. 1840. Type: India, *Koenig s.n.* (BM).
Dalbergia timoriensis DC., Prod. 2: 417. 1825.
Brachypterum scandens (Roxb.) Benth. ex Wight, Ic. Pl. Ind. Or. 275. 1840; Thw., Enum. Pl. Zeyl. 93. 1859.
Deguelia timoriensis (DC.) Taub. in Pflanzenfam. 3(3): 345. 1894.
Derris timoriensis (DC.) Pittier, Contr. U.S. Natl. Herb. 20: 41. 1917.

Liana, to about 30 m high; young stems puberulent, glabrescent; leaves 9–11(–19)-foliolate; stipules small, caducous; leaflets subcoriaceous, oblong to obovate-oblong, (1.5–)2.5–6.5 cm long, 1–2(–3) cm wide, obtuse, acute, breviacuminate or, sometimes, emarginate, the base rounded to subacute, the upper surface glabrous, nitid, the lower surface puberulent or minutely appressed-pubescent; inflorescences axillary, racemose, about 25–45 cm long; bracts minute, deltoid-ovate, caducous; bracteoles ovate, at base of calyx, about 1 mm long; flowers 9–10 mm long; calyx sericeous, subtruncate, 3 mm long, 2.5–3 mm in diameter; petals white to pink or lavender, the vexillum and wings glabrous or nearly so, the keel puberulent toward the apex; ovary 6–8-ovulate; fruit essentially sessile, sericeous, glabrescent, 1–5-seeded, oblong, acute at apex and base, 2.5–8 cm long, about 9–13 mm wide with a wing about 1.5 mm wide along one margin; seeds reniform, dark brown, subnitid, compressed, about 9–10 mm long, 6 mm wide; chromosome number 2n = 26.

Distr. Ceylon, India, Thailand, Malaysia, southern China, and northern Australia, at low elevations.

Vern. Ala vel, bo-kalavel, kala-wel (S); Kalungu kodi, tekil, welan-tekel (T).

Specimens Examined. WITHOUT LOCALITY: *Macrae 12* (BM); *s. coll. s.n.* (P). JAFFNA DISTRICT: Between Paranthan & Mullaitivu, *Jayasuriya 28* (US). ANURADHAPURA DISTRICT: Kekirawa-Anuradhapura Road, *Amaratunga 1386* (PDA); Nochchiyagama-Nikawewa, *Jayasuriya 8* (US); Palugaswewa, *Simpson 9167* (BM); Mihintale, *Simpson 8521* (BM). POLON-NARUWA DISTRICT: Trikonamadu, *Balakrishnan 370* (SFV, US). PUT-TALAM DISTRICT: South of Bangadeniya, 4 mi. N. of Chilaw, *Fosberg & Jayasuriya 52827* (SFV, US). KURUNEGALA DISTRICT: Dunkande, Arankele, *Jayasuriya & Balasubramaniam 530* (US); Nikaweratiya, *Amaratunga 321* (PDA); Kuliyapitiya, *Amaratunga 992* (PDA). MATALE DISTRICT: Road around Sigiriya Rock, *Nowicke, Fosberg & Jayasuriya 371* (US); Dambulla, Pelawehera Tank, Trincomalee Road, *Worthington 1049* (BM, Worthington Herb.); Mahawela, *Amaratunga 1797* (PDA). BATTICALOA DISTRICT: Batticaloa, *Gardner, C.P. 1492* (BM, BR, K, P, PDA); Batticaloa-Kalkudah Road, mile post 3, *Kundu & Balakrishnan 203* (US). AMPARAI DISTRICT: Pottuvil-Panama Road, *Cooray 69073004R* (SFV, US). BADULLA DISTRICT: Hembarawa, 1 mi. S, *Jayasuriya 402* (US). MONERAGALA DISTRICT: Inginiyagala, 25 Aug. 1954, *Sylva s.n.* (PDA); Muppane, 17 Mar. 1927, *de Silva s.n.* (PDA).

4. Derris canarensis (Dalz.) Baker in Hook. f., Fl. Br. Ind. 2: 246. 1878; Gamble, Fl. Pres. Madras 387. 1918; Alston in Trimen, Handb. Fl. Ceylon 6: 87. 1931.

Pongamia canarensis Dalz., Hooker's J. Bot. Kew Gard. Misc. 2: 37. 1850. Type: *Dalzell* (?) *s.n.,* India, "Canara prope Garsuppa", (K).

Derris ovalifolia Wight & Arn. var., Benth. in Miq., Pl. Jungh. 1: 252. 1852; Thw., Enum. Pl. Zeyl. 92. 1859; Benth., J. Linn. Soc. Bot. 4, suppl.: 113. 1860.

Derris oblonga Benth., J. Linn. Soc. Bot. 4, suppl.: 112. 1860; Baker in Hook. f., Fl. Br. Ind. 2: 242. 1879; Trimen, Handb. Fl. Ceylon 2: 93, tab. 29. 1894; Cooke, Fl. Pres. Bombay 405. 1901.

Deguelia canarensis ("Baker") Taub., Bot. Centralbl. 47: 386. 1891.

Deguelia oblonga (Benth.) Taub., Bot. Centralbl. 47: 387. 1891.

Liana, to about 25 m high; young stems puberulent, glabrescent; leaves 9–17-foliolate; leaflets membranaceous to subcoriaceous, lanceolate-oblong to obovate-oblong, (3–)6–13 cm long, 2–4.5 cm wide, breviacuminate, blunt at the apex, the base acute to subcuneate, the upper surface glabrous at maturity, the lower surface moderately to sparsely pubescent with minute, appressed hairs; inflorescences racemose, about 7–10 cm long; bracts minute, ovate; bracteoles lanceolate-deltoid, 1–1.5 mm long, 1 mm wide, immediately at base of calyx; flowers fragrant, 12–15 mm long; calyx reddish, moderately appressed-pubescent, about 3 mm long, 3 mm in diameter, the teeth deltoid, about 1 mm long; petals white to pink or violet, glabrous; fruit elliptic, glabrous, 1-seeded, 3.5–5 cm long, 1.8–2.2 cm wide, recurved-mucronate at the apex, narrowed at the base, narrowly winged along one margin.

Distr. Ceylon and the western coast of India, in moist evergreen forests at about 300–1300 m elevation, rare.

Vern. Diya kala-wel, kalu kala-wel (S).

Specimens Examined. KANDY DISTRICT: Hantane, 2300 ft, *Gardner 476* (BM). KANDY & KEGALLE DISTRICTS: Hantane, *Gardner*, Kitulgala ("Kittoolgala") Mar. 1853, *Thwaites, C.P. 1493* (BM, BR, P, PDA). MATALE DISTRICT: Between Pallegama and Ranamure, riparian vegetation of Halmini-Oya, 300 m elev., *Jayasuriya 340* (US).

5. Derris trifoliata Lour., Fl. Cochinch. 433. 1790; Merr., Comm. Lour. Fl. Cochinch., Trans. Amer. Philos. Soc., n.s. 24 (2): 206. 1935; Polhill, Fl. Trop. East Africa, Papil. 74, fig. 14, 1971. Type: *Loureiro*, China, Canton (holotype P).

Dalbergia heterophylla Willd., Sp. Pl. 3: 901. 1802. Type: *Roxburgh* ?, India, Willdenow Herb. no. *13073* (lectotype, fruiting specimen, B-Willd; syntype, flowering specimen, B-Willd.).

Robinia uliginosa Roxb. ex Willd., Sp. Pl. 3: 1133. 1802. Type: *Roxburgh* ?, India (holotype B-Willd. ?; isotype BM as *Galedupa uliginosa* Roxb.).

Galedupa uliginosa (Roxb. ex Willd.) Roxb., Hort. Beng. 53. 1814; Roxb., Fl. Ind. 3: 243. 1832.

Pongamia uliginosa (Roxb. ex Willd.) DC., Prod. 2: 416. 1825; Wight & Arn., Prod. 262. 1834.

Derris uliginosa (Roxb. ex Willd.) Benth. in Miq., Pl. Jungh. 1: 252. 1852; Thw., Enum. Pl. Zeyl. 92. 1859; Benth., J. Linn. Soc. Bot. 4, suppl.: 107. 1860; Baker in Hook. f., Fl. Br. Ind. 2: 245. 1878; Trimen, Handb. Fl. Ceylon 2: 92. 1894; Alston in Trimen, Handb. Fl. Ceylon 6: 87. 1931; Macmillan, Trop. Pl. & Gard., ed. 5. 370. 1962.

Deguelia uliginosa (Roxb. ex Willd.) Baill., Bull. Mens. Soc. Linn. Paris 1: 444. 1885.

Deguelia trifoliata (Lour.) Taub., Bot. Centralbl. 47: 388. 1891; Taub. in Pflanzenfam. 3 (3): 345. 1894.
Derris heterophylla (Willd.) Backer ex Heyne, Nutt. Pl. Ned.-Ind., ed. 2, 2: 806. 1927.
Derris affinis Benth. in Miq., Pl. Jungh. 1: 252. 1852.
Derris uliginosa var. *loureiri* Benth., J. Linn. Soc. Bot. 4, suppl.: 108. 1860.
Derris forsteniana Blume ex Miq., Fl. Ind. Bat. 1: 144, tab. 3. 1855.

Liana, to about 15 ´m high; young stems glabrous or nearly so; leaves 3–7-foliolate; stipules deltoid-ovate, 2–3 mm long, caducous; stipels setaceous, about 1 mm long, caducous; leaflets coriaceous, ovate to lanceolate or elliptic, 3.5–15 cm long, 2–7.5 cm wide, breviacuminate, the base rounded to subcordate, the surfaces glabrous or nearly so; inflorescences paniculate, subracemose, (6–)10–28 cm long, terminal and axillary; bracts minute, deltoid to lanceolate or ovate, 0.5–1 mm long, usually caducous; bracteoles ovate-deltoid, minute, immediately below calyx or 1–2 mm below; flowers 9–12 mm long; calyx 2–2.5 mm long, 2.5 mm in diameter, subtruncate, ciliolate but otherwise glabrous or nearly so; petals cream-white to lavender or pinkish, glabrous; fruit elliptic to subreniform, puberulent, glabrescent, sessile, 1 or 2-seeded, 3–4.5(–5.5) cm long, 2–3.5 cm wide, with a wing 1–2 mm wide along one margin; seeds brown, reniform, compressed, about 1.5–2 cm long, 1–1.5 cm wide; chromosome numbers 2n = 22, 24.

Distr. In mangrove swamps, Ceylon, India, southern China, Malaysia, Polynesia, Micronesia, East Africa, and Australia.

Uses. The stem fibres of this species are said to be used in making rope and fishing lines. The bark and roots are mentioned by some authors, including Trimen, as being used as fish poison. Alston, however, on the herbarium sheet of his collection *547*, states, "not used for fish-p. The plant used is Tekelankoddi". According to Sudhir Kumar Das, "Medic., Econ. & Useful Pl. of India," undated, p. 36, "*Derris uliginosa*—Large Climber. Bark is highly insecticide and anthelmintic, is taken for destroying intestinal worms and thrown in ponds for killing fishes and aquatics."

Vern. Kala-wel (S); tekil, tilankoddi, uppu thailan-kodi (T).

Specimens Examined. WITHOUT EXACT LOCALITY: *Macrae 465* (BM); "Jaffna, Trincomalee, Batticaloa, *Gardner,* Caltura, 1853, G.D.A.", *Thwaites C.P. 1494* (BM, P, PDA, US). PUTTALAM DISTRICT: Near Puttalam, Aug. 1883, *s. coll. s.n.* (PDA); Chilaw, *Simpson 8139* (BM). TRINCOMALEE DISTRICT: "Koddi-ar", Aug. 1885, *s. coll. s.n.* (PDA); Trincomalee, 5 Dec. 1927, *de Silva s.n.* (PDA), *Alston 547* (PDA). BATTICALOA DISTRICT: "Kochchikada" (Kokkadichchalai ?), *Simpson 7957* (BM). COLOMBO DISTRICT: Negombo, *Simpson 7911* (BM, PDA), July 1930, *de Silva s.n.* (PDA); Thalahena, *Waas 723* (US). COLOMBO OR KALUTARA DISTRICT: Between Colombo and Panadura, side road off A2, *Comanor 990*

(US). KALUTARA DISTRICT: Near Beruwela, March 1887, *s. coll. s.n.* (PDA). GALLE DISTRICT: Galle, *Gardner 244* (BM).

In addition to the five known native species of *Derris* the following three have been introduced into Ceylon.

Derris elliptica (Wall.) Benth., J. Linn. Soc. Bot. 4, suppl.: 111. 1860; Macmillan, Trop. Pl. & Gard. ed. 5, 370, 459. 1962.

Pongamia elliptica Wall., Pl. As. Rar. 3: 20, tab. 237. March 1832; Wight, Ic. Pl. Ind. Or. tab. 420. 1841.
Galedupa elliptica Roxb., Fl. Ind. 3: 242. Oct.–Dec. 1832.
Deguelia elliptica ("Benth.") Taub., Bot. Centralbl. 47: 387. 1891.

Macmillan (l.c.) states that the bark and flowers of *D. elliptica* are used as fish poison, the juice as arrow-poison, and the roots in insecticidal preparations, known as "derris powder".

Derris microphylla (Miq.) Jackson, Index Kew. 1: 332. 1893.

Brachypterum microphyllum Miq., Fl. Ind. Bat. suppl. 296. 1859.
Derris dalbergioides Baker in Hook. f., Fl. Br. Ind. 2: 241. 1874; Worthington, Ceylon Trees 175. 1959; Macmillan, Trop. Pl. & Gard. ed. 5, 84. 1962.
Deguelia dalbergioides (Baker) Taub., Bot. Centralbl. 47: 386. 1891.

Derris robusta (DC.) Benth., J. Linn. Soc. Bot. 4, suppl.: 104. 1860; Trimen, Handb. Fl. Ceylon 2: 92. 1894.

Dalbergia robusta DC., Prod. 2: 417. 1825; Wight, Ic. Pl. Ind. Or. 244. 1840.
Dalbergia krowee Roxb., Fl. Ind. 3: 229. 1832.
Deguelia robusta ("Benth.") Taub., Bot. Centralbl. 47: 388. 1891.

5. AGANOPE

Miq., Fl. Ind. Bat. 1: 151. 1855. Lectotype species: *A. floribunda* Miq. a synonym of *A. thyrsiflora* (Benth.) Polhill (Kew Bull. 25(2): 266. 1971).

Derris Lour. section *Aganope* (Miq.) Benth., J. Linn. Soc. Bot. 4, suppl.: 103. 1860.
Deguelia Aubl. section *Aganope* (Miq. emend. Benth.) Taub. in Pflanzenfam. 3(3): 345. 1894.
Ostryoderris Dunn, Bull. Misc. Inform. 1911: 363. 1911. Lectotype species: *O. impressa* Dunn (Polhill, Kew Bull. 25 (2): 267. 1971).

Lianas or semi-scandent trees; leaves about 9–11-foliolate, alternate, imparipinnate; stipules present but caducous; stipels present or absent; leaflets opposite or subopposite; inflorescences axillary, or terminal, paniculate, thyrsoid; bracts and bracteoles often caducous; flowers with calyx campanulate, 5-dentate or sub-truncate; petals white, sometimes greenish or pinkish; stamens 10, diadelphous 9 : 1 with the vexillar filament free, the anthers uniform, dorsifixed; ovary sessile or short-stipitate, 1–10-ovuled; style incurved; stigma minute, terminal; fruit in-dehiscent, somewhat compressed, oblong, sometimes constricted between seeds, narrowly winged along one or both margins; seeds subreniform or oblong-obovoid with hilum terminal or subterminal; chromosome number 2n = 22.

A genus of six species in Africa, southern Asia, New Guinea, and the Philippines. Only one species is known from Ceylon.

Aganope heptaphylla (L.) Polhill, Kew Bull. 25: 268. 1971.

Sophora heptaphylla L., Sp. Pl. 373. 1753, excl. syn. Pluk., non auct. post. Type: Ceylon, *Hermann* (lectotype *Hermann herb. no. 2: 8* BM; syntype *Hermann herb. 2: 80,* BM).

Pongamia sinuata Wall., Cat. no. 5911. 1832, nom. nud.; ex Benth., J. Linn. Soc. Bot. 4, suppl.: 113. 1860, nomen in synon.

Pongamia grandifolia Grah. in Wall., Cat. no. 5882. 1832, nom. nud.; ex Benth., J. Linn. Soc. Bot. 4, suppl.: 113. 1860, nomen in synon.

Pterocarpus diadelphous Blanco, Fl. Filip. 563. 1837; Merr., Sp. Blanco. 186. 1918. Type: Philippines, *Blanco,* type not known.

Derris sinuata Benth. ex Thw., Enum. Pl. Zeyl. 93. 1859; Benth., J. Linn. Soc. Bot. 4, suppl.: 113. 1860; Baker in Hook. f., Fl. Br. Ind. 2: 246. 1876; Trimen, Handb. Fl. Ceylon 2: 94. 1894. Syntypes: Ceylon, Batticaloa, *Gardner s.n.* in 1846 (PDA); *Gardner 5911* (K); Colombo, Sept. 1852, Sept. 1856; B [illegible, in Kalutara District ?], Dec. 1853, *Thwaites C.P. 1491* (BM, BR, K, P, PDA).

Deguelia sinuata (Thw.) Taub., Bot. Centralbl. 47: 388. 1891.

Derris diadelpha (Blanco) Merr., Philipp. J. Sci. 5: 103. 1910.

Derris heptaphylla (L.) Merr., Interp. Rumph. Herb. Amb. 273. 1917; Alston in Trimen, Handb. Fl. Ceylon 6: 87. 1931.

Derris exserta Craib, Bull. Misc. Inform. 1927: 383. 1927. Type: Thailand, Muang Loei, *Kerr 8785* (holotype K).

Liana: stems glabrous; stipules and stipels not seen; leaves 5–7-foliolate; leaflets coriaceous, elliptic to lanceolate-elliptic or ovate, 5–10.5 cm long, 2.5–7 cm wide, obtuse, sometimes emarginate, the base rounded, the surfaces glabrous; inflorescences paniculate, elongated, usually terminal, sometimes axillary; bracts and bracteoles caducous, not seen; flowers 14–17 mm long; calyx puberulent, 3–4 mm long, 3–4 mm in diameter, subtruncate; petals greenish white or pinkish; ovary 5–6-ovulate; fruit coriaceous, glabrous, somewhat falcate, 5–9(–20) cm long, essentially sessile, 1–4-seeded, about 2.5–3 cm wide, and about 1 cm thick, sometimes constricted between the seeds, rostrate at the apex, narrowed toward the base, the wing along one margin, 1–2 mm wide; seeds reddish-brown, nitid, subreniform, about 2.5 cm long, 1.3 cm wide, 5 mm thick, the hilum orbicular, about 1 mm in diameter, lateral, subterminal.

Distr. In forests and mangrove swamps, Ceylon, India (Bengal), Southeast Asia, New Guinea, and the Philippines.

Specimens Examined. WITHOUT LOCALITY: India ? Ceylon ?, *Amherst 1404* (BM). WITHOUT EXACT LOCALITY: *Beddome 2447* (BM); *Jonville s.n.* (BM). BATTICALOA DISTRICT: Kalladi, E. of Lady Manning Bridge, low elevation, *Jayasuriya et al. 706* (US).

Other members of the Dalbergieae, chiefly from tropical America, have been introduced into Ceylon, as ornamentals or as specimens in the botanical gardens. Herbarium vouchers are known for the following species:

Andira inermis (W. Wright) DC. subsp. **inermis** DC., Prod. 2: 475. 1825. Peradeniya, Royal Botanic Gardens, *Worthington 1186* (Worthington herb.). Kandy, *Rudd 3058* (PDA, SFV, US); *Worthington 1103* (Worthington herb.), *5278* (Worthington herb.), *6520* (Worthington herb.).

Lonchocarpus cyanescens (Schum. & Thonn.) Benth., J. Linn. Soc. Bot. 4, suppl. 96. 1860, an African species, Peradeniya, R.B.G., Aug. 1887, *s. coll.* (PDA), 1888, *s. coll.* (PDA).

Lonchocarpus glabrescens Benth., Hooker's J. Bot. Kew Gard. Misc. 2: 33. 1850. Peradeniya, R.B.G., March 1888, *s. coll.* (PDA), 26 July 1926, *s. coll.* (PDA).

Lonchocarpus latifolius (Willd.) H.B.K., Nov. Gen. Sp. Pl. 6: 383. 1824. Peradeniya, R.B.G., 26 June 1919. *J.M. Silva s.n.* (PDA).

Lonchocarpus sericeous (Poir.) DC., Prod. 2: 260. 1825. Peradeniya, R.B.G., 26 June 1920, *s. coll.* (PDA).

Platymiscium polystachyum Benth. in Seem., Bot. Harald 121, tab. 21. 1853, 1854. Peradeniya, R.B.G., *Rudd 3325* (PDA, SFV, US); 29 March 1926, *J.M. Silva s.n.* (PDA); April 1885, *s. coll. s.n.* (PDA), March 1900, *s. coll. s.n.* (PDA).

Tribe PHASEOLEAE

(by Richard H. Maxwell*)

DC., Mém. Leg. 1825–1827. Type: *Phaseolus* L.; Benth., Comm. Leg. Gen. 1837; Benth. in Benth. & Hook. f., Gen. Pl. 1. 1865; Hutchinson, Gen. Fl. Pl.

*Indiana University Southeast, New Albany, Indiana, USA.

1. 1964; Verdcourt, Kew Bull. 24(2): 253–293, 24(3): 380–440, 507–568. 1970; Verdcourt, Kew Bull. 25(1): 70–146. 1971; Verdcourt, Fl. Trop. East Africa, Papil. 503–807. 1971; Baudet, Prod. Class. Gen. Papilionaceae-Phaseoleae. Bull. Jard. Bot. État 48: 183–220. 1978; Lackey, Iselya 1: 19–53. 1979; Verdcourt, Man. New Guinea Leg. 422–554. 1979; Lackey in Polhill & Raven, Adv. Leg. Syst. Part 1. 301–327. 1981.

Mostly vines, high-climbing woody lianas, or smaller twiners, or prostrate forms, occasionally shrubs or trees, mostly perennials, rarely herbaceous. Leaves alternate, mostly pinnately trifoliolate, rarely 1–many-foliolate; stipules and stipels usually present. Inflorescences mostly axillary, racemose or pseudoracemose (fasciculate racemose), rarely paniculate, rarely 1–few-flowered. Flowers papilionaceous; calyx imbricate, 5–lobed with the upper lobes partially connate, or 4-lobed, the upper lobes more or less entire apically; vexillary petal usually reflexed, the lamina frequently emarginate, occasionally inflexed biauriculate basally, with callosities, or appendages, the wings usually free, the keels mostly coherent distally, gibbous, occasionally rostrate, incurved or twisted. Stamens 10, diadelphous, the vexillary free, or pseudomonadelphous, with the vexillary filament attached medianly to the staminal sheath; anthers mostly 10, perfect, uniform, occasionally dimorphic; pistil mostly ascending or geniculate, occasionally straight; ovary with a basal disc, 1–numerous-ovulate; style with the basal portion occasionally flat, or swollen, the upper style occasionally flat, hard, twisted, or coiled, glabrous or pubescent, the stigma mostly terminal. Fruit 2-valved, usually dehiscent, compressed, linear or oblong, 1 to about 20-seeded. Seeds usually hard, reniform, ovoid or somewhat oblong, varying greatly in size, the hilum mostly small, usually somewhat elliptic to short-oblong, rarely long-linear to encircling about 4/5 of the seed, occasionally with arillate tissue. Seedlings with opposite first leaves.

CONSPECTUS OF SUBTRIBES AND GENERA

a Calyx and leaflet undersurfaces lacking gland-dots; bracteoles present
  b Vexillum greatly exceeding the keel length, or keels greatly exceeding the vexillum; lianas, coarse vines, occasionally trees; 1–few-seeded, occasionally many-seeded, rarely numerous-seeded
    . . . . Subtribe **Erythrininae** Benth. (1837)
    1 Trees
      2 Vexillum longest; fruit 1–few–many-seeded . . . .**1. Erythrina** (8 spp. ?)
      2 Keels longest; fruit 1-seeded. . . . . **4. Butea** (1 sp.)
    1 Lianas, vines
      3 Calyx obliquely truncate. . . . . **2. Strongylodon** (1 sp.)
      3 Calyx symmetrical
        4 Fruit samaroid, without stinging hairs; wings about equal the vexillum length. . . . .
        . . . . **5. Spatholobus** (1 sp.)
        4 Fruit not samaroid, frequently with stinging hairs; wings greatly exceeding the vexillum length
        . . . . **3. Mucuna** (5 spp. ?)

bb Vexillum and keels about equal length; lianas, vines, rarely herbs, never trees; several–many-seeded, rarely 1–few seeded

  c Inflorescences pseudoracemose ....Subtribe **Diocleinae** Benth. (1837)

    1 Calyx appearing bilabiate, the upper lobes much larger than the lower. ....
    ....**7. Canavalia** (7 spp. ?)

    1 Calyx appearing 4–5-lobed

      2 Seed hilum linear, encircling nearly ½ to ⁴/₅ the seed. .... **6. Dioclea** (2 spp.)

      2 Seed hilum elliptic, ovoid, or short-oblong

        3 Upper style involute, bearded, the stigma subterminal. ...... **8. Pachyrhizus** (2 spp.)

        3 Upper style ascending or straight, glabrous, the stigma terminal

          4 Calyx lobes exceeding the tube length; fruit smooth between the seeds..............
          .... **9. Galactia** (1 sp.)

          4 Calyx lobes about equal the tube; fruit exocarp deeply grooved between the seeds. ...
          ....**10. Calopogonium** (1 sp.)

cc Inflorescences not racemose or racemose lacking tubercles

  d Flowers large, resupinate, 2–5.5 cm long. .... Subtribe **Clitoriinae** Benth. (1837)

    1 Vexillum spurred; calyx campanulate.......................**17. Centrosema** (2 spp.)

    1 Vexillum spurless; calyx funnelform.........................**18. Clitoria** (2 spp.)

dd Flowers mostly medium or small, non-resupinate

  e Upper & lower style similar in shape; the upper style & stigma lacking pubescence (except *Pueraria*); keel petals untwisted, mostly erostrate; flowers very small to medium, mostly shades of blue, violet. .... Subtribe **Glycininae** Benth. (1837)

    1 Calyx 4–5-lobed

      2 Anthers dimorphic, 5+5; flowers about 5 mm long. .... **14. Teramnus** (2 spp.)

      2 Anthers uniform; flowers 10 to about 15 mm long

        3 Stigma bearded; leaflets frequently lobed ....**11. Pueraria** (2 spp. ?)

        3 Stigma glabrous; leaflets unlobed

          4 Flowers about 10 mm long; seeds 1–4

            5 Pseudoracemose, 2 or more flowers at a node. ....**12. Neonotonia** (1 sp.)

            5 Racemose, each node 1-flowered .... **13. Glycine** (2 spp. ?)

          4 Flowers to about 15 mm long; seeds 5–many .... **15. Shuteria** (1 sp.)

    1 Calyx obliquely truncate. .... **16. Dumasia** (1 sp.)

ee Upper & lower style dissimilar, the upper style and/or stigma pubescent; keel petals frequently twisted, occasionally rostrate; flowers mostly medium, frequently yellow.................
  ....Subtribe **Phaseolinae** Benth. (1837)

    1 Fruit with 4 longitudinal wings............................**19. Psophocarpus** (1 sp.)

    1 Fruit lacking 4 wings

      2 Seed hilum linear, with conspicuous white arillate tissue.............**20. Lablab** (1 sp.)

      2 Seed hilum ovoid, elliptic, or short-oblong, lacking conspicuous arillate tissue

        3 Inflorescences few-flowered; keel petals untwisted

          4 Flowers rose or crimson to bluish-purple; pollen fine or coarse reticulate

            5 Upper style villous adaxially................................**21. Dipogon** (1 sp.)

            5 Upper style mostly glabrous.............................**22. Dolichos** (1 sp.)

          4 Flowers yellow; pollen tuberculate or spinulose...........**23. Macrotyloma** (3 spp.)

        3 Inflorescences many-flowered; keel petals usually twisted

          6 Pollen finely or coarsely reticulate; leaflets occasionally lobed

            7 Vines, or erect herbs; flowers mostly yellow, rarely reddish, pinkish; leaflets mostly ovate
            ....**24. Vigna** (13 spp. ?)

            7 Somewhat shrubby, erect; flowers maroon; leaflets narrowly elliptic...............
            ....**25. Macroptilium** (1 sp.)

          6 Pollen smooth, or very finely sculptured; leaflets unlobed......**26. Phaseolus** (2 spp.)

aa Calyx & leaflet undersurfaces gland-dotted; bracteoles absent..........................
....Subtribe **Cajaninae** Benth. (1837)
  1 Fruit exocarp deeply grooved between the seeds
    2 Leaflets with 7–8 pairs of primary lateral veins; seeds usually lacking arillate tissue.....
....**27. Cajanus** (1 sp.)
    2 Leaflets with 3–5 pairs of primary lateral veins; seeds arillate......**28. Atylosia** (4 spp.)
  1 Fruit exocarp more or less non-constricted between the seeds
    3 Fruit 3–4-seeded.......................................**29. Dunbaria** (2 spp.)
    3 Fruit 1–2-seeded
      4 Leaves unifoliolate or digitately trifoliolate
        5 Leaflets mostly ovate, trifoliolate or unifoliolate; shrubs; seed hilum oval, lacking arillate
tissue. ....**30. Flemingia** (4 spp. ?)
        5 Leaflets linear, unifoliolate; erect, small plants; seed hilum linear, with conspicuous arillate
tissue...**32. Eriosema** (1 sp.)
      4 Leaves pinnately trifoliolate
        6 Erect shrubs, fruit 2-seeded....................................**30. Flemingia**
        6 Mostly vines, or shrubs with trailing branches; 1–2-seeded......................
....**31. Rhynchosia** (12 spp. ?)

The ''?'' in (spp. ?) indicates that either further collecting may reveal more species present in Ceylon, or a doubt as to species ranking. Species numbers within the genera include naturalized species as well as some cultivated species whose status is questionable. The tribe includes several genera of considerable economic importance, such as *Canavalia, Phaseolus,* and *Vigna.* Helpful references are given with such genera. Because of the large proportion of introduced, naturalized and ornamental species, the pertinent references following accepted names are not always to Ceylon records.

Within the subtribes of tribe Phaseoleae, I am following the genus placement of Bentham (1837) modified by Lackey (1981). Recent, easily accessible illustrations are noted in species references wherever possible. Lackey's sketches of tribe floral dissections (Iselya, 1979) and some fruit and seeds (Adv. Leg. Syst. Part 1. 1981) are very helpful.

Types given as ''not seen'' are usually followed by specimens cited as ''Based on.'' These specimens provided me additional material for my descriptions. I relied heavily on specimens from Coimbatore (COIMB) when Ceylon material was sparse. ''MH'' refers to the Madras Herbarium which was moved to COIMB with other material from the Southern Circle of India.

See Lackey (1981) for a discussion of pseudoracemose inflorescences. I use the term tubercle to describe a conspicuous outgrowth at the pseudoracemose inflorescence rachis node into which 2 or more flower pedicels are inserted.

See Grear & Dengler (Brittonia 28: 281–288. 1976) for a discussion of seed appendages. I did not find it necessary to make fine distinctions regarding growth appendages around the seed hilum in order to segregate genera or species for the Flora.

I am following the realignment of Verdcourt within the subtribe Phaseolinae.

See Verdcourt (Kew Bull. 24(3): 384–390 & 519–525. 1970) for illustrations of the very important gynoecial configurations and keys to the subtribe genera and subgeneric categories. In my gynoecial description of the style portion of the simple pistil, I frequently refer to "lower and upper style." This is an arbitrary division of the single style with the lower part proximal to the ovary, the upper part distal to the ovary and usually terminated by the stigma. This division is useful in describing the configuration of the entire style, and the shape and pubescence of portions of the style. Style characters are necessary in many cases to separate genera and species within the Phaseolinae.

Lackey (Iselya 1(4): 167–186. 1980) has published 72 floral dissection sketches of American Phaseolinae, which include many Ceylon taxa.

I gratefully acknowledge the preliminary manuscript and herbarium work of Dr. Velva E. Rudd and the opportunity, through her, to participate in the Flora project. I also thank Jean Maxwell for her many hours of word processing and proofreading. Any mistakes are mine.

I am also grateful to the directors of the herbaria cited, especially those whose loans were used as reference sets.

KEY TO THE GENERA

1 Trees; flowers large, 4–8 cm long, showy, shades of red, orange or orange-scarlet
  2 Fruit mostly torulose in the seeded portion, evenly distributed, several–many-seeded; stems and vegetative structures frequently armed; vexillary petal largest, longest........**1. Erythrina**
  2 Fruit greatly compressed, 1 distal seed; stems and vegetative structures unarmed; keel petals largest, longest...**4. Butea** (*B. monosperma*)
1 Vines, lianas, occasionally small shrubs, rarely erect herbs; flowers usually less than 4 cm long (except 3. *Mucuna*, 17. *Centrosema*, 18. *Clitoria*), occasionally red or shades of orange, frequently other colours
  3 Bracteoles present; plants lacking conspicuous gland dots; flowers variously coloured, the vexillum with or without darker striae
    4 Wing and keel petals greatly exceeding the vexillum, keels with a hardened apex or beak; calyx and/or fruit frequently with stinging hairs; flowers 3–7.5 cm long, frequently showy......
    ....**3. Mucuna** (incl. subg. *Stizolobium*)
    4 All petals about equal or wings much shorter than the vexillum, keels without a hardened apex; stinging hairs absent; flowers mostly smaller, showy or inconspicuous
      5 Fruit 1–2-seeded; inflorescences pendant, long pedunculate, or paniculate and many-branched with samaroid fruit
        6 Wing petals much shorter than the vexillum, the keels falcate; flowers over 2 cm long, crowded, pendant at the ends of long peduncles; seed hilum linear; fruit round or ellipsoid, 1–2-seeded
        ....**2. Strongylodon** (*S. siderospermus*)
        6 Petals of about equal length, the keels straight; flowers about 9 mm long; inflorescences paniculate, many-branched; seed hilum elliptic, minute; fruit somewhat oblong, samaroid, the apex narrower, 1-seeded........................**5. Spatholobus** (*S. parviflorus*)
      5 Fruit several–many–numerous-seeded (except 1–few in *Dioclea* p.p.); inflorescences erect, mostly pseudoracemose, racemose or not as above
        7 Calyx appearing bilabiate, lower lobes smaller than the upper; flowers usually resupinate
        ....**7. Canavalia**

7 Calyx 4–5-lobed, or the rim obliquely truncate; flowers non-resupinate (except 17. *Centrosema*, 18. *Clitoria*)

  8 Calyx obliquely truncate; seeds black, attached to the placenta after fruit dehiscence; flowers yellow . . . . **16. Dumasia** (*D. villosa*)

  8 Calyx distinctly 4–5-lobed; seeds mostly other than black, not attached to the placenta after fruit dehiscence; flowers variously coloured

    9 Flowers showy, frequently resupinate, 2–6 cm long, bracteoles large; leaflets 3 or 5–9-foliolate; inflorescences frequently short, few-flowered and axillary with paired flowers

      10 Leaflets 3-foliolate; calyx campanulate; vexillary petal slightly exceeding the wings, the back spurred; fruit linear, both sutures swollen; upper style flat, wide apically, minutely pubescent at the apex . . . . **17. Centrosema**

      10 Leaflets 3, or 5–9-foliolate; calyx cylindrical; vexillary petal greatly exceeding the wings, lacking a spur; fruit linear or oblong, upper suture swollen; upper style bearded longitudinally . . . . **18. Clitoria**

    9 Flowers inconspicuous, or if showy usually smaller than above, always non-resupinate; bracteoles small (except *Dioclea virgata*); leaflets 3-foliolate; inflorescences frequently elongate, rarely short, few-flowered and axillary with paired or unpaired flowers

      11 Seed hilum linear, nearly 1/2 to 4/5 encircling; pseudoracemose, the rachis with large tubercles; upper style and stigma glabrous; anthers 10, uniform, or dimorphic, 6 + 4 . . . . **6. Dioclea**

      (*D. javanica*, the liana, probably extirpated. *D. virgata*, a smaller vine, introduced, but probably abandoned)

      11 Seed hilum oval, elliptic or short-oblong, much less than ½ encircling (except 20. *Lablab*, linear and arillate); upper style and stigma glabrous or pubescent; anthers 10, uniform (except frequently 8 or 9 in 10. *Calopogonium*), or dimorphic, 5 + 5 in 14. *Teramnus*

        12 Flowers small, 5–12 mm long

          13 Flowers whitish, shades of blue-purple, or pink-reddish; upper and lower styles more or less uniform, glabrous; mostly small, twining, climbing vines

            14 Anthers uniform; flowers 6–12 mm long; fruit mostly oblong to linear-oblong, occasionally linear, about 4 mm wide, lacking a beak or the beak downcurved

              15 Fruit 5–11-seeded

                16 Calyx 5-lobed; anthers mostly 8, 9; fruit with conspicuous deep transverse grooves between the seeds; flowers shades of blue-violet . . . .
. . . . **10. Calopogonium** (*C. mucunoides*)

                16 Calyx 4-lobed; anthers 10; fruit smooth; flowers pink to reddish

                  17 Leaflets rounded, occasionally emarginate apically, broadly lanceolate to oblong elliptic, papyraceous; calyx lobes about 2x the tube length; inflorescence rachis tuberculate, not florate almost to the base; bracts inconspicuous; fruit exocarp finely, adpressed pubescent; seed colour variable, usually other than black . . . . **9. Galactia** (*G. striata*)

                  17 Leaflets obtuse apically, orbicular or ovate, the terminal occasionally rhomboid, membraneous; calyx lobes about equal the tube; inflorescence rachis atuberculate, florate almost to the base; bracts conspicuous, persistent; seeds black . . . . **15. Shuteria** (*S. vestita*)

              15 Fruit mostly 2–5-seeded

                18 Calyx 4-lobed; flowers 2 or more at a node; climbing native vine . . . .
. . . . **12. Neonotonia** (*N. wightii*)

                18 Calyx 5-lobed; flowers 1 at a node; erect, hirsute, cultivated annual . . .
. . . . **13. Glycine** (Soya Bean)

            14 Anthers dimorphic, 5 + 5; flowers about 5 mm long; fruit dark, linear, about 3 mm wide, with an upturned beak . . . . **14. Teramnus**

13 Flowers yellow; upper and lower styles dissimilar, the upper pubescent; small prostrate, trailing or weakly climbing vines

   19 Weak, climbing vine in the ground cover; mostly upland; leaflets broadly ovate, occasionally elliptic; fruit about 15 × 4 mm, 1–2-seeded; stipules produced, somewhat bilobed; keel petals untwisted .... **24. Vigna**, sect. *Vigna*, p.p. (*V. hosei*)

   19 Prostrate or sprawling, many-branched vines from a single taproot; mostly seacoast, lowlands; leaflets usually lobed; fruit 2.5–5.5 × 0.3–0.4 cm, 6–14-seeded; stipules mostly peltate, or lanceolate and non-produced; keel petals twisted................. **24. Vigna** p.p.

See *V. trilobata* & *V. aconitifolia*

12 Flowers larger than 12 mm long

  20 Keel petals and/or the style more or less straight, untwisted, except the style involute in 8. *Pachyrhizus*

    21 Flowers whitish, bluish, shades of purple or pink; stigma not surrounded by a ring of hairs, except in 19. *Psophocarpus*, 22. *Dolichos* & 21. *Dipogon*; terminal leaflets frequently rhomboid or triangular, occasionally lobed

      22 Stipules produced below insertion; fruit with 4, broad longitudinal wings, frequently to 18 cm long or longer; stigma surrounded basally by long hairs..................
.... **19. Psophocarpus** (*P. tetragonolobus*, cultivated)

      22 Stipules non-produced; fruit lacking 4 broad wings, never to 18 cm long, except occasionally in 8. *Pachyrhizus*; stigma glabrous or with a ring or tuft of hairs

        23 Fruit deeply grooved transversely between the seeds; upper style involute, the stigma a subterminal sphere; wing auricles long, extending back parallel to the claw; leaflets large, the terminal leaflets rhomboid or triangular, frequently lobed, the margins dentate or entire.................................... **8. Pachyrhizus** (cultivated)

        23 Fruit without distinct transverse grooves; upper style erect, non-involute, the stigma terminal; wing auricles reduced; leaflets smaller, frequently similar in shape, entire, mostly unlobed, occasionally lobed

          24 Fruit slender, 3–5 mm wide, long-linear, straight, 7–20-seeded, about 8 cm long; seeds small about 3 × 2 × 1.5 mm; flowers crowded; style glabrous or a few hairs around the stigma; vexillary petal lacking appendages; leaflets occasionally lobed ......
.... **11. Pueraria**

          24 Fruit wider, frequently oblong, falcate, mostly few–several-seeded, rarely to 8 cm long; seeds larger; flowers mostly uncrowded on the rachis or somewhat clustered; upper style pubescent; vexillary petal appendaged; leaflets lobed or unlobed

            25 Coarse, climbing vines; flowers mostly violet-purple; leaflets about 8 × 5 cm; inflorescence rachis distinctly tuberculate; keel petals bend 90° in the middle; stigma large, glabrous; seed hilum linear, about 6 mm long, 1/2 encircling, with white arillate tissue................................ **20. Lablab** (*L. purpureus*)

            25 Smaller vines; flowers mostly rose pink, crimson pink to bluish purple; leaflets about the same size as above or smaller; inflorescence rachis atuberculate; keel petals oblong-falcate or semiorbicular; stigma small, with hairs; seed hilum short-oblong, 2–3 mm long, arillate tissue poorly developed or lacking

              26 Leaflets about 4 × 3 cm; upper style villous adaxially; introduced cultivated ornamental.................................... **21. Dipogon** (*D. lignosus*)

              26 Leaflets 5.5–7.0 × 2.5–5.5 cm, often lobed; upper style glabrous; native....
.... **22. Dolichos** (*D. trilobus*)

    21 Flowers yellow; stigma surrounded by a ring of hairs; leaflets oblong-oval to lanceolate, the terminal leaflets occasionally rhomboid, triangular, rarely lobed; fruit oblong to linear, 2.4–6.5 × 0.6–1.2 cm (pollen tuberculate or spinulose).........**23. Macrotyloma**

  20 Keel petals and/or style twisted, stigma mostly glabrous, not surrounded by a ring of hairs

27 Fruit linear, 7–12 × 0.3–0.4 cm, to about 20-seeded; flowers maroon; shrubby, erect; upper style incurved, bent into a squarish hook . . . .

. . . . **25. Macroptilium** (*M. lathyroides*)

27 Fruit not as above, mostly few–several-seeded; flowers mostly yellow or whitish, shades of blue-purple, occasionally reddish; upper style not as above

  28 Stipules non-produced; keel petals spiralling 1–5 turns; pollen smooth or finely reticulate; leaflets unlobed; style filiform, curved at least 1 turn; garden cultivars, rarely escapes

. . . . **26. Phaseolus**

  28 Stipules produced, bilobed or spurred basally, or non-produced peltate; keel petals various, rarely spiral more than 1-turn or not at all; pollen finely or coarsely reticulate; leaflets unlobed or lobed; lower style filiform or ribbon-like, the upper enlarged, hard or thick, not curved as above; cultivated, escapes and native . . . . **24. Vigna**

3 Bracteoles absent; leaflets, calyx, fruit or most of plant gland-dotted; flowers mostly yellow or orange, vexillum usually with conspicuous darker striae

29 Leaflets unifoliolate or trifoliolate, mostly ovate, occasionally lanceolate or suborbicular; seed hilum ovoid, elliptic or short-oblong, with or without arillate tissue; plants erect shrubs, or climbing, trailing or occasionally prostrate vines

  30 Fruit exocarp deeply grooved transversely, or obliquely between the seeds, septate, (1–)3–7(–8)-seeded; ovary 3–many-ovulate

    31 Entire plant gland-dotted; leaflets with 7–8 pairs of primary lateral veins; seeds usually lacking arillate tissue; small shrubs; cultivated . . . . **27. Cajanus** (*C. cajan*)

    31 Leaflets and calyx usually gland-dotted; leaflets with 3–5 pairs of primary lateral veins; seeds mostly arillate; vines (except *Atylosia trinervia*, a small shrub); native

      32 Upper style glabrous; inflorescences few–several-flowered; seeds with linear, white, 2-lipped arillate tissue; flowers 5 to about 17 mm long (except 2 cm long in *A. trinervia*, a small shrub); fruit deeply grooved between the seeds . . . . **28. Atylosia**

      32 Upper style pubescent or glabrous; inflorescences usually several-flowered; seeds with circular arillate tissue, or a piece of rim aril; flowers 1.4–3.0 cm long; fruit faintly grooved between the seeds . . . . **29. Dunbaria**

  30 Fruit exocarp smooth, usually non-septate; seeds 1–2(–3); ovary 2-ovulate, rarely 3

    33 Shrubs 1–2 m tall; leaves usually digitately trifoliolate, occasionally unifoliolate or pinnately trifoliolate; bracts frequently broad, foliaceous, persistent . . . . **30. Flemingia**

    33 Mostly vines, climbing, trailing, prostrate or smaller shrubs, with trailing branches; leaves pinnately trifoliolate; bracts not as above . . . . **31. Rhynchosia**

29 Leaflets linear, unifoliolate in Ceylon; seed hilum linear, 2-lipped arillate; erect plants to about 20 cm tall; inflorescences congested . . . . **32. Eriosema** (*E. chinense*)

1. ERYTHRINA

L., Sp. Pl. 706. 1753; L., Gen. Pl. ed. 5, 315. 1754; Hutchinson, Gen. Fl. Pl. 1: 432. 1964 (cum syn.); Verdcourt, Fl. Trop. East Africa, Part 4, subf. 2: 541. 1971; Krukoff & Barneby, Lloydia 37(3): 332–459. 1974. Lectotype species: *E. herbacea* L., fide Krukoff; other authors cite *E. corallodendrum* L.

Hypaphorus Hassk., Hort. Bogor. 197. 1858.
Micropteryx Walp., Linnaea 23: 739. 1850. Lectotype: *M. poeppigiana* Walp.
  = *Erythrina poeppigiana* (Walp.) O.F. Cook.
Duchassaingia Walp., Linnaea 23: 741. 1850. Lectotype: *D. glauca* Walp.
  = *Erythrina fusca* Lour.

Corallodendron Kuntze, Rev. Gen. Pl. 1: 171. 1891.

Trees or shrubs, rarely perennial herbs; stems frequently armed. Leaves pinnately trifoliolate, sometimes absent at flowering; stipules persistent or deciduous; stipels usually fleshy, glandular. Inflorescences usually single, unbranched, terminal or axillary, racemose; bracts and bracteoles deciduous or caducous. Flowers 4–9 cm long, solitary or 2–5-fasciculate, showy; calyx tube spathaceous, bilabiate, obliquely truncate or the lobes variable; petals usually red, occasionally orange, yellow or cream, the vexillum reflexed or spreading, simple, greatly exceeding the wings and keels, the wings free; stamens monadelphous or the vexillary free, the anthers uniform; ovary stipitate, 2–many-ovulate, variously pubescent, occasionally with stellate hairs, the style somewhat geniculate or ascending, glabrous distally, the stigma small, capitate, glabrous. Fruit indehiscent, or variously dehiscent, linear-oblong to linear, straight, slightly falcate, contorted or coiled, usually torulose, occasionally somewhat elliptic and follicular, 1–14-seeded; sutures erect, the upper usually with closely parallel ribs to either side. Seeds mostly ellipsoid, horizontal, variable in colour, the hilum oblong or elliptic.

A genus of about 100 species found in the tropics and subtropics of both hemispheres.

This genus has been extensively studied for years by B.A. Krukoff, but with a New World emphasis. In 1974, Krukoff and Barneby reorganized the genus on a worldwide basis and my treatment is within this framework. Barneby's (1974) sketches are especially valuable and are noted in the following species.

KEY TO THE SPECIES

1 Pubescence of ovary, calyx and young fruit never stellate; calyx tube expanding, unsplit or split less than about ½
  2 Leaflets mostly elliptic; seeds dark with blackish mottling; keel lamina at least ½ the vexillum length, vexillum long-clawed, the claw ¼–⅓ the vexillum length . . . . **1. E. fusca**
  2 Leaflets mostly ovate, triangular or deltoid; seeds mostly shades of red, occasionally dark; keel lamina less than ½ the vexillum length or rarely about ½; vexillum short-clawed
    3 Vexillum ovate, obovate to oblong-obovate, to about twice as long as wide; keel lamina to about ½ the vexillum length; calyx 2-lipped, split about halfway; fruit follicular or samaroid, slightly constricted between the seeds or not; seeds mostly dark brown, yellowish, reddish-purple or blackish
      4 Keel petals much longer than the wings; fruit follicular, the valves thin, papery; seeds evenly distributed; leaflets yellow-waxy beneath (ceriferous); stipules lanceolate, about 5 mm long
. . . .**2. E. suberosa**
      4 Keel petals shorter than the wings; fruit samaroid, basal half indehiscent and without seeds, distal half with seeds, the valves not papery; leaflets nonceriferous; stipules large, cup-like
. . . .**3. E. subumbrans**
    3 Vexillum lanceolate, much more than twice as long as wide; keel lamina about 1/5 the vexillum length; calyx rim obliquely truncate or split ⅓–½; fruit torulose or slender moniliform; seeds red, reddish-orange or orange with black
      5 Calyx rim obliquely truncate, unsplit; seeds without black hila

6 Leaflets ovate, acuminate, much longer than wide; vexillum lanceolate or narrowly elliptic; keel petals fused; calyx tube cylindrical; fruit about 10-seeded; seeds red without black mottling ....**4. E. rubinervia**
6 Leaflets wider than long; vexillum spathulate; keel petals free; calyx tube campanulate; fruit about 7-seeded; seeds red with a black line extending toward the chalazal end......... ....**5. E. mitis**
5 Calyx rim truncate, usually split on one or both sides, becoming bilabiate; seeds scarlet or reddish-orange, with black hila.................................**6. E. lysistemon**
1 Pubescence of ovary, calyx and young fruit stellate; calyx tube split almost to the base on one side only
7 Leafless or with leaves at anthesis; fruit 15–30 cm long, prominently ribbed, straight or slightly curved, usually 6–9-seeded, more or less smooth or not deeply constricted between the seeds; leaflets subdeltoid, usually over 10 cm wide at maturity, the bases truncate to rounded; flowers red; leaflet lamina green and yellow variegated or not variegated..........**7. E. variegata**
7 Leafy at anthesis; fruit less than 15 cm long, without prominent ribs, curved, usually 1–4-seeded, deeply constricted between the seeds; leaflets ovate-rhomboid or triangular, usually under 10 cm wide at maturity, the bases broadly cuneate; flowers orange-red; leaflet lamina not variegated ....**8. E. velutina**

Of the 8 species considered, only *Erythrina fusca* and *E. variegata* are native. *Erythrina velutina, E. subumbrans, E. suberosa,* and *E. mitis* were apparently introduced as plantation shade trees. *E. lysistemon* and *E. rubinervia* were introduced as ornamentals.

The descriptions of non-native species rely heavily on the work of Krukoff (1939) and Krukoff and Barneby (1974) as well as the references cited.

1. Erythrina fusca Lour., Fl. Cochinch. 427. 1790; Alston in Trimen, Handb. Fl. Ceylon 6: 79, 80. 1931; Worthington, Ceylon Trees 166. 1959; Verdcourt, Fl. Trop. East Africa, Papil. 547. 1971; Krukoff, J. Arnold Arbor. 53(1): 130. 1972; Krukoff & Barneby, Lloydia 37(3): 340. t. p. 341. 1974 (cum syn.); Verdcourt, Man. New Guinea Leg. 424. 1979; Dwyer & D'Arcy, Ann. Missouri Bot. Gard. 67(3): 693. 1980. Type: Based on *Gelala aquatica* Rumph., Herb. Amboin. 2: 235, t. 78. 1750, fide Krukoff & Barneby (1974), not seen.

Erythrina glauca Willd., Ges. Naturf. Freunde Berlin, Neue Schriften 3: 428. 1801; Benth. in Mart., Fl. Bras. 15(1): 173. 1859; Amshoff in Pulle, Fl. Suriname 2(2): 190. 1939; Krukoff, Brittonia 3: 224. 1939. Type: Venezuela, Caracas. Based on notes by Bredemeyer, fide Krukoff (1939).
Erythrina ovalifolia Roxb., Hort. Beng. 53. 1814, nomen; Roxb., Fl. Ind. 3: 254. 1832; Wight, Ic. Pl. Ind. Or. 1, t. 247. 1840; Thw., Enum. Fl. Zeyl. 89. 1859; Baker in Hook. f., Fl. Br. Ind. 2: 189. 1876; Trimen, Handb. Fl. Ceylon 2: 64. 1894; Ridley, Fl. Malay Penins. 1: 579. 1922. Type: India, Calcutta, *Roxb.*, drawing 972 (K), fide Verdcourt (1971), not seen.

Large trees to about 20 m tall, with a short trunk and spreading crown; branches armed. Leaves present at flowering, the petioles to about 19 cm long, the rachis ½ to ⅕ the petiolar length, glabrescent or glabrous; leaflets mostly elliptic, oval or ovate, about 15 × 7 cm, somewhat coriaceous, glabrous above, with dense, short adpressed hairs beneath or glabrous, rounded or obtuse apically, the bases similar, primary lateral veins in 7–9 pairs; stipules not seen; stipels about 1 mm long, persistent. Inflorescences erect, with ferruginous hairs, soon glabrescent, florate more than ½ the length; bracts and bracteoles ovate, to about 2.5 × 2 mm. Flowers 1–2–5 at the node, about 5 cm long, the pedicels 5–20 mm long; calyx broadly campanulate, asymmetrically split, to about 17 mm long on the keel side, with a spur, to about 14 mm long on the vexillum side, lacerate, ferruginous pubescent, the hairs deciduous; petals orange or dark red, the vexillum rounded-rhomboid, 3–5 cm long, glabrous, the claw about 15 mm long, the wings obliquely obovate, about 25 × 15 mm, the keels ovate-falcate, about 30 × 15 mm; ovary long-stipitate, several–many-ovulate, the style slightly ascending. Fruit linear, somewhat woody, torulose, about 30 × 1.5 cm, with a mostly upcurved beak, about 2 cm long, the stipe about 1.5 cm long, immature with dense, adpressed, ferruginous hairs, then glabrescent, 5–10-seeded. Seeds oblong-ellipsoid, 12–18 × 5–8 mm, opaque, dark brown or blackish, with black mottling, the hilum elliptic, about 7 mm long. Chromosome number 2n = 42.

Distr. West Indies, Central America, tropical South America, India, Malaysia, Indonesia, the Philippines and other Pacific islands.

Ecol. Both Krukoff and Verdcourt believe the species is dispersed by ocean currents. Krukoff (1972) states the species thrives in a variety of habitats, but in low swampy ground the trees reach huge dimensions. Trimen (1894) notes the species rather common in the low country, by the banks of rivers and streams. Flowering is from January through April.

Uses. *Erythrina ovalifolia* is on the Seeds List (Dept. Agr., Royal Botanic Gardens, Peradeniya, Ceylon 1958) as a plantation shade tree (?). Krukoff & Barneby (1974) report the species used extensively in the New World as a shade tree for cacao and coffee.

Vern. Yak-erabadu, Yak-errabadoogas (S).

Specimens Examined. DISTRICT UNKNOWN: Haragama (?), July 1856, *Thwaites* (?) *C.P. 3499* (PDA); 1861, *Thwaites C.P. 3499* (BM), *Thwaites* (?), *C.P. 3499* (K). 1863, *Thwaites C.P. 3499* (P). POLONNARUWA DISTRICT: beyond Polonnaruwa, *Rudd & Balakrishnan 3156* (K, NY, PDA, US). PUTTALAM DISTRICT: A 3, near bridge over Deduru-oya, *Maxwell & Jayasuriya 831* (NY, PDA, US). KURUNEGALA DISTRICT: Kurunegala, *Alston 1453* (PDA); Wenduru Wewa, *Cramer 3061* (US); Talgodapitiya, *Amaratunga 1459* (PDA). MATALE DISTRICT: Nalanda Rest House, *Worthington 3424* (BM); streamside, Demata Oya, Nalanda, *Worthington 3431* (K). KANDY DISTRICT: Peradeniya, *Haigh 9170* (NY), *9322* (NY); Haragama, *Jayasuriya et al. 1411* (US); Hakkinda, *F.W. de Silva, s.n. 25.1.29* (PDA); Mahaweli ganga, *Worthington 69* (BM); Getambe, *Worthington 2377* (K); Peradeniya, *Worthington 4180* (K); Pallekelle, Kandy, *Worthington 5585* (K); Ma Oya, 10 mi. E. of Kandy, *Worthington 6574* (K); Kandy, resthouse, *Worthington 7067* (K); Getambe,

Peradeniya, *Worthington 7068* (K); Wattegama, *Worthington 7069* (K). TRIN-COMALEE DISTRICT: Somapura, *Jayasuriya & Sumithraarachchi 1614* (US); Trincomalee, *Worthington 1370* (K).

Note. *Erthyrina fusca* is the most widespread species of the genus.

2. Erythrina suberosa Roxb., Fl. Ind. 2: 253. 1832; Wight & Arn., Prod. 1: 260. 1834; Baker in Hook. f., Fl. Br. Ind. 2: 189. 1876; Trimen, Handb. Fl. Ceylon 2: 64. 1894; Gamble, Fl. Pres. Madras 1(2): 353. 1918; Ridley, Fl. Malay Penins. 1: 578. 1922; Krukoff, J. Arnold Arbor. 20: 230. 1939 & 53(1): 131. 1972; Krukoff & Barneby, Lloydia 37(3): 351, t. p. 352. 1974 (cum syn.). Type: Circars, India. *Wallich 5959* (K), not seen, fide Wight & Arn. (1834).

Erythrina superosa Roxb., Hort. Beng. 53. 1814, nomen.
Erythrina sublobata Roxb., Fl. Ind. 3: 254. 1832; Wight & Arn., Prod. 1: 261. 1834.
Erythrina alba Roxb. ex Wight & Arn., Prod. 261. 1834, nomen, Roxb. in E.I.C. Mus. t. 104. 1832.
Erythrina maxima Roxb., ex Wight & Arn., Prod. 261. 1834, nomen, Roxb. in E.I.C. Mus. t. 105. 1832.
Erythrina suberosa Roxb. var. *sublobata* (Roxb.) Baker in Hook. f., Fl. Br. Ind. 2: 190. 1876.
Erythrina glabrescens (Prain) R.N. Parker, Indian Forester 46: 647. 1920.

Deciduous trees with corky, deeply cracked bark; branches unarmed or heavily armed. Leaf rachis slightly shorter than the petiole, usually both with white prickles; stipules lanceolate, about 5 mm long; leaflets somewhat reniform or deltoid and rounded, sometimes sinuately lobed, wider than long, or rhomboidal, the upper surface glabrous, the lower woolly pubescent, rounded to obtuse apically, truncate or broadly cuneate basally. Bracts lanceolate, about 3 mm long, caducous. Flowers about 4 cm long, the pedicels 2–5 mm long; calyx tube about 5 mm long, campanulate, splitting to become bilabiate, glabrous; petals bright scarlet, the vexillum broadly oblanceolate to elliptic, to about 3.8 × 2 cm, sessile, the wings minute, about 6 mm long, the keels about 1.8 cm long; ovary few-ovulate. Fruit linear-falcate, follicular, chartaceous, oblong to linear-falcate, non-torulose, to about 15 cm long, 2–5-seeded, with spongy packing between the seeds. Seeds subreniform or ellipsoid, about 8 × 6 × 5 mm, dark, reddish-brown, the hilum elliptic, about 3 × 1.5 mm. Chromosome number $2n = 42$.

Distr. India, Bhutan, Burma, Nepal, Thailand, and Vietnam, fide Krukoff & Barneby (1974).

Specimen Examined. NUWARA ELIYA DISTRICT: Estate at Maskeliya, *Worthington 2724* (BM mixture).

Notes. The follicular, chartaceous fruit in the packet indicates a species of sect. *Suberosae* Krukoff (Krukoff and Barneby, 350. 1974).

The Seeds List (1958) has *E. suberosa* as a shade tree, with seeds available from March through May.

Description based on *Wallich 5959* (BM isotype), *5959C* (BM).

3. Erythrina subumbrans (Hassk.) Merr., Philipp. J. Sci. 5: 113. 1910; Backer & Bakh. f., Fl. Java 1: 628. 1963; Krukoff, J. Arnold Arbor. 53(1): 130. 1972; Krukoff & Barneby, Lloydia 37(3): 355, t. p. 356. 1974 (cum syn.). Type: Java, not seen.

Erythrina lithosperma Miq., Fl. Ind. Bat. 1(1): 209. 1855, non Blume, Cat. Gew. Buitenzorg 92. 1823, nomen; Baker in Hook. f., Fl. Br. Ind. 2: 190. 1876; Trimen, Handb. Fl. Ceylon 2: 64. 1894; Ridley, Fl. Malay Penins. 1: 579. 1922; Alston in Trimen, Handb. Fl. Ceylon 6: 80. 1931; Amshoff in Pulle, Fl. Suriname (2): 191. 1939.
Hypaphorus subumbrans Hassk., Hort. Bogor. Descr. 198. 1858.
Erythrina mysorensis Gamble, Fl. Pres. Madras 1(2): 354. 1918, ined.; Gamble, Kew Bull. 1919: 222. 1919.

Trees, the trunk and main branches usually armed; stems terete, glabrous. Leaves present at flowering, the rachis ⅓–¼ the petiolar length; stipules large, cuplike; leaflets ovate, glabrous, the terminal suborbicular to triangular, occasionally rhomboidal, to about 13 × 12 cm, acute apically, cuneate to rounded basally, with 6–8 pairs of primary lateral veins. Inflorescences numerous-flowered, the rachis tuberculate; bracts and bracteoles linear-lanceolate, caducous. Flowers about 3 cm long, the pedicels about 3 mm long; calyx tube about 4 mm long, appearing 2-lobed, glabrous, lobes 2.5–5.0 mm long; petals orange, the vexillum broadly oblanceolate to ovate, slightly reflexed, glabrous, about 30 × 15 mm, the wings broadly oblanceolate, about 12 × 5 mm, equalling the keels, the keels obovate; vexillary stamen filament lightly fused with the staminal sheath towards the base; ovary short-stipitate, to about 8-ovulate, puberulent, the style straight, pubescent, the upper style with glandular hairs. Fruit narrowly oblong, somewhat samaroid, 8–12 × 1.2–2 cm, yellowish, glabrous, the proximal flat part about 2 mm thick, lacking seeds, the distal seedbearing part about 4 mm thick, 1–5-seeded; upper suture with narrow, adnate ribs, the lower swollen. Seeds subreniform, compressed, about 9 × 6 × 3 mm, reddish-purple, the hilum oval-orbicular, about 1 mm long. Chromosome number 2n = 42.

Distr. Southern India, Burma, Thailand, Laos, Vietnam, Malaya, Sumatra, Java, Borneo, Sarawak, the Philippines, Celebes, Moluccas, Fiji, and the Samoan Islands. The species has been introduced into the New World.

Ecol. The species is found in the hill country and flowers June through January. Krukoff & Barneby (1974) state that these medium-sized trees are "confined to low and middle elevations in light forests, open places and secondary forests."

Uses. According to Stockdale, Petch & Macmillan (The Royal Botanic Gardens, Peradeniya, Ceylon, H.W. Cave & Co., Colombo 1922), the well-known dadap was introduced in 1852 from Java for green manure in tea cultivation. *Erythrina lithosperma* was carried on the Seeds List (1958) as a shade tree, with seeds available March through May.

Vern. Dadap of the Malays. Eramudu (S); Muruinja (T).

Specimens Examined. MATALE DISTRICT: Wahacotte, *Amaratunga 1578* (PDA). KANDY DISTRICT: Andiatenne, Kadugannawa, *Worthington 1423* (K); Peradeniya, Bot. Gard., *Haigh 9171* (NY); Peradeniya, *de Silva 468* (PDA), *469* (PDA); A 3, near Pussellawa, *Mueller-Dombois 67082404* (PDA, US); Upper Hantana, moist patana, *Rudd & Balakrishnan 3051* (K, PDA, US), seedling, *Rudd & Balakrishnan 3051A* (US), *3053* (K, PDA, US); Dodanwela, Peradeniya, *Worthington 4184* (BM, K); Mariawatte, Gampola, *Worthington 6537* (K). NUWARA ELIYA DISTRICT: between Mulhalkele & Uda Pussellawa, *Maxwell 1009* (NY, PDA, US). RATNAPURA DISTRICT: past Rakwana to Lauderdale, *Maxwell et al. 976* (K, PDA, US). BADULLA DISTRICT: vicinity of Welimada, *Maxwell & Jayasuriya 901* (PDA, US).

Notes. The sterile, flattened, indehiscent, proximal part of the fruit is a unique character within the genus. This is the only species in sect. *Hypaphorus* (Hassk.) Krukoff.

4. Erythrina rubinervia H.B.K., Nov. Gen. Sp. 6: 434. 1824; Krukoff, Brittonia 3(2): 307. 1939; Krukoff & Barneby, Lloydia 37(3): 391, t. p. 391. 1974; Dwyer & D'Arcy, Ann. Missouri Bot. Gard. 67(3): 696. 1980. Type: Colombia, S.A., *Humboldt & Bonpland 1787* (Photo NY), not seen.

Small tree or shrub; branches armed. Leaves present at flowering, petioles 8–30 cm long, the rachis about ½ as long, the rachis, petioles and petiolules soon glabrescent; leaflets to about 20 × 12 cm, chartaceous, soon glabrescent or glabrous, acuminate apically, usually cuneate basally, the primary lateral veins in about 10 pairs, the terminal leaflets usually rhombic-ovate. Bracts caducous; bracteoles narrowly elliptic, 0.8–1.1 × 0.3 mm. Flowers about 5.5 cm long, the pedicels 3–9 mm long; calyx cylindrical or tubular-campanulate, the rim somewhat obliquely truncate, the tube to about 2 cm long on the vexillary side, to about 3 cm long on the lower side, pubescent becoming glabrescent or glabrous; vexillary petal scarlet, narrowly elliptic, about 5.0 × 1.0 cm, the cuneate claw about 5 mm long, the wings obliquely elliptic, subauriculate, abruptly acute apically, about 9 × 3 mm, slightly smaller than the keels, about 1/5 the vexillum length, the claw about 1 mm long, the keels semiobovate, auriculate, abruptly acute apically; stamens monadelphous; ovary stipitate, densely pubescent with short spreading hairs, about 12-ovulate. Fruit somewhat woody, torulose, 18–36 × about 1.5 cm, deeply constricted between the seeds, the stipe 3–5 cm long, the beak about 4 cm long, many-seeded. Seeds 10–12 × 6–8.5 mm, scarlet. Chromosome number $2n = 42$.

Distr. Panama, Venezuela, Colombia, Ecuador, Peru, Bolivia. Cultivated in Ceylon.

Specimen Examined. KANDY DISTRICT: Peradeniya Gardens, *s. coll. s.n.* (NY).

Notes. Ecologically the native South American species is confined to higher elevations, between 1000–2500 m, and was probably introduced as an ornamental.

5. Erythrina mitis Jacq., Hort. Schoenb. 2: 47. 1797; Krukoff, Brittonia 3(2): 265. 1939 (cum syn.); Krukoff, Amer. J. Bot. 28: 683–691. 1941; Krukoff & Barneby, Lloydia 37(3): 398, t. p. 398. 1974. Type: Northern Venezuela, Jacq., Hort. Schoenb. 2: pl. 216. 1797, not seen.

Erythrina umbrosa H.B.K., Nov. Gen. & Sp. 6: 434. 1824; Trimen, Handb. Fl. Ceylon 2: 64. 1894. Type: Caracas, Venezuela, *Humboldt 685* (F, GH, NY, photos), not seen.

Small trees; armed or unarmed. Leaves usually present at flowering, the petioles 14–35 cm long, the rachis, petioles and petiolules sparsely pubescent when young, soon glabrous; stipules not seen; leaflets thin-chartaceous, rhombic-ovate or broadly ovate, to about 18 × 15 cm, sparsely pubescent when young, then glabrous, usually long acuminate apically, truncate to cuneate basally, primary lateral veins in 6–9 pairs. Flowers to about 8 cm long, the pedicels about 5 mm long; calyx campanulate, thin, chartaceous, appearing obliquely truncate, to about 15 mm long on the vexillary side, to about 20 mm long on the lower side, sparsely pubescent when young, soon glabrescent; petals scarlet, the vexillum oblanceolate, 4.4–7.8 × 1.1–1.8 cm, rounded or emarginate, the wings about 6–8 × 3.5 mm, the keels free, rounded, about 5–6 × 3.5 mm; ovary long-stipitate, with villous pubescence, several–many-ovulate. Fruit torulose, somewhat woody, the stipe 2.0–3.5 cm long, the seed-bearing portion to about 26 × 1.5 cm, the beak 2–5 cm long, about 7-seeded. Seeds oblong-ellipsoid, about 11 × 8 mm, red, occasionally scarlet and black, the hilum elliptic, with a black line extending about 6 mm towards the chalazal end.

Distr. Northern Venezuela. Introduced into Ceylon.

Uses. Trimen notes the *E. umbrosa* H.B.K. (=*E. mitis*) from Central America is frequently planted for shade on cacao estates in Ceylon. Stockdale et al. (1922) note it was introduced in 1881.

Vern. Madre de Cacao (Spanish).

Specimen Examined. NUWARA ELIYA DISTRICT : Estate, Maskeliya, *Worthington 2724* (BM mixture, excluding fruit in packet).

Notes. The specimen above is tentatively placed with *E. mitis*. The young fruit show an exceptionally long stipe, with the seed-bearing portion apparently torulose. Worthington notes the flowers are apricot, and the tree is very thorny, therefore of no use as a shade tree. Obviously more collections are necessary to remove the tentative status of this species as part of the Flora.

6. Erythrina lysistemon Hutchinson, Kew Bull. 9: 422. 1933; Codd, Bothalia 6(3): 507–511, pl. 1–3. 1955; Verdcourt, Fl. Trop. East Africa, Papil. 547. 1971; Krukoff, J. Arnold. Arbor. 53(1): 135. 1972; Krukoff & Barneby, Lloydia 37(3): 406, t. p. 407. 1974. Type: South Africa, Eastern Transvaal, Crocodile River, June 1932, *Smuts 66* (K), not seen.

Small xerophytic tree, to about 7 m tall; branches armed. Leaves absent at flowering; leaflets rhomboid, occasionally almost hastate, 2.1–9.5 × 1.5–8.5 cm, (fide Verdcourt, 1971), both surfaces glabrous, broadly acuminate apically, very broadly cuneate basally, with 3 prominent veins inserted basally. Inflorescences long pedunculate, the peduncle usually exceeding 6 cm long, the rachis and peduncles somewhat tomentose. Flowers about 5 cm long, the pedicels very short; calyx tube about 1.3 cm long, campanulate, cleft 1/3–1/2 on one or both sides, symmetrically or asymmetrically bilabiate at anthesis, pubescent; petals mostly scarlet, the vexillum oblanceolate, about 5 × 2–2.5 cm, striate, somewhat arched (Krukoff & Barneby, 1974), simple, cuneate basally, folded about the stamens, the wings obliquely oblong or oblanceolate, about 13 × 4 mm, the claw about 1 mm long, the keels fused, broadly elliptic, about 10 × 4 mm, broadly invaginate apically, with two small apical teeth, short-clawed, the wings or keels rarely exceeding 15 mm long; vexillary filament free to the base; ovary long-stipitate, many-ovulate, adpressed tomentellous, the style glabrous. Fruit coriaceous, torulose or moniliform, about 8 cm long including the stipe, 1 cm wide, the stipe about 12 mm long, the beak about 9 mm long, glabrescent, about 7-seeded. Seeds ellipsoid, about 8 × 5 mm, scarlet or reddish-orange, the hilum black, oblong-elliptic.

Distr. Tanzania, Moçambique, Swaziland, Malawi, Zambia, Rhodesia, Botswana and South Africa (Transvaal, N. Natal, E. Cape Province).

Ecol. Flowering in June and July in native habitats. Codd (1955) reports dry habitats: scrub forest, coastal sand dunes, dry savannahs and rocky, wooded slopes.

Uses. Apparently introduced as an ornamental.

Specimen Examined. KANDY DISTRICT: Peradeniya Gardens, *s. coll. s.n.* (NY, annotated by Krukoff, 1975).

Notes. Krukoff (1972) states the species is extensively cultivated in Australian gardens.

7. Erythrina variegata L., in Stickman, Dis. Herb. Amboin. 10. 1754; L., Amoen. Acad. 4: 122. 1759; Merr., Interpr. Rumph. Herb. Amboin. 276. 1917; Alston in Trimen, Handb. Fl. Ceylon 6: 80. 1931; Krukoff, J. Arnold Arbor. 22: 226. 1939; Bullock, Kew Bull. 20(2): 294. 1966; Verdcourt, Fl. Trop. East Africa, Papil. 549. 1971; Krukoff, J. Arnold Arbor. 53(1): 132. 1972; Krukoff & Barneby, Lloydia 37(3): 431, t.p. 433. 1974 (cum ext. syn.); Verdcourt, Man. New Guinea Leg. 425, fig. 100. 1979. Type: Rumph., Herb. Amboin. 2: 234, t. 77. 1741, not seen.

Erythrina corallodendrum L. var. *orientalis* L., Sp. Pl. 706. 1753. Type: Rheede, Hort. Mal. 6: 13, t. 7. 1686. Based on *Gelala litorea* Rumph., Herb. Amboin. 2: 230. t. 76. 1741.

Erythrina picta L., Sp. Pl. ed. 2, 993. 1763; Moon, Cat. 52. 1824. Based on *Gelala alba* Rumph., Herb. Amboin. 2: 234, t. 77. 1741.

Erythrina orientalis (L.) Murr., Comm. Fotting. 8: 35. t. 1. 1787; Backer & Bakh. f., Fl. Java 1: 627. 1963.

Erythrina indica Lam., Enc. 2: 391. 1786; Moon, Cat. 52. 1824; Roxb., Fl. Ind. 3: 249. 1832; Wight & Arn., Prod. 1: 260. 1834; Wight, Ic. Pl. Ind. Or. t. 58. 1839; Thw., Enum. Pl. Zeyl. 89. 1859; Baker in Hook. f., Fl. Br. Ind. 2: 188–189. 1876; Trimen, Handb. Fl. Ceylon 2: 63. 1894; Gamble, Fl. Pres. Madras 1(2): 353. 1918; Ridley, Fl. Malay Penins. 1: 578. 1922; Worthington, Ceylon Trees 168. 1959.

Erythrina corallodendrum Lour., Fl. Cochinch. 427. 1790.

Erythrina lithosperma Blume, Cat. Gew. Buitenzorg 92. 1823, nom. nud., non Miq., Fl. Ind. Bat. 1: 209. 1855.

Erythrina variegata L. var. *orientalis* (L.) Merr., Interpr. Rumph. Herb. Amboin. 276. 1917. Based on *E. corallodendrum* L. var. *orientalis* L., Sp. Pl. 706. 1753.

Erythrina rostrata Ridley, Fl. Malay Penins. 1: 580. 1922.

Trees; branches armed. Leafless or with leaves at flowering, the rachis about 4 cm long, 1/2–1/5 the petiolar length, the rachis and petioles with stellate pubescence; stipules apparently deciduous; stipels globose; leaflets broadly ovate to deltoid, to about 11 × 12 cm, papyraceous, the upper surface glabrous, the lower occasionally with stellate pubescence on the veins, acute to obtuse apically, rounded, truncate or slightly cordate basally, the lateral leaflets inequilateral, the primary lateral veins in about 6 pairs. Inflorescences erect, to about 30 cm long, numerous-flowered, florate about 1/2 the length, somewhat tuberculate, the stellate tomentum somewhat fugacious; bracts oblong, about 2–3 × 1 mm, caducous; bracteoles similar to the bracts. Flowers about 6 cm long, the pedicels to about 1 cm long, single or in fascicles of 2 or 3, crowded distally; calyx tube splitting to form a spathe, about 2 cm long, with fugacious stellate tomentum, then glabrous, with numerous, dark striations; petals dark red, the vexillum elliptic to oblanceolate, glabrous, folded to spreading, simple, about 6 × 2 cm, the wings obovate, slightly exceeding the keels, about 2 cm long, the keels free, ovoid; vexillary stamen free; ovary long-stipitate, with stellate pubescence, several–many-ovulate, the style glabrous, slightly ascending. Fruit indehiscent, dark, somewhat woody, turgid, linear to narrowly oblong with the stipe about 2 cm long, the seed-bearing portion about 10–18 × 2.5 cm, the beak about 1 cm long, tomentose then glabrous, somewhat constricted between the seeds, the exocarp coarse-reticulate veined, 3–8 seeded. Seeds oblong, to about 20 × 11 × 9 mm, purplish-red, the hilum elliptic-oblong or somewhat oblanceolate, about 6 mm long. Chromosome number 2n = 42.

Distr. Throughout the tropics of both hemispheres.

Ecol. This widespread species is very common in the dry, low-country and along the coasts where it has been planted extensively and grows into a large tree. Flowering is reported March though April; but Simpson (*8501*) notes "salt resisting tree at the sea margin with flowers 'chinese vermilion,' '' 1-IX-1931. Moon's localities are Kandy and Colombo.

Uses. Worthington (1959) notes the species is popular for fence posts, toys, and household articles. Roxburgh (1832) indicates it is used for shade, and cuttings are used as support for black pepper vines in India.

Vern. Coral tree, Indian Coral Tree, Thorny-dadap (E); Yak-erabodu, Era-mudu, Era-badu, Katu-eramadu, Weta-erabodu (S); Murukku, Mullu-murukku (T); Murunga (S & T). Thwaites gives Errabadoo-gass. Trimen mentions Ela-erabadu on a drawing by Moon at BM.

Specimens Examined. KURUNEGALA DISTRICT: Pothuhera, *Amaratunga 1573* (PDA). JAFFNA DISTRICT: nr. Kayts, living fence posts, *Rudd 3276* (NY, PDA, US). VAVUNIYA DISTRICT: w/o exact locality, *Worthington 4563* (K). POLONNARUWA DISTRICT: Polonnaruwa, sacred area, *Ripley 224* (US); btwn. Wewa & Elahera, *Rudd & Balakrishnan 3157* (K, NY, PDA, US). PUTTALAM DISTRICT: A 3 nr Mundal, trees planted as a hedge, *Maxwell & Jayasuriya 830* (NY, PDA, US). KANDY DISTRICT: Peradeniya Royal Botanic Gardens, *Haigh 9172* (NY); cultivated, "Seed brought from Bombay, Victoria Gardens, escaped growing in bathroom drain,'' *Worthington 7058* (K as *E. alba*), *Worthington 7073 & 7219* (K). TRINCOMALEE DISTRICT: Niroddumunai, sea margin, *Simpson 8501* (BM); Trincomalee, *Gardner C.P. 1464* (PDA); Trincomalee, seashore, *Worthington 3571* (BM, K). BATTICALOA DISTRICT: NE. of Kalkudah, *Mueller-Dombois 68041903* (PDA, US); Kalkudah Bay at beach in first row of vegetation facing ocean, *Mueller-Dombois 68042001* (K, NY, PDA, US); Kalkudah, above beach at rest house, *Rudd & Balakrishnan 3150* (K, NY, PDA, US), *3151* (PDA, US). AMPARAI DISTRICT: nr. Arugam Bay, trees in a hedge, about 15 m tall, *Maxwell & Jayasuriya 756* (NY, PDA, US). HAMBANTOTA DISTRICT: Tangalla, *Rudd 3089* (K, PDA, US); nr Tissamaharama, *Rudd 3096* (K, NY, PDA, US).

Notes. Rudd (*3157*) notes orange flowers. Krukoff (1972) considers the prominent veining on the pods and the large seeds distinctive characters. The dense stellate tomentum is easily rubbed off. The species epithet refers to the leaves, but I have not seen the typical form in Ceylon. See Verdcourt (1979).

Erythrina indica is listed in the Seeds List (1958) with seeds available January and February.

8. Erythrina velutina Willd., Ges. Naturf. Freunde Berlin, Neue Schriften 3: 426. 1801; Benth. in Mart., Fl. Bras. 15(1): 173. 1859; Trimen, Handb. Fl. Ceylon 2: 64. 1894; Fawcett & Rendle, Fl. Jamaica 4(2): 50, fig. 15. 1920; Amshoff in Pulle, Fl. Surinam 2(2): 192. 1939; Krukoff, Brittonia 3(2): 326. 1939;

Krukoff, Amer. J. Bot. 28: 683–691, pl. 2, f. 14a, 14b. 1941; Krukoff & Barneby, Lloydia 37(3): 435, t. p. 436. 1974. Type: Carabobo (near Valencia), Venezuela, described from Bredemeyer's notes, fide Krukoff (1939), not seen.

Large, spreading, short boled tree; branches armed. Leaves present at flowering, the petioles to about 20 cm long, the rachis 1/2 to 1/5 as long, the rachis, petioles, and petiolules with dense stellate tomentum, soon glabrescent; stipules linear, rigid, about 5 mm long (fide Amshoff, 1939), deciduous; stipels about 1 mm long, persistent; leaflets somewhat ovate-rhomboid to triangular, to about 11 × 12 cm, chartaceous, both surfaces with dense stellate tomentum, the upper surface soon glabrescent, rounded to obtuse apically, broadly cuneate to truncate basally, the primary lateral veins in about 6 pairs. Inflorescences erect, stellate tomentose, soon glabrescent, to about 20 cm long, florate about 1/2 the length, numerous-flowered, somewhat tuberculate; bracts ovate, about 2.5 mm long, caducous; bracteoles linear-lanceolate, about 1 mm long, caducous. Flowers to about 6 cm long, the pedicels about 1 cm long, in fascicles of 2 or 3, crowded distally; calyx finally splitting almost to the base to form a spathe, 2–3 cm long, with fugacious stellate tomentum, then glabrous, striate; petals orange-red to vermilion, the vexillum ovate, obovate or oblong-elliptic, emarginate, inequilateral, glabrous, usually strongly reflexed, 3–5.5 × 1.5–3.0 cm, the claw about 6 mm long, wings obliquely obovate or obliquely elliptic, slightly exceeding the keels, about 1/3 the vexillum length, the keels free, obovate; stamens monadelphous; ovary long-stipitate, few to several-ovulate, with stellate pubescence, the style glabrous, slightly ascending. Fruit indehiscent, somewhat woody, turgid, linear to narrowly oblong, torulose, sometimes irregularly constricted between the seeds, upcurved, tomentose, the stipe to about 2 cm long, the seed-bearing portion from 7.5–14 × 1.2–1.7 cm, the beak about 1 cm long, 1–4-seeded; sutures indistinct. Seeds oblong, (11–)14–17 × (6–)8–11 mm, red, the hilum elliptic-oblong, about 6 mm long. Chromosome number 2n = 42.

Distr. The West Indies and tropical South America.

Ecol. Confined to the drier tropics at lower altitudes.

Uses. Trimen (1894) reports *E. velutina* is frequently met with on estates, especially of cacao, planted as shade trees. Krukoff (1939) notes it is often planted in hedges for ornamental purposes in Venezuela. Stockdale et al. (1922) add that the species was introduced into Ceylon in 1881.

Vern. Coral tree, Red bean tree (E).

Specimens Examined. None, to my knowledge. If the species is as common as Trimen indicates, perhaps sheets determined *E. variegata* should be re-examined.

Notes. This species differs from the closely related *E. variegata* in these ways: possessing leaves at flowering, reflexed vexillum of an orange-red colour, the fruit exocarp somewhat torulose, without distinctive reticular venation, fruit with

1–4 red seeds, a somewhat thicker, ovate-rhomboid or triangular shaped leaflet, and perhaps also being found in drier habitats and farther inland than *E. variegata.*

2. STRONGYLODON

Vogel, Linnaea 10: 585. 1836; Verdcourt, Man. New Guinea Leg. 429. 1979. Type species: Hawaiian Islands, *S. ruber* Vog. (=*S. lucidus* (Forst. f.) Seem.).

Climbing vines; stems terete, glabrous. Leaves trifoliolate; stipules and stipels persistent; leaflets usually ovate. Inflorescences fasciculate-racemose, axillary, with lax or pendant peduncles, tuberculate, numerous-flowered; bracts and bracteoles small, caducous. Flowers medium, elongate; calyx 5-lobed, the upper lobes partially connate, all lobes blunt, much shorter than the tube; petals reddish or blue-green, the vexillum recurved or reflexed, the wings adherent to the keels, the keels incurved, acute or beaked apically, about equalling the vexillum and much longer than the wings; stamens diadelphous, the vexillary free, the anthers uniform; ovary stipitate, 1–several-ovulate, the style geniculate, the upper style glabrous, the stigma terminal, capitate. Fruit indehiscent, the shape depending on the seed number, usually ellipsoid, 1–9-seeded, with a long persistent beak; sutures swollen. Seeds fairly large, ovoid, the hilum linear, nearly 1/2 encircling. Chromosome number $2n = 28$.

A genus of about 20 species in Polynesia, Indo-Malaysia, especially the Philippines, Indonesia, Ceylon and Madagascar.

Strongylodon siderospermus Cordemoy, Fl. Ile Reunion 392. 1895, fide Verdcourt (1979). Type not seen.

Strongylodon pseudolucidus Craib, Bot. Mag. t. 8494. 1913. Type: Madagascar, Plate 8494, Bot. Mag., 1913. Grown at Kew from Madagascar seed. Based on Plate 8494.

Strongylodon secundus St. John, Pacific Sci. 26(1): 121–124, fig. 1. 1972 (syn. fide Verdcourt, 1979). Holotype: Solomon Isl., Guadalcanal I. *Kajewski 2493* (BISH), not seen.

Strongylodon ruber auct., non Vogel, Thw., Enum. Pl. Zeyl. 89. 1859; Baker in Hook. f., Fl. Br. Ind. 2: 191. 1876; Trimen, Handb. Fl. Ceylon 2: 64. 1894; Prain, J. Asiat. Soc. Bengal 66(2): 411. 1897.

Strongylodon lucidum auct., non Seemann, Alston in Trimen, Handb. Fl. Ceylon 6: 80. 1931.

Leaf rachis 2–3 cm long, ½–⅓ the petiolar length; stipules acute, 3–4 mm long; stipels linear, 4–5 mm long; leaflets glabrous, to about 10 × 6 cm, usually smaller, acute or acuminate apically, cuneate to rounded basally, usually with 3 prominent veins inserted at the base, and 2–3 pairs of primary lateral veins. Inflorescences with 1–3-flowers at the node; bracts acute, about 2 mm long;

bracteoles ovate, about 2×1 mm. Flowers 2–2.5 cm long, the pedicels 1–1.5 cm long; calyx tube about 6 mm long, glabrous, all lobes ciliate, scarcely 1 mm long; petals reddish-pink, the vexillum ovate, about 2.2 cm long including a short claw, about 9 mm wide, inflexed biauriculate basally, with two appendages, glabrous, the wings somewhat oblong, the lamina about 9×4 mm, the keels slightly falcate, the lamina about 15×4 mm, acute apically, the wings and keels long-clawed with each claw about 6 mm long; ovary 2-ovulate, long-stipitate, glabrous, the style upcurved gradually, the upper style filamentous. Fruit glabrous, compressed slightly, about 8×4 ×1.5 cm, with a stipe about 1.5 cm long, 1–2-seeded; beak downcurved, about 5 mm long. Seeds about 2.5×1.8×1.5 cm, black, smooth, the hilum 3–4 cm long, surrounded by red arillate tissue.

Distr. The Andamans, Christmas Island, northern Australia, New Guinea, New Caledonia, and westward from Ceylon to Madagascar.

Ecol. A littoral species found in the intermediate regions of the low country, it flowers in June.

Specimens Examined. KURUNEGALA DISTRICT: Kurunegala, June 1853, *Thwaites s.n.* (BM, K, PDA). KANDY DISTRICT: Galagedara, *Gardner s.n. C.P. 1479* (BM, K, PDA); "received 1870", *Thwaites C.P. 1479* (GH).

Notes. According to Trimen (1894) the species is very rare in Ceylon (Trimen cites *C.P. 1465,* but *C.P. 1479* is correct).

Judging from a sheet at BM, *Thwaites* (1854) *C.P. 1479* det *Strongylodon zeylanicum* Thw., Thwaites (?) may have felt he had found a new species. In his Enum. Pl. Zeyl., however, he erroneously assigned the plants to *S. ruber* Vogel, a Polynesian species. Thwaites was followed until the correct relationships were worked out by Craib (1913), but see Verdcourt (1979).

3. MUCUNA

Adans., Fam. Pl. 2: 325. 1763; Prain, J. Asiat. Soc. Bengal 66: 404. 1897; Burkart, Leg. Argentinas ed. 2, 390. 1952; Hutchinson, Gen. Fl. Pl. 1: 432. 1964 (cum syn.); Verdcourt, Kew Bull. 24(2): 286. 1970; Verdcourt, Fl. Trop. East Africa, Papil. 561. 1971; Verdcourt, Man. New Guinea Leg. 433–461. 1979; D'Arcy, Ann. Missouri Bot. Gard. 67(3): 728. 1980, nom. cons. Type species: *M. urens* (L.) DC. (*Dolichos urens* L.) type cons.

Stizolobium P. Br., Hist. Jamaica 290. 1756; Prain, J. Asiat. Soc. Bengal 66: 404. 1897; Kuntze, Rev. Gen. Pl. 207. 1891; Piper, USDA Bur. Pl. Industr. Bull. 179: 9. 1910; Burkart, Leg. Argentinas ed. 2, 391. 1952; D'Arcy, Ann. Missouri Bot. Gard. 67(3): 768. 1980. Lectotype: *S. pruriens* (L.) Medik. (*Dolichos pruriens* L.).

Citta Lour., Fl. Cochinch. 2: 456. 1790; Wight & Arn., Prod. 1: 253. 1834. Type: *C. nigricans* Lour. (= *Mucuna nigricans* (Lour.) Jackson.).

Carpopogon Roxb., Hort. Beng. 54. 1814; Roxb., Fl. Ind. 3: 283. 1832; Wight & Arn., Prod. 1: 253. 1834.

Hornera Neck., Elem. Bot. 3: 43. 1790.

Negretia Ruiz & Pav., Prod. 98, t. 21. 1794. Lectotype: *N. elliptica* Ruiz & Pav. (= *Mucuna elliptica* (Ruiz & Pav.) DC.).

Large, woody lianas or smaller herbaceous, climbing, twining vines, rarely erect. Leaves pinnately trifoliolate; stipulate, usually stipellate. Inflorescences pseudoracemose, paniculate or subumbellate, axillary or occasionally cauliflorous, the rachis frequently with flowers crowded at the ends of long, pendant peduncles, tuberculate; bracts small, caducous, bracteolate. Flowers large, showy; calyx appearing 4–5-lobed, the upper lobes connate or somewhat bifid; petals purple, red or greenish-yellow; the vexillum usually about ½ the wing length, the keels about equalling the wings, usually upcurved or beaked, horny apically; stamens diadelphous, the vexillary free, the anthers slightly dimorphic in size and filament attachment, often barbate; ovary few-to-many-ovulate, pubescent, the style long, uniform or swollen apically, sometimes pubescent, ascending or geniculate, the stigma small, capitate, terminal. Fruit dehiscent or indehiscent, ovoid, oblong or linear, somewhat turgid, often with stinging hairs, the exocarp frequently with transverse or oblique lamellae, occasionally smooth, 1–few–many-seeded; sutures with parallel sutural ribs, occasionally winged. Seeds flattened, ellipsoid or somewhat reniform, the hilum long-linear and lacking arillate tissue, or elliptic to short-oblong with rim arillate tissue.

This genus of about 100 species is primarily tropical, but extends into the subtropics.

Since the genus needs worldwide revision, an estimate of the number of species varies. Hutchinson (1964) estimates 160, Verdcourt (1971) about 100, while D'Arcy (1980) estimates about 12 in *Mucuna* and 8 in *Stizolobium*. D'Arcy reestablishes *Stizolobium* in the *Flora of Panama* as a genus. I do not segregate *Stizolobium* as a separate genus here because the flowers are similar and unique, and there is no recent treatment (see Verdcourt, 1971). Dr. Donald Windler, U.S.A., has begun a monograph of the New World species.

KEY TO THE SPECIES

1 Large woody lianas, perennials; fruit 1–3-seeded; seeds ovoid, flattened; hilum linear, nearly ½ to $^4/_5$ encircling; anthers occasionally barbate (subg. *Mucuna*)

  2 Fruit with distinct transverse or oblique lamellae; flowers purplish or violet

    3 Fruit usually 1-seeded, nearly as wide as long; seed hilum about ¾ encircling; flowers about 3.8 cm long; calyx appearing truncate, short lobed, the lobes about 3 mm long; vexillum bright violet, the keels green; leaflets pubescent beneath.................1. M. monosperma

    3 Fruit usually 2-seeded, longer than wide; seed hilum nearly ½ encircling; flowers 6.5–7.5 cm long; calyx appearing bilabiate, the lower lobe acute, about 5 mm long; vexillum deep blue-purple to reddish; leaflets glabrous beneath......:................2. M. atropurpurea

2 Fruit without distinct lamellae (shallow ribs in *M. novoguineensis*); flowers pale green, yellow
or bright red

    4 Flowers pale green or yellow, about 3.5 cm long; leaflets glabrous beneath; each fruit suture
with a broad parallel wing to either side; coastal native species...........**3. M. gigantea**

    4 Flowers bright red, 5–8 cm long; leaflets glabrous or pubescent beneath; fruit sutures lacking
broad wings; introduced ornamental.........................**4. M. novo-guineensis**

1 Smaller annuals or perennial vines; fruit (2–)5–8-seeded; seeds somewhat reniform or rounded-
oblong; hilum elliptic to short-oblong, less than ¼ encircling, with or without rim arillate tissue;
anthers non-barbate (subg. *Stizolobium*)

    5 Calyx and fruit with stinging hairs....................................**5. M. pruriens**

    5 Calyx and fruit without stinging hairs....................**5. M. pruriens** s.l.—cultivated

1. Mucuna monosperma (Roxb.) DC., Prod. 2: 406. 1825; Wight in Hook.,
Bot. Misc. 2: 346, suppl. t. 12. 1831; Wight & Arn., Prod. 1: 254. 1834 (cum
syn.); Wight, Ic. Pl. Ind. Or. t. 35. 1838; Thw., Enum. Pl. Zeyl. 89. 1859;
Baker in Hook. f., Fl. Br. Ind. 2: 185. 1876 (cum syn.); Trimen, Handb. Fl.
Ceylon 2: 61. 1894; Gamble, Fl. Pres. Madras 1(2): 355. 1918. Type: Ind.
Orient., Cat. Calc. of Roxburgh ?, not seen.

Stizolobium rugosum Moon, Cat. 53. 1832, nomen.
Carpopogon monospermum Roxb., Fl. Ind. 3: 283. 1832.

Twining, climbing lianas; stems terete to ridged, with rufous pubescence when
young, becoming glabrescent. Leaf rachis about ¼ the petiolar length; stipules
deciduous; stipels setaceous, 1–3 mm long, persistent; leaflets densely pubescent beneath, usually with 3 prominent veins at the base, primary lateral veins
in 5–6 pairs, the terminal leaflet obovate, oval or broadly elliptic, about 11 ×
7 cm, the lateral leaflets ovate, rounded or abruptly acute apically, rounded or
broadly cuneate basally, inequilateral. Inflorescences axillary or cauliflorous, pendant or lax, the peduncles to about 15 cm long, usually shorter than the leaves,
condensed paniculate, 6–18-flowered, each branch (or tubercle) 2–3-flowered.
Flowers 3.5–5 cm long, the pedicels about 13 mm long; calyx appearing 4-lobed,
the tube about 8 mm long, orange-brown pubescent, the upper lobes connate,
obtuse, about 3 mm long, the lower longest; petals bright violet and greenish,
the vexillum spreading, obovate or suborbicular, about 3 cm long, bright violet,
short-clawed, the wings narrow, falcate, about 4.5 cm long, bright violet, the
keels narrow, about 5 cm long, with an upcurved acuminate beak, greenish, the
wings and keels short-clawed; ovary usually 1-ovulate, short-stipitate, pilose, the
style somewhat filamentous below, pubescent towards the capitate stigma. Fruit
indehiscent, somewhat ovoid, about 6 × 4 × 2.7 cm, the exocarp with reddish,
stinging hairs and oblique lamellae, 1(–2)-seeded; upper and lower sutures with
parallel 5–9 mm wide wings to each side. Seeds suborbicular, compressed, dark
brown, shining, faintly reticulate, about 3 × 2 × 1.5 cm, the hilum linear, flat,
about ¾ encircling. Chromosome number 2n = 22.

Distr. Ceylon and the hills of Burma and India.

Ecol. Thwaites (1859) reports the species in the Central Province up to 3000 ft. Trimen (1894) has it rather rare and found in the upper zone of the moist low country, flowering January through March.

Specimens Examined. WITHOUT EXACT LOCALITY: Herb. Hort. Soc., 1839, *J.S. Mackenzie s.n.* (K). KANDY DISTRICT: Hunasgiriya, 1854, *Thwaites* (?) *C.P. 2662* (PDA). KEGALLE DISTRICT: Kitulgala, Mar. 1853, *Thwaites* (?) *C.P. 2662* (BM). BADULLA DISTRICT: Lunugala, Jan. 1888, *Trimen* (?) *s.n.* (PDA).

Note. Description based on Herb. *Wight 752* (BM, COIMB), *Roxburgh 276*, *LcN 25* (BM).

2. Mucuna atropurpurea (Roxb.) DC. ex Wight & Arn., Prod. 1: 254. 1834 (cum syn.); Thw., Enum. Pl. Zeyl. 89. 1859; Baker in Hook. f., Fl. Br. Ind. 2: 186. 1876; Trimen, Handb. Fl. Ceylon 2: 61. 1894; Gamble, Fl. Pres. Madras 1(2): 355. 1918. Type: India, Travancore, 1813, *Roxburgh s.n.* (BM).

Carpopogon atropurpureum Roxb., Hort. Beng. 54. 1814; Roxb., Fl. Ind. 3: 287. 1832.
Mucuna atro-purpurea DC., Prod. 2: 406. 1825 (sine desc.).

Twining, climbing lianas; stems terete, glabrous, to about 8 m long. Leaf rachis about ¼ the petiolar length; stipules deciduous; stipels setaceous, about 3 mm long, persistent; leaflets ovate or elliptic, about 10.5 × 4.5 cm, abruptly acute apically, rounded basally, both surfaces glabrous, the laterals inequilateral, the primary lateral veins in about 3 pairs, frequently branching about ½ the distance to the margin. Inflorescences pseudoracemose or paniculate, pendant, longer than the leaves, 10–15-flowered, the rachis nodes distant, each node 1–3-flowered; bracts lanceolate, about 7 mm long. Flowers 6–7.5 cm long, the pedicels 10–16 mm long; calyx tube about 12 mm long, sericeous pubescent, the upper lobes partially connate, obtuse-acute or truncate, 2–3 mm long, the laterals acute or somewhat obtuse, the lower longest, about 5 mm long; petals dark or bluish purple to reddish, short-clawed, the vexillum spreading or straight, probably obovate (?), 4–5.5 cm long, the wings narrow, about 6 cm long, the keels narrow, about 6.4 cm long; ovary few-ovulate, dense ferruginous pilose, the style filamentous, the upper portion somewhat thicker. Fruit indehiscent, short-oblong, septate, to about 9.5 cm long (3-seeded), to 13 cm long (4-seeded), 4 cm wide, 2 cm thick, the exocarp with oblique lamellae, covered with ferruginous stinging hairs, (1–)2(–4)-seeded; upper and lower sutures with parallel wings to either side. Seeds suborbicular, except frequently flattened on one margin, slightly compressed, dark, smooth, about 2.1 × 1.8 × 0.9 cm, grey mottled with brownish-black, the hilum linear, slightly raised, about 1.5 mm wide, 2.2 cm long, nearly ½ encircling.

Distr. India and Ceylon.

Ecol. Thwaites (1859) notes the species not uncommon up to 2000 ft. Trimen (1894) reports it rather rare and found in the dry region.

Most authors note the flowers deep or rich blue-purple, but *Jayasuriya 1036* records a red corolla. Flowering is around January and February.

Vern. Bu-chariya, Buchariwa, Ginipus-wel (S); Punnakalichi, Phandatullai (T).

Specimens Examined. WITHOUT EXACT LOCALITY: *Thwaites C.P. 1483* (K, P), 1854, *Thwaites s.n.* (BM). WITHOUT DISTRICT: Northern Province: *A. Clark 5* (PDA). ANURADHAPURA DISTRICT: Ritigala Strict Nat. Res., *Jayasuriya 1036* (US). MATALE DISTRICT: Galboda off Naula-Elahera Rd., *Jayasuriya 303* (US); between Naula and Dambulla, *Robyns 7327* (PDA, US). TRINCOMALEE DISTRICT: 7 mi. NE. of Muttur, *Jayasuriya et al. 665* (PDA, US); Norway Isl., Trincomalee, *Worthington 1251* (K). BATTICALOA DISTRICT: Batticaloa, *Gardner C.P. 1483* (PDA). AMPARAI or MONERAGALA DISTRICT: S. of Senanayake Samudra along dry stream bed, *Maxwell & Jayasuriya 751* (NY, PDA, US).

Notes. The 4-seeded fruit dimensions above are from 1 pod in the seed box at (COIMB). This large, separate fruit was unusual in size and may belong to another species.

3. Mucuna gigantea (Willd.) DC., Prod. 2: 405. 1825; Wight in Hook., Bot. Misc. 2: 357, suppl. t. 14. 1831; Wight & Arn., Prod. 1: 254. 1834; Thw., Enum. Pl. Zeyl. 89. 1859; Baker in Hook. f., Fl. Br. Ind. 2: 186. 1876; Trimen, Handb. Fl. Ceylon 2: 62. 1894; Gamble, Fl. Pres. Madras 1(2): 355. 1918; Ridley, Fl. Malay Penins. 1: 577. 1922; Backer & Bakh. f., Fl. Java 1: 630. 1963; Verdcourt, Kew Bull. 24(2): 287. 1970; Verdcourt, Fl. Trop. East Africa, Papil. 564. 1971; Ohashi & Tateishi, Jap. J. Bot. 51(6): 163, fig. 2, 3. 1976; Verdcourt, Kew Bull. 33(1): 126. 1979; Verdcourt, Man. New Guinea Leg. 443, fig. 106F. 1979. Type: India, Malabar, Rheede, Hort. Mal. 8: 63. 1688, not seen.

Dolichos giganteus Willd., Sp. Pl. 3: 1041. 1802.

Stizolobium giganteum (Willd.) Spreng., Syst. Veg. 3: 250. 1826; Moon, Cat. 53. 1824.

Carpopogon giganteum (Willd.) Roxb., Hort. Beng. 54. 1814; Roxb., Fl. Ind. 3: 287. 1832.

Twining, climbing lianas; stems terete or slightly ridged, glabrescent at maturity. Leaf rachis about ⅓ the petiolar length, stipules minute, acute-acuminate, 3–4 mm long, persistent; leaflets elliptic or broadly ovate, to about 11 × 6 cm, abruptly acute apically, rounded basally, the laterals inequilateral, the surfaces glabrous when mature, the primary lateral veins in 4–5 pairs. Inflorescences condensed, paniculate, pendant, about 22 cm long, many-flowered, the rachis nodes crowded distally, each node about 3-flowered. Flowers 3(–5) cm long, the pedicels to about 2.5 cm long; calyx appearing 5-lobed, the tube 6–7 mm long, sericeous tomentose, the upper lobes partially connate, blunt, obtuse-acute, 2–3 mm long, the lower slightly longer; petals sulphur-yellow, greenish-yellow or pale green, the vexillum obovate to somewhat orbicular, 1.5–3.0 cm long, the wings narrow, nearly equalling the keels, the keels 3–4.5 cm long, rising into a sharp, shining, tooth-like beak; ovary stipitate, usually few-ovulate. Fruit indehiscent, oblong, stipitate, about 10–15 × 4.5 × 1.5 cm, the exocarp smooth, with long, adpressed, chestnut-brown irritant bristles, usually 1–2–few-seeded; sutures with a broad parallel wing to each side. Seeds broadly ovoid, compressed, dark, about 2 × 2 × 1.2 cm, the hilum linear, flat, about 4.5 cm long, about ¾ encircling.

Distr. Old World tropics.

Ecol. Found along the sea coast. Trimen (1894) reports it grown in the Botanic Gardens at Peradeniya, flowering in January.

Vern. Kana-pus-waela (S).

Specimen Examined. BATTICALOA DISTRICT: *Gardner C.P. 3587* (PDA).

Notes. Moon's locality is Bentota, Galle District. I did not see this specimen. *Gardner C.P. 3587* is 1 fruit.

Description based on *Wight 753* (BM, COIMB). There was little additional material available in COIMB upon which to base a complete description. As Verdcourt notes (1971, p. 565), Asian material may have smaller flowers than African, but additional collections correlating fruit and flowers are necessary to confirm this.

Verdcourt (1979) and Ohashi & Tateishi (1976) have named subspecies. See the latter for additional references.

4. Mucuna novo-guineensis Sheff., Ann. Jard. Bot. Buitenzorg 1: 18. 1876; Verdcourt, Man. New Guinea Leg. 450, fig. 104, 107A. 1979 (cum syn.). Type not seen.

High climbing lianas; mature stems glabrous. Leaves trifoliolate, the rachis about 2 cm long, about ⅓ the petiolar length, mostly glabrous or sparsely adpressed, ferruginous pubescent; stipules not seen, probably caducous; stipels setaceous, 1–2 mm long; leaflets thin, acute to long acuminate apically, mostly rounded basally, both surfaces glabrous or glabrescent below, the terminal leaflet elliptic, the laterals ovate-elliptic, inequilateral. Inflorescences pendant, pseudoracemose, mostly cauliflorous, to 60 cm long, florate to the base, numerous (to about 75) flowered, each tubercle 2–4-flowered, glabrous or glabrescent; bracts and bracteoles not seen. Flowers falcate, to about 7 cm long, the pedicels about 1 cm long; calyx a broad cup, about 8 mm long, sparsely pubescent, with long ferruginous, irritant (fide Verdcourt, 1979) hairs basally and a denser tomentum towards the lobes, the margin truncate, or the lobes becoming small teeth to about 1 mm long; petals red, short-clawed, the vexillum spreading, ovate, about 3 cm long, the wings narrow, about 5.5 cm long, the keels narrow, to about 7 cm long, with narrow, elongate beaks; ovary several-ovulate, with long, somewhat sparse, stiff hairs, extending up the style. Fruit (not seen, desc. ex Verdcourt, 1979) linear-oblong, to about 27 × 5.5 cm, the margins slightly constricted between the seeds, the exocarp with shallow almost transverse ribs, about 5-seeded. Seeds 4–5.5 × 4–5.5 × 1.7 cm.

Distr. New Guinea region.

Ecol. Cultivated in Ceylon, blooming July and August in Kandy. Verdcourt (1979) reports it at 0–1800 m in primary and secondary forests in its native distribution.

Uses. Ornamental.

Specimens Examined. KANDY DISTRICT: Royal Botanic Gardens, Peradeniya, *Maxwell 1028* (US), *Rudd 3061* (US); Aniewatte, Mrs. Udawatte's garden, *Meijer 1068* (US).

5. Mucuna pruriens (L.) DC., Prod. 2: 405. 1825; Benth. in Mart., Fl. Bras. 15(1): 169, t. 46–2. 1859; Baker in Hook. f., Fl. Br. Ind. 2: 187. 1876: Trimen, Handb. Fl. Ceylon 2: 62. 1894; Merr., Interpr. Rumph. Herb. Amboin. 277. 1917; Ridley, Fl. Malay Penins. 1: 577. 1922; Fawcett & Rendle, Fl. Jamaica 4(2): 54. 1920; Amshoff, Fl. Suriname 2(2): 195. 1939; Backer & Bakh. f., Fl. Java 1: 629. 1963; Verdcourt, Fl. Trop. East Africa, Papil. 566, fig. 82/7. 1971; Verdcourt, Man. New Guinea Leg. 451. 1979; Verdcourt, Kew Bull. 35: 743–752. 1981. Type: Amboina, Indonesia, Rumph., Herb. Amboin. 5, t. 142. 1750, fide Verdcourt (1971), not seen.

Dolichos pruriens L., in Stickman, Dis. Herb. Amboin. 23. 1754; L., Syst. Nat. ed. 10, 1162. 1759.

Stizolobium pruriens (L.) Medik., in Vorles. Churpf. Phys. Ges. 2: 399. 1787; Moon, Cat. 53. 1824; Bort, USDA Bur. Pl. Industr. Bull. No. 141, Part 3. 1909; Piper & Tracy, USDA Bur. Pl. Industr. Bull. No. 179. 1–26. 1910; Piper, Proc. Biol. Soc. Wash. 30: 51–52. 1917; Burkart, Leg. Argentinas ed. 2, 391. 1952; D'Arcy, Ann. Missouri Bot. Gard. 67(3): 769, f. 46. 1980.

Carpopogon niveum Roxb., Hort. Beng. 54. 1814; Roxb., Fl. Ind. 3: 285. 1832.

Carpopogon pruriens (L.) Roxb., Hort. Beng. 34. 1814; Roxb., Fl. Ind. 3: 283. 1832.

Mucuna nivea (Roxb.) DC., Prod. 2: 406. 1825; Wight & Arn., Prod. 1: 255. 1834; Baker in Hook. f., Fl. Br. Ind. 2: 188. 1876.

Mucuna prurita Wight in Hook., Bot. Misc. 2: 348. 1831; Wight & Arn., Prod. 1: 255. 1834; Thw., Enum. Pl. Zeyl. 89. 1859 (& var. β); Gamble, Fl. Pres. Madras 1(2): 356. 1918; Alston in Trimen, Handb. Fl. Ceylon 6: 79. 1931.

Carpopogon capitatum Roxb., Fl. Ind. 3: 284. 1832.

Mucuna capitata (Roxb.) Wight & Arn., Prod. 1: 187. 1834.

Mucuna utilis Wall. ex Wight, Ic. Pl. Ind. Or. t. 280. 1840.

Mucuna prurita Hook. var. β Thw., Enum. Pl. Zeyl. 89. 1859. Type: *C.P. 3150* (PDA), nomen.

Mucuna pruriens (L.) DC. var. *biflora* (Thw.) Trimen, Cat. 25. 1885; Trimen, Handb. Fl. Ceylon 2: 63. 1894. Type: Batticaloa, *C.P. 3150* (PDA).

Stizolobium capitatum (Roxb.) Kuntze, Rev. Gen. Pl. 1: 207. 1891; Piper & Tracy, USDA Bur. Pl. Industr. Bull. No. 179. 12, pl. 2A. 1910.

Stizolobium niveum (Roxb.) Kuntze, Rev. Gen. Pl. 1: 207. 1891; Piper & Tracy, USDA Bur. Pl. Industr. Bull. No. 179. 15, pl. 4A. 1910.

Mucuna pruriens (L.) DC. var. *utilis* (Wight) Burck, Ann. Jard. Buitenzorg 11: 187. 1893.

Mucuna lyonii Merr., Philipp. J. Sci. Suppl. 1: 197. 1906.

Stizolobium aterrimum Piper & Tracy, USDA Bur. Pl. Industr. Bull. No. 179. 18, pl. 7. 1910.

Stizolobium utile (Wall. ex Wight) Piper & Tracy, USDA Bur. Pl. Industr. Bull. No. 179. 14, pl. 3B. 1910.

Stizolobium pruritum biflorum Piper, Proc. Biol. Soc. Wash. 30: 60. 1917.

Stizolobium cochinchinense (Lour.) Burkart, Leg. Argentinas ed. 2, 391. 1952 (Fl. Cochinch. 456. 1790), fide Verdcourt (1979). See Backer & Bakh. f., Fl. Java 629. 1963 (forma).

Large, twining, climbing vines, somewhat woody, but usually annuals; stems mostly terete, densely pubescent, glabrous at maturity. Leaf rachis usually $1/7–1/9$ the petiolar length; stipules linear, about 4 mm long; stipels setaceous, 2–3 mm long, persistent; leaflets thin, to about 13 × 9 cm, usually obtuse or acute apically, rounded or truncate basally, both surfaces pubescent, with adpressed, white hairs, the terminal leaflets elliptic, obovate or frequently rhomboidal, the lateral leaflets broadly ovate, greatly inequilateral, the lamina above the midrib about ⅓ as wide as below, the primary lateral veins in 5–7 pairs. Inflorescences pendant, rarely erect, 2–30 cm long, few–many-numerous-flowered, tuberculate, each node usually 2–3-flowered, about 1 cm apart; bracts lanceolate, about 8 mm long, caducous; bracteoles lanceolate, about 1 cm long, caducous. Flowers 3–3.5 cm long (3.8–5.4 cm fide Trimen, 1894), the pedicels 2–4 mm long; calyx 4-lobed, the tube 4–7 mm long, with or without stinging hairs, sericeous tomentose, the upper lobes mostly connate, entire, all lobes acute, the lower longest, about 7 mm long; petals purple, lilac or white, the vexillum ovate, 2–2.5 × 1.5–2.0 cm, the wings narrow, 3–3.5 cm long, slightly shorter than the keels, the keels beaked for 5–6 mm; ovary few–several-ovulate. Fruit linear-oblong, somewhat S-shaped, slightly compressed, septate, about 7.5 × 1.4 × 1 cm, the exocarp densely pubescent with erect, fulvous or ferruginous, stinging hairs (rarely in cultivars), 2–8-seeded; upper suture with an extra rib slightly below each sutural rib (variable in cultivars). Seeds ovoid, about 8 × 7 × 2 mm, mostly larger, brown, mottled with black (variable in cultivars), the hilum oblong, about 4 mm long, surrounded by a raised rim, arillate tissue lacking. Chromosome number 2n = 22, 44 (1 count) or 20 (1 count).

Distr. Throughout the tropics and subtropics worldwide.

Ecol. Rather common in the dry and intermediate regions, flowering in January, February, and April.

Uses. The species is listed as a medicinal herb by Parsons (1937). It is also on the Seeds List (1958), with seeds available throughout the year. The List indicates the tender pods are used as a vegetable, but see notes below.

Vern. Cowitch, Cowage, Cowhage (E); Achariya-pala, Weldamiya, Wandurume, Achariya (S); Chunao-avarai, Poonaykali, Punnaikkaali (T). See notes below.

Specimens Examined. WITHOUT EXACT LOCALITY: *Hermann* Mus. *67* in BM; *Thwaites C.P. 3150* (BM). POLONNARUWA DISTRICT: 30.1.46, *Ahmed s.n.* (PDA). KURUNEGALA DISTRICT: Doluwa, *Jayasuriya & Balasubramaniam 532* (US). BATTICALOA DISTRICT: in 1858, *Thwaites* ? *C.P. 3150* (BM isotype, PDA holotype of *M. pruriens* var. *biflora* Trimen), Nov. 1858, *Thwaites* ? *C.P. 3150* (PDA). BADULLA DISTRICT: Lower Badulla Rd., Apr. 1854, *Thwaites* ? *C.P. 3150* (PDA). HAMBANTOTA DISTRICT: Ruhuna Natl. Pk, *Mueller-Dombois & Comanor 67083008* (PDA, US); rd. to Hinwewe Tank, *Nowicke & Jayasuriya 403* (US); near Sithulpahuwa, *Wirawan 795* (US).

Notes. Although several different classifications have been published, I consider *M. prurita* and *M. pruriens* one species with the cultivated forms, mostly lacking stinging hairs, derived. I agree with Burkart (1952) that the variation among the cultivars is more or less equivalent to that among races or varieties of *Phaseolus vulgaris* (Green or French bean). Verdcourt (1981) has named several new species in subg. *Stizolobium*, but it is certainly doubtful that all the cultivated species named by Piper & Tracy (1910), based on fruit and seed characters, deserve species rank. See Burkart (1952) for additional discussion and references.

The probable Ceylon taxa, regardless of rank, are as follows:

1 Calyx usually, & fruit always, with stiff, stinging hairs; fruit hairs light yellowish-brown to rusty brown; flowers purple; plants wild, native
  2 Inflorescences usually 6–30 cm long, the longer pendant, many–numerous-flowered . . . .
  . . . .**M. pruriens** subsp. **pruriens** var. **pruriens.** *
  2 Inflorescences usually about 2.5 cm long, erect, mostly 2-flowered . . . .
  . . . .**M. pruriens** var. **biflora**
1 Calyx always & fruit almost always lacking stinging hairs; fruit glabrous or with variable pubescence; flowers purple or white; plants introduced, cultivated
  3 Flowers white; fruit exocarp glabrous at maturity; seeds light grey; inflorescences exceeding or equal to the leaves . . . . .Lyon bean (**M. nivea, Stizolobium cochinchinense,** pods and seeds used, a questionable table vegetable, but probably used in curries)

---

*The name *M. prurita* appeared on many sheets at (COIMB), but I could not separate *M. pruriens* as noted above. See Piper & Tracy (1910), and Piper (1917) for the seed and pod characters they believe are significant to delimit the two species.

3 Flowers dark shades of purple; fruit exocarp pubescent at maturity; seeds blackish or light grey
mottled with brown; inflorescences usually not exceeding the leaves
   4 Seeds mostly solid black, occasionally faintly mottled; fruit 8–15 cm long, with prominent
   longitudinal exocarp ridges..............................**M. pruriens** var. **utilis** group
      5 Fruit villous, blackish pubescent; inflorescences short, subumbellate, 5–6-flowered......
      ...**M. capitatum** (India, the young pods used in curries after the pubescence is removed)
      5 Fruit pubescence not as above; inflorescences longer, the nodes more distant, many-flowered
         6 Fruit pubescence erect or ascending, yellowish; seeds dull black with ferruginous mottling
         .... **S. utile**
         (=*M. pruriens* var. *utilis* ?, forage & green manure, the pubescence may be irritating)
         6 Fruit pubescence adpressed, minute, white; seeds shining, solid black ....
         ....Mauritius or Bengal bean (**S. aterrimum**, forage & green manure)
   4 Seeds ash-grey, mottled with brown or black; fruit 5–6 cm long, exocarp with obscure longitudinal
   ridges................................ Florida Velvet bean (**S. deeringianum** Bort,
   forage & green manure, the fruit may have a few stinging hairs)

Currently subg. *Stizolobium* cultivars are used mainly as forage and green
manure, and incidentally as ornamental vines.

Roxburgh's (1832) favourable comparison of *Carpopogon niveum* with
Phaseolus vulgaris (Green or French bean) is now discounted. The age of the
pods and seeds, the food preparation method, even the individual person's
physiology affect the bean's palatability and after effects. The variability of the
poisonous principles has not been entirely eliminated in the cultivated varieties.

The roots, seeds, and hairs of *M. pruriens* were all used medicinally. A common application was as a vermifuge. Most of the literature maintains young, tender
pods are cooked and eaten as a vegetable (*M. pruriens*).

Maxwell et al. 970 was purchased from a man who grew it off the Sinharaja
Forest Road near Kudawe Village. I believe this was the Lyon Bean (*M. nivea*).
We were told that after the packing inside the pod was removed, the seeds and
pod were boiled and eaten.

4. BUTEA

Roxb. ex Willd., Sp. Pl. 3: 917. 1802, nom. cons.; Blatter, J. Indian Bot. Soc.
8: 133–138. 1929; Lackey in Polhill & Raven, Adv. Leg. Syst. 312. 1981. Type
species: *B. monosperma* (Lam.) Taubert, typ. cons.

Plaso Adans., Fam. Pl. 2 : 325. 1763. Type: 'H.M. 6. t. 16 & 17'.

Trees or large twining, woody lianas; young stems pubescent. Leaves
trifoliolate; stipulate and stipellate; leaflets obovate to orbicular. Inflorescences
fasciculate-racemose or paniculate, axillary or terminal, many-flowered; bracts
and bracteoles lanceolate, caducous. Flowers large, showy; calyx broadly campanulate, 5-lobed, the upper lobes deeply emarginate or connate, then appearing
4-lobed; petals exerted, reddish, usually shaded with orange and silver, about
equal in length or keels slightly exceeding the vexillum, the vexillum recurved,

ovate or lanceolate, simple, the wings falcate, adherent to the keels, the keels semilunate; stamens diadelphous, the vexillary free, the anthers uniform; ovary sessile or shortly stipitate, 2-ovulate, the style long-geniculate, the upper style glabrous, the stigma terminal, small, capitate, glabrous. Fruit partially dehiscent around the seed or indehiscent, tomentose, oblong or broadly linear, obtuse apically, 1-seeded; sutures swollen. Seed in the distal end of the fruit, obovate, compressed, the hilum small.

A genus of about three species, found in Ceylon, Indochina, Malaysia, and Java.

Butea monosperma (Lam.) Taub. in Pflanzenfam. 3(3): 365. 1894; Alston in Trimen, Handb. Fl. Ceylon 6: 80. 1931; Backer & Bakh. f., Fl. Java 1: 628. 1963; Verdcourt, Man. New Guinea Leg. 461, fig. 108. 1979.

Erythrina monosperma Lam., Enc. 1: 391. 1783.
Butea frondosa Koen. ex Roxb., Asiat. Res. 3: 469. 1792, nom. invalid. (sine descr. gen.); Willd., Sp. Pl. 3: 917. 1802, nom. illeg. (*E. monosperma* in syn.); Moon, Cat. 52. 1824; Roxb., Fl. Ind. 3: 244. 1832; Wight & Arn., Prod. 1: 261. 1834; Thw., Enum. Pl. Zeyl. 89. 1859; Beddome, Fl. Sylv. t. 176. 1872; Baker in Hook. f., Fl. Br. Ind. 2: 194. 1876; Bentley & Trimen, Med. Pl. Pt. 31, t. 79. 1878; Trimen, Handb. Fl. Ceylon 2: 66. 1894; Gamble, Fl. Pres. Madras 1(2): 357. 1918. Based on Koenig, Roxburgh collections from Coromandel and Ind. Orient (BM).

Trees; young stems terete. Leaf rachis about ⅓ the petiolar length; stipules linear-lanceolate, caducous; stipels setaceous, about 2 mm long; terminal leaflets 10–12 × 8–10 cm, obtuse and occasionally emarginate apically, cuneate basally, with 6–8 pairs of primary lateral veins prominent beneath. Inflorescences fasciculate-racemose, paniculate from some nodes, each tubercle 1–3 flowered. Flowers 4–5 cm long, the pedicels 2–2.5 cm long; calyx tube about 10 mm long, tomentose, urceolate, the upper lobes partially connate, all lobes acute, about 3 mm long; petals tomentose, scarlet-orange, the vexillum ovate, about 5 cm long, partially reflexed or spreading, basally inflexed biauriculate, the keels falcate, beaked; ovary villose, the style exserted beyond the keel at anthesis, the lower style bearded. Fruit 13.5–15.5 × 4–4.5 × about 0.5 cm, stipitate, light brown, obtuse or rounded apically, the pedicel attached at the upper suture. Seeds suborbicular, flat, about 3 × 2.2 × 0.2 cm, the hilum suboval, 2–3 mm long. Chromosome number 2n = 18.

Distr. Ceylon, India, and Indochina.
Ecol. The species occurs in the open in dry regions of Ceylon. Flowering is reported from early June through August, but the Gunawardene collection below from Peradeniya flowered 21.3.30.

Uses. Trimen (1894) notes the exudate hardens into the "Bengal Kino" of commerce. Gamble (1918) adds that the lac insect is grown upon the "Kino" gum. Also an ornamental.

Vern. Gas-kela (S); Parasu, Murrakan (T).

Specimens Examined. LOCALITY UNKNOWN: 1854, *Thwaites C.P. 1465* (BM); *Walker s.n.* (NY, "Campbell" on label). JAFFNA DISTRICT: Jaffna *Gardner C.P. 1465* (K, PDA). ANURADHAPURA DISTRICT: Mihintale, in compound of Dagoba, *Cramer 4094, 4147* (US). POLONNARUWA DISTRICT: E. of Polonnaruwa, marker 57-3, *Theobald & Grupe 2319* (US); Rd. to Batticaloa, marker 57, *Waas 630* (US); Polonnaruwa, *Worthington 86* (BM, K), *Ripley 173* (US, sterile). MATALE DISTRICT: Sigiriya Rock, *Nowicke et al. 340* (NY, P, US); Sigiriya, *Worthington 1177* (K). KANDY DISTRICT: Peradeniya, Royal Botanic Gardens, 21.3.30, *D.W.K. Gunawardene s.n.* (PDA), *Worthington 1321 & 1796* (K). AMPARAI DISTRICT: Padiyatalawa, Talawa country, *Worthington 6305* (K). BADULLA DISTRICT: Lunugala, S. of Bibile in savannah, *Worthington 6851* (K). MONERAGALA DISTRICT: Bibile, *Kostermans 24393* (US), 12.7.24, *J.M. Silva s.n.* (PDA), *Worthington 4052 & 5115* (K).

Notes. Moon (1824) also lists *B. superba* Roxb. for Ceylon with the locality of Kandy and the Sinhala name "Wel-Kela." He characterizes *B. frondosa* (= *B. monosperma*) as downy-branched and *B. superba* as smooth branched. Gamble (1918) notes *B. superba* is a large climbing shrub, with the lowest calyx tooth equal to the side ones, the base of the pods narrowed, with the breadth of the pod 1 inch. *Butea frondosa* (= *B. monosperma*) is an erect tree, with the lowest calyx-tooth much shorter than the laterals, and the base of the pod rounded, almost semi-cordate, with the width 1.5 inch. Alston (1931) adds a key with *B. superba* "Climber; racemes 12 in long" (30 cm) vs. *B. monosperma*, "erect tree; racemes 6 in long" (15 cm).

I have seen no *B. superba* specimens from Ceylon.

Blatter (1929) considers *Spatholobus* a section of *Butea*, but because of inflorescence and floral differences I maintain both at genus rank. Lackey (1981) places *Spatholobus* next to *Butea* in subtribe Erythrininae Benth.

5. SPATHOLOBUS

Hassk., Fl. 25: p. 2, Beibl. 52. 1842; Prain, J. Asiat. Soc. Bengal 66(2): 412. 1897; Blatter, J. Indian Bot. Soc. 8: 137. 1929. Type species: Malaysia, *S. littoralis* Hassk.

Climbing lianas; stems terete or slightly angular. Leaves large, pinnately trifoliolate; stipulate and stipellate. Inflorescences paniculate, numerous-flowered; bracts and bracteoles small, caducous. Flowers small, subsessile or borne on short pedicels; calyx 4–5-lobed, the upper lobes connate, entire or slightly emarginate; petals red, purple or white, clawed, about equal, the vexillum ovate, simple, the wings free, obliquely oblong, the keels somewhat oblanceolate; stamens diadelphous, the vexillary free, the anthers uniform; ovary 2-ovulate, the style uniform, sharply upcurved, the stigma terminal, small, capitate. Fruit oblong or broadly linear, frequently rounded at the apex and base, samaroid with 1 apical seed, the basal portion compressed and indehiscent, the apical portion somewhat dehiscent, tomentose; both sutures swollen. Seed subreniform to obovate.

A genus of about 15 species in India, tropical Asia, Malaysia and the Philippines.

Spatholobus parviflorus Roxb. ex Kuntze, Rev. Gen. Pl. 205. 1891; Hu, J. Arnold Arbor. 5: 228. 1924. Type: Mountainous parts of Rajamundree Circar, grown from seed sent to Calc. Bot. Gard. by Captain Alexander Denten in 1797, fide Roxburgh (1832), not seen.

Butea parviflora Roxb., Hort. Beng. 53. 1814, nomen; Roxb., Fl. Ind. 3: 248. 1832; Wight & Arn., Prod. 1: 261. 1834; Wight, Ic. Pl. Ind. Or. t. 210. 1839; Blatter, J. Indian Bot. Soc. 8: 137. 1929.

Butea sericophylla Wall., Cat. 5441. 1831–1832, nomen. Based on *Wallich 5441* (BM).

Spatholobus roxburghii Benth. in Miq., Pl. Jungh. 238. 1852, nomen; Baker in Hook. f., Fl. Br. Ind. 2: 193. 1876; Trimen, Handb. Fl. Ceylon 2: 66. 1894 (note); Prain, J. Asiat. Soc. Bengal 66(2): 412. 1897; Gamble, Fl. Pres. Madras 1(2): 358. 1918.

Large lianas. Leaf rachis 1–1.5 cm long, the petiole to 8 times as long; stipules small, acute; stipels about 1 mm long, acute; leaflets broadly obovate-ovate, to about 20 × 13 cm, obtuse or abruptly acute apically, broadly cuneate or rounded basally, silky pubescent, the lateral leaflets inequilateral, the primary lateral veins in about 8 pairs, prominent beneath. Inflorescences axillary, brown pubescent; bracts linear, about 3 mm long; bracteoles lanceolate, about 1.5 mm long. Flowers 8–10 mm long, the pedicels 1–2 mm long; calyx fulvous tomentose, the upper lobes emarginate, all lobes acute, about 3 mm long, equalling the tube; petals white, the vexillum about 8 mm long, with a claw about 3 mm long, the wings about 6 mm long, with a claw about 4 mm long, the keels somewhat similar in shape to the wings, long-clawed. Fruit 10–12.5 cm long, the basal portion 3.5–4 cm wide, the apical portion narrower, 3–4 mm thick at the seed, ferruginous tomentose, the pedicel 1.5–2 cm long, attached at the upper margin. Seed flat, dark, about 14 × 9–10 × 2–2.5 mm, the hilum elliptic-oblong, about 1 mm long, without arillate tissue.

Distr. South India and Burma.

Vern. Pilacchi valli (T), fide Gamble (1918).

Specimens Examined. LOCALITY UNKNOWN: *Walker 1331* (K). KAN-
DY DISTRICT: Peradeniya, Royal Botanic Gardens, *Rudd 3062* (K, NY, US).

Notes. As Trimen (1894) points out, considering this species naturalized based
on the Walker specimen is problematic. The Rudd collection, January 30, 1970,
is in fruit. Roxburgh (1832) reports the flowers of the type white, as do most
authors except Baker (1876) who states "Corolla bright red . . ."

Description based on *Wallich 5440* (K), *Wight 914* (NY).

6. DIOCLEA

H.B.K., Nov. Gen. & Sp. 6: 437, t. 576. 1824; Benth. in Mart., Fl. Bras. 15(1):
160. 1859; Maxwell, Ph.D. Thesis, S. Ill. Univ., Carbondale, Ill. 1969 (Revision New World); Maxwell, Ann. Missouri Bot. Gard. 67(3): 662. 1980. Type
species: Colombia, S. A., *D. sericea* H.B.K.

Hymenospron Spreng., Syst. Veg. ed. 16, 283. 1825–28.

High climbing woody lianas, or smaller, twining, climbing vines; stems pubescent. Leaves trifoliolate; stipules produced below insertion, or non-produced, persistent; stipellate; leaflets usually ovate. Inflorescences usually axillary,
pseudoracemose, erect, numerous-flowered, tubercles distinct, usually clavate,
each several-flowered; bracts and bracteoles caducous, or persistent. Flowers of
medium size; calyx with the upper lobes partially or entirely connate; petals shades
of reddish-purple, purple to blue, the vexillum reflexed, complex, glabrous or
pubescent, the wings free, the keels variable, the wings and keels about the same
length, shorter than the vexillum; stamens monadelphous, the vexillary fused to
the staminal sheath, the anthers uniform or dimorphic; disc present; ovary
few–many-ovulate, the style geniculate, the upper style glabrous, the stigma terminal, capitate, glabrous. Fruit indehiscent or dehiscent, usually oblong, compressed to flat, septate, 1- to about 15-seeded; upper suture with a parallel rib
or wing on each valve. Seeds variable in shape, usually compressed or flattened,
vertical, the hilum linear, nearly ½ to ⁴/₅ encircling.

A diverse genus of about 50 species centred in tropical America. A few species
of the larger lianas are found throughout the tropics of the Old World. At least
one of the smaller vines, *D. virgata*, has been introduced through botanic gardens
into Asia.

KEY TO THE SPECIES

1 Large, coarse, high-climbing lianas; stipules produced below insertion; calyx tube pubescent, upper lobe bifid; petals carnose; keel petals with obtuse beaks, the upper margin entire; fruit large,
somewhat turgid, 2–3-seeded; seeds globose, the hilum about ¾ encircling...**1. D. javanica**

1 Smaller vines, twining, climbing, occasionally in dense clumps; stipules non-produced; calyx tube
glabrous, upper lobes connate, entire; petals membranous; keel petals erostrate, the upper margin
fimbriate towards the claw; fruit smaller, oblong, flat, 7- to about 15-seeded; seeds rounded-oblong,
flat, the hilum nearly ½ encircling.......................................**2. D. virgata**

1. Dioclea javanica Benth. in Miq., Pl. Jungh. 236. 1852; Ridley, Fl. Malay Penins. 1: 570. 1922; Alston in Trimen, Handb. Fl. Ceylon 6: 81. 1931; Backer & Bakh. f., Fl. Java 1: 631. 1963; Verdcourt, Man. New Guinea Leg. 465, fig. 109. 1979. Type: Java, Junghuhn's herbarium at (L), not seen.

Parrana rubra Rumph., Herb. Amboin. 5: 9, t. 5. 1755; Merr., Interpr. Rumph. Herb. Amboin. 280. 1917. Based on *Dolleschal s.n.* (W), Fl. Amboinensis exsiccata.

Dolichos hexandrus Roxb., Hort. Beng. 55. 1814, nom. nud. Based on Roxb. Icon. in King, Duthie & Prain, Ann. Royal Bot. Gard. Calcutta 9(1): 30, pl. 40. 1906.

Dolichos coriaceus Grah. in Wall., Cat. 5562. 1831–32, nom. nud. Based on *Wallich 5562* (G, K). Malaysia.

Lepidamphora volubilis Zoll. ex Miq., Fl. Ind. Bat. 1: 217. 1855, nom. nud. Based on *Zollinger Herb. 763–967* (P). Java.

Dioclea fergusonii Thw., Enum. Pl. Zeyl. 412. 1864. Type: Near Colombo, Ceylon, *Ferguson s.n.* (BM & P isotypes, PDA holotype).

Dioclea reflexa auct. mult., non Hook. f., Baker in Hook. f., Fl. Br. Ind. 2: 196. 1876; King, Duthie & Prain, Ann. Roy. Bot. Gard. Calcutta 9(1): 30, pl. 40. 1906; Koorders in Exk. Fl. 403. 1912.

High climbing lianas; younger stems with dense fulvous-ferruginous, villous pubescence. Leaf rachis ⅓–¼ the petiolar length; stipules acute to acuminate, produced; stipels setaceous, 1–2 mm long; leaflets ovate to oval-elliptic, to about 18 × 8 cm, acute apically, frequently with an abrupt acuminate tip about 1.5 cm long, rounded basally, pubescent on both sides, with about 8 pairs of primary lateral veins prominent beneath. Inflorescences stout, 30–85 cm long; bracts linear-lanceolate, about 8 mm long, somewhat reflexed, caducous or semipersistent; bracteoles flabellate, persistent. Flowers 1.5–2.5 cm long, the pedicels about 3 mm long; calyx tube about 6 mm long, the upper lobes partially connate, bifid, the lower lobe about 7 mm long; petals pale violet, of about equal length, somewhat carnose, glabrous, the keels with obtuse beaks; anthers of the staminal sheath alternately dimorphic, the vexillary perfect; ovary about 3-ovulate, villous, the lower style somewhat swollen. Fruit indehiscent, somewhat oblong, frequently somewhat arcuate, to about 12 × 5 × 1.2 cm, with cottony packing between the seeds, light brown, glabrescent, 2–3-seeded; upper margin flat, the upper suture with a narrow wing about 5 mm to either side, the lower suture naviculate, knife-like, the pedicel attached at the upper suture. Seeds suborbicular, compressed, dark, about 3 × 2.6–3.0 × 1.3 mm, the hilum about 8 mm long, about ¾ encircling. Chromosome number unknown. The closely related *D. reflexa* Hook. f. is 2n = 22.

Distr. Assam, Andaman Islands, India, Malaysia and Indonesia to New Guinea. To my knowledge there has been only one collection from Ceylon and

our examination of the locality in 1972 indicates survival of this species is doubtful due to destruction of the natural habitat.

Ecol. High, climbing liana of the low moist region.

Specimens Examined. COLOMBO DISTRICT: Kaduwela (fide Trimen,
1894), *Ferguson s.n.* (BM & P isotypes, PDA holotype & isotypes, *C.P. 3817*
at the top of the sheets).

Notes. The Old World species of *Dioclea* are not clearly understood and the
nomenclature is tangled. The West African *D. reflexa* Hook. f., possessing
5+5 dimorphic anthers, large, strongly reflexed, semi-persistent bracts, a somewhat elliptic fruit, with an erect wing to either side of the upper suture, is not
the same as the Asian species depicted above.

I believe there are several species of Asian *Dioclea* and that *D. fergusonii*
(= *Dolichos hexandrus*) may be separate from *Dioclea javanica,* based primarily on fruit characters. I presume the *Mucuna hexandra* of Mabberley (Taxon,
29: 605. 1980) is also based on the 6 perfect and 4 imperfect anther condition
described above. If the synonymy is valid and taxonomy correct, *Dioclea hexandra* (Ralph) Mabberley could become the valid name for *Dioclea fergusonii*
(= *Dolichos hexandrus*) with *Dioclea javanica* as a possible separate taxon.

In my opinion, *Dioclea reflexa* Hook. f. is a completely different taxon. The
taxon described above is not conspecific with any tropical New World species
known to me.

Description based on *Junghuhn 108* (K).

2. Dioclea virgata (Rich.) Amshoff, Meded. Bot. Mus. Herb. Rijks. Univ. Utrecht
52: 69. 1939; Maxwell, Ann. Missouri Bot. Gard. 67 (3): 672. 1980. Type: French
Guiana, Herb. *Desvaux* (P as *Mucuna virgata* Desv.).

Dolichos virgatus Rich., Actes Soc. Hist. Nat. Paris 111. 1792 (fide Amshoff,
1939). Type unknown.

Mucuna virgata (Rich.) Desv., Ann. Sci. Nat. Bot. 1: 423. 1826 (fide Amshoff
1939). Type: Herb. *Desvaux* (P).

Dioclea lasiocarpa Mart. ex Benth., Comm. Leg. Gen. 2: 69. 1837; Benth. in
Mart., Fl. Bras. 15(1): 166, pl. 44. 1859; Baker in Hook. f., Fl., Br. Ind.
2: 196. 1876. Types: Bahia, Brazil, *Martius 2016* (M, lectotype var. α), *Martius s.n.* (BR paratypes, M paratypes var. α); Pará, Brazil, *Martius s.n.* (M
var. β); Rio Jequitinhonha, Minas Gerais, Brazil *Pohl 3222* (W holotype of
var. γ, W isotype).

Canavalia bracteolata Merr., J. Straits Branch Roy. Asiat. Soc. 86: 313. 1922.
Type: Sandakan, British North Borneo, *Ramos 1151* (US lectotype).

Twining, climbing vines, or gregarious clumps in the open, frequently forming dense thickets. Leaf rachis 4(–10) mm long, much less than ¼ the petiolar
length; stipules triangular, about 2.5 mm long; stipels setaceous; leaflets ovate,
to about 7 × 4 cm, acute to acuminate apically, somewhat cordate basally, sparsely

pubescent to glabrescent above, pubescent below, with about 5 pairs of primary lateral veins. Inflorescences slender, 40–80 cm long; bracts acute, to about 3 mm long; bracteoles broadly ovate to suborbicular, about 8 × 6 mm, covering the flower bud before anthesis, caducous. Flowers about 2.5 cm long, the pedicels about 8 mm long; calyx tube about 8 mm long, glabrous, the upper calyx lobes fused, the lower about 8 mm long; petals reddish-purple, the vexillum obovate, membranous, puberulent, longest, the wings usually with a spur, the keels shortest, somewhat oblanceolate, erostrate, the upper margin fimbriate toward the claw; anthers uniform, 10 perfect; ovary 10–15-ovulate, villous, the lower style not swollen. Fruit dehiscent, about 11 × 2 × 0.4 cm, brown, with erect, stiff, somewhat fugacious, ferruginous hairs, with papery packing between the seeds, 6–15-seeded; upper suture with an erect wing to either side, the lower convex. Seeds subreniform to oblong-rounded, flat, about 11 × 7 × 2 mm, usually reddish-brown with darker mottling, the hilum about 9 mm long, slightly less than ½ encircling.

Distr. Native to South and Central America, extending into Southeast Mexico. Introduced through botanical gardens into the Old World tropics, it has become naturalized only in North Borneo, to my knowledge.

Ecol. Found primarily in moist tropical regions at forest edges or in open woodland, but extending high up moist ravines and out into drier regions along stream beds.

Uses. The species was probably introduced into Ceylon as an ornamental. It and some of its forms are being tested in Colombia, S.A., for grazing palatability.

Specimen Examined. KANDY DISTRICT: Peradeniya, Royal Botanic Garden, w/o date, *s. coll. s.n.* (PDA, cult. folder).

Notes. The vexillum has a dark purple spot above the yellow of the target area.

7. CANAVALIA

DC., Prod. 2: 403. 1825; DC., Mem. Leg. 375. 1826; Hutchinson, Gen. Fl. Pl. 1: 427. 1964; Sauer, Brittonia 16(2): 106–181. 1964 (Revision, cum syn.); St. John, Israel J. Bot. 19: 161–219. 1970; Verdcourt, Fl. Trop. East Africa, Papil. 571–577. 1971; D'Arcy, Ann. Missouri Bot. Gard. 67(3): 562–571. 1980. Type species: *C. ensiformis* (L.) DC. (*Dolichos ensiformis* L.), nom. cons.

Canavali Adans., Fam. Pl. 2: 325, 531. 1763, nom. rejic.
Wenderothia Schlecht., Linnaea 12: 330. 1838. Type: *C. villosa* Benth. (*W. discolor* Schlecht.).
Neurocarpum Hook. & Arn., Bot. Beech. Voy. 286. 1838.

Climbing, twining or trailing vines, occasionally erect herbs in cultivation, mostly perennials; stems terete, usually pubescent. Leaves alternate, pinnately trifoliolate; stipules small; stipellate; leaflets mostly ovate to suborbicular. Inflorescences axillary, erect or pendant, fasciculate-racemose, usually numerous-flowered, each node about 3-flowered; bracts and bracteoles small, caducous. Flowers resupinate; calyx 4–5-lobed, the upper lobes large, connate or each obtuse, rounded, the lower and lateral lobes much smaller, crowded, acute; petals whitish, pink, rose or shades of violet-purple, somewhat carnose, about equal in length, the vexillum reflexed, emarginate, orbicular, elliptic or obovate, biauriculate, bicallose, the wings narrow, free, the keels wider than the wings, upcurved, frequently beaked; stamens monadelphous, the anthers 10, uniform; ovary linear, several–numerous-ovulate, sessile or stipitate, pubescent, the style ascending or geniculate, the upper style glabrous, the stigma capitate, terminal or slightly subterminal. Fruit usually dehiscent, linear, oblong-linear, or oblong, compressed or somewhat turgid, straight or arcuate, several–numerous-seeded; upper suture with 2 closely parallel ribs, usually with an extra rib on each valve, 2–6 mm below each sutural rib. Seeds somewhat oblong to elliptic, or reniform, hard, the hilum oblong-linear to long-linear, $^1/_5$ to ½ encircling, without arillate tissue.

About 52 species in the tropics and subtropics of each hemisphere. A few are cultivated for food in the tropics.

KEY TO THE SPECIES

1 Leaflets oval, orbicular, frequently emarginate; common tropical seacoast species; seed hilum about 7 mm long, encircling about $^1/_5$ the seed . . . . **1. C. rosea**
1 Leaflets broadly ovate or ovate-elliptic, apices acute or acuminate; lowland or upland species, only *C. cathartica* occasionally on open beaches; seed hilum 9–20 mm long, encircling ¼ to about ½ the seed
  2 Flowers pink, rose or shades of violet-purple; fruit 8–30 × 2.5–4.5 cm; seed hilum to about 15 mm long; wild or cultivated
    3 Fruit 8–12 × about 4.5 cm, the extra rib about 5 mm below the sutural rib, tardily dehiscent or indehiscent; upper calyx lobes much shorter than the tube; wild species in coastal habitats
      . . . .**2. C. cathartica**
    3 Fruit 13–30 × about 4 cm, the extra rib 2–4 mm below the sutural rib, usually spirally dehiscent; upper calyx lobes about equal or slightly shorter than the tube; wild or cultivated species, upland or lowland, frequently along rivers
      4 Leaflets pubescent beneath, with long, dense hairs; seeds small, about 13 × 8 × 6 mm; lateral calyx lobes acute, much shorter than the lower lobe . . . .**3. C. mollis**
      4 Leaflets glabrous beneath, or glabrescent, with short, sparse hairs, occasionally pubescent, with short, dense hairs in *C. virosa*; seeds larger, 16–21 × 12–15 × 8–10 mm; lateral calyx lobes obtuse, about equal or slightly shorter than the lower lobe
        5 Leaflets broadly ovate; seeds clear brown or dark brown with dark mottling; fruit mostly 10–20 cm long, occasionally smaller; native or occasionally cultivated
          6 Leaflets with short, blunt apices; fruit with the extra rib about 4 mm from the sutural rib; calyx 10–14 mm long; vexillum to about 3 cm long; seed hilum 11–13 mm long; fruit 8–17 × 2.5–3.0 cm . . . .**4. C. virosa**
          6 Leaflets with long, acuminate apices; fruit with the extra rib about 2 mm from the sutural rib; calyx about 15 mm long; vexillum to about 3.5 cm long; seed hilum about 15 mm long; fruit to about 20 × 3.5 cm . . . .**5. C. gladiolata**

1. Canavalia rosea (Sw.) DC., Prod. 2: 404. 1825; Verdcourt, Fl. Trop. East Africa, Papil. 576, 577. 1971; Verdcourt, Man. New Guinea Leg. 475. 1979. Types not seen. See Sauer (1964) and Verdcourt (1971) for their discussions of the types and nomenclature.

Dolichos maritimus Aubl., Hist. Pl. Guian. F. 765. 1775. Type not seen.
Dolichos obtusifolius Lam., Enc. Meth. 2: 295. 1786.
Dolichos roseus Sw., Prod. Veg. Ind. Occ. 105. 1788.
Dolichos rotundifolius Vahl., Symb. Bot. 2: 81. 1791; Moon, Cat. 53. 1824; Roxb., Fl. Ind. 3: 302. 1832.
Canavalia maritima (Aubl.) Thouars in Desv., J. Bot. 1: 80. 1813; Amshoff in Pulle, Fl. Suriname 2(2): 211. 1939; Backer & Bakh. f., Fl. Java 1: 633, 651. 1963; Sauer, Brittonia 16(2): 163. 1964; D'Arcy, Ann. Missouri Bot. Gard. 67(3): 567, fig. 9. 1980.
Canavalia obtusifolia (Lam.) DC., Prod. 2: 404. 1825; Wight & Arn., Prod. 1: 253. 1834; Benth in Mart., Fl. Bras. 15(1): 178, t. 48. 1859; Thw., Enum. Pl. Zeyl. 88. 1859; Baker in Hook. f., Fl. Br. Ind. 2: 196. 1876; Trimen, Handb. Fl. Ceylon 2: 68. 1894; Ridley, Fl. Malay Penins. 1: 572. 1922.
Canavalia podocarpa Dunn in Piper and Dunn, Kew Bull. 137. 1922; Alston in Trimen, Handb. Fl. Ceylon 6: 81. 1931. Type: *Wight 748* (K).
Canavalia lineata auct., Prain, J. Asiat. Soc. Bengal 66(2): 418. 1897; Gamble, Fl. Pres. Madras 1(2): 360. 1918.

Coarse vines, trailing or climbing, stems sparsely pubescent, becoming glabrescent. Leaf rachis 1.5–4.0 cm long, about ½ the petiolar length; stipules lanceolate, deciduous, about 4 mm long; stipels small, deciduous; leaflets coriaceous, elliptic to suborbicular, to about 12 × 7 cm, obtuse or emarginate apically, broadly cuneate, rounded to truncate basally, glabrous or sparsely pubescent above, glabrous to densely pubescent below, the primary lateral veins in about 7 pairs. Inflorescences erect or pendant, the rachis 4–18 cm long, tuberculate, peduncles 10–20 cm long, several-flowered. Flowers about 3 cm long, the pedicels 3 mm long; calyx tube about 8 mm long, pubescent, the upper lobes 4–5 mm long; vexillary petal somewhat orbicular to elliptic, about 2.5 × 2 cm; ovary many-ovulate, the style slender, upcurved. Fruit dehiscent, oblong, compressed, 6.5–15 × 2.2–3.0 cm, shortly beaked, with adpressed pubescence, finally glabrescent, the exocarp smooth or wrinkled, about 7-seeded, the extra rib about 2 mm below the sutural rib. Seeds ellipsoid or ovoid, dark, smooth, shining, slightly compressed, 1.5–2.0 × 1.0–1.4 × 0.5–1.1 cm, the hilum linear, about 8 mm long.

Distr. Pantropical on beaches and with littoral vegetation; rarely further inland. The seeds are buoyant.

Ecol. All collections from Ceylon with habitat data are seashore. Collectors note the flowers red, bright rose, mauve to reddish purple, and blooming throughout the year. Trimen (1894) reports the flowers are sometimes white.

Uses. Trimen (1894) reports the seeds are eaten boiled. Ridley (1922) reports the seeds edible when young, making a good pease-porridge and that Malays eat the flowers raw. Possible use as a beach binder is noted in Madras.

Vern. Mudu-awara (S).

Specimens Examined. WITHOUT EXACT LOCALITY: Verlos Bay (= Vandeloos Bay, Batticaloa ?) *Gardner C.P. 1484* (PDA); *Thwaites s.n.* (K), *Thwaites C.P. 1484* (BM). PUTTALAM DISTRICT: Wilpattu Natl. Pk. Palugaturai, trailing on beach, *Cooray 69092806R* (BM, NY, US), *Fosberg et al. 50921* (US), *Robyns 7335* (US), *Wheeler 12102* (US); Wilpattu, Kollankanatta Beach, *Wirawan et al. 949* (BM, NY, PDA, US). TRINCOMALEE DISTRICT: Irrakkakandi, *Comanor 775* (BM, US); 13 mi. N. of Trincomalee, sandy beach, *Davidse 7556* (MO, US). BATTICALOA DISTRICT: Mankeni, on sea beach, *Cramer 4777* (US); 8 mi. S. of Batticaloa, sea coast on loose sand, *Townsend 73/271* (US). AMPARAI DISTRICT: Pottuvil, *Bernardi 16002* (US); Arugam Bay, low dunes back of beach, *Fosberg & Sachet 53033* (NY, US), *Maxwell & Jayasuriya 753* (NY, PDA, US). COLOMBO DISTRICT: in 1858, *W. Ferguson s.n.* (PDA).

Note. Both names, *C. rosea* (Sw.) DC. and *C. maritima* (Aubl.) Thouars are commonly used. The nomenclature is more difficult than the taxonomy as the leaflet shape and habitat make identification easy.

Description based on *Wallich 5532, Wight 748* (K).

2. Canavalia cathartica Thouars, J. Bot. (Desvaux) 1: 81. 1813 (as "Canavali"); Sauer, Brittonia 16(2): 158. 1964; St. John, Israel J. Bot. 19: 216, fig. 21. 1970; Verdcourt, Fl. Trop. East Africa, Papil. 574, fig. 84. 1971; Verdcourt, Man. New Guinea Leg. 471, fig. 111. 1979. Lectotype: Rheede, Hort. Mal. 8: 87, fig. 45. 1688, selected by Sauer (1964), but see Verdcourt's reasons for altering the type (1971).

Lablab microcarpus DC., Prod. 2: 402. 1825. Based on Rumphius "Cacara laut," Herb. Amboin. 5: 390, t. 141/1. 1747.

Canavalia turgida Grah., ex A. Gray, Bot. U.S. Expl. Exped. 1: 440. 1854; Miq., Fl. Ind. Bat. 215. 1855; Prain, J. Asiat. Soc. Bengal 66(2): 417. 1897 (p.p.); Ridley, Fl. Malay Penins. 1: 573. 1922. Type: not seen. Based on *Wallich 5534* (K).

Canavalia ensiformis (L.) DC. var. *turgida* Baker in Hook. f., Fl. Br. Ind. 2: 196. 1876. Based on *Wallich 5534* (K).

Canavalia microcarpa (DC.) Piper, Proc. Biol. Soc. Wash. 30: 176. 1917; Merr.,
    Interpr. Rumph. Herb. Amboin. 280. 1917; Backer & Bakh. f., Fl. Java
    1: 633, 651. 1963. Type not seen.
Canavalia obtusifolia, non (Lam.) DC., Gamble, Fl. Pres. Madras 1(2): 360.
    1918.

Coarse, climbing, vines; stems sparsely pubescent or glabrous. Leaf rachis
to about 4 cm long, usually less than ⅓ the petiolar length; stipules lanceolate
to oblong, about 2.5 mm long; leaflets broadly ovate to suborbicular, from
9.5–15.0 × 6.5–10.0 cm, abruptly acute to somewhat acuminate apically, broadly
cuneate to truncate, or somewhat rounded basally, both surfaces sparsely pubescent or glabrous, with 5–7 pairs of primary lateral veins. Inflorescences lax, pendant, numerous-flowered, 15–40 cm long, tuberculate; bracteoles obtuse, about
1 mm long. Flowers 2.5 to about 4 cm long, the pedicels about 2 mm long; calyx
tube 9–11 mm long, puberulent, the upper lobes rounded, about 3 mm long; petals
pink, purple, or light violet, the vexillum obovate or suborbicular, 2.5–3.5 cm
long, the wings and keels narrow, somewhat oblanceolate, claws about 7 mm
long; ovary 7–8-ovulate, pubescent, the style ascending, glabrous. Fruit oblong,
tardily dehiscent or indehiscent, 7–12 × 3.0–4.5 × about 1 cm, with a short,
upturned beak, somewhat turgid, 7–8-seeded, the lower margin naviculate, with
the extra rib 4–6 mm below the sutural rib. Seeds oblong-ellipsoid, about 17 ×
10 × at least 7 mm, reddish-brown, the hilum linear, about 12 mm long.

Distr. East Africa, tropical Asia, India, Malaysia through Polynesia, Ryukyu
Islands, N. Queensland in Australia, and the New World tropics.
    Ecol. The species is found back from the beaches with *C. rosea,* then into
the coastal lowland scrub and forest edges. It should flower year around.
    Uses. None found, but I assume the specific epithet refers to cathartic pro-
perties from some part of the plant.
    Vern. None known, except the general "Wild Bean" (E).
    Specimens Examined. COLOMBO DISTRICT: vines in slough near
pumping station on Kelani Ganga, *Maxwell 1035* (NY, PDA, US); 4 mi. E. of
Kaduwela, *Maxwell 1036* (PDA, US); Thalahena, near lagoon edge, *Waas 731*
(US). GALLE DISTRICT: Koggala, border of lake, *Cramer 3523, 4548* (US);
Bonavista, *Kundu & Balakrishnan 496* (US); Koggala, SE. of Galle, *Rudd 3084*
(PDA, US); across bay from Closenberg Hotel, *Sohmer et al. 8893* (NY).
    Notes. *Canavalia cathartica* was placed in Ceylon on the authority of Sauer
(1964) and a re-examination of the specimens cited above. He cites in Ceylon
s. coll. 61 (L) & *Walker 1450* (U), which I did not see. I see no reason to doubt
the presence of the species in Ceylon and suggest specimens of the common *C.
virosa* should be re-examined. *Canavalia cathartica* can be distinguished from
C. virosa in possessing shorter, wider, mostly indehiscent fruit, upper calyx lobes
⅓ to ¼ the tube length, pendant inflorescences and possibly larger and more
orbicular leaflets.

Verdcourt (1971) notes that hybrids between *C. rosea* and *C. cathartica* may be common in some parts of the world.

Description based on *Wallich 5534* (K), *Wight 736A* (BM).

3. Canavalia mollis Wall. ex Wight & Arn., Prod. 1: 253. 1834; Sauer, Brittonia 16(2): 157. 1964. Type: India, *Wight 735* (NY, P syntype).

Canavalia ensiformis (L.) DC. var. *mollis* (Wall. ex Wight & Arn.) Baker in Hook. f., Fl. Br. Ind. 2: 196. 1876. Type: *Wallich 5533* (BM).
Canavalia virosa (Roxb.) Wight & Arn. var. *mollis* (Wall. ex Wight & Arn.) Gamble, Fl. Pres. Madras 1(2): 359. 1918.
Canavalia virosa auct. Thw., Enum. Pl. Zeyl. 89. 1859.

Climbing or trailing vines; leaflets ovate, about 15 cm long, acute or acuminate apically, rounded basally, sparsely pubescent above, sparse to dense below, the primary lateral veins in about 6 pairs. Inflorescences numerous-flowered; bracteoles obtuse, about 1 mm long. Flowers to about 3.5 cm long, the pedicels about 1 mm long; calyx about 13 mm long, constricted behind an apiculate tip, pubescent; vexillary petal about 3 cm long. Fruit spirally dehiscent, linear-oblong to oblong, compressed, about 15 × 2.5 × 0.8–1.0 cm, dark brown, many-seeded, the extra rib about 4 mm below the sutural rib. Seeds rounded, oblong-elliptic, slightly compressed, about 13 × 8 × 6 mm, dark brown with black mottling, the hilum linear, about 10 mm long.

Distr. India, Ceylon and Indonesia.

Ecol. Frequently found near rivers at low elevations to 1700 m. Flowering mostly during the dry season.

Specimen Examined. WITHOUT EXACT LOCALITY: in 1861, *Thwaites C.P. 2782* (BM, det. by J. Sauer, 1962).

Notes. The species is added to the Ceylon flora on the authority of the monographer, Sauer (1964). Thwaites (1859) places *C.P. 2782* under *C. virosa*.

The distinguishing characters in Sauer's (1964) key are: upper calyx lip nearly equalling the tube, and the lower calyx teeth extended, narrow, acute, the laterals largely covered by the lowest tooth.

4. Canavalia virosa (Roxb.) Wight & Arn. in Wight, Cat. 44. 1833; Wight & Arn., Prod. 1: 253. 1834; Thw., Enum. Pl. Zeyl. 89. 1859 (p.p. ?); Gamble, Fl. Pres. Madras 1(2): 359. 1918; Alston in Trimen, Handb. Fl. Ceylon 6: 81. 1931; Backer & Bakh. f., Fl. Java 1: 633. 1963; Sauer, Brittonia 16(2): 152. 1964; Verdcourt, Fl. Trop. East Africa, Papil. 573. 1971. Type: Sauer (1964) found no type. Roxburgh's drawing *2085* (K lectotype, fide Verdcourt (1971), not seen).

Dolichos virosus Roxb., Fl. Ind. 3: 301. 1832 (excl. syn.); Moon, Cat. 52. 1824, nomen ?

Canavalia ensiformis (L.) DC. var. *virosa* (Roxb.) Baker in Hook. f., Fl. Ind. 2: 196. 1876; Trimen, Handb. Fl. Ceylon 2: 67. 1894.

Coarse vines, strong climbers or trailing, to 15 m long; stems pubescent, becoming glabrescent. Leaf rachis 1.5–3.5 cm long, about ⅓ the petiolar length; stipules about 2 mm long; leaflets ovate, 6–16 × 4–12 cm, acute or acuminate apically, rounded basally, sparsely to densely pubescent, the venation prominent on both sides, the primary lateral veins in about 7 pairs. Inflorescences 15–40 cm long, the rachis tuberculate, numerous-flowered; bracteoles obtuse, about 2 mm long. Flowers about 3 cm long, the pedicels 2–5 mm long; calyx tube about 7 mm long, sparsely pubescent, the upper lobes truncate, about 5 mm long; petals pink, lilac or purple, the vexillum somewhat orbicular, with 2 small, rounded appendages in the target area, the wings and keels oblanceolate, about 2.4 cm long; ovary about 8-ovulate, short-stipitate, the lower style flat, pubescent for about 10 mm, upcurved, the upper style about 7 mm long, glabrous. Fruit dehiscent, oblong, compressed, 10–15 × about 3.5 cm, about 7-seeded, the extra rib about 4 mm below the sutural rib. Seeds ovoid-ellipsoid, compressed, about 2 × 1 × 0.8 cm, brownish or dark brown, the hilum linear, about 1.3 cm long. Chromosome number 2n = 22.

Distr. Widespread through tropical Africa, Malagasy Republic, S. Arabia, India, Ceylon, and south into Java and probably Indonesia.

Ecol. The species is common in the moist low country and is found up to 900 m. It blooms December through June.

Uses. Roxburgh (1832) reports the plant poisonous. Baker (1876) notes the seeds are said to be bitter.

Vern. Moon reports Moodu-awara (S). A collector's label notes Wal-awara (S); this name also applies to *C. ensiformis*.

Specimens Examined. WITHOUT DISTRICT: Puttuvil (Pottuvil ?, AMPARAI DISTRICT), 21.3.27, *J.M. Silva s.n.* (PDA). JAFFNA DISTRICT: between Elephant Pass and Paranthan, *Rudd 3294* (K, PDA, US). ANURADHAPURA DISTRICT: Anuradhapura, *Kundu & Balakrishnan 338* (US). PUTTALAM DISTRICT: Wilpattu Natl. Pk., edge of Kali Villu, *Maxwell & Jayasuriya 801* (NY, PDA, US); Daluwa, along cart track to beach, *Maxwell & Jayasuriya 819* (PDA, NY, US); Wilpattu, *Mueller-Dombois et al. 69043001* (NY, PDA, US); S. of Puttalam, *Simpson 8180* (BM); Wilpattu, *Wirawan et al. 870* (US), *1059* (US). KANDY DISTRICT: Peradeniya, Royal Botanic Gardens, *Alston 1075* (PDA). NUWARA ELIYA DISTRICT: Hewaheta, *Alston 1480* (PDA). TRINCOMALEE DISTRICT: Katiraiveli, Trinco, *Alston 588* (PDA). BATTICALOA DISTRICT: S. of Batticaloa, *Simpson 8291* (BM). AMPARAI DISTRICT: Arugam Bay, 21.3.27, *J.M. Silva s.n.* (PDA). COLOMBO DISTRICT: Seeduwa (Siduwa ?), a climber on tall mangroves, *Amaratunga 2075* (PDA). BADULLA DISTRICT: Mi. 54 between Mahiyangana-Padiyatalawa, *Jayasuriya 2115* (US); road from Bibile to Mahiyangana,

Maxwell 1026 (NY, US). HAMBANTOTA DISTRICT: Ruhuna Natl. Pk. (Yala), Buttawa Bungalow area, *Comanor 368* (PDA, US); near Yala, *Mueller-Dombois & Comanor 67062308* (PDA, US); Ruhuna Natl. Pk., *Cooray 69011710R* (US); Yala, near Guard's hut, *Cooray 70032104R* (US).

Notes. Thwaites (1859) and Trimen (1894) cite *C.P. 2782,* but a sheet has been annotated *C. mollis* by Sauer (1964), the monographer.

Alston (1931) notes in raising *C. ensiformis* var. *virosa* to *C. virosa,* its shorter, 5–6-seeded, oblong, pod about 10 × 3.7 cm, with the hilum ⅓ encircling.

5. Canavalia gladiolata Sauer, Brittonia 16(2): 148. 1964. Type: Burma, *Lace 5464* (K holotype).

Canavalia gladiata (Jacq.) DC. var. *spodosperma* Voigt, Hort. Sub. Calc. 235. 1845. Type not located, fide Sauer (1964).

Climbing, twining vines. Leaflets broadly ovate, chartaceous, about 11 × 8 cm, acute to acuminate apically, rounded basally, sparse pubescent to glabrescent on both sides, with about 6 pairs of primary lateral veins. Inflorescences 9–20 cm long, the rachis 1–3 cm long, 5–10-flowered, bracts not seen; bracteoles obtuse, about 1.5 mm long. Flowers about 4 cm long, the pedicels about 1.5 mm long; calyx about 15 mm long, the upper lobes shorter than the tube, the upper edge constricted behind a non-apiculate tip, the lowest lobe slightly subulate, the laterals obtuse, all lobes about 2 mm long, with short, sparse, white pubescence; vexillary petal about 3.5 cm long. Fruit spirally dehiscent, compressed, linear to linear-oblong, 13–20 × about 3.2 cm, pale tan, 5–8-seeded (?), the extra rib about 2 mm below the sutural rib. Seeds (ex desc. Sauer, 1964) oblong, moderately compressed, about 20 × 12 × 8 mm, clear brown with dark mottling, buoyant, but quickly permeable, the hilum about 15 mm long.

Distr. Primarily Southeast Asia, north into China and west into India.

Ecol. The species climbs on thickets and trees along river banks and mountain slopes from low elevations to at least 1700 m, and flowers in summer and fall (Sauer, 1964).

Uses. Sauer reports that natives occasionally eat the pods but the species is not in cultivation.

Specimen Examined. WITHOUT EXACT LOCALITY: *Reynaud s.n.* (P, 2 sheets det. by Sauer).

Notes. A doubtful addition to the Flora. The specimens are not in Sauer's monograph (1964). I do not have Reynaud as a Ceylon collector, and the 2 sheets cited may be from Burma rather than Ceylon, although Ceylon is on the label.

6. Canavalia ensiformis (L.) DC., Prod. 2: 404. 1825; Baker in Hook. f., Fl. Br. Ind. 2: 193. 1876 (p.p. & excl. vars.); Trimen, Handb. Fl. Ceylon 2: 67. 1894 (p.p. excl. syn. & var.); Gamble, Fl. Pres. Madras 1(2): 359. 1918 (p.p.);

Alston in Trimen, Handb. Fl. Ceylon 6: 80, 81. 1931; Amshoff in Pulle, Fl. Suriname 2(2): 212. 1939; Burkart, Leg. Argentinas ed. 2, 409. 1952; Sauer, Brittonia 16(2): 142. 1964 (cum syn.); Verdcourt, Fl. Trop. East Africa, Papil. 572. 1971; Verdcourt, Man. New Guinea Leg. 473. 1979; D'Arcy, Ann. Missouri Bot. Gard. 67(3): 565. 1980. Type: Jamaica, Spanish Town, *Sloane Herb.* 3, 67 (BM holotype), not seen, fide Verdcourt (1971).

Dolichos ensiformis L., Sp. Pl. 725. 1753.

Vines, climbing or trailing perennials, though usually cultivated as a bushy annual; stems sparsely pubescent or glabrous. Leaf rachis 1–3.5 cm long, about ⅓ the petiolar length; stipules deciduous; leaflets elliptic or ovate-elliptic, 6–20 × 3–11 cm, obtuse or abruptly acute-acuminate apically, cuneate basally, glabrous or sparsely pubescent on both surfaces. Inflorescences 15–50 cm long, numerous-flowered, tuberculate; bracteoles obtuse, about 2 mm long. Flowers about 3 cm long, the pedicels 2–5 mm long; calyx tube 6–7 mm long, sparsely pubescent, the upper lobes somewhat truncate, about 5 mm long, the lower about 2.5 mm long, slightly exceeding the laterals; petals rose, lavender to purple, the vexillum about 2.7 cm long. Fruit linear-oblong, 15–35 × 3–3.5 cm, numerous-seeded, the extra rib about 5 mm below the sutural rib. Seeds rounded-oblong, compressed, white, 14–21 × 10–15 × 7–10 mm, the hilum linear, 6–9 mm long. Chromosome number 2n = 22.

Distr. Cultivated pantropically for food. It is an infrequent escape. A prehistoric American Indian domesticate (Sauer, 1964).

Ecol. Alston (1931) reports the species as cultivated and naturalized in Ceylon. The escapes (wild ?) are perennial vines.

Uses. Cited by Parsons (List of Medicinal Herbs, Dept. Agr. Bull. 91. 1937). The seeds are edible, but primary use is as a cover crop or green manure.

Vern. Horse-bean, Jack bean, Sabre-bean, Sword bean (E); Awara, Bu-wal-awara, Wal-awara (S); Koliavarai, Segapu (T). Some of these names may also apply to the next species, *C. gladiata,* another cultivated plant.

Specimens Examined. DISTRICT UNKNOWN: Henaratgoda, 8.5.17, *s. coll. s.n.* (PDA). JAFFNA DISTRICT: Jaffna, *Gardner C.P. 1481* (PDA). ANURADHAPURA DISTRICT: Kekirawa, "wild", Aug. 1885, *s. coll. s.n.* (PDA). KANDY DISTRICT: Peradeniya Gardens, Jack Bean, 25.4.17, *J.M. Silva s.n.* (PDA).

Note. Description based on *Wight 750e* (K).

7. Canavalia gladiata (Jacq.) DC., Prod. 2: 404. 1825; Wight & Arn., Prod. 1: 253. 1834; Wight, Ic. Pl. Ind. Or. 3: t. 753 ? 1843–45; Benth. in Mart., Fl. Bras. 15(1): 178. 1859 (p.p.); Thw., Enum. Pl. Zeyl. 88. 1859; Merr., Interpr. Rumph. Herb. Amboin. 281. 1917; Ridley, Fl. Malay Penins. 1: 572. 1922; Piper & Dunn, Kew Bull. 134. 1922; Alston in Trimen, Handb. Fl. Ceylon 6: 81. 1931;

Amshoff in Pulle, Fl. Suriname 2(2): 211. 1939; Burkart, Leg. Argentinas ed. 2, 409, 410. 1952; Sauer, Brittonia 16(2): 149. 1964 (cum syn. & vars.); Verdcourt, Fl. Trop. East Africa, Papil. 572. 1971; Verdcourt, Man. New Guinea Leg. 474. 1979; D'Arcy, Ann. Missouri Bot. Gard. 67(3): 567. 1980. Lectotype: Jacquin, Ic. Rar. 3, t. 560. 1787, not seen. Plant grown in Vienna, origin undetermined.

Dolichos gladiatus Jacq., Ic. Pl. Rar. 3: t. 560. 1787; Moon, Cat. 53. 1824 (incl. vars.); Roxb., Fl. Ind. 3: 300. 1832 (p.p.).

Canavalia maxima Thouars, J. Bot. (Desvaux) 1: 80. 1913 (p.p., Rheede's Baramareka). Lectotype: Rheede, Hort. Mal. 8: 85, t. 44. 1688, fide Sauer, 1964, not seen.

Canavalia ensiformis auct., Baker in Hook. f., Fl. Br. Ind. 2: 195. 1876 (p.p.); Trimen, Handb. Fl. Ceylon 2: 67. 1894 (p.p. ?); Gamble, Fl. Pres. Madras 1 (2): 359. 1918 (p.p.).

Climbing, trailing vines to about 4 m long, cultivated annuals or rarely escaped perennials. Leaf rachis 2.5–5 cm long, ½ to ⅓ the petiolar length; stipules about 2 mm long; leaflets ovate, 8–20 × 5–12 cm, long-acuminate apically, rounded to broadly cuneate basally, sparsely pubescent on both surfaces, primary lateral veins in about 6 pairs. Inflorescences 20–35 cm long, numerous-flowered, tuberculate; bracteoles obtuse, about 1 mm long. Flowers about 3.5–4 cm long, the pedicels about 2 mm long; calyx tube about 10 mm long, glabrous or glabrescent, the upper lobes about 5–7 mm long; petals white (lavender, D'Arcy (1980)), the vexillum suborbicular, 3–3.5 cm long. Fruit usually spirally dehiscent, linear-oblong to linear, compressed, 20–40–(60 ?) × 3.5–5 cm, about 15–(20)-seeded, an extra rib about 2 mm below the sutural rib. Seeds oblong-elliptic, 21–35 × 14–20 × 12–14 mm, reddish to clear brown, frequently with dark mottling, white in var. *alba,* the hilum linear, 15–20 mm long. Chromosome number $2n = 22, 44$.

Distr. Widely dispersed pantropically through cultivation. Escapes are rare. An old far Eastern domesticate (Sauer, 1964).

Ecol. Found in the lowlands and medium elevations, it flowers all year in the tropics, and in summer and fall in marginal regions (Sauer, 1964).

Uses. Cultivated for food, the young pods and seeds are commonly cooked and eaten. Also occasionally grown as an ornamental vine, fetish plant or cover crop (Sauer, 1964).

Vern. Awara, Awara-Sudu (white-seeded var.), Awara-Ratu (red-seeded var.) (S). Some of the English names, such as Sword bean, attributed to *C. ensiformis* probably refer to this species also.

Specimens Examined. VAVUNIYA/ANURADHAPURA DISTRICT: between Vavuniya and Medawachchiya, a colony, probably escaped from cultivation, (flowers & fruit) 16 Mar. 1970, *Rudd 3254* (NY, PDA, US).

Notes. Species *C. virosa, C. ensiformis* and *C. gladiata* have been confused in Ceylon, but I am assuming that Moon's (1824) *Dolichos gladiatus* var. α, characterized as white, does have white flowers and may be *C. ensiformis*, which also has white seeds. A variety mentioned by Roxburgh (p. 301, 1832) with white flowers, the legume about 2 feet long, often containing as many as twenty large, white, arillate seeds is certainly not *C. ensiformis*. Perhaps Roxburgh's variety is *C. gladiata* var. *alba* (Makino) Hisauchi (Sauer, 1964), a white-seeded variety named from Japan.

8. PACHYRHIZUS

Rich. ex DC., Prod. 2: 402. 1825; Clausen, Cornell Univ. Agric. Exp. Sta. Mem. 264. 1945; Hutchinson, Gen. Fl. Pl. 1: 439. 1964 (cum syn.). Type species: *Dolichos erosus* L., *P. erosus* (L.) Urban, typ. cons.

Cacara Thouars, Dict. Sci. Nat. 6: 35. 1805 (1806 ?), nom. rejic.

Annual or perennial, climbing, twining, herbaceous vines, frequently with large tap-roots. Leaves pinnately trifoliolate; stipulate and stipellate; leaflets ovate, the terminal often rhomboidal, triangular, frequently lobed. Inflorescences axillary, fasciculate-racemose, usually tuberculate, few–numerous-flowered; bracts and bracteoles linear or setaceous, caducous. Calyx 5-lobed, the upper lobes partially connate, emarginate, largest, obtuse, the lateral and lower lobes much smaller, about equal, acuminate or acute, smaller than or about equalling the tube; petals blue, occasionally white, the vexillum obovate, inflexed biauriculate basally, the wings oblong, somewhat falcate to oblanceolate, the keels somewhat ovate, incurved, about equalling the wings; stamens diadelphous, the vexillary usually free, the anthers 10, uniform; ovary subsessile, several–numerous-ovulate, the style upcurved and involute terminally, mostly bearded, flattened towards the apex, the stigma minute, subterminal. Fruit linear or linear-oblong, dehiscent, compressed, septate, the exocarp frequently deeply constricted transversely between the seeds. Seeds ovoïd, compressed, the hilum small, oval or elliptic, without arillate tissue.

A genus of about six species in the tropics of both hemispheres, probably of New World origin. Spread through cultivation for the edible tuberous roots.

See "Yam Bean" in "Tropical Legumes: Resources for the Future", pp. 21–27, 1979, Nat. Acad. Sci., Washington D.C., U.S.A.

KEY TO THE SPECIES

1 Leaflet lobes dentate, rarely entire, the bases very broadly cuneate or truncate; wing auricle about ⁴/₅ the claw length; fruit 8–14 cm long, hairs over 1 mm long or glabrescent; widest seed dimension 5–11 mm..**1. P. erosus**
1 Leaflets unlobed, entire, the bases medium cuneate; wing auricle about ½ the claw length; fruit 6.5–30 cm long, hairs to 1 mm long or glabrescent; widest seed dimension 11–14 mm......
....**2. P. tuberosus**

1. Pachyrhizus erosus (L.) Urban, Symb. Ant. 4: 311. 1905; Merr., Interpr. Rumph. Herb. Amboin. 132, fig. 2. 1917; Clausen, Cornell Univ. Agric. Exp. Sta. Mem. 264. 1945; Burkart, Leg. Argentinas ed. 2, 413. 1952; Backer & Bakh. f., Fl. Java 1: 643. 1963; Verdcourt, Man. New Guinea Leg. 477, fig. 112. 1979; D'Arcy, Ann. Missouri Bot. Gard. 67 (3): 744, fig. 40. 1980 (cum syn.). Type: ? Herb. Linn (LINN 900.11), not seen, fide D'Arcy (1980).

Dolichos erosus L., Sp. Pl. 2: 726. 1753.
Dolichos bulbosus L., Sp. Pl. ed. 2, 1021. 1763; Roxb., Fl. Ind. 3: 309. 1832.
Pachyrhizus angulatus Rich. ex DC., Prod. 2: 402. 1825; Wight & Arn., Prod.
    1: 251. 1834 (cum syn.); Benth. in Mart., Fl. Bras. 15(1): 199. 1859; Baker
    in Hook. f., Fl. Br. Ind. 2: 207. 1876; Ridley, Fl. Malay Penins. 1: 569. 1922.

Strong, high-climbing vines; stems mostly terete, with sparse, adpressed pubescence. Leaf rachis 2–3 cm long, about ⅓ the petiolar length, with sparse, adpressed to ascending hairs; stipules lanceolate or acute, about 3 mm long; stipels setaceous, about 2.5 mm long, persistent; both leaflet surfaces with sparse, adpressed pubescence, the terminal leaflets to about 10 × 11 cm, the upper margin frequently with shallow sinuses and acute, blunt or dentate lobes, blunt-acute apically, cuneate or truncate basally, the lateral leaflets broadly ovate, usually shallowly lobed, inequilateral, 3 prominent veins inserted basally. Inflorescences axillary, 20–60 cm long (fide Backer, 1963), mostly smaller, many–numerous-flowered, florate ½ or less the length, ferruginous to canescent pubescent, the tubercles globose, somewhat elongate, each about 3-flowered; bracteoles linear-lanceolate, about 1 mm long. Flowers borne on pedicels about 6 mm long; calyx tube 4–6 mm long, with adpressed canescent or fulvous hairs, the upper lobe 5–6 mm long, the lateral and lower lobes acute, about 4 mm long; petals blue, bluish-violet (or reddish, fide Baker, 1876), membranous, apparently glabrous, the vexillum obovate-elliptic, about 17 × 15 mm, with a semicircular ridge of tissue connecting the auricles above the claw, the claw 4–5 mm long, the wings about 15 × 5 mm, the auricle incurved, acuminate, about 4 mm long, the claw about 5 mm long, the keels broadly oblanceolate, 12–14 × 4–5 mm, the claw about 8 mm long; ovary subsessile, about 6 mm long, villous, about 11-ovulate. Fruit about 9 × 1.2 × 0.3 cm, with fairly long, adpressed, fulvous or ferruginous hairs over 1 mm long, older pods becoming glabrescent, about 10-seeded. Seeds about 7 × 6.5 × 3 mm, dark brown, shining, with the hilum margin flattened, the hilum about 3.5 mm long. Chromosome number 2n = 22.

Distr. Found in the tropics of both hemispheres. Roxburgh (1832) states the plant is an exotic, introduced from the Malay Islands and reached the Islands via South America and the Philippines. Merrill (1917) adds that Spanish colonists brought the bean to the Philippines from Mexico.

Ecol. Yam beans grow best in a moist tropic climate.

Uses. The roots are eaten, either raw or boiled. The mature seeds are very poisonous. Ridley reports the young pods are eaten; however, Tropical Legumes (1979) states that only the tubers are safe to eat and further that leaves, stems, roots, ripe pods and seeds possess insecticidal properties and can also be toxic to man. In Central America, as in the Philippines and Asia, immature pods are sometimes eaten, but care and experience are necessary to avoid the toxic effects. This report indicates yam beans are a very practical tropical food resource that has not been utilized sufficiently; it gives research contacts worldwide. Balansa (*Balansa 2280* (US), Tonkin, village of Dong-Tom, Vietnam) notes the seeds are used in small doses as a vermifuge.

Vern. Yam bean (E).

Specimens Examined. KANDY DISTRICT: Oct. 1888, *s. coll. s.n.* from Java, flowers blue, 2 sheets; Royal Botanic Gardens nursery, Nov. 1888, *s. coll. s.n.* Jan. 1889, *s. coll. s.n.* (PDA, all in cultivar folder.).

Notes. Backer & Bakh. f. (1963) note that the cultivars sometimes escape but disappear in time.

In addition to the specimens cited, specimens from India, SE. Asia and China were used for the description.

2. Pachyrhizus tuberosus (Lam.) Spreng., Syst. Veg. ed. 6, 4(1): 281. 1827; Spreng, Bull. Misc. Inform. 62, fig. 1. 1889; Clausen, Cornell Univ. Agric. Exp. Sta. Mem. 264. 1945; Burkart, Leg. Argentinas ed. 2, 416. 1952. Type not seen.

Dolichos tuberosus Lam., Enc. 2: 296. 1786.

Strong, high-climbing vines; stems terete, pubescent. Leaf rachis from ½ to ¼ the petiolar length, sparsely pubescent; stipules lanceolate or acute, about 5 mm long (?); stipels setaceous, about 2.5 mm long, persistent; both leaflet surfaces with short, sparse, adpressed pubescence, the terminal leaflets deltoid to triangular, without distinct lobes, about 13 × 11 cm, acute and frequently narrowing apically, cuneate basally, the lateral leaflets broadly ovate, smaller than the terminal, greatly inequilateral, with 3 prominent veins inserted basally. Inflorescences to about 40 cm long, numerous-flowered, florate about ½ the length, the tubercles usually elongate, incurved, the lower nodes to about 8-flowered, the upper about 3-flowered; bracteoles linear, about 1.5 mm long. Flowers about 2.8 cm long, the pedicels about 6 mm long; calyx tube with adpressed, ferruginous pubescence, the upper lobe about 6 mm long, the lateral and lower lobes acute, 3.5–4 mm long; petals lavender, blue or purple, occasionally white, the vexillum obovate-elliptic, about 18 × 16 mm, with a semicircular ridge of tissue connecting the auricles above the claw, the claw about 5 mm long, the wings oblanceolate, about 17 × 6 mm, the auricles acuminate, about 3.5 mm long, the claw about 6 mm long, the keels obliquely oblong, about 15 × 7 mm, somewhat beaked or the apex acute, the claw about 8 mm long, the upper margin of the wings and keels sparsely ciliate; ovary subsessile, about 6 mm long, villous, about 7-ovulate. Fruit spirally dehiscent, about 8.5 × 1.2 cm, with mostly adpressed, ferruginous pubescence, the hairs about 1 mm long (the small-fruited hairy variety, fide Clausen, 1945). Seeds not seen. Chromosome number 2n = 22.

Distr. Mostly the New World tropics. Introduced into the Old World tropics, but apparently not as common as *P. erosus*. Believed to be native to the Amazon head waters regions of South America and parts of the Caribbean, the species was distributed from the Royal Botanic Gardens at Kew to tropical British Colonies (Tropical Legumes, 1979).

Ecol. Grows best in the moist tropics, but, as with *P. erosus,* the species will grow cultivated in many different environments.

Uses. Cultivated for the roots and green beans. The root is used in curry. Other uses are the same as for *P. erosus*; however, the tubers are not as large.

Vern. Yam bean (E).

Specimens Examined. KANDY DISTRICT: Royal Botanic Gardens, Nov. 1887, *s. coll. s.n.* "Fls. pure-white," (PDA), *s. coll. s.n.* "*Dolichos tuberosus* from Kew*", (PDA).

Notes. In addition to the specimens cited, the description was based on material from Goiás, Brazil; Huigra, Ecuador (det. by Clausen); and Leon Province, Ecuador (det. by Clausen).

The species was growing in the Herb or Students' Garden at the Royal Botanic Gardens at Peradeniya in 1972.

Although the species was introduced into Ceylon in the 1880's, to my knowledge, it has not become naturalized.

9. GALACTIA

P. Br., Civ. Natr. Hist. Jamaica 298. 1756; Urban, Symb. Ant. 2: 307. 1900–1901; Hutchinson, Gen. Fl. Pl. 1: 430. 1964 (cum syn.); Burkart, Darwiniana 16: 663–796. 1971. Type Species: *G. pendula* Pers.

Odonia Bertol., Lucub. Herb. 35. Bonnae. 1822.

Climbing, twining, trailing vines or small erect shrubs. Leaves usually pinnately, occasionally digitately, trifoliolate or 1-foliolate, rarely pinnately 5–7-foliolate; stipulate and stipellate; leaflets elliptic, ovate to suborbicular. Inflorescences axillary, from few-flowered on very short peduncles to many flowered, fasciculate-racemose from distinct tubercles; bracts and bracteoles usually persistent. Calyx 4-lobed, the upper lobes connate, entire; petals whitish through shades of pink to purple, the vexillum ovate to suborbicular, simple, the wings free, the keels obliquely oblong; vexillary stamen free or connate medianly, the anthers 10, uniform; ovary sessile, many-ovulate, the style usually ascending, slender, uniform, exserted, the upper style glabrous, the stigma capitate, small, glabrous. Fruit dehiscent, oblong, to linear-oblong, flat, straight, many-seeded; sutures slightly swollen or indistinct, usually with a small beak. Seeds small, the hilum elliptic to oblong.

A pantropical genus of perhaps 50–80 species extending into the temperate regions. The vast majority of species are found in the New World.

Galactia striata (Jacq.) Urban, Symb. Ant. 2: 320. 1900; Fawcett & Rendle, Fl. Jamaica 4: 56. 1920; Burkart, Darwiniana 16: 715, fig. 16, 17. 1971 (cum syn.); D'Arcy, Ann. Missouri Bot. Gard. 67(3): 700, fig. 28. 1980. Type: Hort. Vindob. I: 32, t. 76. 1770, from specimens cultivated in Hort. Vindob. & Monac., Herb. Jacq., fide Urban., sheet upon which tab. 76 was based (BM).

Glycine striata Jacq., Hort. Bot. Vindob. 1: 32, t. 76. 1771? See Burkart, p. 719. 1971.
Glycine tenuiflora Klein ex Willd., Sp. Pl. 3(2): 1059. 1802; DC., Prod. 2: 241. 1825; Moon, Cat. 53. 1824; Roxb., Fl. Ind. 3: 319. 1832.
Teramnus tenuiflorus (Willd.) Spreng., Syst. Veg. 3: 235. 1826.
Galactia tenuiflora (Klein ex Willd.) Wight & Arn., Prod. 1: 206. 1834 (sensu vars.); Wight & Arn., Prod. 1: 206. 1834; Benth. in Mart., Fl. Bras. 15(1): 143. 1859 (incl. vars., p.p.); Thw., Enum. Pl. Zeyl. 88. 1859; Baker in Hook. f., Fl. Br. Ind. 2: 192. 1876; Trimen, Handb. Fl. Ceylon 2: 65. 1894; Gamble, Fl. Pres. Madras 1(2): 356. 1918; Nooteboom & Van Steenis, Reinwardtia 5(4): 431. 1961; Backer & Bakh. f., Fl. Java 1: 631. 1963; Verdcourt, Fl. Trop. East Africa, Papil. 578. 1971; Verdcourt, Man. New Guinea Leg. 478, fig. 113. 1979. Photo of the holotype at B seen at K. India, Pondicherry. Also based on *Wallich 5511, Wight Cat. 875* (NY).
Galactia villosa Wight & Arn., Prod. 1: 207. 1834; Gamble, Fl. Pres. Madras 1(2): 357. 1918. Type: *Wight 939* (K).
Galactia striata (Jacq.) Urban var. *tenuiflora* (Klein ex Willd.) Burkart, Darwiniana 16: 721, fig. 19, 20. 1971.

Small vines, trailing or climbing, perennial from a woody root-stock; stems slender, pubescent. Leaves trifoliolate, subsessile, the rachis 5–15 mm long, the petioles 2–5 cm long; stipules linear or acute, about 3 mm long, persistent; stipels setaceous, about 1 mm long, persistent; leaflets broadly lanceolate to oblong-elliptic, to 2.5–3.5 × 1.7–2.2 cm, acute, rounded or sometimes emarginate apically, rounded to somewhat cordate basally, paler beneath, glabrescent above, sericeous pubescent beneath, with about 6 pairs of primary, lateral veins. Inflorescences lax, to about 10 cm long (2, 3 times longer ?, other authors), the tubercles distant, each 1–3-flowered; bracts at the base of the tubercles about 2 mm long; bracteoles narrowly ovate, about 2 mm long. Flowers 6–10 mm long, the pedicels about 1 mm long; calyx tube about 3 mm long, pubescent, all lobes lanceolate to acuminate, 4–6 mm long, the lower longest; petals reddish or pink, the vexillum obovate, about 10 × 6 mm, short-clawed, the wings and keels obliquely oblong, about equal in length, about 8 mm long, the claw about 2 mm long; stamens diadelphous, the vexillary free; ovary linear, 8–10-ovulate, sericeous pubescent, the style ascending, glabrous. Fruit linear to linear-oblong, compressed, sericeous pubescent, becoming glabrous, about 3–5 × 0.6 × 0.2 cm, non-septate, 4–10-seeded; upper suture with closely parallel ribs, the lower swollen. Seeds ovoid-elliptic, compressed, oblique, variously coloured, to about 4 × 3 × 1 mm, the hilum somewhat elliptic, about 1 mm long. Chromosome number 2n = 20.

Distr. Pantropical and extending through the subtropics into some temperate regions.

Ecol. Trimen reports the species as rather common in the upper zone of the low country: Kotmale, Hantane, Matale. He also reports the flowers as pink.

Uses. The name *G. striata* is common in agronomy literature pertaining to field trials in tropical and subtropical regions of the New World. The uses are many: forage in submarginal grazing regions, high priority nitrogen fixation, and as a cover crop for green manure, hay, or silage. The species is usually grown in a legume mixture or legume-grass mixture. See Matos and Alcantara, Zootecnia 14(1): 51–57. 1976, *Galactia striata,* a promising legume for central Brazil.

Specimens Examined. NUWARA ELIYA, KANDY AND MATALE DISTRICTS: Kotmale, Hantane, Matale, *Gardner C.P. 1468* (PDA); *Thwaites C.P. 1468* (K); *s. coll. C.P. 1468* (PDA, mixture, fruit only).

Notes. I follow Burkart (1971) in considering *Galactia tenuiflora* (Klein ex Willd.) Wight & Arn. sensu, *Glycine tenuiflora* Klein ex Willd., *Wight 875,* and *Wallich 5511* types similar to *Galactia striata.* From the basis of the variety *tenuiflora* Burkart maintains *G. longifolia* (Jacq.) Benth., including Bentham's *G. tenuifolia* var. *eriocarpa,* in synonymy. Burkart does not mention *G. villosa* which could be maintained based on *Wight 939* and *Wallich 5509b, c,* and *5510* p.p. at species or varietal rank as Verdcourt (1971) suggests. I consider the Ceylon specimens to be *G. striata* var. *tenuiflora.*

10. CALOPOGONIUM

Desv., Ann. Sci. Nat. Bot., ser. 1, 9: 423. 1826; Hutchinson, Gen. Fl. Pl. 1: 429. 1964. Type species: *C. mucunoides* Desv.

Stenolobium Benth., Ann. Wiener Mus. 2: 125. 1837.

Twining, climbing vines. Leaves trifoliolate; stipulate and stipellate; leaflets mostly ovate. Inflorescences axillary, fasciculate-racemose, tuberculate, few–many-flowered; bracts and bracteoles small, mostly caducous. Flowers medium to small; calyx 5-lobed or the upper two lobes partially connate; petals mostly shades of blue and violet, the vexillum reflexed, biauriculate basally, the wings and keels adherent, about equal in length; stamens diadelphous, the vexillary free, the anthers uniform, mostly 8–9; ovary many-ovulate, sessile, the style slightly geniculate, the upper style glabrous, the stigma terminal, capitate, glabrous. Fruit dehiscent, hirsute, linear or linear-oblong, compressed, usually septate between the seeds and constricted externally, several-seeded; sutures erect, slightly swollen. Seeds dark, globose or oblong, compressed, small, the hilum small, oval.

A genus of six to eight species native to tropical America.

Calopogonium mucunoides Desv., Ann. Sci. Nat. Bot., ser. 1, 9: 423. 1826; Amshoff in Pulle, Fl. Surinam 2(2): 196. 1939 (cum syn.); Burkart, Leg. Argentinas ed. 2, 413. 1952; Backer & Bakh. f., Fl. Java 1: 630. 1963; Verdcourt, Fl. Trop. East Africa, Papil. 577. 1971; Verdcourt, Man. New Guinea Leg. 481, fig. 114. 1979; D'Arcy, Ann. Missouri Bot. Gard. 67(3): 560. 1980. Type: French Guiana (P, fide Amshoff, Meded. Bot. Mus. Herb. Rijks. Univ. Utrecht 52: 65. 1939), not seen.

Stenolobium brachycarpum Benth. in Seem., Bot. Her. 109. 1853; Benth. in Mart., Fl. Bras. 15(1): 140. 1859. Based on *Seeman 215* (K), Panama.

Low twining or trailing vines; stems terete, pilose. Leaf rachis about 1 cm long, $1/6$–$1/10$ the petiolar length, the petiole and rachis pilose; stipules acute to lanceolate, about 4 mm long, persistent; stipels linear-filamentous, 3–4 mm long; leaflets ovate, elliptic or rhomboid-ovate, about 6 × 4 cm, acute to obtuse apically, broadly cuneate to rounded basally, both surfaces sparsely pubescent, the lateral leaflets oblique, with about 4 pairs of primary lateral veins. Inflorescences to about 25 cm long, the rachis condensed or the nodes distant and elongate, ferruginous pilose, each node 2–6-flowered; bracts lanceolate; bracteoles lanceolate, about 5 mm long. Flowers about 7 mm long, subsessile; calyx about 6 mm long, pilose, the upper lobes partially connate, about equalling the tube, the lower lobes acuminate, much longer than the tube; petals short-clawed, the vexillum obovate, about 9 × 5–6 mm, emarginate, inflexed biauriculate, blue, the wings narrowly oblong, nearly equalling the vexillum, the keels incurved, oblong, auriculate; ovary 6–8-ovulate, villous. Fruit straight or falcate, about 2.5 × 0.4 × 0.3 cm, septate, light brown, hirsute ferruginous, the exocarp deeply constricted between the seeds, about 6-seeded; sutures indistinct. Seeds somewhat oblong, 2–4 × 2.5–3.0 × 2 mm. Chromosome number 2n = 36.

Distr. The species is native to tropical South America, Central America and to some of the West Indies. It was introduced into tropical Africa and Asia, and apparently into Ceylon about 1926 where it has become naturalized.

Ecol. Flowers January through June. Roadsides of the wet zone.

Uses. Cover crop to control erosion, green manure, and possibly as forage for cattle.

Specimens Examined. KANDY DISTRICT: Peradeniya Exp. Sta., 26.1.26, *Alston s.n.* (PDA); Kandy, *Fairchild & Dorsett 373* (US). COLOMBO DISTRICT: Moratuwa, Katubedda Rd., *Cooray 70012303R* (PDA, US). RATNAPURA DISTRICT: A 4, marker 86, *Comanor 1119* (NY, PDA, US). GALLE DISTRICT: Bona Vista, *Cramer 2907* (PDA).

Notes. The Ceylon collections show short inflorescences with densely clustered flowers; however, tropical American specimens often have long inflorescences with few-flowered nodes.

11. PUERARIA

DC., Ann. Sci. Nat. Bot., Ser. 4: 97. 1825; Benth., J. Linn. Soc. Bot. 9: 121–125. 1867. Type species: *P. tuberosa* (Roxb.) DC., India.

Neustanthus Benth. in Miq., Pl. Jungh. 234. 1852.

Twining, climbing or trailing vines. Leaves trifoliolate; stipulate and stipellate; leaflets ovate with a lobing tendency, the terminal leaflets frequently rhomboid. Inflorescences usually axillary, fasciculate-racemose, long pedunculate, many–numerous-flowered, tuberculate; bracts and bracteoles caducous or subpersistent. Flowers medium; calyx 4–5-lobed, the upper lobes frequently connate, usually longest, all lobes obtuse to acuminate; petals reflexed, usually blue or purplish, biauriculate basally, the wings and keels about the same length, frequently adherent, the keels sometimes beaked; stamens usually monadelphous, the vexillary filament attached towards the base, sometimes free, the anthers uniform; ovary many-ovulate, the style geniculate, the upper style glabrous, the stigma terminal, capitate, sometimes with a few hairs. Fruit dehiscent, long-linear, somewhat terete, pubescent, septate or occasionally continuous between the seeds, many-seeded; sutures indistinct or slightly swollen. Seeds small, compressed, subreniform or transversely oblong, the hilum small, elliptic.

A genus of about 15 species of tropical Asia, China, Japan, New Guinea and Polynesia.

Pueraria phaseoloides (Roxb.) Benth., J. Linn. Soc. Bot. 9: 125. 1867; Baker in Hook. f., Fl. Br. Ind. 2: 199. 1876; Prain, J. Asiat. Soc. Bengal 66(2): 420. 1897; Prain, Bengal Pl. 1: 396. 1903; Ridley, Fl. Malay Penins. 1: 571. 1922; Backer & Bakh. f., Fl. Java 1: 632. 1963; Verdcourt, Man. New Guinea Leg. 485, fig. 115. 1979; Lackey in Polhill & Raven, Adv. Leg. Syst. Part 1. 2: 318. 1981. Type: Grown at Calcutta from seeds sent by Kerr from Canton, China, fide Roxburgh, and *Roxburgh* drawing *1890* (K), fide Verdcourt (1971), not seen.

Dolichos phaseoloides Roxb., Fl. Ind. 3: 316. 1832.
Dolichos viridis Ham. in Wall., Cat. 5559 & 5559B. 1831–1832, nom. nud.
Phaseolus decurrens (= *P. decurvus* sphalm.) Grah. in Wall., Cat. 5612. 1831–1832, nom. nud.
Neustanthus phaseoloides (Roxb.) Benth. in Miq., Pl. Jungh. 234. 1852.
Neustanthus javanicus Benth. in Miq., Pl. Jungh. 235. 1852. Type: Java, Merapi, R. Kuning, *Junghuhn s.n.* (K holotype).
Pueraria javanica (Benth.) Benth., J. Linn. Soc. Bot. 9: 125. 1865; Burkart, Leg. Argentinas ed. 2, 407, fig. 122. 1952.

Pueraria phaseoloides (Roxb.) Benth. var. *javanicus* (Benth.) Baker in Hook. f., Fl. Br. Ind. 2: 199. 1876; Verdcourt, Fl. Trop. East Africa, Papil. 596, fig. 87. 1971.

Herbaceous vines; stems terete, pubescent. Leaves sparsely pubescent, the rachis ½–⅓ the petiolar length; stipules acute, non-produced, about 3 mm long, persistent; stipels linear-filamentous, about 5 mm long; lateral leaflets broadly ovate, greatly oblique, to about 10 × 8 cm, the terminal leaflets to about 8 × 11 cm, acute apically, cuneate basally, pubescent above and below, sometimes pilose below, the primary lateral veins in 4–5 pairs, frequently branching before the margins, prominent beneath. Inflorescences erect, axillary; bracts lanceolate, about 2 mm long; bracteoles lanceolate, about 2.5 mm long, subpersistent. Flowers about 1.5 cm long, the pedicels about 2 mm long; calyx tube about 4 mm long, the upper lobes connate, emarginate, longest, the lateral lobes obtuse, the lower acute, all lobes shorter than the tube; petals white with purple or lavender markings, the vexillum obovate, inflexed biauriculate, about 15 mm long, slightly exceeding the wings and keels, the keels beaked; stamens monadelphous; ovary 15–20-ovulate, the stigma bearded. Fruit spirally dehiscent, slightly upcurved distally, about 8 cm long, 4 mm wide, 1.5 mm thick, septate, 7–17-seeded; sutures indistinct. Seeds subreniform, about 3 × 2 × 1.25 mm, horizontal, orange or reddish brown, the hilum about 1 mm long. Chromosome number variously reported as 2n = 20, 22 or 24.

Distr. India, East Asia, Malaysia, and Indonesia. Introduced and naturalized into Ceylon and the New World tropics.

Ecol. Flowers November through March.

Uses. Ground cover and erosion control.

Specimens Examined. KANDY DISTRICT: Exp. Sta., Peradeniya, 15.1.25, *s. coll. s.n.* (PDA), 26.1.26, *s. coll. s.n.* (PDA); Hataraliayadda, in rubber field, *Amaratunga 93* (PDA, US) ; Peradeniya, *J.M. de Silva 124* (NY); NW. of Galagedera, *Rudd 3315* (K, NY, PDA, US); Galagedera, *Rudd 3334* (PDA, US); Doluwa, nr. Kandy, *Rudd & Albert 3159* (NY, PDA, US). COLOMBO DISTRICT: Labugama Forest Reserve, *Rudd & Jayasinghe 3326* (PDA, US). KALUTARA DISTRICT: Pelawatte, *Cramer 2807* (PDA, US). KEGALLE DISTRICT: Karawanella, on rubber land, a cover crop, *Amaratunga 1236* (PDA); Deraniyagala, *Simpson 9144* (BM). GALLE DISTRICT: Kanawa Estate, Bentota, *s. coll. s.n.* (PDA, in cult. folder).

Notes. I have seen no exsiccatae of the Kudzu vine, *P. lobata* (Willd.) Ohwi, from Ceylon; but I did observe sterile plants of this species growing wild along the roadside in Uva Province, Moneragala District, marker 134 west of Wellawaya on 1 June, 1972. Kudzu vine has larger, more frequently lobed leaflets and a much shorter, wider fruit than *P. phaseoloides*.

Pueraria can be distinguished by the large leaflets, the long, linear, many-seeded fruit and the small oblong seeds with small elliptic hila.

Lackey (1981) observes that the genus needs revision, but that three groups can be easily recognized. One group contains *P. phaseoloides* and *P. subspicata* Benth. This group is characterized by 4 or more flowers at a node, non-produced stipules, distinct upper calyx teeth and terete pods.

Description based on *Wallich 5559, 5559B* (BM) and *Junghuhn 210, 151* (K).

12. NEONOTONIA

Lackey, Phytologia 37: 209–212. 1977; Lackey, Taxon 27: 560. 1978; Hymowitz & Newell in Polhill & Raven, Adv. Leg. Syst. Part 1. 319. 1981. Type species: *N. wightii* (Arn.) Lackey.

Notonia Arn. in Wight & Arn., Prod. 1: 449. 1834, non *Notonia* DC. in Guillemin, Arch. Bot. (Paris) 2: 518. 1833.

Johnia Arn. in Wight & Arn., Prod. 1: 449. 1834, non *Johnia* Roxb., Fl. Ind. 1: 168. 1832.

Bujacia E. Meyer, Comm. Pl. Afr. Aust. 127. 1836, p.p. Type: *B. gampsonychia* (= *Teramnus labialis,* fide Lackey (1977)).

Glycine sect. Javanica sensu Harms in Engler, Pfl. Afr. 3(1): 654. 1915, p.p.

Glycine subg. Glycine sensu F.J. Herm. (revision), Techn. Bull. USDA. 1268: 24. 1962.

Glycine subg. Bracteata Verdcourt, Taxon 15: 34. 1966; Newell & Hymowitz, Amer. J. Bot. 65(2): 168–179. 1978.

Twining, procumbent perennial vines. Leaves trifoliolate; stipulate and stipellate; leaflets ovate to elliptic. Inflorescences in axillary pseudoracemes, each node 2–several-flowered, without distinct tubercles; bracts and bracteoles usually lanceolate. Flowers small; petals whitish, pink to blue or purple, long-clawed, glabrous; upper calyx lobes mostly united, entire; vexillary petal spreading, simple, somewhat biauriculate, obovate to orbicular, the wings and keels about equal in length, the vexillum longest; stamens monadelphous or the vexillary separating with age, the anthers uniform, glabrous; ovary several–many-ovulate, the style short, glabrous, incurved, the stigma small, capitate, terminal, glabrous. Fruit dehiscent, linear or oblong, flattened or subterete, few-seeded; sutures slightly swollen or indistinct. Seeds ovoid-oblong, horizontal, the hilum minute, elliptic, usually with a small, scale-like persistent aril.

A genus of one, or possibly two, species native to the Old World, centered in Africa and India extending into Southeast Asia.

Neonotonia wightii (Grah. ex Wight & Arn.) Lackey, Phytologia 37(3): 210. 1977; Lackey in Polhill & Raven, Adv. Leg. Syst. Part 1: 319. 1981; Lackey, Iselya 2(1): 11–12. 1981. Syntypes: Southern India, *Wight 871* (K), *872* (COIMB, K); *Heyne* in Wall. Cat. No. 5528, *Wallich 5528* (BM, K).

Notonia wightii Wight & Arn., Prod. 1: 208. 1834.

Johnia wightii Wight & Arn., Prod. 1: 449. 1834.

Glycine bujacia Benth., Comm. Leg. Gen. 62. 1837.

Soja wightii Grah., Walp. Ann. 4: 552. 1858; Thw., Enum. Pl. Zeyl. 88. 1859. Based on *Wallich 5530* (BM, K).

Glycine wightii (Grah. ex Wight & Arn.) Verdcourt, Taxon 15(1): 35. 1966; Verdcourt, Fl. Trop. East Africa, Papil. 528. 1971; Verdcourt, Man. New Guinea Leg. 493. 1979.

Glycine javanica auct. mult., non L., Moon, Cat. 53. 1824; Baker in Hook. f., Fl. Br. Ind. 2: 183. 1876; Trimen, Handb. Fl. Ceylon 2: 59. 1894; Gamble, Fl. Pres. Madras 1(2): 351. 1918; F.J. Herm., Techn. Bull. USDA. 1268: 29–36, figs. 8–10. 1962.

Climbing vines to about 4 m long; stems somewhat angular, pubescent. Leaf rachis and petiole 5–8 cm long; stipules acute-acuminate, 5–7 mm long, persistent, stipels setaceous, about 4 mm long, persistent; leaflets acute or obtuse apically, cuneate, rounded, truncate to cordate basally, both surfaces pubescent, the terminal to about 10.5 × 8 cm, the laterals to about 9.5 × 7 cm, usually smaller than the terminal, the lamina inequilateral. Inflorescences to 30 cm long, florate 8–12.5 cm; bracts and bracteoles linear. Flowers 4–8 mm long, the pedicels about 2 mm long or subsessile, in fascicles along the rachis, densely congested towards the apex; calyx 4-lobed, the tube about 2 mm long, with long tawny hairs outside, sericeous inside, all lobes prominently linear-lanceolate, about twice the tube length; petals dull purplish-white, the vexillum straight, 4–6 mm long, whitish, the claw about 2 mm long, the wings oblanceolate, with a distinct, sharp auricle, the keels oblong-falcate; vexillary stamen usually united to the staminal sheath; ovary sessile, about 3–5-ovulate, densely pubescent. Fruit oblong, compressed, 2–2.7(–3.5) cm long, 4–6 mm wide, 2 mm thick, septate, with adpressed or spreading hairs, 3–5-seeded; sutures indistinct. Seeds oblong-truncate, about 3.5 × 2.5 × 1.5 mm, black, shiny. Chromosome number 2n = 22 or 44.

Distr. The Ceylon taxon occurs in India, Ceylon, Malaya, Java, and New Guinea.

Ecol. Flowers November and December. Trimen notes the species is common in the low country, where it appears sometimes cultivated as a pulse.

Vern. Goradiya (S).

Specimens Examined. LOCALITY UNKNOWN: *Macrae 322* (BM); in 1854, *Thwaites C.P. 1477* (BM, P); *Walker s.n.* (PDA); *Col. & Mrs. Walker s.n.* (K). KANDY DISTRICT: Peradeniya, Nov. 1882, *Trimen ? C.P. 1477* (PDA). NUWARA ELIYA DISTRICT: Hanguranketa ?, Dec. 1851, *Thwaites ? C.P. 1477* (PDA). BADULLA DISTRICT: below Ella, *Fosberg & Sachet 53184* (US).

Notes. F.J. Herm. (1962) in his revision of *Glycine* divided '*G. javanica*' into several subspecies and varieties.

Verdcourt (1966) discovered the type of *Glycine, G. javanica* L. to be a *Pueraria.* Subsequently *Glycine* was conserved with *G. clandestina* as the type.

Glycine javanica auct. mult., non L., sensu F.J. Herm. (1962), was encompassed by *G. wightii* (Grah. ex Wight & Arn.) Verdcourt (1966, 1971).

Lackey (1977) subsequently segregated the *G. wightii* taxa, including *Glycine petitiana* (A. Rich.) Schweinf., into the genus *Neonotonia.* Some cytological and chemical evidence has been produced to justify this separation (Hymowitz & Newell, 1981). (See also *Glycine.*)

No nomenclatural changes are necessary in following Lackey, as *Neonotonia wightii* subsp. *wightii* var. *wightii* is the only taxon found in Ceylon.

My description is primarily based on specimens examined in the Madras Herb. (COIMB). I found the flowers to be closer to the 4 mm end of the range, but as F.J. Herm. (1962) and Verdcourt (1971) note, variation is extensive. The corolla is whitish, the vexillum may have a mauve spot, and the wings a purplish tinge, turning reddish or bright orange-red upon drying.

13. GLYCINE

Willd., Sp. Pl. 3(2): 1053. 1802, nom. cons; F.J. Herm., (Revision) Techn. Bull. USDA. 1268: 9. 1962 (as *Glycine* L. p.p.); Verdcourt, Taxon 15(1): 34. 1966; Verdcourt, Kew Bull. 24(2): 256. 1970; Newell & Hymowitz, Amer. J. Bot. 65(2): 168–179. 1978; Hymowitz & Newell in Summerfield & Bunting, Adv. Leg. Sci. 251. 1980 (& sect. 5. *Glycine*); Lackey in Polhill & Raven, Adv. Leg. Syst. Part 1. 318. 1981. Type species: *G. clandestina* Wendland.

Soja Moench, Meth. Pl. 153. 1794 ("Soia"), index ("Soja") noms. rejic.; Lackey, Taxon 27: 560. 1978; Lackey, Taxon 31: 541. 1982. Type: *Soja hispida* Moench (*Dolichos soja* L.).

Twining perennial vines (except *G. max* an annual, erect herb). Leaves trifoliolate, pinnate or digitate; stipulate and stipellate. Inflorescences axillary, racemose, each node 1-flowered, or rarely in sessile fascicles in the lower axils, without distinct tubercles; bracteate and bracteolate. Flowers small, white or blue or purplish, long-clawed; calyx 5-lobed; vexillary petal glabrous, suborbicular to obovate or rhomboid, the wings narrow, somewhat adherent to the keels, the keels nearly as long as the wings; stamens monadelphous or the vexillary separating with age, the anthers uniform, glabrous; ovary several–many-ovulate, the style short, slender, somewhat incurved, glabrous, the stigma small, capitate, terminal, glabrous. Fruit dehiscent, linear or oblong, subterete, the exocarp pubescent, nonconstricted between the ⸜eds, 2–8-seeded; sutures slightly swollen or indistinct. Seeds ovoid-oblong, horizontal, the hilum minute, elliptic, usually with a small, scale-like persistent aril.

With Hermann's (1962) subg. *Glycine* moved to *Neonotonia* Lackey, subgenus *Leptocyamus* (Benth.) F.J. Herm. becomes subg. *Glycine* which includes the type of the genus. Those six species are primarily Australian, but extend to the S. Pacific Islands, Formosa, the Philippines and Micronesia.

The second subgenus, *Soja* (Moench) F.J. Herm., of about three "species," includes the cultivated soybean and occupies an Asian distribution.

Glycine max (L.) Merr., Interp. Rumph. Herb. Amboin. 274. 1917; F.J. Herm., Techn. Bull. USDA. 1268: 39, fig. 12. 1962 (cum syn.); Burkart, Leg. Argentinas ed. 2, 382, fig. 133. 1952; Verdcourt, Kew Bull. 24(2): 256. 1970; Verdcourt, Man. New Guinea Leg. 492, fig. 117. 1979. Lectotype: Cliff. specimen in Linn. Herb. chosen by Piper and Prain, not seen.

Cadelium Rumph., Herb. Amboin. 5: 388, t. 140. 1747–1750.

Phaseolus max L., Sp. Pl. 725. 1753; L., Fl. Zeyl. No. 280. 1747; Moon, Cat. 52. 1824; Trimen, Handb. Fl. Ceylon 2: 72. 1894 (nomen = *Vigna radiata* (L.) Wilczek); Prain, J. Asiat. Soc. Bengal 66(2): 422. 1897.

Dolichos soja L., Sp. Pl. 727. 1753; L., Fl. Zeyl. No. 534. 1747; Roxb., Fl. Ind. 3: 314. 1832.

Soja hispida Moench, Meth. 153. 1794; Wight & Arn., Prod. 1: 247. 1834.

Glycine soja Benth., J. Linn. Soc. Bot. 8: 266. 1833, non *Glycine soja* Sieb. & Zucc. in Abh. Math.-Phys. Kl. Königl. Bayer. Akad. Wiss. 4(2): 199. 1845; Baker in Hook. f., Fl. Br. Ind. 2: 184. 1876 (nomen, fide Prain); Backer & Bakh. f., Fl. Java 1: 625. 1963.

Glycine hispida (Moench) Max., Bull. Acad. Imp. Sci. Saint-Pétersbourg 18: 398. 1873; Baker in Hook. f., Fl. Br. Ind. 2: 184. 1876 (sensu *G. hispida*, fide Prain); Prain, J. Asiat. Soc. Bengal 66(2): 403. 1897.

Soja max (L.) Piper, J. Amer. Soc. Agron. 6: 84. 1914.

Annual herbs to about 1 m tall; stems erect, bushy, terete, with spreading hairs. Leaf rachis 6–16 mm long, 1/5 or less the petiolar length; stipules acute, 3–7 mm long, persistent; stipels setaceous, 1–3.5 mm long, persistent; leaflets broadly ovate, suborbicular, oval or elliptic-lanceolate, 6–10 × about 3 cm, acute or occasionally obtuse apically, cuneate or rounded basally, pubescent with lax hairs to glabrescent. Inflorescences axillary, shorter than the leaves, few-flowered, usually 1 flower at a node; bracts lanceolate, striate, 4.5–5.5 mm long; bracteoles linear-lanceolate or setaceous, 2–3 mm long. Flowers about 7 mm long, the pedicels less than 1 mm long; calyx tube about 3 mm long, with long ascending hairs, all lobes acute-acuminate, the upper deeply emarginate, the lower longest, nearly equalling the tube; petals white, pink or bluish to purple, the vexillum squarish, the lamina about 4 × 4 mm, the claw blunt, about 1 mm long, the wings obovate to oblong, the lamina about 3 × 2 mm, with a distinct auricle, the claw about 2 mm long, the keels somewhat oblong, the claw about 2 mm long; vexillary stamen usually united to the staminal sheath; ovary sessile, about 3-ovulate, pubescent. Fruit tardily dehiscent, turgid, from 2–7 cm long, 8–15 mm wide, 6 mm thick, hirsute, septate, the exocarp somewhat depressed between the seeds,

2–3-seeded; sutures indistinct, the upper usually ending in a short beak. Seeds 6–11 × 5–8 × 4 mm, yellow, tan, olive, green or black, occasionally bicoloured, the hilum less than 1 mm long. Chromosome number $2n=40$, some counts 38 or 80.

Distr. Probably a native of Asia, cultivated worldwide from tropical to temperate regions, occasionally found as an escape, but apparently never persisting in the wild state.

Vern. Soy bean, Soya bean (E); Boo-mae (S).

Specimen Examined. COLOMBO DISTRICT: *Hermann* Mus. *22* in BM.

Notes. See Prain (1897), Piper (1914), F.J. Herm. (1962), Backer & Bakh. f. (1963), and Hymowitz & Newell (1980) for discussions of the assignment of names to cultivated and wild taxa. See also Summerfield & Bunting [eds.], Adv. Leg. Sci. 251–335. 1980, for references and current work on soya beans.

Moon's (1824) list states 'cultivated' and I assume Hermann's collection was either cultivated or an escape.

Description based on *Hermann,* Mus. 22 in BM.

14. TERAMNUS

P. Br., Hist. Jamaica 290. 1756; Verdcourt, Kew Bull. 24(2): 263–284. 1970. Type species: *T. volubilis* Swartz.

Glycine L. p.p., Wight & Arn., Prod. 1: 208. 1834, non *Glycine* Willd., nom. cons.

Twining, climbing vines; stems pubescent. Leaves trifoliolate; stipulate and stipellate; leaflets usually elliptic. Inflorescences axillary, pseudoracemose, many–numerous-flowered, without tubercles; bracts and bracteoles linear-lanceolate or lanceolate, persistent. Flowers minute, in pairs or fascicles; calyx upper lobe bifid or entire; petals white, pink or purplish, of about equal length, the vexillum spreading, non-reflexed, the wings sometimes with pubescent auricles, sometimes with a distal tooth-like structure from the lamina, adherent to the keels; stamens monadelphous, the anthers dimorphic, 5 + 5; ovary many-ovulate, the style short, thick, persistent, slightly upcurved, glabrous, the stigma capitate, glabrous. Fruit dehiscent, linear, compressed, dark, the beak upcurved, several to many-seeded; upper suture occasionally raised, the lower slightly swollen or indistinct. Seeds oblong, truncate on each end, the hilum oval or elliptic.

A pantropical genus of about eight species.

KEY TO THE SPECIES

1 Hairs of the fruit sparsely adpressed or fruit glabrescent; seeds smooth; upper calyx lobes usually bifid. . . . . **1. T. labialis**

1 Hairs of the fruit erect strigose or spreading, about 1 mm long; seeds roughened; upper calyx lobes usually connate. . . . . **2. T. mollis**

1. Teramnus labialis (L. f.) Spreng., Syst. Veg. 3: 235. 1826; Benth., J. Linn. Soc. Bot. 8: 265. 1865; Baker in Hook. f., Fl. Br. Ind. 2: 184. 1876; Trimen, Handb. Fl. Ceylon 2: 60. 1894 (p.p.); Gamble, Fl. Pres. Madras 1(2): 352. 1918; Ridley, Fl. Malay Penins. 1: 570. 1922; Backer & Bakh. f., Fl. Java 1: 626. 1963; Verdcourt, Kew Bull. 24(2): 266. 1970 (cum syn.); Verdcourt, Fl. Trop. East Africa, Papil. 535. 1971; Verdcourt, Man. New Guinea Leg. 496, fig. 119. 1979. Type: *Herb. Linnaeus 901.14* (LINN holotype), not seen, fide Verdcourt (1971).

Glycine labialis L. f., Sp. Pl. 325. 1781; Roxb., Fl. Ind. 3: 318. 1832; Wight & Arn., Prod. 1: 208. 1834; Wight, Ic. Pl. Ind. Or. t. 168. 1839; Thw., Enum. Pl. Zeyl. 88. 1859.
Glycine parviflora Lam., Enc. 2: 738. 1798; Moon, Cat. 53. 1824.

Stems slender with adpressed to ascending canescent pubescence. Leaf rachis 4–12 mm long, the petioles 2.0–4.5 cm long, sparsely pubescent; stipules acuminate, about 2 mm long, persistent; stipels lanceolate, about 1 mm long, persistent; leaflets usually elliptic, occasionally ovate or elliptic-lanceolate, rarely suborbicular or orbicular, pale, to about 7 × 3.5 cm, usually acute or occasionally obtuse or rounded apically, usually rounded basally or similar to the apex, glabrous or with sparse adpressed pubescence above, adpressed canescent pubescence beneath. Inflorescences 1–9 cm long, usually single, occasionally 2 or more in a cluster; bracts linear-lanceolate, striate, about 1 mm long; bracteoles linear, about 2 mm long. Flowers about 4 mm long, the pedicels less than 2 mm long; calyx tube to about 2 mm long, with long pubescence, all lobes acuminate, the longest about equalling the tube; vexillary petal obovate, simple, about 4 mm long, the claw about 1 mm long, white, purplish (or pink?), the wings with the lamina narrow-oblong, about 2.5 mm long, the claw about 1.4 mm long, the keels oblanceolate, the lamina about 2.5 mm long, the claw 1.5 mm long; inner alternate anthers small, sterile; ovary sessile, the stigma terminal. Fruit to about 4.5 cm long, 3 mm wide, 1.5 mm thick, with adpressed pubescence, (3–)7–12-seeded; both sutures slightly swollen, the upper margin with a persistent upcurved beak. Seeds smooth, yellow-brown to dark brown, about 2.5 × 1.5 × over 1 mm, the hilum minute, about 0.5 mm long, frequently with a persistent piece of rim-aril. Chromosome number 2n = 28 (& 20).

Distr. Tropical and South Africa, India, Ceylon, Malaya, Indo-China, Philippines, New Guinea, Guam, E. Indian Ocean islands, West Indies, and Guyana.

Ecol. Common in the low country. The species flowers from November through at least June.

Vern. Wal Kollu (S).

Specimens Examined. LOCALITY UNKNOWN: *Thwaites C.P. 1471* (K). JAFFNA DISTRICT: Causeway btwn. Jaffna and Kayts, *Rudd 3274* (K, PDA, US); btwn. Jaffna and Elephant Pass, *Rudd 3288* (K, PDA, US).

ANURADHAPURA DISTRICT: btwn. Dambulla and Kekirawa, *Rudd 3253* (K, PDA, US). POLONNARUWA DISTRICT: Sacred area, *Dittus WD70062701* (US). MATALE DISTRICT: Dambulla, *Amaratunga 520* (PDA), *Simpson 9801* (BM); btwn. Nalanda and Dambulla, *Alston 2405* (PDA); btwn. Naula and Dambulla, *Robyns 7328* (PDA, US); Nalanda, *Simpson 9230* (BM). KANDY DISTRICT: Peradeniya, *Gardner C.P. 1471* (PDA). MONERAGALA DISTRICT: nr. Wellawaya, *Maxwell & Jayasuriya 760* (PDA, US); btwn. Koslanda and Wellawaya, *Rudd & Balakrishnan 3209* (PDA, US). MATARA DISTRICT: nr. Matara, *Rudd 3088* (PDA, US). HAMBANTOTA DISTRICT: Ruhuna Natl. Pk., *Cooray 6901121R* (US), *70032513R* (PDA, US), *Mueller-Dombois 69010717* (US); rd. to Kataragama, *Rudd 3098* (K, PDA, US).

2. Teramnus mollis Benth., J. Linn. Soc. Bot. 8: 265. 1865; Gamble, Fl. Pres. Madras 1(2): 352. 1918; Alston in Trimen, Handb. Fl. Ceylon 6: 79. 1931; Alston, Kandy Fl. 30, fig. 155. 1938; Verdcourt, Kew Bull. 24(2): 276. 1970. Type: India, Madras: Courtallum, *Herb. Wight 222* (K lectotype).

Glycine mollis Wight & Arn., Prod. 1: 209. 1834 (p.p.); Thw., Enum. Pl. Zeyl. 88. 1859 (p.p.).
Teramnus labialis (L.f.) Spreng. var. *mollis* Baker in Hook. f., Fl. Br. Ind. 2: 184. 1876; Trimen, Handb. Fl. Ceylon 2: 60. 1894 (forma).

Stems slender, with long ascending hairs; stipules acuminate, persistent; stipels lanceolate, persistent; leaflet elliptic to broadly lanceolate, the terminal elliptic to somewhat rhomboid, to 15 cm long (fide Gamble) but usually smaller, acute apically, rounded basally, both surfaces canescent pubescent, dense beneath. Inflorescences clustered, densely villous; bracts striate. Flowers probably to about 4 mm long, the pedicels less than 2 mm long; calyx appearing 4-lobed, with long pubescence; vexillary petal obovate, the wings narrow-oblong, long-clawed, the keels oblanceolate, long-clawed; ovary sessile, the stigma terminal. Fruit to about 4.5 × 0.4 cm, villous, with ascending spreading, tawny hairs over 1 mm long, about 8-seeded; both sutures slightly swollen, the upper margin ending in an upcurved beak. Seeds roughened when mature, dark, the hilum minute, elliptic, frequently with a persistent piece of rim-aril. Chromosome number 2n = 28.

Distr. India, Ceylon, East Pakistan, Burma and Indonesia.
Ecol. Common in the low country, and probably flowering with *T. labialis*, November through June.
Vern. Probably Wal Kollu (S), the same as *T. labialis*.
Specimens Examined. LOCALITY UNKNOWN: 1819, *Moon s.n.* (BM); *Thwaites C.P. 1471* (BM p.p., GH); *Walker "N. 308"* (K, mixture); *s. coll. s.n.* "Kandy" as *Glycine sp. nov.* (BM). KANDY DISTRICT: Patiyagama, *Rudd & Jayasinghe 3248* (PDA, US).
Notes. This genus is easily distinguished in Ceylon by its small flowers on

an atuberculate rachis, the 5 + 5 dimorphic anther condition, and the linear fruit with upturned beaks.

Species within the genus are not easily separable using conventional characters (Verdcourt 1970, 1971). I agree after trying to separate *T. mollis* from *T. labialis* using Ceylon and COIMB specimens. The best *T. mollis* character seems to be the long, spreading hairs on the fruit, not at all like 'typical' *T. labialis*. Besides the characters in the Key to the Species, Alston (1931) notes *T. mollis* possesses longer, lanceolate-elliptic leaflets, and shorter, more crowded, densely villous inflorescences than *T. labialis*.

15. SHUTERIA

Wight & Arn., Prod. 1: 207. 1834, nom. cons.; Van Thuan, Adansonia Ser. 2. 12(2): 291–305. 1972 (Revision). Type species: *S. vestita* Wight & Arn., typ. cons.

Twining vines. Leaves trifoliolate; stipules and stipels persistent; leaflets orbicular or ovate. Inflorescences axillary, racemose, many-flowered, tubercles absent; bracts acute to lanceolate, striate, persistent; bracteoles acuminate, persistent. Flowers small; calyx 4-lobed, the upper lobes entire, connate, the lower somewhat longer; petals pinkish, red to purple, the vexillum spreading, striate, simple, the wings and keels equal in length, slightly shorter than the vexillum; stamens diadelphous, the vexillary free, the anthers uniform; ovary several–many-ovulate, the style geniculate, the upper style glabrous, the stigma terminal, capitate, glabrous. Fruit dehiscent, linear-oblong, flat, with a persistent down-curved beak, 5–11-seeded; upper and lower sutures slightly swollen. Seeds dark, reniform or ovoid, flat, the hilum orbicular, minute.

A genus of five species of India, Ceylon, and extending to East Asia, the Philippines, Malaysia, and Indonesia.

Shuteria vestita Wight & Arn., Prod. 1: 207. 1834; Wight, Ic. Pl. Ind. Or. t. 165. 1839; Thw., Enum. Pl. Zeyl. 88. 1859; Baker in Hook. f., Fl. Br. Ind. 2: 181. 1876 (vars. non var. *involucrata*); Trimen, Handb. Fl. Ceylon 2: 58. 1894; Prain, J. Asiat. Soc. Bengal 66(2): 401. 1897; Gamble, Fl. Pres. Madras 1(2): 350. 1918; Backer & Bakh. f., Fl. Java 1: 625. 1963; Van Thuan, Adansonia ser. 2. 12(2): 291–305, pl. 1–5. 1972. Type: India, Dindygul Hills, *Wallich 5512* (K).

Glycine vestita Grah. in Wall., Cat. 5512. 1831, nom. nud.
Shuteria glabrata Wight & Arn., Prod. 1: 207. 1834.
Shuteria densiflora Benth. in Miq., Pl. Jungh. 232. 1852.
Shuteria vestita Wight & Arn. var. *densiflora* (Benth.) Baker in Hook. f., Fl. Br. Ind. 2: 182. 1876.
Shuteria vestita Wight & Arn. var. *glabrata* Baker in Hook. f., Fl. Br. Ind. 2: 182. 1876.

Climbing or creeping vines; stems slender, terete, ferruginous hirsute when young; stipules lanceolate, about 4 mm long; stipels linear or setaceous, about 2 mm long, exceeding the petiolules; leaflets membranous, 2–3.5 × 1.5–3 cm, pubescent, obtuse apically, cuneate to rounded basally, the terminal sometimes rhomboidal. Inflorescences 2–10 cm long, florate almost to the base; bracts lanceolate, about 2 mm long or longer; bracteoles 2 mm long. Flowers about 10 mm long, the pedicels about 2 mm long; calyx tube about 3 mm long, hirsute, all lobes acuminate, the lower longest, nearly equalling the tube; vexillary petal obovate, about 10 mm long, glabrous, with the claw short, the wings and keels suboblong, the laminas 4–6 × 2 mm, the claws about 3 mm long; ovary about 6-ovulate, hirsute, the upper style about 2 mm long. Fruit about 3 cm long, 4–5 mm wide, 2 mm thick, straight or slightly falcate, light brown, hirsute becoming glabrescent, non-septate, (2–)5–6(–7)-seeded. Seeds 2.5–3.0 × 1.5–2.0 × 1 mm, horizontal. Chromosome number $2n = 22$.

Distr. Throughout the genus range.

Ecol. Flowers August (and perhaps as early as June) through March. Found in the upland regions (1200 to 2100 m).

Specimens Examined. LOCALITY UNKNOWN: March 1836, *s. coll.,* "Hb. Wight," (K); *Gardner C.P. 661* (PDA); 1819, *Moon s.n.* (BM); *Thwaites* ? *C.P. 661* (PDA); 1854, *Thwaites C.P. 661* (BM); *Col. Walker s.n.* (PDA, several sheets). KANDY DISTRICT: Gampola, *Amaratunga 1536* (PDA). NUWARA ELIYA DISTRICT: Below Hakgala, March 1921, *A de A (de Alwis* ?) *s.n.* (PDA); Maskeliya, *C.F. & R.J. van Beusekom 1524* (PDA, US); Ramboda Falls, *Fairchild & Dorsett 379* (US); "Newera Ellia," Jan. 1846, *Gardner s.n.* (K); "N. Ella, Ramboddi" (Ramboda ?), *Gardner 661* (PDA); McDonald's Valley, below Hakgala, *Rudd & Balakrishnan 3165* (K, MO, PDA, US); Hakgala, climbing in jungle, *Simpson 9126* (BM); Patana, way to Ft. MacDonald, 8.10.06, *A.M.S. (Smith* ?) *s.n.* (PDA); Patana, Hakgala, 1.10.06, *A.M.S. (Smith* ?) *s.n.* (PDA), *J.C.W. (Willis* ?) *s.n.* (PDA); Maturata, Sept. 1850, *(Thwaites* ?) *C.P. 661* (BM, PDA). BADULLA DISTRICT: Haputale, *A. de Alwis 11* (PDA); Bolgandavela, Uma oya, *J.M. Silva 234* (PDA); Passara, Uva, Jan. 1888, *(Trimen* ?) *s.n.* (PDA).

Notes. This species is easily distinguished by the persistent bracts, being florate almost to the base of the peduncle and the blackish seeds with minute orbicular hila. Ceylon specimens examined have smaller fruit than reported by the authors cited. The flower colour varies throughout the species range. Van Thuan (1972) reports the corolla red; Backer and Bakh. f. (1963) report the standard white or pale pink with reddish violet veins, and the wings and keels reddish violet. The axillary inflorescences may be single or in clusters of 2 or 3.

16. DUMASIA

DC., Ann. Sci. Nat. Bot. 4: 96. 1825. Type species: *D. villosa* DC.

Twining, herbaceous vines; stems terete. Leaves trifoliolate; stipules and stipels persistent; leaflets ovate. Inflorescences axillary, racemose, several–numerous-flowered, atuberculate; bracts and bracteoles small, persistent. Flowers small to medium, the pedicel attachment under the calyx; calyx obliquely truncate, without prominent lobes; petals shades of yellow, long-clawed, glabrous, the vexillum spreading, simple, slightly exceeding the wings and keels, the latter about equal lengths; stamens diadelphous, glabrous, the vexillary free, the anthers uniform; ovary few–many-ovulate, the lower style bent basally, flattened medially, glabrous or sometimes pubescent, the upper style geniculate at the base, glabrous, the stigma capitate, glabrous. Fruit dehiscent, linear-oblong, torulose, with a small persistent beak, few–many-seeded; upper and lower sutures slightly swollen or indistinct. Seeds subglobose, the hilum oblong-orbicular to elliptic, without an aril.

A genus of about eight species of the Old World tropics.

Dumasia villosa DC., Ann. Sci. Nat. Bot. 4: 97. 1825; DC., Mem. Leg. 257, t. 44. 1826; Verdcourt, Fl. Trop. East Africa, Papil. 513, fig. 74. 1971; Verdcourt, Man. New Guinea Leg. 488, fig. 120. 1979. Type: Nepal, *Wallich* (G holotype, K isotype), fide Verdcourt (1971), not seen.

Perennial vines; stems to about 3 m long, pubescent or glabrous-glabrescent. Leaf rachis 5–20 mm long, 1/4–1/5 the petiolar length; stipules linear, 2–5 mm long; stipels setaceous, 1–2 mm long; leaflets to about 6 × 5 cm, densely pubescent to glabrous, obtuse, rounded or slightly emarginate apically, broadly cuneate or truncate basally. Inflorescences usually equal to or shorter than the leaves, 4–30-flowered; bracts linear-lanceolate, about 2 mm long; bracteoles linear, about 1 mm long. Flowers about 1.5 cm long, borne on sub-basal pedicels 2–4 mm long; calyx tube 7–9 mm long, glabrous or pubescent; vexillary petal broadly oblanceolate, emarginate, basally biauriculate, the lamina about 14 mm long, the claw about 6 mm long, the wings and keels somewhat oblong, the lamina about 4 × 2–3 mm, the claws about 10 mm long; vexillary stamen wider at the base; ovary about 4-ovulate, glabrous or pubescent, the upper style widening into a flat plate about 2 mm below the stigma. Fruit somewhat flattened, 2–4 × about 0.6 × 0.5 cm, glabrous or pubescent, light brown, 1–4-seeded; upper suture slightly swollen, the lower indistinct. Seeds about 6 × 4 × 4 mm, black, the hilum about 1.5 mm long. Chromosome number 2n = 20.

Distr. A montane species of East Africa, Malagasy Republic, India, Ceylon, China, Taiwan, and parts of Malaysia and Indonesia.

KEY TO THE VARIETIES

1 Fruit densely hairy; stems pubescent with erect hairs; leaflets densely pubescent below......
....**1.** var. **villosa**
1 Fruit glabrous; stems glabrous or with a few adpressed hairs; leaflets glabrous or sparsely pubescent below...**2.** var. **leiocarpa**

1. var. villosa Baker in Hook. f., Fl. Br. Ind. 2: 183. 1876; Gamble, Fl. Pres. Madras 1(2): 350. 1918; Backer & Bakh. f., Fl. Java 1: 624. 1963; Verdcourt, Fl. Trop. East Africa, Papil. 513, fig. 74. 1971.

Dumasia pubescens DC., Mem. Leg. 257, t. 44. 1826. Type: *Wallich,* not seen.
Dumasia congesta Grah. in Wall., Cat. 5524. 1832, nom. nud.
Dumasia congesta Wight & Arn., Prod. 1: 206. 1834. Type: Nepal, *Wallich 5524* (K), not seen. Based on Wight, Ic. Pl. Ind. Or. t. 445. 1840–1843, *Wight 749* (COIMB).

Stems usually covered with erect, ferruginous hairs. Leaflets sparsely pubescent above, adpressed to ascending hairs beneath about 1 mm long. Calyx tube sparsely pubescent; ovary and lower style pubescent. Fruit densely covered with canescent-fulvous hairs about 0.5 mm long.

Distr. A montane species of Nepal, India, East Africa, Java and the Malagasy Republic.

Notes. Trimen reports that the ordinary form of the species has densely hairy stems and leaves which are very pubescent beneath; but this condition has not been found in Ceylon. Verdcourt characterizes var. *villosa* as having densely hairy pods. I have seen no specimens from Ceylon that can be assigned with certainty to the typical variety.

2. var. leiocarpa (Benth.) Baker in Hook. f., Fl. Br. Ind. 2: 183. 1876; Trimen, Handb. Fl. Ceylon 2: 58. 1894.

Dumasia leiocarpa Benth. in Miq., Pl. Jungh. 2: 231. 1852. Type: not seen.
Dumasia villosa DC. var. β Thw., Enum. Pl. Zeyl. 88. 1859. Based on *C.P. 663* (BM, K, P), Ceylon.

Stems glabrous or with a few adpressed hairs. Leaflets glabrous or very sparsely pubescent beneath. Calyx tube, ovary, lower style, and fruit all glabrous.

Distr. Ceylon, India, Malaysia and Java.

Ecol. Flowering February and March. Common in upper montane zone, fide Trimen.

Specimens Examined. LOCALITY UNKNOWN: *Thomson s.n.* (K); 1854, *Thwaites C.P. 663* (BM, K, P 3 sheets); *Col. Walker s.n.* (PDA), *Walker s.n.* (P), *79* (K); *Col. & Mrs. Walker s.n.* (K, P). KANDY DISTRICT: Hanguranketa, *Simpson 9207* (BM). NUWARA ELIYA DISTRICT: Newahette (Hewaheta?), Feb. 1883, *s. coll.* (Trimen?) *s.n.* (PDA); Rd. to Ambawela, Mar. 1922, *de Alwis s.n.* (PDA); "N. Ellia," *Gardner 210* (BM, K); *s. coll.,* Herb. R. Wight, "in hb. Kew" (Gardner?) *210* (PDA); jungle margin, Hakgala, *Simpson 9043* (BM); jungle back of Hakgala, *s. coll.,* 1.3.06, *s.n.* (PDA); 26.2.06, *A.W.S.? s.n.* (PDA); Ramboddi (Ramboda?), 1851, *Thwaites? s.n.* (PDA).

Notes. The genus is easily distinguished by the oblique-truncate calyx, the pedicel attachment, and the few dark seeds in the torulose fruit.

Description based on *Wallich 5523, 5523a, 5523b* (BM).

17. CENTROSEMA

(DC.) Benth., Comm. Leg. Gen. 53. 1837, nom. cons.; Hutchinson, Gen. Fl. Pl. 1: 446. 1964; Barbosa-Fevereiro, Rodriguesia 42: 159–219. 1977 (Revision, Brazil). Type species: *C. brasilianum* (L.) Benth. (*Clitoria brasiliana* L.), typ. cons.

Clitoria sect. III, *Centrosema* DC., Prod. 2: 234. 1825.
Bradburya Raf. in Kuntze, Rev. Gen. Pl. 164. 1891, non *Bradburya* Raf., Fl. Ludov. 104. 1817.

Twining, climbing vines. Leaves trifoliolate; stipulate; stipellate; leaflets usually ovate. Inflorescences racemose, mostly congested, axillary, 1–several-flowered; bracteate; bracteoles conspicuous. Flowers large; calyx campanulate, 5-lobed, the upper lobes partially connate; petals shades of blue or violet, the vexillum longest, slightly reflexed, usually spurred basally, variously pubescent, the wings free, narrow, the keels semiorbicular; vexillary stamen free or slightly adherent, the anthers uniform; ovary many-ovulate, the style upcurved, somewhat flattened or spathulate distally, the stigma terminal, pubescent beneath. Fruit spirally dehiscent, linear, septate, compressed, acuminate apically, terminating in a long, straight, persistent beak, many-seeded; sutures thickened to either side and with narrow ribs. Seeds oblong to orbicular, horizontal, the hilum oblong, or somewhat oval-elliptic.

A native tropical American genus of 35–50 species.

KEY TO THE SPECIES

1 Bracteoles about equal the calyx length; terminal leaflets broadly lanceolate, ovate or elliptic, rounded basally; lower calyx lobe linear, greatly exceeding the rest; fruit about 6 mm wide . . . .
. . . . **1. C. pubescens**
1 Bracteoles greatly exceeding the calyx length; terminal leaflets rhomboidal or somewhat oblong, broadly cuneate basally; lower calyx lobe acute, slightly exceeding the rest; fruit about 12 mm wide
. . . . **2. C. plumieri**

1. Centrosema pubescens Benth., Comm. Leg. Gen. 55. 1837; Benth. in Mart., Fl. Bras. 15(1): 131, t. 34, fig. II. 1859; Fawcett & Rendle, Fl. Jamaica 4: 45, fig. 12. 1920; Alston in Trimen, Handb. Fl. Ceylon 6: 83. 1931; Amshoff in Pulle, Fl. Suriname 2(2): 18. 1939; Backer & Bakh. f., Fl. Java 1: 624. 1963; Verdcourt, Fl. Trop. East Africa, Papil. 520, fig. 76. 1971; Verdcourt, Man. New Guinea Leg. 501, fig. 122. 1979; D'Arcy, Ann. Missouri Bot. Gard. 67(3):

576. 1980. Type: *Keerle* in Herb. Martius, Tlalpuxahua, Mexico (M holotype), fide Verdcourt (1971), not seen.

Clitoria plumieri Tuss., Fl. Ant. 4: 79, t. 27. 1827, non Turp.

Centrosema berteriana Vogel, Linnaea 10: 598. 1836, non DC. Type: Probably (B), destroyed.

Bradburya pubescens (Benth.) Kuntze., Rev. Gen. Pl. 1: 164. 1891.

Periandra berteriana auct., non Benth., Trimen, Handb. Fl. Ceylon 2: 75. 1894.

Stems slender, terete, about 3 m long, sparsely pubescent; stipules ovate, 2–4 mm long; stipels filamentous, flat, about 2 mm long. Leaflets lanceolate to broadly ovate, glabrescent above, pubescent beneath, acute to obtuse, or occasionally rounded apically, rounded basally, the terminal 4–8 × 1.5–4.5 cm, larger than the laterals. Inflorescences 2.5–6.0 cm long, the peduncles mostly slightly longer than the petioles; bracts ovate-lanceolate, about 5 mm long, deciduous; bracteoles ovate, about 9 mm long, persistent. Flowers 2–3.5 cm long, the pedicels about 7 mm long; calyx tube 2–3 mm long, glabrous, the upper lobe bifid, shorter than the laterals, the lower longest, linear, about 8 mm long; petals violet, short-clawed, the vexillum orbicular, to about 3 cm long, emarginate, puberulent outside, the wings somewhat falcate, oblanceolate, about 3 cm long, pubescent, the auricle curved back, about 2 mm long, the keels semiorbicular, weakly auriculate, pubescent along the lower margin; vexillary stamen free; style basally pubescent, becoming flattened distally and glabrous, the stigma with a fringed margin. Fruit glabrescent, about 12 cm long, 6 mm wide, 2.5 mm thick, reddish brown, compressed, the beak about 1 cm long, 12–22-seeded. Seeds about 4 mm long, reddish brown with darker mottling, the hilum about 1.5 mm long. Chromosome number $2n = 20$.

Distr. A common roadside species of tropical America now found throughout the tropics worldwide.

Ecol. The species has been introduced into Ceylon and is established throughout the Island. Flowering is through April, October and January.

Specimens Examined. KURUNEGALA DISTRICT: Pilessa, between Kurunegala and Mawatagama, *Rudd 3314* (K, PDA, US); road to Kandy, 4 mi. SE. of Kurunegala, *Fosberg & Jayasuriya 52713* (US); Hiripitiya, *Maxwell & Jayasuriya 779* (PDA, US). KANDY DISTRICT: Aniewatta, *Rudd & Balakrishnan 3060* (K, PDA, US); Kandy, *Rudd 3050* (K, PDA, US); Kandy, Katukelle, Piachaud Gardens, *Rudd & Fernando 3319* (K, PDA, US); Peradeniya, *Comanor 319* (K, PDA, US), *475* (K, PDA, US), *Maxwell & Jayasuriya 841* (PDA, US), *847* (PDA, US). KALUTARA DISTRICT: Pelawatte, *Cramer 2804* (PDA, US). BADULLA DISTRICT: Welimada, *Maxwell & Jayasuriya 887* (PDA, US); between Passara and Lunugala, *Maxwell & Jayasuriya 1022* (PDA, US). MONERAGALA DISTRICT: 2 mi. W. of Diyaluma Falls, *Maxwell & Jayasuriya 762* (PDA, US).

2. Centrosema plumieri (Pers.) Benth., Comm. Leg. Gen. 54. 1837; Benth. in Mart., Fl. Bras. 15(1): 127. 1859; Trimen, Handb. Fl. Ceylon 2: 75. 1894; Fawcett & Rendle, Fl. Jamaica 4: 44. 1920; Urban, Symb. Ant. 8: 302. 1920–1921; Alston in Trimen, Handb. Fl. Ceylon 6: 83. 1931; Amshoff in Pulle, Fl. Suriname 2(2): 184. 1939; Pittier, Bol. Técn. Minist. Agric. No. 5: 72, pl. 44. 1944; Backer & Bakh. f., Fl. Java 1: 624. 1963; Verdcourt in Fl. Trop. East Africa, Papil. 520. 1971; Verdcourt, Man. New Guinea Leg. 499. 1979; D'Arcy, Ann. Missouri Bot. Gard. 67(3): 575. 1980. Type: Based on a plate by Plumier in Desc. Pl. Amer. 94, t. 108. 1757, not seen.

Clitoria plumieri Turp. ex Pers., Syn. Pl. 2: 303. 1807, non Tuss., 1827.
Bradburya plumieri (Pers.) Kuntze., Rev. Gen. Pl. 1: 64. 1891.

Stems robust, terete, glabrous; stipules narrowly ovate, striate, about 7 mm long; stipels filamentous, about 6 mm long; terminal leaflets the larger, about 8 × 7.5 cm, somewhat rhomboid or oblong, abruptly acute apically, cuneate basally, the lateral leaflets rounded to somewhat truncate basally, densely pubescent beneath. Inflorescences about 6 cm long, the peduncles usually shorter than the leaves, the rachis somewhat tuberculate; bracts orbicular, about 7 mm long; bracteoles elliptic-ovate, striate, about 15 mm long, persistent, almost twice as long as the calyx. Flowers about 4.5 cm long, the pedicels about 5 mm long; calyx tube glabrous, the upper lobes very short, the laterals slightly longer, the lower longest, acute, about 8 mm long; petals whitish, reddish or purple inside, short-clawed, the vexillum orbicular, about 3.5 cm long, pubescent outside, the wings somewhat falcate, oblanceolate, about 3 cm long; vexillary stamen free; ovary 12–14-ovulate, the style flattened distally, the stigma with a fringed margin. Fruit about 13 cm long, 12 mm wide, 4–6 mm thick, glabrous, the beak about 3 cm long, about 12-seeded. Seeds about 7 × 6 × 3 mm, the hilum about 4 mm long. Chromosome number $2n = 20$.

Distr. West Indies, Mexico, Central and tropical South America. Introduced into the Old World Tropics.

Ecol. Flowering April and September (Alston, 1931). The species was introduced into Ceylon. Trimen (1894) notes the species is naturalized in many places in the low country, as about Colombo.

Specimens Examined. KANDY DISTRICT: Peradeniya, Sept. 1887, *s. coll.* (Trimen ?) (PDA); April 1891, *s. coll.* (Trimen ?) (PDA); 1 Dec. 1924, *s. coll.* (PDA); 15 Jan. 1925, *s. coll.* (PDA).

Notes. The specimens examined were in the cultivar folders. I am uncertain as to the reasons for the introduction, but in all probability they were either for cover crops or ornamentals.

See Barbosa-Fevereiro (1977) for supplemental illustrations.

18. CLITORIA

L., Sp. Pl. 753. 1753; Benth., J. Linn. Soc. Bot. 2: 33–44. 1858; Hutchinson, Gen. Fl. Pl. 1: 446. 1964; Fantz, Monograph, Ph.D. Thesis, Univ. Fla., U.S.A., 1977; Fantz, Sida 8(1): 90–94. 1979. Type species: *C. ternatea* L.

Ternatea Tourn. ex L., Syst. Nat. ed. 1. 1735; H.B.K., Nov. Gen. Sp. 6: 415. 1823.
Neurocarpum Desv., J. Bot. (Desvaux) 1: 119. 1813, 2: 75. 1814.

Twining, climbing vines or small, erect woody shrubs; pubescence of uncinate (hooked) hairs common. Leaves trifoliolate or 5–7(–9)-foliolate; stipulate and stipellate; leaflets ovate, elliptic, or orbicular. Inflorescences axillary, few-flowered, racemose; bracts 2–6, persistent or caducous; bracteoles 2, striate, persistent. Chasmogamous flowers large, showy, resupinate by twisting of the pedicels; calyx funnelform, 5-lobed, all lobes acute, acuminate; petals various shades of white, blue violet, the vexillum longest, slightly reflexed, spurless, the wings adherent to the keels, the keels shortest; stamens diadelphous, the vexillary free, the anthers uniform; ovary stipitate, several–many-ovulate, densely pubescent, the style straight, bearded, usually twisted, flattened, the stigma terminal, usually pubescent basally. Fruit spirally dehiscent, linear or oblong, compressed, occasionally subterete, stipitate, usually with a persistent downcurved beak, several-seeded; upper suture thickened, occasionally with the ribs 3–5 mm from the upper suture, the lower suture thickened or indistinct. Seeds ovoid or elliptic-oblong, the hilum short, linear or oval-orbicular.

A genus of about 60 species mostly native to tropical America.

See Fantz (Ann. Missouri Bot. Gard. 67(3): 582. 1980) for a description of cleistogamous flowers. My descriptions are of chasmogamous flowers.

KEY TO THE SPECIES

1 Vines; leaves 5–7(–9)-foliolate; fruit flat, 7–15 cm long, the upper ribs adnate to the suture; inflorescences reduced, 1-flowered...**1. C. ternatea**
1 Small erect shrubs; leaves 3-foliolate; fruit subterete, 3–5 cm long, the upper ribs about 3 mm to each side of the upper suture; inflorescences occasionally elongate, 1–few-flowered.......
....**2. C. laurifolia**

1. Clitoria ternatea L., Sp. Pl. 753. 1753; L., Fl. Zeyl. No. 283. 1747; Moon, Cat. 53. 1824; Roxb., Fl. Ind. 3: 321. 1832; Wight & Arn., Prod. 1: 205. 1834; Benth. in Mart., Fl. Bras. 15(1): 118, t. 31, f. 1. 1859; Thw., Enum. Pl. Zeyl. 88. 1859 (as *Clitorea*); Baker in Hook. f., Fl. Br. Ind. 2: 208. 1876; Trimen, Handb. Fl. Ceylon 2: 75. 1894; Gamble, Fl. Pres. Madras 1(2): 365. 1918; Amshoff in Pulle, Fl. Suriname 2(2): 177. 1939; Backer & Bakh. f., Fl. Java 1: 623. 1963; Verdcourt in Fl. Trop. East Africa, Papil. 515, fig. 75. 1971; Verdcourt, Man. New Guinea Leg. 502, fig. 123. 1979; Fantz, Ann. Missouri Bot.

Gard. 67(3): 592. 1980. Type: Herb. Linnaeus (LINN 902.1), not seen.

Ternatea vulgaris H.B.K., Nov. Gen. Sp. 6: 415. 1824.

Vines; stems about 2 m long, terete. Leaves pinnately 5–7(–9)-foliolate, the rachis 3–4 cm long; stipules linear-lanceolate, about 4 mm long; stipels setaceous, about 2 mm long; leaflets ovate to broadly elliptic, 2–4.5 × 1.2–2.5 cm, with sparse, strigulose pubescence, obtuse to sometimes rounded apically, cuneate to rounded basally. Inflorescences much shorter than the leaves, 1-flowered, the peduncles 6–10 mm long; bracts at the base of the pedicels 2–3 mm long, frequently unequal; bracteoles orbicular, 6–10 mm long. Flowers inverted, 3.5–6.5 (?) cm long, the pedicels about 5 mm long; calyx tube about 11 mm long, very sparsely pubescent, all lobes acuminate, of about equal length, 7–9 mm long; petals with the outer edges of the vexillum and wings white to very light violet, or from white to deeper shades of blue and purple in most cultivated forms, the vexillum obovate, emarginate, somewhat reflexed, 3.5–4.5 cm long, puberulent distally, the wings oblong, ½–⅔ the vexillum length, the claw about 8 mm long, the keels shorter than the wings, rostrate, long-clawed; ovary pubescent, the upper style slightly upcurved, pubescent to densely villous apically, the stigma mostly terminal, somewhat cupped, bearded. Fruit sparsely pubescent or glabresent, flat, linear, 7–15 cm long, 6 mm wide proximally to 12 mm wide distally, about 3 mm thick, the exocarp light brown, 8–12-seeded. Seeds ovoid-oblong, about 6 mm long, horizontal, the hilum orbicular. Chromosome number 2n = 16.

Distr. A native of the Old World tropics (Fantz, 1980), now found throughout tropical America, introduced and naturalized elsewhere. A frequent escape from cultivation.

Ecol. Trimen (1894) cites the species as commonly found in the dry regions and perfectly wild, while in the moist regions it is usually cultivated. Flowering is throughout the year. The flowers of cultivars are often much longer than their wild counterparts.

Uses. The species is found on the Seeds List (1958) for purchase as an ornamental blooming throughout the year. The double flowered and white forms are frequently cultivated. *C. ternatea* is also listed as a medicinal herb by Parsons (1937).

Vern. Nil-katarodu, Katarodu-wel (S); Karuttappu (T).

Specimens Examined. WITHOUT LOCALITY: *G. Thomson s.n.* (K). VAVUNIYA DISTRICT: Highway to Mannar near Parayanalankulam, *Theobald & Krahulik 2858* (US). PUTTALAM DISTRICT: Nr. Nochchiyagama, *Bernardi 15361* (US); Wilpattu Region, Puttalam lagoon edge, *Fosberg & Jayasuriya 52810* (US); ocean shore west of Pomparippu, *Maxwell & Jayasuriya 804* (NY, PDA). KURUNEGALÅ DISTRICT: Melsiripura, *Amaratunga 1338* (PDA). MATALE DISTRICT: 10 April, 1924, *Petch s.n.* (PDA). KANDY DISTRICT: Gannoruva, cult., *G.A. Ramanayake & J.E.S. 10040* (PDA); Galagedera,

Amaratunga 954 (PDA), *Rudd 3316* (MO, PDA, US); Kandy, *Rudd with Balakrishnan 3063, 3064* (K, MO, PDA, US); Peradeniya, cult., 10.6.27, *Livera s.n.* (PDA), cult., *J.M. Silva 828* (PDA), *Amaratunga 651* (PDA). BATTICALOA DISTRICT: Batticaloa, *Gardner 1485* (PDA). COLOMBO DISTRICT: Kalagedihena, *Amaratunga 1844* (PDA); Kochchikade, *Simpson 7959* (BM); Mt. Lavinia, *Amaratunga 2068* (PDA). KEGALLE DISTRICT: Dumbuluwa, *Amaratunga 918* (PDA). RATNAPURA DISTRICT: cult., Carney, *Maxwell et al. 952* (NY, PDA). MONERAGALA DISTRICT: Kataragama, *Rudd 3106* (K, MO, PDA, US). HAMBANTOTA DISTRICT: Hambantota, Dec. 1882, (*Trimen ?*) *s.n.* (PDA); Ruhunu (Yala) Natl. Pk., Karaugaswala, *Mueller-Dombois & Cooray 67121046* (PDA, US); Patanagala, *Wirawan 699* (MO, PDA, US).

Note. Description based on Ceylon, *Hermann* Mus 3: 13, 3: 20, in BM.

2. Clitoria laurifolia Poir. in Lam., Enc. Suppl. 2: 301. 1811; DC., Prod. 2: 235. 1825; Backer & Bakh. f., Fl. Java 1: 623. 1963. Type not seen.

Neurocarpum laurifolium (Poir.) Desv. in Ham., Prod. Fl. Ind. Oc. 51. 1825.
Neurocarpum cajanaefolium Presl, Symb. Bot. 1: 17, t. 9. 1830–1831.
Clitoria cajanaefolia (Presl) Benth., J. Linn. Soc. Bot. 2: 40. 1858; Benth. in Mart., Fl. Bras. 15(1): 121. 1859; Baker in Hook. f., Fl. Br. Ind. 2: 209. 1876.

Small shrubs; stems erect, somewhat ridged, 1–1.5 m tall. Leaves mostly trifoliolate, subsessile; stipules acute, about 4 mm long, persistent; stipels linear to filamentous, about 3 mm long, persistent; leaflets elliptic to broadly lanceolate, about 5 × 3 cm, pubescent, emarginate or rounded apically, cuneate basally, adpressed pubescent beneath, nitidous above. Inflorescences usually 1–few-flowered, the rachis 2–5 cm long, bracts at the base of the pedicels acute, frequently unequal, 2 mm long; bracteoles obovate, about 8 × 5 mm. Flowers about 5 cm long, the pedicels about 5 mm long; calyx tube about 12 mm long, the upper lobes obtuse, the rest acute, about 7 mm long; petals light violet, the vexillum suborbicular, emarginate, about 4.5 × 4.5 cm, the claw 3 mm long, the wings obliquely oblong, about 18 mm long, the claw 12 mm long, slightly exceeding the keel length, the keels falcate, rostrate, with the beaks slightly twisted, long-clawed; ovary about 8-ovulate, the style geniculate, the upper style fringed with 2 adaxial rows of hairs, cupped, the stigma flat-capitate. Fruit glabrous, subterete, 3–4.5 × about 1 × 0.7 mm, with a persistent, downcurved beak, septate, about 7-seeded; upper suture with parallel ribs about 4 mm to either side. Seeds ovoid, about 6 × 5 × 2–3 mm, transverse, sticky, the hilum ovoid orbicular, 1 mm long. Chromosome number 2n = 24.

Distr. Throughout tropical America, introduced into Asia and Africa, where it occasionally becomes naturalized.

Ecol. Naturalized in the wet zones, occasional to locally abundant, and apparently blooming throughout the year.

Uses. The species is carried on the Seeds List (1958) as *C. cajanifolia* Benth. probably as an ornamental. It has been cultivated at Peradeniya and Simpson (*8619*, BM) notes it was formerly used as a green manure.

Vern. Andanahiriya (Alston).

Specimens Examined. KANDY DISTRICT: Kandy, *Worthington 4967* (K). COLOMBO DISTRICT: Tarakuliya, S. of Negombo, *Comanor 1149* (PDA, US); SE. corner of Colombo Airport, *Maxwell 1032* (NY, PDA, US). RAT-NAPURA DISTRICT: Hills W. of Weddagala, *Fosberg 56599* (US); Kalawana, *Maxwell et al. 972* (PDA, US); Kuruwita, 5.4.28, *Alston s.n.* (PDA); near Rat-napura, culvert 57/6, *Maxwell & Fernando 984* (NY, PDA, US); Ratnapura-Matugama Road, culvert 20/4, *Cooray 70020807R* (NY, PDA, US). GALLE DISTRICT: Batapola, *Simpson 8619* (BM); Gonapinnuwala, *Cramer 2696* (PDA, US).

Notes. The flowers are light violet, with the back of the vexillum white, a darker violet inside, with light yellow streaks in the target area.

Description based on Presl, Symb. Bot. 1: 17, t. 9. 1830.

19. PSOPHOCARPUS

DC., Prod. 2: 403. 1825; Hutchinson, Gen. Fl. Pl. 1: 442. 1964 (cum syn.); Verdcourt, Fl. Trop. East Africa, Papil. 602. 1971; Verdcourt & Halliday, Kew Bull. 33(2): 191–227. 1978 (Revision, cum syn.), nom. cons. Type species: *P. tetragonolobus* (L.) DC. (*Dolichos tetragonolobus* L.)

Usually vines. Leaves 1-foliolate or pinnately trifoliolate; stipules and stipels persistent, the stipules produced below insertion. Inflorescences axillary, fasciculate-racemose, tuberculate, few–several-flowered; bracteate and bracteolate. Calyx 4–5-lobed, the upper lobes connate, entire or emarginate; petals blue or purple, the vexillum suborbicular, biauriculate, with appendages, glabrous, the wings free, somewhat ovate, the keels obliquely oblong, somewhat beaked; vex-illary stamen free or fused to the staminal sheath, the anthers uniform; ovary linear, ridged, few–numerous-ovulate, the style geniculate, the upper style beard-ed, bent at the apex, the stigma penicillate. Fruit oblong to linear, dehiscent, 4-winged, somewhat septate, few–numerous-seeded. Seeds ovoid or oblong, the hilum short, oval or elliptic, with or without arillate tissue.

An Old World tropical genus of nine species.

Psophocarpus tetragonolobus (L.) DC., Prod. 2: 403. 1825; Wight & Arn., Prod. 1: 252. 1834; Baker in Hook. f., Fl. Br. Ind. 2: 211. 1876; Backer & Bakh. f., Fl. Java 1: 644. 1963; Verdcourt & Halliday, Kew Bull. 33(2): 196, fig. 3. 1978 (cum syn.); Verdcourt, Man. New Guinea Leg. 533, fig. 130. 1979. Type: *Lobus quadrangularis* Rumph., Herb. Amboin. 5: 374, t. 133. 1747, not seen.

Dolichos tetragonolobus L., Syst. Nat. ed. 10, 1162. 1759; Moon, Cat. 53. 1824;
  Roxb., Fl. Ind. 3: 305. 1832.
Dolichos ovatus Grah. in Wall., Cat. 5540. 1831, nom.
Lobus quadrangularis Rumph., Herb. Amboin. 5: 374, t. 133. 1747–1750, fide
  Merr., Interpr. Rumph. Herb. Amboin. 286. 1917.

Twining, climbing vines; stems terete, glabrous. Leaves pinnately trifoliolate,
the rachis 2–5 cm long, the petiole 3–4 times as long, both glabrous; stipules
lanceolate, about 5 mm long above insertion, 1–4 mm below, striate, glabrous;
stipels lanceolate, 3–5 mm long; leaflets ovate, thin, 6–14.5 × 3–9 cm, acuminate
apically, broadly cuneate or rounded or sometimes truncate basally, both surfaces glabrous, gland-dotted beneath, primary lateral veins in about 7 pairs, the
terminal leaflet frequently rhombic or deltoid. Inflorescences about 15 cm long,
each tubercle 1–2-flowered; bracts lanceolate, ovate, striate, about 2 mm long,
glabrous, persistent; bracteoles ovate-orbicular, striate, about 3 mm long, glabrous,
persistent. Flowers about 3 cm long, the pedicels 8–12 mm long; calyx tube to
about 8 mm long, glabrous, the upper lobe shallowly emarginate, obtuse, the
laterals acute, the lower longest, acuminate, about 5 mm long; vexillary petal
reflexed, to about 3 cm long, the wings to about 18 mm long, with an auricle
almost equalling the claw, the keels slightly shorter than the wings; vexillary
stamen usually slightly fused; ovary few–many-ovulate. Fruit linear, to about
18 × 1 cm, glabrous, each wing 5–15 mm wide, with crenate margins, about
14-seeded. Seeds ovoid, dark, about 6 × 5 mm, the hilum oblong-oval, about
3 mm long, without arillate tissue. Chromosome number 2n = 18.

Distr. Origin unknown. Cultivated in the tropics of both hemispheres.
Vern. The Winged Bean, Princess Bean (E); Dara-dambala (S).
Specimens Examined. DISTRICT UNKNOWN: Gonahena, 7.5.17, *s.
coll. s.n.* (PDA); Gannoruwa, "strong climbing bean, with pale blue flowers,"
Amaratunga 2031 (PDA). KANDY DISTRICT: Peradeniya, Royal Botanic
Gardens, Feb. 1898, *s. coll. s.n.* (PDA).
Notes. I include this species on the authority of Moon who lists two varieties,
the white, Sudu-dambala, and the blue, Nil-dambala. While apparently still grown
in gardens for food; it has not, to my knowledge, become naturalized in Ceylon.
A food plant of outstanding promise, its leaves, shoots, flowers, pods and seeds
can be eaten. Certain varieties in Burma and Papua, New Guinea, also produce
potato-like tubers with very high protein.
See "The Winged Bean: A High Protein Crop for the Tropics." Nat. Acad.
of Sci., Washington, D.C., U.S.A., 43 pp., 1975, and "Tropical Legumes:
Resources for the Future," Nat. Acad. Sci., Washington, D.C., U.S.A., 331
pp. 1979. Verdcourt and Halliday (1978) also supply numerous references.
Description based on Hort. Calc. HT *5440 C, D* (K, Herb. Benth. 1854) and
Bourne s.n. (K).

20. LABLAB

Adans., Fam. Pl. 2: 325. 1763; Verdcourt, Kew Bull. 24(3): 380, 409. 1970; Verdcourt, Fl. Trop. East Africa, Papil. 696. 1971; Verdcourt, Taxon 21(4): 533. 1972; Westphal, Taxon 24: 189–192. 1975; Verdcourt in Summerfield & Bunting, Adv. Leg. Sci. 45–48. 1980. Type species: *L. purpureus* (L.) Sweet.

Dolichos sensu auct., non sensu Verdcourt 1971; Int. Code Bot. Nomenclature 352. 1978.

Twining, climbing vines. Leaves tr foliolate; stipules and stipels persistent; leaflets ovate or triangular, the terminal frequently rhomboid. Inflorescences fasciculate-racemose, erect, axillary, tuberculate, many-flowered; bracts caducous; bracteoles persistent. Flowers medium-sized; calyx somewhat 4-lobed, the upper lobes usually entirely connate, the lower longest; petals violet, purple to reddish-purple, occasionally white, the vexillum reflexed with 2 appendages or bicallose, glabrous, the wings free, the keels long-beaked; stamens usually diadelphous, the vexillary free, the anthers 10, uniform; ovary few–several-ovulate, the style geniculate, flattened, the upper style bearded adaxially, straight, the stigma terminal, fairly large, capitate, glabrous. Fruit semilunate, oblong-falcate to oblong, compressed with a small downcurved beak, few–several-seeded; sutures swollen, usually without distinct ribs or wings, the upper suture frequently verrucose. Seeds subreniform or ovoid, compressed, the hilum linear, white arillate.

A widespread monotypic tropical genus. The species has been subdivided into many infraspecific taxa because of its variability under extensive cultivation.

Lablab purpureus (L.) Sweet, Hort. Brit. ed. 1, 481. 1827; Verdcourt, Kew Bull. 24(3): 409. 1970; Verdcourt, Fl. Trop. East Africa, Papil. 696, fig. 104. 1971 (subsp. *uncinatus*); Verdcourt, Man. New Guinea Leg. 537, fig. 131. 1979; D'Arcy & Poston, Ann. Missouri Bot. Gard. 67(3): 714, fig. 31. 1980. Type: not found, see Verdcourt (1971).

[*Cacara alba* Rumph., Herb. Amboin. 5: t. 136, 137, 141(?). 1747–1750.]
Dolichos lablab L., Sp. Pl. 725. 1753; Roxb., Fl. Ind. 3: 305. 1832; Benth. in Mart., Fl. Bras. 15(1): 197, t. 51, fig. 2. 1859; Baker in Hook. f., Fl. Br. Ind. 2: 209. 1876 (excl. p.p. syn.); Trimen, Handb. Fl. Ceylon 2: 76. 1894; Piper, USDA Bull. 318. 1915; Gamble, Fl. Pres. Madras 1(2): 367. 1918; Alston in Trimen, Handb. Fl. Ceylon 6: 83, 84. 1931; Backer & Bakh. f., Fl. Java 1: 643. 1963. Type: *Phaseolus niger lablab* of Alpini from Egypt, 1592, fide Verdcourt (1971).
Dolichos purpureus L., Sp. Pl. ed. 2, 1021. 1763; Sm., Exot. Bot. 29, t. 74. 1806; Moon, Cat. 53. 1824; Roxb., Fl. Ind. 3: 306. 1832 (var.).

Dolichos albus Lour., Fl. Cochinch. 534. 1790; Moon, Cat. 53. 1824; Roxb.,
   Fl. Ind. 3: 306. 1832 (var.).
Lablab vulgaris Savi in Nuovo Giorn. Lett. 8: 116, fig. 8/a–c. 1824; Wight in
   Hook., Bot. Misc. 2: 352, suppl. t. 15. 1831; Wight & Arn., Prod. 1: 250.
   1834; Wight, Ic. Pl. Ind. Or. 1: t. 57. 1838 & 1: t. 203. 1839; Thw., Enum.
   Pl. Zeyl. 90. 1859.
Lablab cultratus DC., Prod. 2: 402. 1825; Wight & Arn., Prod. 1: 251. 1834;
   Trimen, Handb. Fl. Ceylon 2: 76. 1894.
Dolichos lignosus auct. non L., Sp. Pl. 726. 1753; Roxb., Fl. Ind. 3: 307. 1832;
   Prain, J. Asiat. Soc. Bengal 6(2): 430. 1897; Alston in Trimen, Handb. Fl.
   Ceylon 6: 84. 1931.

Vines, usually perennial, about 3 m long; stems terete to slightly angled, sparsely strigulose to glabrescent. Leaf rachis ½–¹/₁₀ the petiolar length; stipules acute,
about 4 mm long; stipels lanceolate, about 4 mm long; leaflets about 8 × 5 cm,
glabrescent, acute-acuminate apically, usually with 3 prominent veins inserted
basally, the primary lateral veins in 3 or 4 pairs, the terminal leaflet sometimes
lobed, the laterals oblique. Inflorescences florate ⅓ to ½ or greater the length,
the tubercles distant, occasionally elongate; bracts soon caducous; bracteoles ovate-
orbicular, about 4 mm long. Flowers reflexed, about 1.5 cm long, the pedicels
about 2 mm long; calyx tube about 3 mm long, carnose, the upper lobe broadly
obtuse, the rest acute, the lower about 2 mm long; vexillary petal suborbicular,
about 12 mm long, wider than long, basally inflexed biauriculate, with 2 double-
eared appendages in the target area, the wings obovate, the keels slender-falcate,
upcurved 90 degrees in the middle, fringed adaxially on the distal edges; ovary
few-ovulate, tomentose, the upper style with the edges incurved, villous. Fruit
dehiscent, 4–5 × about 2 ×0.3 cm, the beak about 5 mm long, 2–4-seeded; upper suture with narrow ribs, the lower indistinct. Seeds extremely variable in
size, about 7 × 5 × 2 mm, vertical to oblique, the hilum about half encircling.
Chromosome number 2n = 20, 22, or 24.

   Distr. Worldwide in the tropics.
   Ecol. Flowers January through June in the high country and probably all year
in the low country.
   Uses. "Natives in Ceylon report fresh flowers from the roadside forms are
mixed with chili peppers and salt, fried and eaten." (For additional information
see "Tropical Legumes: Resources for the Future," Nat. Acad. Sci., Washington,
D.C., U.S.A., 1979. This report emphasizes the variability of Lablab bean (over
200 genotypes are known) and that the center of research has been India. Varieties
are used as food: young pods, dried seeds, leaves and flowers, sprouts and protein concentrate from the seeds; forage: grazing, as hay and mixed with grasses
to improve pasture quality; a green manure, erosion control, nitrogen fixer, cover
crop in plantations and a second crop in rice fields after harvest.
   Vern. Moon gives Ratu-peti-dambala for the purple form and Sudu-peti-

dambala for the white (S). Trimen gives Tatta-payaru and Minni (T). He notes Pal-avarai has been used for the wild form and Dambala for the cultivated varieties. Also used: Ho-dhambala, Kos-ata-dhambala, Kiri-dambala (S); Motchai (T); and Bonavist bean, Hyacinth bean, Wild bean, and Lablab bean (E).

Specimens Examined. LOCALITY UNKNOWN: *Hermann* Mus. *60* in BM; 1861, *Thwaites C.P. 1470* (BM). COLOMBO DISTRICT: Henaratgoda, flrs. white, cultivar folder, 8.5.17, *s. coll. s.n.* (PDA). JAFFNA DISTRICT: Jaffna, 1846, *Gardner C.P. 1470* (PDA). MANNAR DISTRICT: Illupai-Kadunai, abundant, Feb. 1890, *s. coll. s.n.* (PDA). ANURADHAPURA DISTRICT: Maddawachchiya to Anuradhapura, *Simpson 9366* (BM). PUTTALAM DISTRICT: Wilpattu Natl. Pk., ocean shore W. of Pomparippu, *Maxwell & Jayasuriya 806* (NY, PDA, US). KANDY DISTRICT: Getambe, 9.7.27, *Alston 1410* (PDA); Peradeniya, Royal Botanic Gardens (all cultivated), May 1887, *s. coll. s.n.* (PDA), Feb. 1898, *s. coll. s.n.* (PDA), *Amaratunga 2040, 2041* (PDA), 25.2.21, *JMS s.n.* (PDA). NUWARA ELIYA DISTRICT: Hakgala, cultivated, 21.10.17, *s. coll. s.n.* (PDA); Maturata, collector illegible, 1857, *(Thwaites ?) C.P. 1470* (PDA); Hewaheta, *Alston 1007* (PDA), roadside, *Balakrishnan NBK 426* (NY, PDA, US); McDonald's Valley below Hakgala, *Rudd & Balakrishnan 3177* (PDA, US); Ramboda, *Simpson 8904* (BM). TRINCOMALEE DISTRICT: Trincomalee, cultivated, 1862, *Glenie C.P. 1470* (PDA). BADULLA DISTRICT: Welimade, April 1921, ADA *(A. de Alwis ?) s.n.* (PDA); Bandarawela, Craig estate jungle, 25.3.06, *A.M.S. s.n.* (PDA); A5, about 1 mi. N. of Palugawala, along roadside, *Maxwell 1016* (NY, PDA, US); cultivated in gardens, *Maxwell 1017, 1018, 1019, 1020* (PDA, US). HAMBANTOTA DISTRICT: Ruhuna Natl. Pk., Mahasilawa, *Cooray 70032603R* (PDA, US); Butuwa, sand dunes, arid zone, *Worthington 5219* (K).

Notes. Trimen (1894) remarks about the variability of the pods, flowers and seeds. Plants growing at Peradeniya in the Royal Botanic Gardens during the summer of 1972 all had white flowers.

My description is based on the wild and garden forms which are twining, rather than the field forms which are erect and bushy, have an unpleasant odor and cannot be used as a vegetable, and on Herb. *Wight 747* (BM, NY).

Alston (1931) in following Prain (1897), introduces the name *Dolichos lignosus* L. with *D. lablab,* but see under *Dipogon* and the references above for discussions of the complicated synonymy and nomenclature.

I am considering *Lablab* easily separable from the rest of the subtribe primarily because of gynoecial characters (geniculation and pubescence of the style, and the large, capitate stigma), as well as the slender, upcurved beaks of the keel petals and the linear, white arillate tissue of the seed.

21. DIPOGON

Liebm., Index Sem. Hort. Acad. Hauniensi 27. 1854; Liebm., Ann. Sci. Nat.

Bot. ser. 4, 2: 374. 1854; Verdcourt, Kew Bull. 24(3): 406. 1970 (cum syn.). Type species: *D. lignosus* (L.) Verdcourt.

Verdcourtia Wilczek, Bull. Jard. Bot. Etat 36: 250. 1966; Verdcourt, Taxon 17(5): 537. 1968.

Small twining or trailing vines. Leaves trifoliolate; stipules and stipels persistent; leaflets triangular, acute apically, broadly cuneate or truncate basally. Inflorescenses axillary, usually clustered in a subumbellate manner at the apex, few-flowered; bracts and bracteoles lanceolate, persistent. Flowers small; calyx 4-lobed, the upper lobe entire, the lower longest; petals rose pink, variable in cultivated forms, glabrous, the vexillum obovate, with 2 basal appendages, the wings oblanceolate, long-clawed, the keels semiorbicular, beaked, about equalling the wings, shorter than the vexillum, long-clawed; vexillary stamen free, the anthers uniform; ovary 5–8-ovulate, the lower style glabrous and swollen towards the base, geniculate, the upper style villous adaxially, the stigma with a ring of hairs. Fruit spirally dehiscent, linear-oblong to oblong-falcate, glabrous, 5–7-seeded; upper suture with adnate ribs, the lower swollen. Seeds ovoid, the hilum linear-oblong, raised, white.

A monotypic genus apparently native and widespread in South Africa, and cultivated in Australia, South America and elsewhere.

Dipogon lignosus (L.) Verdcourt, Taxon 17: 537. 1968; Verdcourt, Kew Bull. 24(3): 406. 1970 (cum syn.); Lackey in Polhill & Raven, Adv. Leg. Syst. Part 1. 317, fig. 13. 1981. Type: Grown from seed sent to Copenhagen from Brazil, holotype not preserved.

Dolichos lignosus L., Sp. Pl. 726. 1753; L., Hort. Cliff. 360, t. 20. 1737; Smith, Spicileg. Bot. 19, t. 21. 1792; Sims, Bot. Mag. 11: t. 380. 1797; Wight & Arn., Prod. 1: 251. 1834 (nota, non *Lablab cultratus*); Freeman, Bot. Gaz. 66: 512–523, figs. 3, 5, 6, 7. 1918; Burkart, Leg. Argentinas ed. 2, 422, fig. 128 j–k. 1952.
Dolichos gibbosus Thunb., Prod. Fl. Cap. 130. 1800.
Verdcourtia lignosa (L.) Wilczek, Bull. Jard. Bot. Etat 36: 254, t. 5, 6. 1966.

Stems ridged, with very sparse ascending to adpressed pubescence, the white rigid hairs about 0.5 mm long. Leaves membranous, the petioles sparsely pubescent, from 3–12 cm long, the rachis ⅓ as long; stipules conspicuous, lanceolate, about 6 mm long, 1.5 mm wide at the base, membranous, striate, glabrous to sparsely ciliate, non-produced; stipels linear-lanceolate, about 3 mm long, glabrous to sparsely ciliate; leaflets glabrescent above, glabrous below, with a pair of prominent primary lateral veins inserted basally, with 1 or 2 weaker pairs above, the terminal leaflets ovate to broadly rhomboid, about 4 × 3 cm, broadly cuneate basally, the lateral leaflets inequilateral, 2.5–3.0 × 2.0–2.5 cm, rounded basally.

Inflorescences 1.5–14 cm long, 2–6-flowered; bractlets lanceolate, about 4 mm long, striate, ciliate, persistent at the base of the pedicels; bracts about 2.5 mm long, with a prominent midrib, ciliate; bracteoles with a rounded apex, about 2.5 mm long, ciliate. Flowers about 1.3 cm long, the pedicels about 5 mm long; calyx tube about 2.5 mm long, membranous, glabrous, the upper lobes connate, the lateral lobes blunt, about 1 mm long, the lower acute, about 1.5 mm long, all lobes canescent and ciliate along the margins; petals rose pink, reddish-purple, later turning blue, the vexillum with a blue basal spot, about 12 × 10 mm, clawed, the wings about 9 mm long, the claw of wings and keels about 5 mm long; pistil geniculate; ovary with white ciliate hairs along the upper margin or glabrescent, the upper style villous adaxially for about 3 mm. Fruit to about 3.5 × 0.7 cm, septate. Seeds mostly reddish-brown with darker mottling, about 4 × 3 × 2.5 mm, the hilum about 3 mm long. Chromosome number 2n = 22.

Distr. Same as genus. Not naturalized in Ceylon to my knowledge. Apparently endemic to South Africa, it is naturalized in Queensland and New South Wales, Australia. Verdcourt (1970) also reports Azores, Madeira, California (USA), Chile and Uruguay. Burkart (1952) reports the species (as *Dolichos lignosus*) growing in Buenos Aires with a white-flowered variety.

Uses. Ornamental.

Specimen Examined. LOCALITY UNKNOWN: 1845, *G. Thomson s.n.* (K, with the label, "Leguminosae, *Dolichos* illegible 10/65" (cult.)).

Notes. Wight & Arn. (1834) realized *Dolichos lignosus* L. was not the same as the *D. lignosus* of Roxburgh (1832), and placed it in their *Lablab cultratus*.

Thomson s.n., dated 1845, is missing in Thwaites (1859) and the *Lablab cultratus* DC. in Trimen (1894) follows Wight & Arn. as a variant of *Lablab purpureus*. Gamble (p. 367, 1918) follows Prain (1897) as does Alston (1931). I could find no reference within the Ceylon literature to *Dipogon* and assume if the *Thomson s.n.* collection was noticed it was assigned variant status under *Lablab purpureus*.

Description based on Hort. Chelsea Gard. *3711* (BM).

22. DOLICHOS

L., Sp. Pl. 725. 1753; L., Gen. Pl., ed. 5, 324. 1754; Verdcourt, Kew Bull. 24(3): 380, 411. 1970; Verdcourt, Taxon 21(4): 533. 1972; Verdcourt in Summerfield & Bunting, Adv. Leg. Sci. 45. 1980, typ. cons. Type species: *D. trilobus* L. (= *Dolichos falcatus* Willd.).

Climbing or prostrate vines, erect herbs or shrubs. Leaves mostly trifoliolate; stipulate and stipellate; leaflets usually ovate, sometimes rhomboidal. Inflorescences axillary, fasciculate-racemose, often condensed subumbellate, or flowers solitary, usually few-flowered; bracts and bracteoles usually lanceolate, persistent. Flowers small to medium; calyx 4–5-lobed, the upper lobes connate, entire or bifid, all lobes about equal in length; petals yellow, reddish to bluish-purple, glabrous, about equal in size, the vexillum reflexed, somewhat orbicular, usually wider than long, biauriculate basally, with vertical lamelliform append-ages in the target area, the wings obovate or oblong, the keels untwisted, fre-quently beaked; stamens diadelphous, the vexillary free, the anthers uniform; ovary few–several-ovulate, the style upcurved, variously swollen and thickened, glabrous, the stigma capitate, with a ring of hairs. Fruit narrow-oblong, frequently upcurved, compressed, non-septate, several-seeded; sutures without conspicuous ribs or wings. Seeds ovoid or subreniform, somewhat compressed, the hilum short-oblong to suborbicular, arillate tissue poorly developed or absent.

A genus of about 60 species centered in Africa but extending to India and eastern Asia.

Dolichos trilobus L., Sp. Pl. 726. 1753 (p.p.); Verdcourt, Taxon 17: 170. 1968; Verdcourt, Kew Bull. 24(3): 380 & 422. 1970; Verdcourt, Fl. Trop. East Africa, Papil. 678. 1971. Type: Herb. Pulkenet (BM syntypes), not seen. See Verdcourt, Taxon 21(4): 533. 1972.

Phaseolus madraspatensis Plukenet, Almagest. Bot. 292, t. 214, fig. 3. 1692.
Dolichos falcatus Klein in Willd., Sp. Pl. 3: 1047. 1802; Roxb., Fl. Ind. 3: 311.
    1832; Wight & Arn., Prod. 1: 249. 1834; Thw., Enum. Pl. Zeyl. 90. 1859;
    Baker in Hook. f., Fl. Br. Ind. 2: 211. 1876; Trimen, Handb. Fl. Ceylon
    2: 77. 1894; Gamble, Fl. Pres. Madras 1(2): 366. 1918; Backer & Bakh.
    f., Fl. Java 1: 644. 1963. Type: E. India, *Klein s.n.* (K isotype ?).
Dolichos tuberosus Roxb. in E.I.C., mus. t. 289 (excl. syn. Lour. & Plukenet,
    fide Wight & Arn. (1834)).

Small twining, climbing or trailing vines; stems slender, angled, sparsely pubescent. Leaf rachis about 1/4 the petiolar length; stipules broadly lanceolate, about 6 mm long, shortly produced, persistent; stipels linear-lanceolate, about 1 mm long, persistent; leaflets often lobed, 5.5–7.0 × 2.5–5.5 cm, acuminate or acute apically, cuneate to rounded basally, usually with three prominent veins inserted basally, the surfaces somewhat gland-dotted, scabrous, glabrescent, the terminals triangular or rhomboidal, the laterals oblique. Inflorescences 8–40 mm long, atuberculate, each node 1–2-flowered; bracts linear, 1 mm long; bracteoles mostly 4-striate, 2 mm long, attached to the pedicels. Flowers about 13 mm long, the pedicels about 2 mm long; calyx membranous, the upper lobe acute, the lower longest, about 2 mm long, about equalling or shorter than the tube, sparsely pubes-cent; vexillary petal lamina about 10 mm long, short-clawed, crimson pink to bluish-purple, the appendages rounded, short, the wings free, the lamina about 10 mm long, the claw about 2 mm long, the keels oblong-falcate, without pouches or beaks; ovary 5–8-ovulate, apparently glabrous, the style constricted, the lower style thickened and flat basally, becoming tenuous distally, the upper style grooved, the stigma somewhat flat with hairs to about 1 mm long. Fruit dehiscent, glabrous, falcate-oblong, 5.5–6.0 cm long, 9 mm wide, 2–3 mm thick, with a persistent downcurved beak, 4–6-seeded; upper suture closely paralleled by thin inconspicuous ribs, the lower swollen. Seeds orbicular-oblong, 5–7 × 4 × 2 mm, dark brown, the hilum short-oblong, 2 mm long. Chromosome number 2n = 20.

Distr. Africa, Arabia, India, Ceylon, China, and the Philippines.

Ecol. Flowers January through March. It is found in the low country and to 600 m, especially in the dry region.

Vern. Wal-dambala (S).

Specimens Examined. LOCALITY UNKNOWN: *Thwaites C.P. 1472* (BM, K); *Col. Walker s.n.* (K). JAFFNA DISTRICT: Pallawarayankaddu, Feb. 1890, *s. coll. s.n.* (PDA); between Paranthan and Pooneryn, *Rudd & Balakrishnan 3299* (PDA, US). MANNAR DISTRICT: Nochchikulam, *Jayasuriya et al. 617* (US); Madhu Rd., *Jayasuriya et al. 619* (US); 25 mi. from Vavuniya on Mannar Rd., *Simpson 9341* (BM). POLONNARUWA DISTRICT: 1 mi. N. of Dimbulagala, *Jayasuriya et al. 714* (PDA, US); between Kalkudah and Polonnaruwa, *Rudd & Balakrishnan 3155* (K, NY, US). KANDY DISTRICT: Haragama, 10.1.28, *Alston s.n.* (PDA), *Jayasuriya & Balasubramaniam 447* (PDA, US); Peradeniya, Royal Botanic Gardens, 29.1.22, *J.M.S. s.n.* (PDA); Hantane, *Gardner s.n.*, *C.P. 1472* (PDA). BATTICALOA DISTRICT: Batticaloa, Mar. 1858, *s. coll. s.n.* (PDA).

Notes. This species is distinguished from close relatives by its glabrous style, constricted at the base, the untwisted keel petals, the protruding, non-lamelliform appendages on the face of the vexillum, and the non-tuberculate pollen. Specimens labelled *D. falcatus* Klein from other parts of the orient possess longer, many-flowered inflorescences which usually do not exceed the leaves and have definite tubercles.

See Verdcourt (1970) for a discussion of the characters which separate *Dolichos* s.l. into several genera.

Verdcourt (1970) divides the species into subspecies, and indicates the Ceylon material, pods 2.5–7.5 cm long and leaflets to 6.5 × 5.2 cm, belongs to subsp. *trilobus*. *Dolichos trilobus* subsp. *trilobus* is divided into 2 varieties: var. *trilobus* with ovate or rhomboid leaflets is in Ceylon, while var. *stenophyllus* Verdc. with leaflets oblong-lanceolate to linear-lanceolate is only found in Tanganyika (Tanzania).

Description based on Herb. *Wight 743* (K, NY) and Madras material (COIMB).

23. MACROTYLOMA

(Wight & Arn.) Verdcourt, Kew Bull. 24(3): 322, 400. 1970; Verdcourt, Fl. Trop. East Africa, Part 4, subf. 2: 581. 1971; Verdcourt, Taxon 27: 219. 1978, nom. cons. Lectotype: *M. uniflorum* (Lam.) Verdcourt (*Dolichos uniflorus* Lam.).

Kerstingiella Harms, Ber. Deutsch. Bot. Ges. 262: 230, t. 3. 1908.
Dolichos L. group Macrotyloma Wight & Arn., Prod. 1: 248. 1834.
Dolichos L. subg. *Macrotyloma* (Wight & Arn.) Baker in Hook. f., Fl. Br. Ind.
2: 210. 1876.

Small climbing, prostrate or occasionally erect herbs. Leaves pinnately
trifoliolate, occasionally unifoliolate; stipulate and stipellate; leaflets oblong-oval
to lanceolate. Inflorescences axillary, fasciculate or appearing racemose, solitary
to many-flowered; bracts acuminate to broadly ovate; bracteolate. Flowers small
to medium; calyx 4–5-lobed, the upper lobes connate, entire or slightly bifid;
petals whitish, yellow or light greenish-yellow, all clawed, about equal in length,
the vexillum orbicular or elliptic, with 2 long, linear, laminar appendages,
glabrous, the wings narrow, the keels wider than the wings, slightly falcate,
untwisted, without beaks; stamens diadelphous, the vexillary free, the anthers
uniform, with tuberculate or spinulose pollen; ovary few–many-ovulate, the lower
style usually slender, bent or ascending, often hairy, the upper style flattened
or thick, usually glabrous, the stigma terminal, capitate, surrounded by a ring
of hairs. Fruit dehiscent, oblong to linear, compressed, upcurved, non-septate,
usually with a persistent beak, 2–6-seeded. Seeds oblong-oval to ovoid, compressed, the hilum oblong-linear to elliptic, usually without well developed arillate
tissue.

A tropical African and Asian genus of about 25 species, some cultivated for
food and fodder.

KEY TO THE SPECIES

1 Lower and lateral calyx lobes long acuminate; fruit hairy; flowers about 1 cm long; cultivated,
  erect or suberect annual . . . . **1. M. uniflorum**
1 Lower and lateral calyx lobes acute; fruit glabrescent to glabrous; flowers as above or 1.2–1.5(–2)
  cm long; native twining perennial
    2 Leaflets somewhat membranous; fruit 5–6-seeded . . . . **2. M. axillare**
    2 Leaflets somewhat coriaceous; fruit usually 2–3-seeded . . . . **3. M. ciliatum**

1. Macrotyloma uniflorum (Lam.) Verdcourt, Kew Bull. 24(3): 322, 379–390,
401, 404–406. 1970; Verdcourt, Fl. Trop. East Africa, Papil. 583. 1971. Type:
Plant grown at the Jardin du Roi, Paris, from seed obtained by Sonnerat in India
(P-LA, holotype), fide Verdcourt, not seen.

Dolichos uniflorus Lam., Enc. 2: 299. 1786; Wight & Arn., Prod. 1: 248. 1834;
    Thw., Enum. Pl. Zeyl. 90. 1859; Backer & Bakh. f., Fl. Java 1: 644. 1963.
Glycine villosa nomen ? Moon, Cat. 53. 1824.
Kerstingiella uniflora (Lam.) Lackey, Phytologia 38: 229. 1978.
Dolichos biflorus auct. non L., Roxb., Fl. Ind. 3: 313. 1832; Baker in Hook.
    f., Fl. Br. Ind. 2: 210. 1876; Trimen, Handb. Fl. Ceylon 2: 77. 1894 (var.
    glaber Thw.); Gamble, Fl. Pres. Madras 1(2): 366. 1918; Alston in Trimen,
    Handb. Fl. Ceylon 6: 83, 84. 1931.

Twining, climbing or partially erect vines. Leaf petioles 0.8–6.8 cm long, the rachis 2.5–10 mm long; stipules lanceolate, 4–8 mm long; leaflets thin, ovate-rhomboid, obovate or elliptic, 1–5 × 1–3 cm, much larger in Indian material (7 × 6.5 cm), rounded to acute apically, rounded basally, both surfaces glabrescent to long hairy pubescent, the lateral leaflets inequilateral. Inflorescences about 1.5 cm long, or subsessile, the flowers occasionally solitary, usually 2–3 fasciculate; bracts linear, about 2 mm long; bracteoles linear, about 3 mm long. Flowers to about 1.5 cm long, the pedicels 1–7 mm long; calyx tube 3 mm long, pubescent, the upper lobes triangular, the remaining lobes lanceolate-acuminate, the apices extended, filiform, 5–7 mm long, petals yellow or greenish yellow, occasionally whitish, the vexillum obovate-oblong, about 10 × 6 mm, usually with a purple spot, short-clawed, the wings and keels greenish-yellow, the keels nearly equalling the wings; ovary usually few-ovulate, densely pubescent, the lower style very slender, bent, the upper style thick, curved upwards into almost a complete circle (not a spiral turn), glabrous. Fruit straight or falcate, 4–6 cm long, 6–8 mm wide, narrowing basally, acuminate at the apex, densely pubescent, then glabrescent, 4–7-seeded; sutures prominent, the upper occasionally verrucose. Seeds dark or variously coloured, oblong or subreniform, 3–4.2 × 2.8–3.5 × 1.5–2 mm, the hilum elliptic, about 1 mm long. Chromosome number 2n = 20, 22.

Distr. India, Africa, Ceylon, and Java. Widely cultivated in Asia.

Ecol. Trimen (1894) reports the species (*D. biflorus*) occasionally cultivated in the dry regions as a cultivated form of *D. uniflorus*. Flowering is November through April.

Uses. Forms or varieties of the species are extensively cultivated in the tropics as a cover crop. The seeds may also be eaten during hard times.

Vern. Horse Gram, Madras Gram (E); Kollu (S) and (T).

Specimens Examined. DISTRICT UNKNOWN: "Kolla" Cult-Dumbara, Mar. 1893, *s. coll. s.n.* (PDA); *Thwaites 1475* (K); *Thwaites C.P. 1468* (BM, mixture with *C.P. 1468, = Galactia*); Homagama, *Alston 1237* (PDA); Dolosbage, *Gardner C.P. 1475* (PDA). KURUNEGALA DISTRICT: Kurunegala Rock, Dec. 1883, *s. coll. s.n.* (PDA). MATALE DISTRICT: Dambulla garden, 1900, *s. coll. s.n.* (PDA); "Damboul," Mar. 1828, *s. coll. C.P. 1475* (PDA); Dambulla, *Amaratunga 536* (PDA); Sigiriya, *Amaratunga 859* (PDA). KANDY DISTRICT: Peradeniya (?), kitchen garden, cult., 1900, *s. coll. s.n.* (PDA). COLOMBO DISTRICT: Colombo, 1858, *W.W. Ferguson s.n.* (PDA, as *C.P. 1475*). KEGALLE DISTRICT: Ambepussa, *Amaratunga 1468, 1723* (PDA).

Notes. Roxburgh (1832) reports the species as an erect annual found only in a cultivated state. Wight & Arnott (1834) indicate the *Dolichos biflorus* of Roxburgh is very probably *D. uniflorus* Lam. and treat it as such. Baker (1876) uses *D. biflorus* and under this name treats *D. biflorus* as stems "lengthened out and twining" with *D. uniflorus* as the suberect annual form.

Trimen (1894) maintains *D. uniflorus* Lam. var. *glaber* Thwaites, ms. (nomen), may not be *D. uniflorus,* but is certainly *D. axillaris* E. Meyer. He maintains *D. biflorus* as the annual cultivated form of *D. uniflorus.*

I maintain, following Verdcourt (1970, 1971) there are at least 3 species of *Macrotyloma* in Ceylon, with *M. uniflorum* consisting of many varieties and forms. Verdcourt (1971) segregates var. *uniflorum* on the basis of pods 6–8 mm wide. This is horse gram, native of India.

Trimen is correct in that a part of his material is *D. axillaris.* The combination *D. axillaris* E. Mey. var. *glaber* E. Mey. was made in 1836 and I have placed Thwaites ms. combination in synonymy. A part of the *C.P. 1475* sheets come out as *D. axillaris* var. *glaber.*

Description based on type photographs of *Dolichos uniflorus* Lam. (*D. biflorus* auct., non L.) cultivated at Leiden, specimen in Herb. Meerburgh (Leiden), photo of type in Herb. Lamarck, "de M. Sonnerat, an Jardin R. Glycine" (all at K); also based on *Wight 741* (BM).

2. Macrotyloma axillare (E. Mey.) Verdcourt, Kew Bull. 24(3): 402, pl. 8/14, 395. 1970; Verdcourt, Fl. Trop. East Africa, Part 4, subf. 2: 586. 1971; Verdcourt, Man. New Guinea Leg. 530, fig. 129. 1979; Lackey in Polhill & Raven, Adv. Leg. Syst. Part 1. 317, fig. 4.12. 1981. Types: South Africa, Natal, *Drege* & Durban (Port Natal), *Drege* (both K isosyntypes), not seen.

Dolichos axillaris E. Mey. var. *glaber* E. Mey., Comm. Pl. Afr. Austr. 1: 144. 1836.

Dolichos uniflorus Lam. var. *glaber* (E. Mey.) Thw. ex Trimen, Handb. Fl. Ceylon 2: 76. 1894.

Macrotyloma axillare (E. Mey.) Verdcourt var. *glabrum* (E. Mey.) Verdcourt, Kew Bull. 24(3): 402. 1970.

Climbing or trailing vines; stems perennial, to about 3.5 m long, glabrescent. Leaf rachis 2–11 mm long, about 1/5 the petiolar length; stipules oval-ovate, 2–5 mm long; leaflet lamina thin, ovate-lanceolate or subrhombic, about 2.5 cm long, acute apically, broadly cuneate to rounded basally, glabrous. Inflorescences in 2–4(–10)-flowered fascicles; bracts linear-lanceolate, 2–3 mm long, persistent; bracteoles linear, 2.5–3 mm long, persistent. Flowers 1.3–2 cm long, the pedicels 2–3 mm long; calyx tube about 2 mm long, pubescent, all lobes lanceolate, 2.5–5 mm long; petals pale yellow, the vexillum with a crimson or purplish spot in the target area, oblong-elliptic, 1–2.4 × 0.6–1.5 cm (fide Verdcourt), emarginate, short-clawed, the appendages about 5 mm long, the wings and keels greenish to yellow; ovary several-ovulate, pubescent. Fruit about 6.5 × 0.7 cm, pubescent or glabrescent (glabrous, Trimen, 1894), 5–6-seeded; upper suture ending in a beak, 0.5–1.5 cm long. Seeds brown to dark red, with sparse to very dense black mottling, oblong-oval, 3–4.2 × 2.5–3 × 0.6–1.5 mm, the hilum minute, rim-aril obsolete or lacking. Chromosome number 2n = 20.

Distr. Africa to Madagascar, Mauritius, Ceylon, and New Guinea to Australia.

Ecol. Trimen (1894) cites rocky places in the low country to 600 m (as *D. uniflorus* var. *glaber*). Verdcourt (1971) found the species in grassland, bushland and open forest, from 0 to 2520 m.

Specimens Examined. LOCALITY UNKNOWN: *Thwaites C.P. 1475* (K). COLOMBO DISTRICT: *Ferguson 2/80* (K, type folder, 2 sheets, drawing ("sp. n."), letter, both sheets det. *D. axillaris* var. *glabra* by Brenes (Brenan?) XI. 1951); Colombo, *Ferguson-Thwaites C.P. 1475* (K, as type material *D. uniflorus* var. *glaber* Thwaites ex Trimen).

Notes. Verdcourt (1971, p. 586) characterizes var. *glabrum* as possessing flowers 1.2–1.5(–2) cm long and stems with sparse, rarely dense adpressed hairs.

3. Macrotyloma ciliatum (Willd.) Verdcourt, Kew Bull. 24(3): 404. 1970. Type not seen.

Dolichos ciliatus Willd., Sp. Pl. 3: 109. 1802; Wight & Arn., Prod. 1: 249. 1834;
    Thw., Enum. Pl. Zeyl. 90. 1859; Baker in Hook. f., Fl. Br. Ind. 2: 210.
    1876; Trimen, Handb. Fl. Ceylon 2: 77. 1894; Gamble, Fl. Pres. Madras
    1(2): 367. 1918.
Dolichos prostratus Koenig Mss. in Roxb., Fl. Ind. 3: 310. 1832.

Twining, climbing perennial vines; stems slender, terete, glabrous or with short, spreading hairs. Leaf rachis 4–15 mm long, 1/4 to 1/8 the petiolar length; stipules ovate to acute, about 5 mm long, striated, rigid, persistent; stipels setaceous, about 1 mm long; leaflet laminas thick, coriaceous, ovate, oblong-oval or broadly elliptic, 2.5–6.5 × 2.2–4.5 cm, acute to obtuse apically, rounded basally, sometimes pubescent on the veins beneath, otherwise glabrescent or glabrous, the margins ciliate with white silky hairs, the veins prominent beneath. Inflorescences short-axillary, from 1 to about 1.5 cm long, racemose or somewhat cymose, 3–6-flowered; bracts oval to oblong-lanceolate, striate, persistent; bracteoles oblong-lanceolate, striate, persistent. Flowers about 12 mm long, the pedicels 1–4 mm long, calyx tube about 2 mm long (?), all lobes lanceolate-setaceous, the upper lobes bifid; petals pale yellow, the vexillum obovate, slightly exceeding the wings and keels. Fruit upper margin straight or somewhat upcurved, the lower naviculate, acuminate apically, greatly compressed, 2.4–4 cm long, 6–10(–12) mm wide, 3–4 mm thick, glabrous, 2–4-seeded; upper sutures ending in a sharp, usually downcurved beak, 5–7 mm long. Seeds about 3 mm long (fide Trimen) or about 7 × 4–5 × 2 mm, ovoid, dark brown, smooth, the hilum oblong-linear, about 2 mm long, white, slightly raised.

Distr. India and Ceylon.

Ecol. Warm, dry regions. Rare. Flowering in January (Trimen, 1894) and February.

Specimens Examined. LOCALITY UNKNOWN: *Thwaites C.P. 1469* (K). JAFFNA DISTRICT: Elephant Pass, Feb. 1890, *Gardner C.P. 1469* (PDA). HAMBANTOTA DISTRICT: Ruhuna Natl. Pk., Buttawa Bungalow, *Comanor 371* (US).

Note. Description based on Herb. *Wight 740* (A, COIMB, K).

24. VIGNA

Savi, Nuovo Goirn. Lett. 8: 113. 1824; Piper, Contr. US Natl. Herb. 22(9): 663. 1926; Burkart, Leg. Argentinas ed. 2, 417–436. 1952; Marechal, Bull. Jard. Bot. Etat 37: 461. 1967; Verdcourt, Kew Bull. 24(3): 507–567. 1970; Marechal et al. Taxon 27: 199. 1978; Marechal et al., Boissiera 28: 1–273. 1978; Verdcourt, Man. New Guinea Leg. 515–529. 1979; Verdcourt in Summerfield & Bunting, Adv. Leg. Sci. 45. 1980; Lackey & D'Arcy, Ann. Missouri Bot. Gard. 67(3): 791. 1980. Marechal, Mascherpa & Stainier in Polhill & Raven, Adv. Leg. Syst. Part 1. 329. 1981. Type species: *Vigna luteola* (Jacq.) Benth.

See *Vigna,* Section 7, in Summerfield & Bunting, Adv. Leg. Sci. 393–468. 1980.

Haydonia Wilczek, Bull. Jard. Bot. Etat 24: 405. 1954.

Twining, climbing vines, occasionally erect herbs, without uncinate hairs. Leaves trifoliolate; stipules variable, usually produced below insertion; stipels persistent; leaflets broadly lanceolate, ovate, narrow rhomboidal or occasionally suborbicular. Inflorescences pseudoracemose or subumbellate, axillary or terminal, somewhat tuberculate or glandular at insertion of the pedicels, few–many-flowered; bracts caducous or deciduous, sometimes reduced, occasionally persistent; bracteoles usually lanceolate, caducous, occasionally persistent. Flowers medium to relatively large and showy; calyx 5-lobed, the upper lobes partially connate, all lobes acute to somewhat setaceous; petals yellow, greenish-white to shades of blue, purple and pink, the vexillum reflexed, mostly orbicular, with 2 or 4 appendages, rarely without appendages or without callosities on the face, the keels truncate, obtuse or with a beak, occasionally incurved and twisted; stamens diadelphous, the vexillary free, the anthers uniform; ovary several–numerous-ovulate, the style upcurved, variously twisted, the lower style slender, glabrous, the upper style thickened, incurved, with long hairs, the stigma subterminal, beaked or hoof-shaped. Fruit usually dehiscent, linear, occasionally very long or somewhat linear-oblong to oblong, slightly compressed or terete, with a persistent beak, usually septate, several–numerous-seeded; sutures indistinct or slightly swollen. Seeds usually oblong or subreniform, frequently truncate on the ends, small- to medium-sized, the hilum elliptic or oblong, occasionally with arillate tissue.

A pantropic genus of about 150 species, currently divided into several infrageneric categories.

KEY TO THE SPECIES

1 Stipules peltate, lanceolate or ovate, produced below insertion
  2 Keel petals untwisted, or slightly twisted, not incurved more than 180 degrees, keel pockets absent; style with the thickened part straight, or gently curved, the beak mostly lacking, occasionally with a short beak
    3 Stipules to 4 mm long, bilobed at the base; flowers yellow; leaflets mostly unlobed; native
      4 Fruit 1–2-seeded, c. 1.5 cm long; stems prostrate in the ground cover, or weakly climbing; flowers c. 1 cm long; upland . . . .**1. V. hosei**
      4 Fruit 4–8(–12)-seeded, usually exceeding 4 cm long; stems climbing, twining; flowers exceeding 1 cm long; coast, lowlands
        5 Leaflets broadly elliptic to somewhat orbicular; seeds 2–6(–8); stipules ovate, c. 3 mm long; sea coast habitat, above the high tide mark into the scrub . . . .**2. V. marina**
        5 Leaflets ovate, ovate-elliptic to somewhat lanceolate; seeds mostly 6–9; stipules ovate-lanceolate, 3–4 mm long, bilobed below insertion; lowlands . . . .**3. V. luteola**
    3 Stipules 8–28 mm long, somewhat lanceolate, produced, constricted at insertion; flowers yellow, white, greenish to purplish; leaflets mostly lobed; cultivated in Ceylon . . . .
      . . . .**4. V. unguiculata**
  2 Keel petals twisted, keel pocket present; styles beaked or with a lobe-like appendage
    6 Stipules lanceolate, distinctly or obscurely bilobed; style curved, the apex thick, with a short, lobe-like appendage; stigma subterminal . . . .**5. V. vexillata**
    6 Stipules peltate (except *V. aconitifolia*); style with a slender beak
      7 Stems frequently clumped from a taproot, mostly prostrate or sprawling, occasionally weakly climbing
        8 Stipules lanceolate or lanceolate-peltate, 5–7 mm long, occasionally appearing non-produced; leaflets usually deeply dissected, occasionally the lobes lanceolate to linear-lanceolate; uplands ? . . . .**6. V. aconitifolia**
        8 Stipules peltate; leaflets lobed, rarely unlobed, never deeply dissected; lowlands, sea coast, edges of paddy fields . . . .**7. V. trilobata**
      7 Stems erect, mostly stronger, twining, climbing vines, occasionally erect herbs
        9 Seeds rough, with ridges or riblets; keel pocket short; leaflets narrow-ovate towards lanceolate; fruit tomentose, or with bristly, ferruginous hairs, glabrescent in *V. radiata* var. *glabra*
          10 Stems, leaves and fruit densely pubescent
            11 Fruit with short hairs; seeds lacking a distinct rim aril
              12 Stems mostly erect; leaflets mostly unlobed . . . .**8a. V. radiata** var. **radiata**
              12 Stems twining, prostrate; leaflets frequently lobed . . .**8b. V. radiata** var. **sublobata**
            11 Fruit with long hairs; seeds with a raised rim aril . . . .**9. V. mungo**
          10 Stems, leaves and fruit glabrescent . . . .**8c. V. radiata** var. **glabra**
        9 Seeds smooth; keel pocket very long; leaflets broadly ovate to suborbicular; fruit glabrous when mature . . . .**10. V. umbellata**
1 Stipules non-produced; keel without pockets; style lacking beaks
  13 Keels twisted; style coiled 3–5 complete turns
    14 Keels twisted 5–7 turns, lacking pockets; leaflets not deeply dissected; stems twining
      15 Fruit 15–20-seeded; seeds black, about 8 mm long, the hilum white, calyx lobes obtuse; flowers mostly shades of purple; cultivated ?, rare . . . .**11. V. caracalla**
      15 Fruit 9–15-seeded; seeds brown, 5.5–7 mm long; calyx lobes lanceolate; flowers mostly dark pink; common in moist regions . . . .**12. V. adenantha**
    14 Keels twisted, much less than 5–7 turns, keel pocket present; leaflets frequently deeply dissected; stems clumped from a taproot, sprawling, occasionally climbing . . . . . **6. V. aconitifolia**
  13 Keels untwisted, slightly twisted, or somewhat incurved; style erect or variously bent, never coiled 3–5 turns

16 Leaflets ovate, ovate-rhomboid, disarticulate; fruit linear, 7–10 cm long, c. 3.5 mm wide, seeds c.10; upper style hoof-shaped; upland . . . .**13. V. grahamiana**

16 Leaflets broadly elliptic to somewhat orbicular, persistent; fruit linear-oblong, 5–7 cm long, c. 8 mm wide; seeds 2–6(–8); upper style never hoof-shaped; littoral plants . . . .
. . . .**2. V. marina**

1. Vigna hosei (Craib) Backer in Backer & Van Slooten, Geillustreend Handb. Jav. Theeonkruiden 153. 1924; Backer & Bakh. f., Fl. Java 1: 642. 1963; Verdcourt, Kew Bull. 24(3): 533. 1970; Verdcourt, Man. New Guinea Leg. 517. 1979. Type: Selangor, Kuala Lumpur, introduced from Sarawak as cutting to Mr. Hose, Director Agriculture, H2335/1913 (K holotype), not seen, fide Verdcourt (1970).

Dolichos hosei Craib., Bull. Misc. Inform. 1914: 76. 1914; Alston in Trimen, Handb. Fl. Ceylon 6: 84. 1931.

Vigna oligosperma Backer, nom. nud.; W. M. van Helten, Pract. Mede. Alge. Proef. Landb. Buiten. No. 16: 58, 61. 1924; Anon., Man. of Green Manuring, Dept. Agric., Peradeniya, Ceylon 83, 114. 1931, fide Verdcourt (1970).

Twining, trailing, or somewhat prostrate vines, about 1 m long; stems slender, ridged, with a few long hairs. Leaf rachis about 6 mm long, 1/10 the petiolar length; stipules bilobed at the base or hastate, striate, produced, about 3 mm long, glabrous; stipels acuminate, about 1 mm long; leaflets thin-papyraceous, broadly ovate or occasionally elliptic, 3.5–4.5 × 1.8–2.8 cm, acute or obtuse apically, broadly cuneate or rounded basally, both surfaces glabrous or sparse strigose beneath, with 3 prominent veins inserted at the base, the laterals slightly inequilateral. Inflorescences axillary, 0.8–7.0 cm long, few-flowered, with retrorse hairs, usually much shorter than the petioles; bracts not seen, caducous (?); bracteoles linear-lanceolate, 1 mm long, hyaline, persistent. Flowers 7–10 mm long, the pedicels about 2 mm long; calyx tube urceolate, 2 mm long, the upper lobes connate, obtuse, the laterals acute, the lowest subulate, all lobes about 1 mm long, ciliate; petals yellow, the vexillum broadly orbicular, about 9 × 11 mm, emarginate, basally inflexed biauriculate, glabrous, with 2 small oblique appendages in the target area, the claw less than 1 mm long, the wings obovate, about 8 × 4 mm, with a small spur, the claw about 2 mm long, the keels semiorbicular, untwisted, about 7 × 3.5 mm, the upper margin almost straight, acute apically, the left keel with a large transverse (?) pocket, the claw about 2 mm long; ovary 2-ovulate, the lower style tenuous, about 1 mm long, the upper style about 5 mm long, with dense hairs extending about 2.5 mm on the inside to the apex, the apex somewhat truncate, the stigma lobed. Fruit oblong, turgid, nonseptate (?), dehiscent, about 15 × 4 × 4 mm, puberulent, (1)–2-seeded. Seeds rounded-oblong, about 5 × 3 × 3 mm, horizontal, dark, shining, smooth, the hilum elliptic-oblong, whitish, about 2 mm long, without arillate tissue. Chromosome number 2n = 20.

Distr. Native to Java and Borneo. Found also in Malaysia, Indonesia and Tanzania.

Ecol. This inconspicuous small vine has escaped from the estates where it was introduced and has now become established in the hill country where it probably flowers year round.

Uses. Alston (1931) notes this small creeping species is now grown extensively as a cover crop under old rubber in Ceylon.

Specimens Examined. KANDY DISTRICT: Peradeniya, Exp. Sta., 26.1.26, *Alston s.n.* (PDA); Gampola, Orwell Estate, *Alston 715* (K, PDA); nr. Kandy, *Hepper 4492* (K, PDA, US); 3 mi. S. of Gampola on A5, *Maxwell 855* (NY); hills above Kandy, *Maxwell & Jayasuriya 846* (NY, PDA, US), *848* (NY, PDA, US), *Worthington 7229* (K, US). NUWARA ELIYA DISTRICT: Maskeliya, *C.F. & R.J. van Beusekom 1531* (US).

Notes. *Vigna hosei* and the next two species fall into subg. *Vigna*, Sect. *Vigna*. See Verdcourt (p. 526, 1970) for Sect. *Vigna* characters.

2. Vigna marina (Burm.) Merr., Interpr. Rumph. Herb. Amboin. 285. 1917; Alston in Trimen, Handb. Fl. Ceylon 6: 82. 1931 (cum syn.); Backer & Bakh. f., Fl. Java 1: 642. 1963; Verdcourt, Fl. Trop. East Africa, Papil. 626. 1971; Verdcourt, Man. New Guinea Leg. 520, fig. 127. 1979. Type: Molucca Is., Amboina. Based on Rumph. Herb. Amboin. 5: 391. t. 141/2. 1750, not seen.

Phaseolus marinus Burm., Ind. Alter Univ. Herb. Amboin. (18). 1769.
Dolichos luteus Sw., Prod. Veg. Ind. Occ. 105. 1788; Moon, Cat. 53. 1824; Thw., Enum. Pl. Zeyl. 90. 1859 (as *D. luteus* Sw. ?). Type: Jamaica, *Swartz,* not seen.
Vigna anomala Walp., Rep. 1: 779. 1842.
Vigna retusa Walp., Rep. 1: 778. 1842; Ridley, Fl. Malay Penins. 1: 568. 1922.
Vigna lutea (Sw.) A. Gray, Bot. Wilkes Exped. 1: 452. 1854; Baker in Hook. f., Fl. Br. Ind. 2: 205. 1876; Prain, J. Asiat. Soc. Bengal 66(2): 428. 1897.
Vigna luteola Benth. ms. ex Thw., Enum. Pl. Zeyl. 90. 1859, nomen. Based on *C.P. 1482* (BM, K, PDA).
Vigna luteola auct., non Benth. in Mart., Fl. Bras. 15(1): 194, t. 50/2. 1859 (at least p.p.); Trimen, Handb. Fl. Ceylon 2: 73. 1894; Baker in Hook. f., Fl. Br. Ind. 2: 205. 1876.

Trailing or prostrate vines; stems terete, sparsely pubescent or glabrous at maturity. Leaf rachis about 11 mm long, 1/3 the petiolar length; stipules acute, obscurely or non-produced, 2–3 mm long, glabrous, persistent; stipels lanceolate, about 2 mm long; leaflets elliptic, obovate or suborbicular, about 5 × 4 cm, obtuse or rounded apically, cuneate or rounded basally, with adpressed short, bristly hairs on both surfaces or glabrous. Inflorescences axillary, 6–8 cm long, usually exceeding the leaves, the rachis somewhat atuberculate, florate 1/2–1/3 the length, pubescent, few flowered; bracts not seen; bracteoles lanceolate, about 1.5 mm long. Flowers about 12 mm long, the pedicels about 4 mm long; calyx tube urceolate, about 2 mm long, glabrous, the upper lobes connate, the laterals acute, the lower longest, somewhat subulate, to about 2 mm long, ciliate; petals yellow, the vexillum obovate, subsessile, about 12 × 14 mm, rounded or slightly emarginate, basally inflexed biauriculate, glabrous, with 2 small oblique append-ages in the target area, the wings obliquely obovate, rounded, about 7 mm wide, about equalling the vexillum, the claw about 2 mm long, the keels semiorbicular, shorter than the vexillum, untwisted, about 6 mm wide, without pockets, the claw about 2 mm long, the beak obtuse; upper style with dense hairs on the inside, the apex extended, hoof-like, the stigma subterminal. Fruit linear-oblong, turgid, 5–7 × about 0.8 × 0.5 cm, subseptate, the exocarp slightly depressed between the seeds, glabrous, about 6-seeded; sutures swollen. Seeds quadrate, horizon-tal, about 4.5 × 4 × 3 mm, black, shining, smooth, the hilum linear-oblong, about 2.5 mm long, without arillate tissue. Chromosome number 2n = 22.

Distr. Tropical Asia and America, South Africa, Polynesia and Australia.

Ecol. A widely distributed vine occurring above the high tide mark and into the coastal scrub at sea level. According to the only collections with dates, flower-ing in natural habitats is in February and July, but it probably occurs all year.

Vern. Field Bean (E); Karal Li-me (S); Kodippayaru (T).

Specimens Examined. KANDY DISTRICT: Royal Botanic Gardens, Peradeniya, cultivated, *Alston 779 & 822* (K, PDA), *1236* (PDA). AMPARAI DISTRICT: Pottuvil, *Bernardi 16011* (US). COLOMBO DISTRICT: nr. seacoast, Colombo, July 1838, *Ferguson s.n., C.P. 1482* (PDA), 1861, *Thwaites C.P. 1482* (BM, K); Mt. Lavinia, *Rudd 3075* (K, PDA, US). GALLE DISTRICT: Galle (?), *Gardner C.P. 1482* (PDA).

Notes. Moon (1824), under *Dolichos luteus,* cites Jamaica, Caltura. Thwaites (1859), uses *V. luteola* Benth. ms., and notes ''An *Dolichos luteus,* Sw ?'' Baker (1876) uses *V. repens* Baker (nom. illegit.), *V. lutea* A. Gray and *V. luteola.* Bentham (1859) cites *Thwaites,* Ceylon, under *V. luteola.* Trimen (1894) uses *V. luteola* Benth. ex Thw.

Alston (1931) introduces the name *V. marina* Merr. for Trimen's *V. luteola* Benth. ex Thw., placing most of Trimen's synonymy under *V. marina* and segregating *V. luteola* Benth. Alston also states he believes Baker's (1876) cita-tion of the Thwaites specimens is probably a mistake with *V. marina* or *Phaseolus calcaratus* being taken for *V. luteola.* Alston gives no indication that specimens of *V. luteola* have been found in Ceylon and Abeywickrama (Ceylon J. Sci., Biol. Sci. 2(2): 173. 1959) does not include the name *V. luteola* in his Provisional Check List.

The only relatively new specimen I examined was that of Rudd. The Alston collections were of specimens grown at the Royal Botanic Gardens from seeds gathered at Matara in 1925 & 1926. I am following Alston (1931) in using the name *V. marina* for all the specimens examined in Ceylon.

Verdcourt (1971) uses both *V. marina* and *V. luteola.* Lackey and D'Arcy (1980) use *V. luteola* and *V. retusa* (Meyer) Walp. They state the name is based on the interpretations of the Burman figure with Merrill's interpretations matching their *V. luteola* and Marechal's (1978) matching their *V. retusa*. I cannot match Lackey and D'Arcy's (1980) nomenclature and taxonomy in the Flora of Panama to Verdcourt's (1971) in Fl. Trop. East Africa or to the specimens I examined in Ceylon. The specimens examined share characteristics of the 3 "taxa."

3. Vigna luteola (Jacq.) Benth. in Mart., Fl. Bras. 15(1): 194. t. 50/2. 1859; Verdcourt, Kew Bull. 24(3): 509, 513, 515, 517, 519, 526. 1970; Verdcourt, Fl. Trop. East Africa, Papil. 625, 627 (note). 1971 (cum syn.); Verdcourt, Man. New Guinea Leg. 518. 1979. Type: based on Jacq., Hort. Vind. t. 90, drawn from plant grown in Vienna from tropical American seed (BM ? holotype, LINN 900-4 ? isotype), fide Verdcourt (1971).

Dolichos repens L., Pl. Jam. Pugill. 19. 1759. Type: Jamaica, *P. Browne*, not
   found, fide Verdcourt (1971).
Dolichos luteolus Jacq., Hort. Vind. 1: 39. t. 90. 1770. (BM isotype).
Vigna repens (L.) Kuntze, Rev. Gen. 212. 1891, non Baker in Hook. f., Fl. Br.
   Ind. 2: 205. 1876, nom. illegit.

Coarse vines; stems rooting at the nodes, pubescent to glabrescent. Leaf rachis 5–12 mm long, 1/4–1/6 the petiolar length; stipules ovate-lanceolate, 3–4 mm long, produced-bilobed below insertion; stipels probably lanceolate; leaflets ovate, ovate-elliptic or ovate-lanceolate, 2.5–10.0 × 0.4–5 cm, acute or acuminate apically, rounded or cuneate basally, both surfaces sparsely pubescent. Inflorescences axillary, 6–45 cm long, the rachis about 2 cm long, few-flowered; bracts ovate-lanceolate, 1.5–2 mm long, deciduous; bracteoles ovate-lanceolate, 1-nerved, 1.5–2 mm long. Flowers 1.5–2.5 cm long, the pedicels 4–9 mm long; calyx tube 3–4 mm long, pubescent or glabrescent, the upper lobes connate, rarely somewhat bifid, all lobes deltoid, 2–4 mm long; petals yellow or greenish, sometimes tinged with pink, the vexillum obovate, 1.3–2.5 × 1.2–2.6 cm, emarginate, glabrous, subsessile, the wings about equalling the vexillum, obliquely obovate, rounded, about 7 mm wide, the claw about 2 mm long, the keels semiorbicular, shorter than the vexillum, untwisted, about 6 mm wide, without pockets, the claws about 2 mm long, the beak short, obtuse; upper style mostly straight or slightly curved with dense hairs on the inside, the apex extended, hoof-like. Fruit reflexed or upcurved, 4–8 cm long, 5–6.5 mm wide, the exocarp very slightly depressed between the seeds, with dense adpressed pubescence, 6–9(–12)-seeded. Seeds oblong, 3–6 × 3–4 × 2–3.5 mm, dark, red-brown or grey-brown speckled with black, the hilum short-oblong, with little or no rim-arillate tissue. Chromosome number $2n = 22$.

Distr. A widespread tropical lowland species found throughout Africa, Asia and America. Verdcourt (1971) notes E. African collections were usually from swampy or seasonally wet grassland, reedy and sandy lake shores and swamp forest at 650–1920 m altitude.

Notes. As noted under *Vigna marina*, *V. luteola* was entered for Ceylon by Thwaites (1859), but Alston (1931) later indicated this was probably in error. I concur with Alston. I have seen no *V. luteola* from Ceylon and the descriptions are from references cited.

In following Verdcourt, I have placed *Dolichos luteus* in synonomy under *Vigna marina* and have retained *V. luteola* with *V. repens* in synonomy. *Vigna marina* and *V. luteola* are very close and, with *V. hosei*, represent the Ceylon species of subg. *Vigna*, sect. *Vigna*.

Whether *V. luteola* is actually present in Ceylon depends on a reexamination of *V. marina* collections or perhaps a worldwide examination of these two taxa, with *V. retusa*, to see if character differences warrant lesser rank.

4. Vigna unguiculata (L.) Walp., Rep. 1: 779. 1842; Piper, USDA Bur. Pl. Industr. Bull. No. 229. 1912 & USDA Bur. Pl. Industr. Circ. 124: 29. 1913; Verdcourt, Kew Bull. 24(3): 542. 1970 (cum syn.); Verdcourt, Fl. Trop. East Africa, Papil. 642. 1971; Verdcourt, Man. New Guinea Leg. 526, fig. 128. 1979; Marechal et al., Taxon 27: 200. 1978 (cum syn.). Type: Specimen grown at Uppsala from seeds received by Linnaeus from Barbados (lecto.), fide Verdcourt (1971).

Dolichos unguiculatus L., Sp. Pl. 725. 1753.
Dolichos sinensis L., Herb. Amboin. 23. 1754; Moon, Cat. 52. 1824; Roxb., Fl. Ind. 3: 302. 1832; Wight & Arn., Prod. 1: 250. 1834 (cum syn. p.p. & vars.).
Phaseolus cylindricus L., Herb. Amboin. 23. 1754.
Dolichos sesquipedalis L., Sp. Pl. ed. 2, 1019. 1763; Moon, Cat. 52. 1824 (& vars.); Roxb., Fl. Ind. 3: 303. 1832.
Dolichos catjang Burm. f., Fl. Ind. 161. 1768; L., Mant. 2: 269. 1771 ('catiang'); Moon, Cat. 53. 1824; Roxb., Fl. Ind. 3: 303. 1832.
Vigna catjang (Burm. f.) Walp., Linnaea 13: 533. 1839; Miq., Fl. Ind. Bat. 1: 188. 1855, nomen; Baker in Hook. f., Fl. Br. Ind. 2: 205. 1876 (as *V. catiang* Endl.); Trimen, Handb. Fl. Ceylon 2: 74. 1894 (as *V. catiang* Endl.); Gamble, Fl. Pres. Madras 1(2): 365. 1918.
Vigna sinensis (L.) Hassk., Cat. Pl. Hort. Bogor 279. 1844; Trimen, Handb. Fl. Ceylon 2: 74. 1894; Alston in Trimen, Handb. Fl. Ceylon 6: 82. 1931.
Vigna sesquipedalis (L.) Fruhw., Anbau Hulsenfr. 254. 1898; Alston in Trimen, Handb. Fl. Ceylon 6: 82. 1931.
Vigna cylindrica (L.) Skeels, USDA Bur. Pl. Industr. Bull. 282: 32. 1913; Alston in Trimen, Handb. Fl. Ceylon 6: 82. 1931.

Variable cultivated species, herbaceous, trailing or climbing vines, occasionally erect, annual or perennial; stems ridged, mostly glabrous. Leaf rachis 0.6–2.5

cm long, ⅓–¹/₆ the petiolar length; stipules constricted at insertion, the distal portion oblanceolate, 6–20 mm long, the proximal portion acute, 2–6 mm long; stipels lanceolate; leaflets ovate, rhomboid or lanceolate, to about 16 × 12.5 cm, acute or obtuse apically, cuneate or somewhat rounded basally, both surfaces glabrous or sparsely pubescent, usually with 3 prominent veins inserted at the base, the terminal leaflets subhastate or 3-lobed basally, the laterals greatly inequilateral, lobed outside towards the base. Inflorescences axillary, to about 40 cm long, florate 1/4–1/7 the length; bracts lanceolate, about 4 mm long, deciduous; bracteoles oblanceolate, about 3 mm long, semipersistent. Flowers 1.5–3.5 cm long, the pedicels about 2 mm long; calyx tube about 4 mm long, all lobes acute, ciliate, the upper lobes emarginate, shorter than the tube; petals white, greenish, yellow or purplish, the vexillum orbicular, basally inflexed biauriculate, the paired appendages rounded or emarginate apically, glabrous, 1.2–3.3 cm long, the claw 5 mm long, the keels obliquely oblong, about equalling the wings, untwisted, the beak obtuse or narrow, without a pocket (?), shortclawed; style constricted at the base, the upper style uniform, hairy inside about 1/2 the length, obliquely truncate at the apex. Fruit somewhat turgid, about 15 cm long to much longer, 7 mm wide, 4 mm thick, glabrous, about 16-seeded. Seeds ovoid or rounded-oblong, about 8 × 7 × 2 mm, variously coloured, the hilum elliptic, about 2 mm long, surrounded by a raised rim aril. Chromosome number 2n = 22, 24.

Distr. Cultivated and grown throughout the tropics as food crops. Probably native to the Old World.

Uses. Food crops.

Vern. In Ceylon, *Vigna catjang* (catjang) and *V. unguiculata* (Cowpeas) are grouped together as: Black-eye bean, Black eye-Pea, Catiang-bean, Cherry bean, Cowpea, Jerusalem Pea, Marble pea and Tonkin-pea (E); Li-me, Mil-me, Mekaral, Wanduru-me (S); Kodip-payam (T). The Yard-long or asparagus bean (E) and Polon-me (S) refer to *V. sesquipedalis*.

Specimens Examined. POLONNARUWA DISTRICT: Between Habarane and Kantalai, plants creeping & twining in grass, *Rudd & Balakrishnan 3124* (PDA, US). RATNAPURA DISTRICT: S. of Weddagala, Kudawe village, *Maxwell et al. 969* (K, PDA, US).

Notes. See Summerfield & Bunting, Adv. Leg. Sci. sect. 7. 1980, for current information, references and research contacts on the cowpeas.

Description based on *Wight 738* (COIMB).

The three cultivated "species" grown in Ceylon are considered subspecies by Verdcourt and others and can be separated as follows:

1 Fruit 7.5–13 cm long, erect or ascending when green; seeds oblong-cylindrical or somewhat reniform, 5–6 mm long, about as wide as thick; habit frequently suberect or erect-bushy, sometimes twining ....**V. unguiculata** subsp. **cylindrica** (Catjang)

1 Fruit 20–90 cm long, pendant or becoming so when green; seeds sub- or elongate-reniform or subglobose, 6–12 mm long, rarely as wide as thick; habit usually twining, occasionally erect
    2 Fruit 20–30 cm long, neither flabby nor inflated when green, fairly compact & without constrictions between the seeds; seeds 6–9 mm long, if subglobose occasionally as wide as thick; twining habit, sometimes erect....................**V. unguiculata** subsp. **unguiculata** (Cowpeas)
    2 Fruit 30–90 cm long, elongated, becoming inflated and flabby when green, the seeds widely separated & with constrictions between the seeds; seeds 8–12 mm long, elongate-reniform, much longer than wide or thick; mostly twining habit...... **V. unguiculata** subsp. **sesquipedalis** (yard-long bean) (From various sources cited above)

This species is in subg. *Vigna,* sect. *Catiang* (DC.) Verdc. See Verdcourt (p. 542, 1970) for the section characters.

5. Vigna vexillata (L.) A. Rich., Hist. Fis. Polit. Nat. 1. Cuba (Spanish ed.) 11: 191. 1845; Benth. in Mart., Fl. Bras. 15(1): 194. 50/1 (p.p.). 1859; Thw., Enum. Pl. Zeyl. 90. 1859; Baker in Hook. f., Fl. Br. Ind. 2: 206. 1876; Trimen, Handb. Fl. Ceylon 2: 74. 1894; Gamble, Fl. Pres. Madras 1(2): 364. 1918; Urban, Symb. Ant. 4: 310. 1903–1911, 8: 312. 1920–1921; Backer & Bakh. f., Fl. Java 1: 643. 1963; Verdcourt, Kew Bull. 24(3): 553, 555, 1970 (cum syn.); Verdcourt, Fl. Trop. East Africa, Papil. 652–655. 1971; Verdcourt, Man. New Guinea Leg. 527. 1979; Lackey & D'Arcy, Ann. Missouri Bot. Gard. 67(3): 801. 1980. Type: Cuba, Havana: *Phaseolus flore adorator vexillo amplopatulo* of Dill., Hort. Elth. t. 234/302. 1732, lectotype, not seen, fide Verdcourt (1971).

Phaseolus vexillatus L., Sp. Pl. 724. 1753.
Phaseolus pulniensis Wight, Ic. Pl. Ind. Or. t. 202. 1839.

Twining or trailing vines; stems to 6 m, with dense pale or dark brown pubescence. Leaf rachis 0.4–3 cm long, about 1/4 the petiolar length, strigose; stipules lanceolate, about 12 mm long, very slightly produced, cordate basally, becoming oblique, striate, persistent; stipels linear, to about 3 mm long, strigose; leaflets usually ovate to lanceolate, rarely slightly lobed, 2.5–16.5 × 0.4–8.3 cm, acute or acuminate apically, rounded or truncate basally, both surfaces pubescent, raised reticulate. Inflorescences axillary, somewhat subumbellate, 4.5–36 cm long, longer than the leaves, the rachis short, 2-6-flowered; bracts not seen; bracteoles linear or somewhat lanceolate, about 4 mm long. Flowers about 2.5 cm long, the pedicels 1–2 mm long; calyx tube 5–7 mm long, pubescent with whitish or brown hairs, the upper lobes deeply emarginate, all lobes acuminate or somewhat linear, usually exceeding the tube length; petals usually pink-violet, the vexillum suborbicular, about 20 × 30 mm, glabrous, without appendages in the target area, the claw about 3 mm long, the wings rounded-oblong, about 22 mm long, the claw about 5 mm long, the keels semiorbicular, about 20 mm long, the beak sharp, twisted, incurved about 180 degrees, left keel with a raised pocket, small basal pockets in both petals (fide Verdcourt), the claw about 4 mm long; upper style widened, flat, curved, thickened apically, with a short lobe-like appendage, the stigma subterminal (fide Verdcourt). Fruit erect, linear, turgid, 4–14 cm long, 2.5–4 mm wide, about 3 mm thick, with erect, bristly brown hairs, 7–18-seeded. Seeds quadrate or oblong-reniform, about 3 × 2.5 × 2 mm, horizontal, brown to dark red with black mottling, the hilum oblong, about 1.5 mm long, arillate tissue weakly developed. Chromosome number 2n = 22.

Distr. Tropical Africa, South Africa, India, SW. Asia and widespread in the tropics of both hemispheres.

Ecol. This species is found in the hill country and moist regions, blooming April through August. Trimen (1894) reports the species as rather common.

Uses. The species is cited as a pasture cover crop, green manure and erosion control crop in "Tropical Legumes: Resources for the Future," p. 34. 1979. The use and possible cultivation of the edible tubers as a root crop is advocated.

Vern. Walima (S) on *Simpson 8246* (BM).

Specimens Examined. LOCALITY UNKNOWN: *Mrs. & Col. Walker s.n.* (K, 2 sheets). KANDY DISTRICT: betw. Madugoda & Hunnasgiriya, *Rudd & Balakrishnan 3239* (PDA, US). NUWARA ELIYA DISTRICT: past Mulhalkele to Uda Pussellawa, *Maxwell 1008* (PDA, US); Maturata, April 1860, *Thwaites ? C.P. 2781* (BM, PDA). RATNAPURA DISTRICT: About 13 mi. NE. of Deniyaya, *Davidse 7891* (MO, US). BADULLA DISTRICT: Canyon below Ella Rest House, *Fosberg & Sachet 53178* (NY, US); near Jangula, *Simpson 8246* (BM).

Notes. This species is the only representative in Ceylon of subg. *Plectropis,* sect. *Plectropis.* See Verdcourt (p. 553, 1970) for section characteristics.

6. Vigna aconitifolia (Jacq.) Marechal, Bull. Jard. Bot. Etat 39: 160. 1969; Verdcourt, Kew Bull. 23(3): 464. 1969 & 24(3): 557. 1970. Type: Jacq. t. 52 ?, not seen.

Phaseolus aconitifolius Jacq., Obs. Bot. 3: 2. t. 52. 1768; Roxb., Fl. Ind. 3: 299. 1832; Wight & Arn., Prod. 1: 247. 1834; Baker in Hook. f., Fl. Br. Ind. 2: 202. 1876; Trimen, Handb. Fl. Ceylon 2: 71. 1894; Prain, J. Asiat. Soc. Bengal 66(2): 421. 1897; Alston in Trimen, Handb. Fl. Ceylon 6: 81. 1931.

Dolichos dissectus Lam., Enc. 2: 300. 1786, excl. ref. Plukenet (= *Vigna trilobata* (L.) Verdcourt, fide Verdcourt (1970)).

Small climbing, trailing or prostrate vines, annuals or perennials; stems very numerous from a single deep taproot, stems frequently with copious long spreading or reflexed hairs. Leaf rachis about 15 mm long, ⅓–⅙ the petiolar length; stipules lanceolate, acuminate, 5–7 mm long, apparently non-produced, or peltate, shortly produced, persistent; stipels filamentous, 2–4 mm long; leaflets 3-lobed or frequently with variable deep sinuses incised almost to the midrib, the lobes lanceolate or linear, 2.5–5.5 × 0.2–2 cm, acute, obtuse, frequently rounded apically, cuneate to very narrow basally, pubescent beneath. Inflorescences axillary, 5–12 cm long, few-flowered; bracts lanceolate-oblong, 3.5 mm long, persistent; bracteoles lanceolate, oblique, about 3 mm long. Flowers about 7 mm long, the pedicels about 1 mm long; calyx tube about 2 mm long, glabrous (?), the upper lobes connate or emarginate, obtuse, the rest triangular, shorter than the tube; petals yellow, the vexillum suborbicular, about 6 × 8 mm, deeply emarginate, glabrous, without appendages in the target area, subsessile, the wings somewhat triangular, about 4 × 4 mm, the claw 1.5 mm long, the keels falcate, about 5 × 2 mm, the beak incurved, coiled, the left keel with a distinct median pocket, the claw about 2 mm long; ovary several-ovulate, the lower style constricted at ovary insertion, the upper style wide, twisted, beaked at the apex, with long hairs for about 2 mm below the stigma. Fruit linear, straight, spirally dehiscent, 2.5–5.5 cm long, 3–4 mm wide, 1.5–3 mm thick, septate, the exocarp slightly depressed between the seeds, 6–9(–14)-seeded; sutures swollen. Seeds horizontal, oblong-rounded, cylindrical, about 2.5 × 1.5 × 1.5 mm, yellow brown or light brown with darker mottling, hard, smooth, the hilum linear-oblong, or elliptic, less than 1 mm long, whitish, without distinct arillate tissue. Chromosome number 2n = 22.

Distr. India, NW. to the Himalayas, and parts of Africa. This species may be as widely spread through Asia as *V. trilobata,* which it resembles. In Ceylon Trimen (1894) gives Jaffna and Batticaloa Districts. Alston (1931) cites Polonnaruwa District.

Ecol. Trimen (1894) reports the species rather common in the dry region, producing its yellow flowers January through August. Probably flowers year round.

Uses. The species is cited in the List of Medicinal Herbs (as *P. aconitifolius* Jacq.). Roxburgh (1832) reports the species much cultivated and used for feeding cattle. The species is the most drought resistant of the cultivated *Vigna,* according to Jain & Mehra (Summerfield & Bunting [eds.], Adv. Leg. Sci. 459. 1980). It is one of the 6 *Vigna* spp. listed as cultivated to varying degrees in Asia.

Vern. Dew-bean, Moth bean, Moth (E); Makushtha (S); Kollu, Tulkapavir (T).

Specimens Examined. All placed with *Vigna trilobata.*

Notes. I observed in my studies at COIMB that leaflet lobing variation can be considerable on the same specimen. The specimens determined *P. aconitifolius* at COIMB have significantly larger leaflets than the Ceylon specimens, and as Roxburgh states the leaflet lobes are linear-lanceolate. There are specimens with the lobes dissected to the base so that they appear almost digitately 5-foliolate. *Vigna aconitifolia*'s other obvious vegetative character, its lanceolate stipule, not at all peltate, and non-produced, separates it from *V. trilobata.* The stipules of the Indian material I examined at COIMB, do not fit the subg. *Ceratotropis* stipule description of Verdcourt (1970), but as Verdcourt notes many problems in *Vigna* remain. Trimen (1894) states that "Dr. Thwaites did not distinguish this from *P. trilobus,* and it is scarcely separable as a species." This and the following 4 species, *V. trilobata, radiata, mungo,* and *umbellata,* fall within subg. *Ceratropis* (Piper) Verdcourt. See Verdcourt (p. 556, 1970) for the subgenus characters.

Vigna aconitifolia may have been introduced into Ceylon since the 1890's but I believe prior to this time the name was assigned to deeply lobed leaflet forms of *V. trilobata.*

Description based on *Wight 728* (COIMB), *759* (NY), and other specimens at COIMB.

7. Vigna trilobata (L.) Verdcourt, Taxon 17: 172. 1968 (cum syn.); Verdcourt, Kew Bull. 24(3): 557, 560. 1970; Verdcourt, Man. New Guinea Leg. 525. 1979. Lectotype: India, Plukenet, Almagest. 292. Phytogr. t. 120, fig. 3 (error for 7). 1694, not seen. See Verdcourt (p. 171, 1968).

Dolichos trilobatus L., Mant. 1: 101. 1767.

Phaseolus trilobatus (L.) Schreber, Nova Acta Acad. Caes. Leop.-Carol. German Nat. Cur. 4: 132. 1770; Baillon, Bull. Soc. Linn. Paris 1: 379. 1883; Leese, Amer. Midl. Naturalist 60: 144. 1958.

Phaseolus trilobus sensu Ait., Hort. Kew. 3: 30. 1789 quoad nomen tantum et auct. mult. non *Dolichos trilobus* L., Sp. Pl. 726. 1753 (p.p.); Moon, Cat. 52. 1824; Roxb., Fl. Ind. 3: 298. 1832; Wight & Arn., Prod. 1: 246. 1834 (& vars.); Wight, Ic. Pl. Ind. Or. t. 94. 1839; Thw., Enum. Pl. Zeyl. 90. 1859; Baker in Hook. f., Fl. Br. Ind. 2: 201. 1876; Trimen, Handb. Fl. Ceylon 2: 71. 1894; Gamble, Fl. Pres. Madras 1(2): 362. 1918; Backer & Bakh. f., Fl. Java 1: 640. 1963.

Small vines, usually prostrate or trailing, never twining nor climbing (?); stems slender, terete, numerous from a single deep taproot, perennial, glabrous or glabrescent. Leaf rachis about 5 mm long, 1/5 or less the petiolar length; stipules distinctly peltate, about 8 mm long, persistent; stipels minute, to 1 mm long; leaflets usually with shallow lobes, rarely unlobed, about 1.5 cm long, 1.5 cm wide, obtuse apically and basally, sparsely pubescent or glabrous, only the midrib prominent beneath or 3-veined basally, the terminal leaflets frequently 3-lobed, the laterals frequently 1-lobed on the outer margin. Inflorescences axillary, 5.5–13 cm long, greatly exceeding the leaves, the rachis 5–10 mm, 2–3-flowered; bracts ovate, about 3.5 × 2 mm, persistent; bracteoles obliquely lanceolate, about 3 mm long. Flowers about 7 mm long, the pedicels about 1 mm long; calyx tube about 2 mm long, glabrous (?), the upper lobes connate or occasionally emarginate, obtuse, the rest triangular, shorter than the tube; petals yellow, the vexillum about 6 × 8 mm, deeply emarginate, glabrous, without appendages in the target area, subsessile, the wings somewhat triangular, about 4 × 4 mm, the claw 1.5 mm long, the keels falcate, untwisted, about 5 × 2 mm, the beak incurved, coiled, the left keel with a distinct, median pocket, the claw about 2 mm long; ovary several-ovulate, the lower style constricted at ovary insertion, the upper style flat, twisted, beaked at the apex, with long hairs for about 2 mm below the stigma, the stigma globose, subterminal. Fruit spirally dehiscent, linear, straight, 3.8–4.8 cm long, about 3 mm wide, 2–3 mm thick, glabrous, septate; 6–13-seeded; sutures swollen. Seeds oblong-rounded, cylindrical, about 3 × 2 × 2 mm, horizontal, brown with darker mottling, hard, smooth, the hilum elliptic, about 1 mm long, whitish, raised. Chromosome number 2n = 22.

Distr. India, NW. to the Himalayas and Burma. Probably spread through cultivation through SW. Asia, to Indonesia and New Guinea.

Ecol. Common in the dry regions of Ceylon and on the seashores. Flowering probably occurs throughout the year.

Uses. Cultivated in India only as a cover crop and for fodder. Tribal people today still gather seeds from wild plants for food as Roxburgh reported in 1832.

Vern. Bin-me, Munwenna (S); Navippayaru, Pachapayaru (T).

Specimens Examined. JAFFNA DISTRICT: Elephant Pass, S. end of Jaffna Lagoon, *Fosberg & Balakrishnan 53569* (US); betw. Jaffna causeway & Kayts, *Rudd 3268* (PDA, US); nr. Kankesanturai, sand, above beach, *Rudd 3280* (PDA, US); betw. Jaffna & Elephant Pass, *Rudd 3286* (NY, PDA, US), 1859, *Thwaites ? C.P. 1474* (P), Feb. 1890, *s. coll.* (PDA), *Thwaites C.P. 1474* (BM, K). MANNAR DISTRICT: Mannar Isl., vic. Pesalai, *Robyns 7340* (US); Mannar Isl., Talaimannar, 17.7.16, *J.M.S. s.n.* (PDA). ANURADHAPURA DISTRICT: nr. Anuradhapura Resthouse, *Maxwell & Jayasuriya 798* (NY, PDA, US); betw. Anuradhapura & Galkulama, *Rudd 3308* (US). POLONNARUWA DISTRICT: Sacred area, *Dittus WD70032602* (US); betw. Polonnarwa & Kalkudah, *Rudd & Balakrishnan 3154* (PDA, US). PUTTALAM DISTRICT: Mannar-Puttalam Rd., Wilpattu Natl. Pk., *Cooray 70020116R* (NY, US), *Davidse & Sumithraarachchi 8210* (MO, PDA, US); Wilpattu, nr. Pallugaturai, *Fosberg et al. 50908* (US); Wilpattu, Kali Villu, *Fosberg et al. 50955* (NY, US), *50972* (US), *Mueller-Dombois 68091107* (PDA, US), *Mueller-Dombois & Cooray 69050112R* (PDA, US), *Wirawan et al. 1007* (NY, PDA); shore west of Sinnapadu, *Fosberg & Jayasuriya 52825* (US); vic. Puttalam Lagoon, nr. beach in sand, *Maxwell & Jayasuriya 822, 824* (NY, PDA, US); Kollankanatta, beach in tall grass, *Mueller-Dombois & Cooray 69042722* (US); W. of Wilpattu, Pallugaturai Beach, *Robyns 7337* (US); Chilaw, *Simpson 8200* (BM); Kalpitiya, *Simpson 9175* (BM). TRINCOMALEE DISTRICT: 1858, *s. coll. C.P. 1474* (PDA); w/o date, *s. coll. C.P. 1474* (PDA). BATTICALOA DISTRICT: sea shore, 1858, *s. coll. C.P. 1474* (PDA); Kalkudah, *Cramer 2741* (US); 8 mi. S. of Batticaloa, sea coast, *Townsend 73/274* (US); Keeli-Kudah, *Waas 2119* (US). AMPARAI DISTRICT: N. of Arugam Bay, *Fosberg & Sachet 53042* (NY, PDA, US); betw. Amparai & Maha Oya, mi. 19–20, *Rudd & Balakrishnan 3226, 3227* (US). COLOMBO DISTRICT: Negombo, July 1930, *J.M. de Silva s.n.* (PDA), seashore, 21.10.49, *J.E. Senaratne s.n.* (PDA); Kochchikade, sandy palm grove,

Simpson 7966 (BM). HAMBANTOTA DISTRICT: Yala Natl. Pk., beach behind Yala campsite, *Comanor 680* (US); Yala, Komawa Wewa, *Cooray 70032512R* (US), dunes at Nahasi Villa, *Davidse 7784* (MO, PDA, US), beach east of Butawa Modera, *Fosberg 50318* (US), mouth of Menik Ganga, *Fosberg 51067* (US), sand beach, next to Uraniya Lagoon, *Nowicke & Jayasuriya 391* (NY, US), *Sohmer et al. 8994* (US).

Notes. Wight & Arn. (p. 246, 1834) divide the taxon into var. α, pubescent or nearly glabrous, with leaflets roundish or entire; var. β, pubescent or nearly glabrous with leaflets deeply 3-lobed; and var. γ, hairy, with leaflets deeply 3-lobed. They feel the last variety was the same as plate 276 of Roxb., in E.I.C., Mus. 1820 (as *Phaseolus aconitifolius*), but not of Fl. Ind. 3: 298. 1832. Prain (1897) and others agree.

Thwaites (1859) states "common amongst grass in the hotter parts of the island, varying greatly in the shape of the leaflets." A sheet at (K) with "Thwaites?" as collector has var. *maritimus* written on it and corresponds to Wight & Arn.'s var. α, but Thwaites did not validly describe the variety. This may be var. *oxalideus* Grah., fide Baker (1876).

Baker (1876) notes that leaflets of the annual, cultivated specimens show less lobing and are rarely entire, while those of perennial, wild specimens are deeply lobed.

Rudd notes that *Rudd & Balakrishnan 3277* is a plant with entire leaflets growing among *Rudd & Balakrishnan 3226,* plants with dissected leaflets. I believe the widespread distribution of *Phaseolus aconitifolius* recorded for Ceylon is due to confusing the deeply lobed leaflet varieties or forms of *V. trilobata* with *V. aconitifolia.*

8. Vigna radiata (L.) Wilczek, Fl. Congo Belge 6: 386. 1954; Verdcourt, Kew Bull. 24(3): 558–60. 1970 (cum syn.); Verdcourt, Fl. Trop. East Africa, Papil. 655–657. 1971; Verdcourt, Man. New Guinea Leg. 523. 1979. Type: Ceylon, *Phaseolus zeylanicus siliquis radiatim digestis* of Dillenius, Hort. Eltham. 315. t. 235/304. 1732 (K lectotype), fide Verdcourt (1971), not seen.

Phaseolus radiatus L., Sp. Pl. 725. 1753; Moon, Cat. 52. 1824; Thw., Enum.
     Pl. Zeyl. 89, 412. 1859; Prain, J. Asiat. Soc. Bengal 66: 422. 1897; Gam-
     ble, Fl. Pres. Madras 1(2): 362. 1918; Alston in Trimen, Handb. Fl. Ceylon
     6: 82. 1931.
Phaseolus sublobatus Roxb., Fl. Ind. 3: 288. 1832; Backer & Bakh. f., Fl. Java
     1: 640. 1963.
Phaseolus glaber Roxb., Fl. Ind. 3: 291. 1832. Type: E. India, *Roxburgh s.n.* (K).
Phaseolus aureus Roxb., Fl. Ind. 3: 297. 1832; Alston in Trimen, Handb. Fl.
     Ceylon 6: 81, 82. 1931; Burkart, Leg. Argentinas ed. 2, 436. 1952; Backer
     & Bakh. f., Fl. Java 1: 640. 1963. Type: Bengal, *Roxburgh* drawing *1604*
     (K), not seen.

Phaseolus trinervius Wight & Arn., Prod. 1: 245. 1834; Thw., Enum. Pl. Zeyl.
   90. 1859; Baker in Hook. f., Fl. Br. Ind. 2: 203. 1876; Trimen, Handb. Fl.
   Ceylon 2: 72. 1894. Type: India, *Wallich 5603* (K).
Phaseolus mungo L. var. *glaber* (Roxb.) Baker in Hook. f., Fl. Br. Ind. 2: 203.
   1876.
Phaseolus calcaratus Roxb. var. *glaber* (Roxb.) Prain, J. Asiat. Soc. Bengal 66:
   424. 1897.
Azukia radiata (L.) Ohwi, Fl. Japan 691. 1953.
Phaseolus max nomen, Trimen, Handb. Fl. Ceylon 2: 72. 1894 (p.p. as 'green
   gram').
Phaseolus mungo auct. mult. non L., Moon, Cat. 52. 1824 (green seeded) ?;
   Roxb., Fl. Ind. 3: 292. 1832.

Erect or climbing, herbaceous annual; stems to about ½ m long, with fulvous
or brown, long, spreading or bristly pubescence. Leaf rachis 1.5–4 cm long, ⅓–
⅐ the petiolar length; stipules peltate, 1–1.8 × 0.3–1 cm, ciliate, persistent;
stipels lanceolate; leaflets elliptic, rhomboid or ovate, 5–16 × 3–12 cm, unlobed
or 2–3-lobed, acuminate apically, broadly cuneate or rounded basally, both surfaces glabrous or pilose, sometimes with 3 prominent veins inserted at the base.
Inflorescences axillary, to about 10 cm long, few to numerous-flowered; bracts
ovate-lanceolate, 4–5 mm long; bracteoles linear-lanceolate or oblong, 4–7 mm
long, striate, somewhat persistent. Flowers about 1.5 cm long, the pedicels 2–3
mm long; calyx tube 3–4 mm long, glabrous, the upper lobes bifid, all lobes
triangular, about 2 mm long, ciliate; petals yellow, greenish yellow, sometimes
pinkish inside, the vexillum suborbicular, about 1.2 × 1.6 cm, emarginate,
glabrous, without appendages in the target area, subsessile, the wings somewhat
triangular, about equalling the vexillum, short-clawed, the keels obliquely-oblong
or falcate, slightly shorter than the wings, the beak incurved through 180°, the
left keel with a distinct median pocket, short-clawed; ovary many-ovulate, the
lower style constricted at ovary insertion, the upper style flat, twisted, with long
hairs below the stigma, beaked at the apex, the stigma globose. Fruit spirally
dehiscent, straight, turgid, 4–9 cm long, 5–6 mm wide, about 4 mm thick, with
dark brown, short, spreading bristly pubescence, septate, (8–)11–14-seeded. Seeds
oblong-rounded, cylindrical, 2.5–4.2 × 2.5–3.2 × 2.3–2.8 mm, horizontal,
greenish, brown or blackish, hard, the testa rough with raised ridges and parallel
riblets between, the hilum elliptic or short-oblong, 1.5–2 mm long, the aril not
developed (fide Verdcourt, 1971). Chromosome number 2n = 22, some counts
44.

Distr. Cultivated crop of ancient origin. Widespread throughout the tropics
and subtropics of both hemispheres.
   Ecol. Mung or Green Gram tolerates a drier environment than *V. mungo*,
Black Gram.
   Uses. See Jain & Mehra in Summerfield & Bunting, Adv. Leg. Sci. 459.

1980. They report *V. radiata* with high palatability: the pods eaten raw or cooked, the whole or split seeds used in soups and other preparations, and the seed sprouts also eaten.

Vern. Sona Mung, Mung bean or Green Gram, Jerusalem pea (E); Mun, Bu-me, Mun-eta (S); Chirup payaru, Chiruppataru, Pani-payir (T).

Specimens Examined. LOCALITY UNKNOWN: *s. coll. C.P. 1476* (K); *Thwaites C.P. 1476* (BM); "received 1870", *Thwaites C.P. 1476* (GH); *Gardner 233, 234* (K); Herb Hooker, *Walker s.n.* (K). DISTRICT UNKNOWN: Central Province, *s. coll. C.P. 3622* (K). PUTTALAM DISTRICT: Wilpattu Natl. Pk., nr. Sadpuda Kallu, *Fosberg et al. 50823* (US). KANDY DISTRICT: Mi. 20 Pussellawa, *Jayasuriya & Tirvengadum 997* (NY, US); Exp. Stat., Peradeniya, a ground cover, *N.D. Simpson 8754, 9204* (PDA). NUWARA ELIYA DISTRICT: Hakgala, patana beyond lab, *s. coll. s.n.* (PDA); Ambawela, Mar. 1921, *H.D.A. s.n.* (PDA); escape from cultivation, Hakgala, *Rudd & Balakrishnan 3167, 3171* (K, NY, PDA, US). AMPARAI DISTRICT: persisting after cultivation, N. of Bakiella, *Fosberg & Sachet 53130* (US). BADULLA DISTRICT: betw. Bandarawela & Haputale, *Rudd & Balakrishnan 3200* (K, PDA, US).

Notes. Following Verdcourt (1970, 1971), I have separated *V. mungo* from *V. radiata,* although *V. mungo* may only be a variety of *V. radiata.* The taxonomy and nomenclature are mixed in the Ceylon literature.

Description based on *Wallich 5603* (K).

KEY TO 3 VARIETIES
(From sources cited above)

1 Stems, leaves and fruit glabrescent ....
.... var. **glabra** (Roxb.) Verdc. (=*Phaseolus glaber* Roxb., *P. mungo* var. *glaber* (Roxb.) Baker, *P. calcaratus* var. *glaber* (Roxb.) Prain.)
1 Stems, leaves and fruit densely pubescent
  2 Stems twining or prostrate, 1 to about 2 m long; leaflets frequently slightly lobed; seeds 10–15, dark, grey or brown; wild or naturalized..
.... var. **sublobata** (Roxb.) Verdc. (=*P. lobatus* Roxb., *P. trinervius* Wight & Arn.)
  2 Stems mostly erect, 0.3–0.7 m tall; leaflets unlobed; fruit dark, 8–10-seeded; seeds green, yellow; cultivated or relic .... var. **radiata** (=*P. radiatus* L., *P. aureus* Roxb.)

Jain & Mehra (1980) feel the base population of *Vigna* spp. similar to *V. sublobata* s.l. leads to the *V. mungo* and *V. radiata* lines, with *V. sublobata* still existing in the wild.

9. Vigna mungo (L.) Hepper, Kew Bull. 11: 128. 1956; Verdcourt, Kew Bull. 24(3): 558. 1970; Verdcourt, Man. New Guinea Leg. 521. 1979. Type: Plukenet, Almagest. 290. 1694, not seen.

Phaseolus mungo L., Mant. 1: 101. 1767; Wight & Arn., Prod. 1: 245. 1834; Baker in Hook. f., Fl. Br. Ind. 2: 203. 1876 (p.p.); Prain, J. Asiat. Soc. Bengal

66(2): 423. 1897 (p.p. max.); Gamble, Fl. Pres. Madras 1(2): 363. 1918; Alston in Trimen, Handb. Fl. Ceylon 6: 82. 1931.

Phaseolus roxburghii Wight & Arn., Prod. 1: 246. 1834.

Phaseolus mungo var. *radiatus* (L.) Baker in Hook. f., Fl. Br. Ind. 2: 203. 1876, non *P. radiatus* L.

Phaseolus radiatus auct., Roxb., Fl. Ind. 3: 296. 1832; Thw., Enum. Pl. Zeyl. 89. 1859 (*C.P. 1473* p.p.?); Prain, J. Asiat. Soc. Bengal 66(2): 422. 1897 (p.p. & var. *grandis* Prain), non L.

Phaseolus max auct., Roxb., Fl. Ind. 3: 295. 1832; Trimen, Handb. Fl. Ceylon 2: 72. 1894 (p.p. & var. *radiatus* (Baker) Trimen), non *Glycine max* (L.) Merr.

The species description is basically the same as for *V. radiata*. Several authors consider *V. mungo* a variety of *radiata*, with the species rank and separation one of convenience and long usage.

Vigna mungo is a suberect cultivar with many branches, the stem and branches dark, often with darker spots (Roxb. 1832, as *Phaseolus max*), with longer hairs than *V. radiata* vars.; leaflets unlobed; fruit with hairs longer than *V. radiata* vars.; seeds somewhat oblong, blackish, the testa rough with raised ridges and with a distinctly raised rim aril surrounding the hilum.

Distr. Same as *V. radiata*.

Ecol. More suitable to a moist environment than Green Gram.

Uses. See under *V. radiata*. *Phaseolus mungo* L. is carried on the Seeds List as an annual with seeds available all the year. I assume it can be cultivated year round wherever suitable. It blooms year round and is rarely found as an escape from crop or garden plots. Seeds of *V. radiata*, Green Gram, are reportedly more palatable than *V. mungo*, Black Gram.

Vern. Urd, Black Gram (E); Ulundu, Mun, Bu-me (S); Uluntu, Chiruppayaru (T).

Specimens Examined. BADULLA DISTRICT: cultivated at Ekiriankumbura, Jan. 1888, *s. coll. s.n.* (PDA). TRINCOMALEE DISTRICT: Trincomalee, *Glenie C.P. 3622* (PDA).

Notes. Wight & Arn. (1834) note the specimen of Flora Zeylanica in Hermann's herbarium is not in flower, is very hairy all over but does not appear erect. Prain (1897) apparently assigns this plant to his *P. radiatus* var. *grandis* (*V. mungo*). Trimen (1894) cites Herm. Mus. 22. Fl. Zeyl. n. 280 under *P. max* (= *V. radiata* p.p. *max*) and Herm. Mus. 47, Fl. Zeyl. n. 281 under *P. max* var. *radiatus* (= *V. mungo*). I could not make a distinction in Herm. Mus. at BM.

10. Vigna umbellata (Thunb.) Ohwi & Ohashi, Jap. J. Bot. 44: 31. 1969; Verdcourt, Kew Bull. 24(3): 560. 1970; Verdcourt, Fl. Trop. East Africa, Papil. 621, 622, 656. 1971 (notes); Verdcourt, Man. New Guinea Leg. 525. 1979. Type: Japan, *Thunberg* Herb. at Uppsala No. *16789* (UPS), not seen.

Dolichos umbellatus Thunb., Trans. Linn. Soc. London 2: 339. 1794.

Phaseolus pubescens Blume, Cat. Gew. Buitenz. 93. 1823; Backer & Bakh. f.,
Fl. Java 1: 640. 1963.

Phaseolus calcaratus Roxb., Fl. Ind. 3: 289. 1832; Wight & Arn., Prod. 1: 245.
1834; Thw., Enum. Pl. Zeyl. 412. 1860; Baker in Hook. f., Fl. Br. Ind. 2:
203. 1876; Trimen, Handb. Fl. Ceylon 2: 73. 1894; Gamble, Fl. Pres. Madras
1(2): 362. 1918; Ridley, Fl. Malay Penins. 1: 566. 1922.

Vigna calcarata (Roxb.) Kurz., J. Asiat. Soc. Bengal 45: 247. 1876.

Azukia umbellata (Thunb.) Ohwi, Fl. Japan 691. 1953.

Small vines; stems flexuose, twining and prostrate at the base, often rooting
at the nodes (fide Trimen, 1894), to about 1.5 m long, sparsely pubescent. Leaf
rachis about 5 mm long, $1/5$–$1/10$ the petiolar length, with deflexed hairs; stipules
peltate to lanceolate-oblong, about 4 mm long, persistent; stipels acicular-linear,
2–3 mm long; leaflets thin, ovate, frequently 3-lobed, 4–13 × 2.2–5.8 cm, acute-
acuminate apically, broadly cuneate, rounded or somewhat cordate basally, both
surfaces sparsely villous, the laterals inequilateral. Inflorescences axillary, usually
single, finally longer than the leaves, the rachis to about 4 cm long, tuberculate,
5–20-flowered; bracts linear-lanceolate, 4–5 mm long, caducous; bracteoles linear,
about 1.5 mm long, semipersistent. Flowers 7–12 mm long, the pedicels to 6
mm long; calyx tube about 4 mm long, glabrous, all lobes triangular, ciliate,
1.5–4 mm long, the lower longest; petals yellow, the vexillum suborbicular, about
1.2 × 1.6 cm, emarginate, glabrous, without appendages in the target area,
subsessile, the wings somewhat obtriangular, about equalling the vexillum, short-
clawed, the keels obliquely-oblong or falcate, slightly shorter than the wings,
with an incurved beak, the right keel with a large, long, horn-like pocket, short-
clawed; lower style constricted at ovary insertion, the upper style flat, narrowing to a sharp apical beak, with long hairs inside below the stigma. Fruit spirally
dehiscent, linear, nearly straight, pendant ?, 4–9 cm long, about 5 mm wide,
2–3 mm thick, green or tan, glabrous, septate, about 6–12-seeded. Seeds ovoid-
subcylindrical, dark brown, about 2.5 × 2 × 2 mm, horizontal, smooth, the
hilum elliptic or rounded, 1.5–2 mm long, with a raised rim aril. Chromosome
number $2n = 22$.

Distr. Throughout India, Malaysia and the Philippines. Cultivated throughout
tropical Asia and parts of Africa.

Ecol. The species is adapted to hot climates and high humidity, ideal for low
altitude tropics. Photoperiod restricts flowering and seed set to equatorial regions.
Trimen (1894) indicates the species common in the grass of the moist region from
1–4000 ft, Gamble (1918) to 7000 ft (both as *P. calcaratus*). Flowering September
through November.

Uses. According to "Tropical Legumes: Resources for the Future," 1979,
the Rice Bean was used in older farming techniques to produce a nutritious seed
crop between rice harvesting and planting, but multiple cropping has eliminated

this practice. This report describes the bean as having a high nutritional value, a wide variety of cultivars available, and, under tropical conditions, simultaneous pod maturity in as little as 60 days. The beans must be cooked before eating, and they lack the strong taste of *Phaseolus vulgaris*.

Specimens Examined. DISTRICT UNKNOWN: Ambagmuwa, Nov. 1854, *s. coll. C.P. 1473* (PDA). KANDY DISTRICT: Govt. stock garden, Peradeniya, 2.11.19, *J.M. Silva s.n.* (PDA). NUWARA ELIYA DISTRICT: Ramboda, Nov. 1854, *s. coll. C.P. 1473* (BM, PDA).

Notes. Trimen (1894) indicates neither *P. calcaratus* (*V. umbellata*) nor *P. trinervius* (*V. radiata* var. *sublobata* (Roxb.) Verdc.) is ever cultivated in Ceylon. Neither species seems common in the Ceylon flora at present and may now exist only as escapes from cultivation.

11. Vigna caracalla (L.) Verdcourt, Kew Bull. 24(3): 552. 1970; Lackey & D'Arcy, Ann. Missouri Bot. Gard. 67(3): 793. 1980. Type: India orient, *L.* Sp. *1017,* not seen.

Phaseolus caracalla L., Sp. Pl. 725. 1753; Moon, Cat. 52. 1824; Wight & Arn., Prod. 1: 244. 1834 (note); Paxton, Bot. Mag. 10: 267. c. 1840; Miq., Fl. Ind. Bat. 195. 1855; Benth. in Mart., Fl. Bras. 15(1): 182. 1859; Piper, Contr. U.S. Natl. Herb. 22: 677. 1926 (cum syn.); Backer & Bakh. f., Fl. Java 1: 639. 1963.

Twining, high-climbing vines; stems perennial, to about 7 m long, glabrous or sparsely pubescent. Leaves large, the petioles to 15 cm long; stipules small, ovate or oblong, non-produced, persistent; stipels small, oblong; leaflets membranous, 10–15 × about 10 cm, acute apically, both surfaces slightly pubescent, the terminal ovate-rhomboid, cuneate basally, the laterals obliquely-ovate, rounded to truncate basally. Inflorescences axillary, several-flowered, 10–12 cm long, florate about ½ the length, each tubercle 1–2-flowered; bracts not seen; bracteoles ovate. Flower buds coiled in 2–3 complete spirals, flowers about 6 cm long, the pedicels about 2 mm long; calyx tube about 12 mm long, glabrous, all lobes obtuse, shorter than the tube, with hyaline margins, the upper lobes deeply emarginate; petals white or yellow with pink or violet wings, the vexillum 4.5–6 cm long, twisted, the wings long-clawed, twisted, the keels narrow, incurved-coiled 4–5(5–7 Verdcourt, 1970) complete turns, the keel pockets absent; upper style with the thick part incurved through about 180–340 degrees (Verdcourt, 1970), without a beak. Fruit pendant, to about 18 × 1.3 cm, glabrous, somewhat torulose, 15–20-seeded. Seeds somewhat globose, brown.

Distr. Native to the American tropics, now widely distributed pantropically throughout the lowlands. In Ceylon Moon gives Colombo, but see note.

Uses. An ornamental.

Vern. Snail-flower (E); Moodu-mae (S).

Specimens Examined. None.

Notes. A questionable addition to the Flora. No specimens from Ceylon have been seen. This species may have been confused with the next, *V. adenantha*. *Vigna caracalla* is the type of Verdcourt's subg. *Cochliasanthus* (Trew) Verdc. See Verdcourt (p. 552, 1970) for the characters of the subgenus.

Marechal et al. (1978) place Verdcourt's subg. in synonymy under *Vigna* sect. *Caracallae* (DC) Marechal in subg. *Sigmoidotropis* (Piper) Verdc. Marechal et al. (1978) move *Phaseolus* sect. *Leptospron* Benth. with the species *P. adenantha* into *Vigna* subg. *Sigmoidotropis* sect. *Leptospron* Marechal et al.

12. Vigna adenantha (G. Meyer) Marechal et al., Taxon 27: 202. 1978; Lackey & D'Arcy, Ann. Missouri Bot. Gard. 67(3): 792. 1980. Type: Guyana, R. Essequibo, *Rodschied* (GOET holotype), fide Verdcourt (1971), not seen.

Phaseolus adenanthus G. Meyer, Prim. Fl. Esseq. 239. 1818; Baker in Hook. f., Fl. Br. Ind. 2: 200. 1876 (cum syn.); Trimen, Handb. Fl. Ceylon 2: 70. 1894; Gamble, Fl. Pres. Madras 1(2): 361. 1918; Burkart, Leg. Argentinas ed. 2: 429, fig. 130 a–f. 1952; Verdcourt, Fl. Trop. East Africa, Papil. 615. 1971 (ex descr. p.p.); Verdcourt, Man. New Guinea Leg. 511. 1979.

Phaseolus rostratus Wall., Pl. As. Rar. 1: 50, t. 63. 1830; Wight & Arn., Prod. 1: 244. 1834; Wight, Ic. Pl. Ind. Or. t. 34. 1838. Type: *Wallich 5610* (BM, K), *5610A, C* (BM).

Phaseolus alatus Roxb., Fl. Ind. 3: 288. 1832. Type: Herb. Hort. Calc., *Roxburgh s.n., 1813* (BM).

Phaseolus truxillensis H.B.K., Nov. Gen. & Sp. 6: 451. 1823; Benth. in Mart., Fl. Bras. 15(1): 186. 1859; Thw., Enum. Pl. Zeyl. 89. 1859.

Phaseolus caracalla auct., Moon, Cat. 52. 1824; non L., Sp. Pl. 725. 1753, = *Vigna caracalla* (L.) Verdcourt.

Vines, twining, climbing perennials to 4 m long; stems glabrous or sparsely pubescent, rooting at the nodes. Leaf rachis 6–20 mm long, ½ to ¹/₆ the petiolar length; stipules oblong-ovate, about 4 mm long, striate; leaflets 2.5–14 × 1.6–8 cm, obtuse to acuminate apically, rounded basally, both surfaces with adpressed sparse pubescence, the laterals inequilateral. Inflorescences pseudoracemose, 10–25 cm long, florate ¼–¹/₅ the length, 6–12-flowered; bracts ovate-oblong, about 2 mm long, deciduous; bracteoles ovate-oblong, 3–4 mm long, striate. Flowers 1.5–3 cm long, the pedicels 2–3 mm long; calyx tube 3–4 mm long, sparsely pubescent, the upper lobes partially connate, the lower lobe lanceolate, about 4 mm long; petals dark pink or white, the vexillum somewhat orbicular, emarginate, to about 2.3 × 2.5 cm, sparsely pubescent or glabrous, the wings obovate, the keels spirally incurved about 3 turns; ovary many–numerous-ovulate, linear, pubescent, the upper style without a thickened portion, with long deflexed hairs towards the apex, the stigma apparently subterminal. Fruit broadly linear, compressed, 7.5–14 × 0.7–1.4 × 0.5 cm, glabrous or sparsely pubescent, 9–15-seeded. Seeds reniform, 5.5–7.0 × 4.5–5.5 × 2.5–3.5 mm, the hilum elliptic, about 1.2 mm long. Chromosome number 2n = 22.

Distr. Pantropical, probably introduced from South America.

Ecol. Trimen (1894) reports the species rather common in the moist regions up to 4000 ft; the dark pink flowers appear in December.

Vern. Snail flower (E); Wal-me (S); Karalsona (T). Moon (1824) gives Moodu-mae (S).

Specimens Examined. LOCALITY UNKNOWN: *Thwaites C.P. 1480* (BM, K); *Col. Walker s.n.* ex Herb. Wight MH# 16112 (COIMB, PDA). DISTRICT UNKNOWN: Colombo or Galle (?), *Gardner C.P. 1480* (PDA). COLOMBO DISTRICT: Delature, roadside, *Simpson 8949* (BM).

Notes. Trimen suggests this may be the plant called *P. caracalla* by Moon (Cat. 52), with the English name "Snail-flower."

Description based on Herb. *Wight 761* (GH, K), *2444* (K), *750* MH# 16113 (COIMB), and ex descr. Verdcourt (1971).

13. Vigna grahamiana (Wight & Arn.) Verdcourt, Kew Bull. 24(3): 562. 1970; Verdcourt, Fl. Trop. East Africa, Papil. 660. 1971 (note). Type: Courtallum, India, *Wight 724* (K).

Phaseolus grahamianus Wight & Arn., Prod. 1: 244. 1834; Thw., Enum. Pl.
　　Zeyl. 90. 1859; Baker in Hook. f., Fl. Br. Ind. 2: 201. 1876; Trimen, Handb.
　　Fl. Ceylon 2: 70. 1894; Gamble, Fl. Pres. Madras 1(2): 362. 1918.
Dolichos subcarnosus Wight & Arn., Prod. 1: 249. 1834. Type: *Wight 737* (K),
　　non Prain, J. Asiat. Soc. Bengal 66(2): 430. 1897.

Twining, climbing vines; stems slender, frequently branching, somewhat ridged, glabrous. Leaf rachis about 1.2 cm long, ½–⅓ the petiolar length; stipules broadly acute, about 3 mm long, glabrous, striate, non-produced; stipels lanceolate, 1–2 mm long; leaflets membranous, both surfaces glabrous, readily disarticulating, with 3 prominent veins inserted at the base, the terminal ovate, rounded-rhomboid, or deltoid, to about 7.5 × 4.5 cm, mostly smaller, acuminate apically, rounded or broadly cuneate basally, the laterals ovate, equilateral, to about 6 × 3.5 cm, mostly smaller, rounded or acute apically, rounded or broadly cuneate basally. Inflorescences axillary, 8–20 cm long, longer than the leaves, florate ½–¼ the length, each tubercle 1–3-flowered; bracts ovate, about 2 mm long, striate, glabrous, caducous; bracteoles minute, rounded, membranous, caducous. Flowers about 1.5 cm long (?), the pedicels about 3.5 mm long; calyx tube about 2 mm long, glabrous, all lobes obtuse or acute, ciliate, the lower longest, about 1.5 mm long; petals reddish (?) or violet purple, short-clawed, the vexillum orbicular, about 1 × 1.5 cm, without appendages, somewhat callous at the base, the wings obovate, about 8 × 7 mm, the keels about 10 mm long, erect distally, with narrow beaks incurved to almost 360°, apparently untwisted (?), without a pocket or the pocket shallow; style with the thick part upcurved, beaked above the stigma, with long hairs below the stigma, the stigma subterminal. Fruit spirally dehiscent, straight or slightly curved, 7.5–10.0 cm long, about 3.5 mm wide, glabrous, septate, with both margins swollen, frequently with a beak 2–3 mm long, about 10-seeded. Seeds oblong, truncate on the ends, about $6 \times 2.5 \times 1$ mm, reddish brown, horizontal, the hilum oblong-linear, about 2 mm long, without arillate tissue. Chromosome number n = 10.

Distr. Courtallum, Madras, India and Ceylon.

Ecol. The species is found in the low country, flowering in November through March.

Specimens Examined. LOCALITY UNKNOWN: "received 1870," *Thwaites C.P. 3586* (GH). VAVUNIYA DISTRICT: Mi. 113, between Medawachchiya and Mannar, *Jayasuriya et al. 587* (US). ANURADHAPURA DISTRICT: Wilpattu Natl. Pk., Marai Villu (?), *Cooray 70020225R* (US). MATALE DISTRICT: "Damboul," (Dambulla?), Mar. 1868, *Thwaites C.P. 3586* (K, PDA). BATTICALOA DISTRICT: by the sea, Batticaloa, Nov. 1858, *Thwaites C.P. 3586*, Mar. 1868, *Thwaites C.P. 3586* (K, PDA). GALLE DISTRICT: Galle (?), 1862, *Thwaites (?) C.P. 3586* (PDA).

Notes. Trimen (1894) describes the flowers as pale bright violet-purple rather than reddish as stated by Baker (1876). The non-produced stipules, the extended and incurved keel beak, and unsculptured pollen separate this species (& subgenus) from *Vigna* proper.

This species is the type of subg. *Macrorhynchus* Verdcourt. See Verdcourt (p. 561, 1970) for subgenus characters.

25. MACROPTILIUM

(Benth.) Urban, Symb. Ant. 9: 457. 1928; Verdcourt, Kew Bull. 24(3): 523. 1970; Marechal et al., Boissiera 28: 151. 1978; Lackey & D'Arcy, Ann. Missouri Bot. Gard. 67(3): 724. 1980. Type species: *M. lathyroides* (L.) Urban.

Phaseolus sect. *Macroptilium* Benth., Comm. Leg. Gen. 76. 1837.

Small vines or erect herbs to about 1 m tall. Leaves pinnately trifoliolate, rarely unifoliolate; stipules non-produced, persistent; stipellate; leaflets mostly narrowly elliptic or somewhat rhomboid-lanceolate. Inflorescences axillary and terminal, erect, elongate, pseudoracemose, usually many-flowered, tuberculate; bracteate and bracteolate. Flowers medium-sized, short-pedicellate; calyx narrowly campanulate, pubescent, the upper lobes deeply emarginate, obtuse, the rest acute, ciliate, much shorter than the tube; petals reddish, violet, purplish or white, the vexillum reflexed, somewhat ovate or orbicular, the wings somewhat rounded, the keels with the beak incurved, without distinct pockets, the wings and keels long-clawed; stamens diadelphous, the anthers uniform; ovary many–numerous-ovulate, the upper style somewhat uniform, thickened, distally erect, then bent into a squarish hook, untwisted, pubescent towards the apex, the stigma terminal. Fruit spirally dehiscent, long-linear, turgid or compressed, non-septate, numerous-seeded; sutures indistinct. Seeds small, round-reniform or oblong-truncate, horizontal, the hilum minute, elliptic, without arillate tissue.

About 20 species, native to the New World tropics. A few have been introduced into the Old World and *M. lathyroides* is naturalized.

Macroptilium lathyroides (L.) Urban, Symb. Ant. 9: 457. 1928; Verdcourt, Fl. Trop. East Africa, Papil. 503. 1971; Verdcourt, Man. New Guinea Leg. 505, fig. 124. 1979; Lackey and D'Arcy, Ann. Missouri Bot. Gard. 67(3): 727, fig. 36. 1980. Type: Not seen.

Phaseolus lathyroides L., Sp. Pl. ed. 2, 1018. 1762; Urban, Symb. Ant. 4: 310. 1903–1911 & 8: 312. 1920–1921 & 9: 457. 1928; Alston in Trimen, Handb. Fl. Ceylon 6: 81. 1931; Backer & Bakh. f., Fl. Java 1: 639. 1963.

Phaseolus semierectus L., Mant. 1: 100. 1767; DC., Prod. 2: 396. 1825; Benth. in Mart., Fl. Bras. 15(1): 189. 1859 (p.p.); Thw., Enum. Pl. Zeyl. 90. 1859; Baker in Hook. f., Fl. Br. Ind. 2: 201. 1876; Trimen, Handb. Fl. Ceylon 2: 70. 1894; Gamble, Fl. Pres. Madras 1(2): 362. 1918; Ridley, Fl. Malay Penins. 1: 567. 1922.

Phaseolus psoraleoides Wight & Arn., Prod. 1: 244. 1834; Wight, Ic. Pl. Ind. Or. t. 249. 1840. Type: *Wight 730* (BM, COIMB).

Somewhat shrubby, erect or occasionally climbing annual with ascending branches; stems terete, glabrous or with long retrorse, pilose, fugacious hairs; stipules linear-lanceolate, about 4 mm long; stipels setose, persistent; leaflets 3–6 × about 1.5 cm, obtuse or acute apically, broadly cuneate to rounded basally, both surfaces glabrous or soon glabrescent. Inflorescences to about 45 cm long, frequently florate about 1/2 the length; bracts long, setaceous; bracteoles setaceous. Flowers about 2.5 cm long, frequently in pairs, the pedicels about 2 mm long; calyx 5-lobed, the tube campanulate, about 5 mm long; petals mostly reddish purple, the vexillum somewhat orbicular, the wings obovate, rounded apically, about 2.3 cm long, expanding beyond the standard, the keels broad-linear, the wings and keels with claws about 5 mm long; stamens glabrous; ovary villous, the style tenuous below, the upper bearded below the stigma. Fruit terete or slightly compressed, 7.5–10.5 cm long, 3–4 mm wide, straight or slightly falcate, about 20-seeded. Seeds oblong-truncate, hard, dark, smooth, about 3.5 × 2 × 2 mm, the hilum about 1 mm long. Chromosome number 2n = 22.

Distr. Found throughout the tropics, this species is probably native to the West Indies and Central America. It has been introduced into Africa and Asia where it has become naturalized.

Ecol. Common in waste ground in the low country as a weed, it flowers year round.

Uses. Verdcourt (1971) reports the species cultivated in trial plots in Africa. The uses considered are probably conventional forage, fodder and soil fertility, rather than human food.

Specimens Examined. LOCALITY UNKNOWN: *Mrs. & Col. Walker s.n.* (K). JAFFNA DISTRICT: nr. Paranthan, *Townsend 73/89* (US). ANURADHAPURA DISTRICT: weeds in paddy fields, *Amaratunga 2187, 2235* (PDA); Madatugama, *Balakrishnan NBK 510* (US); Wilpattu Natl. Pk., *Fosberg et al. 50770* (US); btwn. Anuradhapura & Galkulama, *Rudd 3306* (K, NY, PDA, US); nr. Kekirawa, *Simpson 9225* (BM); btwn. Wilpattu & Anuradhapura, *C.F. & R.J. van Beusekom 1641* (PDA, US). POLONNARUWA DISTRICT: 3 mi. SW. of Elahera, *Davidse 7334* (MO, PDA, US); btwn. Habarane & Polonnaruwa, *Rudd & Balakrishnan 3137* (NY, PDA, US). PUTTALAM DISTRICT: W. of Wilpattu Natl. Pk., *Mueller-Dombois et al. 69043006* (NY, PDA, US). KURUNEGALA DISTRICT: Hiripitiya, *Maxwell & Jayasuriya 780* (NY, PDA, US); Kurunegala, *Worthington 4004* (BM, K). MATALE DISTRICT: "Damboul," Mar. 1862, *s. coll. s.n. C.P. 3609* (PDA); 6 mi. E. of Dambulla, *Davidse & Sumithraarachchi 8118* (MO, PDA, US); N. of Dambulla, *Fosberg & Balakrishnan 53392* (US); btwn. Dambulla & Habarane, *Koyama & Herat 13566* (NY, PDA). KANDY DISTRICT: Peradeniya, Aug. 1858, *s. coll. s.n. C.P. 3609* (PDA), Royal Botanic Gardens, June 1891, *s. coll. s.n.* (PDA), *Amaratunga 641* (PDA). TRINCOMALEE DISTRICT: Trincomalee, *Mrs. Braund s.n.* (PDA), *Worthington 1854* (K). AMPARAI DISTRICT: btwn. Amparai & Kalmunai, *Kostermans 24366* (US); shore of Senanayake Samudra, *Maxwell & Jayasuriya 745* (PDA, US), *Mueller-Dombois & Comanor 67072612* (PDA, US); Wadinagala, *Townsend 731279* (K); btwn. Amparai & Maha Oya, *Rudd & Balakrishnan 3235* (PDA, US). MONERAGALA DISTRICT: E. of Wellawaya, *Hepper & G. de Silva 4738* (K, US). HAMBANTOTA DISTRICT: nr. Ridiyagama, *Cramer 2839* (PDA); nr. Tissamaharama, *Rudd 3095* (K, PDA, US).

Note. Description based on Wight, Ic. Pl. Ind. Or. t. 249, *Wight 730* (BM, COIMB) and *Wight 762* (K).

26. PHASEOLUS

L., Sp. Pl. 723. 1753; L., Gen. Pl. ed. 5, 323. 1754; Irish, Missouri Bot. Gard. 12th Ann. Rep. 81–165. 1901; Tracy, USDA Bur. Pl. Industr. Bull. 109. 1907; Jarvis, Cornell Bull. 260: 149–245. 1908; Piper, Contr. US Natl. Herb. 22(9): 663. 1926; Bailey, Enc. Hort. 3: 2574–77. 1939; Romero, Bol. Inst. Nac. Invest. Agron. 9(21): 557–568. 1949; Burkart, Leg. Argentinas ed. 2, 426–437. 1952; Verdcourt, Kew Bull. 24(3): 507–526. 1970; Verdcourt, Fl. Trop. East Africa, Papil. 613. 1971; Verdcourt in Summerfield & Bunting, Adv. Leg. Sci. 45. 1980; Evans in Summerfield and Bunting, Adv. Leg. Sci. 337. 1980;

Marechal, Mascherpa & Stainer in Polhill & Raven, Adv. Leg. Syst. Part 1. 329–335. 1981. Type species: *Phaseolus vulgaris* L.

Climbing, prostrate or erect herbs with uncinate hairs. Leaves pinnately trifoliolate; stipules non-produced, persistent; stipellate; leaflets ovate or rhombic. Inflorescences axillary, pseudoracemose, somewhat tuberculate, each node few–numerous-flowered; bracteate and bracteolate. Flowers small to medium; calyx 4–5-lobed, the upper lobes bifid or shallowly emarginate; petals yellow to white, scarlet, pink or purple, the vexillum orbicular, biauriculate, reflexed, with or without oblique, transverse appendages in the target area, the wings usually obovate, frequently conforming to the keel spiral, the keels narrow, beaked, spiralling 1 to 5 complete turns; stamens diadelphous, the vexillary filament free, the anthers somewhat uniform, the pollen without wide reticulations; ovary 2–many-ovulate, the lower style slender, the upper style usually thick, hard, spiralling at least 1 complete turn, glabrous or with hairs towards the apex, variously bent, the stigma variable, oblique, subterminal or terminal. Fruit dehiscent, linear, oblong, or falcate, compressed or turgid, 3–many-seeded, frequently with a persistent beak. Seeds reniform or ovoid-oblong, the hilum oblong or elliptic, lacking well-developed arillate tissue.

With the removal of many species to *Vigna,* about 50 species remain in *Phaseolus.* The genus contains important crop species grown in tropical and temperate regions worldwide.

The following descriptions are primarily based on seeds bought in India, United States cultivars, and the specimens cited.

KEY TO THE SPECIES

1 Fruit linear-lanceolate; seeds about 10, rounded-reniform or oblong; flowers about 1.5 cm long
....**1. P. vulgaris**
1 Fruit oblong falcate or oblanceolate; seeds 3–4, orbicular-reniform; flowers to about 1 cm long
....**2. P. lunatus**

1. Phaseolus vulgaris L., Sp. Pl. 723. 1753; Moon, Cat. 52. 1824; Roxb., Fl. Ind. 3: 287. 1832; Wight & Arn., Prod. 1: 243. 1834; Benth. in Mart., Fl. Bras. 15(1): 182. 1859; Baker in Hook. f., Fl. Br. Ind. 2: 200. 1876; Trimen, Handb. Fl. Ceylon 2: 69. 1894; Gamble, Fl. Pres. Madras 1(2): 361. 1918; Burkart, Leg. Argentinas ed. 2, 426. fig. 130, g–l. 1952; Backer & Bakh. f., Fl. Java 1: 639. 1963; Verdcourt, Fl. Trop. East Africa, Papil. 614. 1971; Verdcourt, Man. New Guinea Leg. 513. 1979; Temple & Song in Summerfield & Bunting, Adv. Leg. Sci. 365. 1980. Type: Cultivated at Uppsala, specimen 899/1 (LINN, lectotype, fide Verdcourt, 1970), not seen.

Small climbers to erect herbs, annuals; stems pubescent to glabrescent. Leaf rachis 1–2.5 cm long, about ¼ the petiolar length; stipules lanceolate, about

4 mm long; leaflets ovate or somewhat rhombic, about 10 × 4 cm, acuminate apically, rounded basally, pubescent on both sides, the laterals inequilateral. Inflorescences 1–3-flowered, the peduncles to about 5 cm long; bracts ovate, about 3 mm long; bracteoles ovate-lanceolate, about 6 mm long. Flowers 1–2 cm long, the pedicels 3–10 mm long; calyx tube about 3 mm long, puberulent, the upper 2 lobes partially connate, all lobes triangular, about 1 mm long; petals white, purple, shades of yellow or shades of pink, the vexillum obovate-oblong, about 1.5 × 1 cm, glabrous, the wings obovate, the keels about 2.2 cm long; ovary many-ovulate, linear, the upper style curved at least one complete spiral, with hairs just below the apex, the stigma subterminal. Fruit linear to linear-lanceolate, about 10 × 1 cm, turgid, filled between the seeds, about 10-seeded, frequently beaked. Seeds oblong-ellipsoid or reniform, compressed, 1–2 cm long, 6–12 mm wide, 4–11 mm thick, the hilum elliptic, about 1 mm long. Chromosome number $2n = 22$.

Distr. Cultivated throughout the tropic and temperate regions of the world. Escaping but probably never becoming naturalized. Place of origin unknown.

Vern. Climbing bean, French bean, Haricot bean, Kidney bean, Runner bean (E); Bonchi (Dutch, fide Trimen); Potu-bonchi (S); Bonchi-kai (T). Moon gives the Sinhala dambala for dwarf kidney beans, sudu-dambala for var. α (Canterbury kidney-beans), and gives kolu dambala for var. β (Negro kidney-beans).

Notes. The following seven items were culled from a letter, dated the 12 May, 1927, found in the *Phaseolus* folder at PDA, that had been written to the PDA economic botanist from the systematic botanist; it concerns varieties of *P. vulgaris*.

1. Dwarf var. Gas bonchi (S), Murran-Bonchi (T). "Common bean in cultivation in gardens and fields."

2. Climbing var. Wel-bonchi (S), Podi-Bonchi (T). "Common in gardens, pods indistinguishable from the last."

3. Another variety is noted similar to the last, but with longer pods.

4. Climbing; pods flat. Paturu Bonchi (S), Peria Bonchi (T). French bean (of the market), Peradeniya and Nawalapitiya.

5. Climbing; pods flat, very large. Paturu Bonchi (S), Pinya Bonchi (T). French bean (of the market).

6. Climbing, pods green, round. Murunga bonchi & Drumstick Beans. "Does not grow here as well as upcountry."

7. Climbing, pods yellow. Butter-Bonchi (S) Butter beans. Chiefly upcountry.

I purchased Gas-bonchi (S) in Coimbatore, India, in 1972. The seeds are rounded reniform or oblong, mostly tan to light, reddish-brown, mottled with darker splotches, smooth, shining, the larger 12–15 mm × about 7 × 5 mm; the hilum central, elliptic, white, sometimes very slightly raised, 2 mm long, with 2 small adjoined, raised papillae next to the hilum. Wel-bonchi seeds from India are similar to Gas-bonchi except that they are of a uniform dark brown colour. Murunga-bonchi beans from India are also similar to Gas-bonchi except they are of a uniform light brown or tan colour. Gas-murunga-bonchi seeds are also similar except they are of a darker brown colour mottled with orange-brown splotches. Butter bean seeds from India are similar to Gas-bonchi except for the uniform very dark brown colour.

2. **Phaseolus lunatus** L., Sp. Pl. 724. 1753; Moon, Cat. 52. 1824; Roxb., Fl. Ind. 3: 287. 1832; Wight & Arn., Prod. 1: 244. 1834; Wight, Ic. Pl. Ind. Or. 3: t. 755. 1844; Benth. in Mart., Fl. Bras. 15(1): 181. 1859; Baker in Hook. f., Fl. Br. Ind. 2: 200. 1876; Trimen, Handb. Fl. Ceylon 2: 69. 1894; Gamble, Fl. Pres. Madras 1(2): 361. 1918; Burkart, Leg. Argentinas ed. 2, 426. fig. 130, m. 1952; Backer & Bakh. f., Fl. Java 1: 639. 1963; Verdcourt, Fl. Trop. East Africa, Papil. 615, fig. 95. 1971; Verdcourt, Man. New Guinea Leg. 513. 1979; Rachie et al. in Summerfield & Bunting, Adv. Leg. Sci. 375. 1980. Type: Bengal: *Phaseolus benghalensis scandens . . . striato* of Bergen, Cat. Stirp. Hort. Acad. Viadr. compl. 99 (1744) fide Verdcourt, not seen.

Small climbers or erect herbs, annuals or biennials; stems glabrous or pubescent. Leaf rachis 1–5 cm long, ½ to ¹/₅ the petiolar length; stipules ovate-lanceolate, 2–3.5 mm long; leaflets ovate, 3–15 × 2–10 cm, acute to acuminate apically, rounded or truncate basally, sparsely pubescent or glabrous, the laterals inequilateral. Inflorescences pseudoracemose, florate 2/3 to 1/5 the peduncular length, tubercles few to several-flowered, the peduncles 2–30 cm long; bracts lanceolate, about 1.5 mm long, persistent; bracteoles elliptic or ovate, 1.5–2 mm long, persistent. Flowers about 1 cm long, the pedicels 5–10 mm long; calyx tube 1.5–2.5 mm long, puberulent, the upper lobes emarginate, all lobes triangular, 5–8 mm long; petals white, yellowish or rose, the vexillum rounded or somewhat oblong, 5–7 × 5–8.5 mm, emarginate, sparsely pubescent or glabrous, the keels 1–1.4 cm, spirally incurved about 1.5 turns; ovary 2–4-ovulate, elliptic, pubescent, the lower style geniculate, the upper style bearded below the sharply beaked apex, the stigma region subterminal. Fruit oblong-falcate or oblanceolate, glabrous or pubescent, 5–10 × 1.2–2.2 cm, compressed, with a short apiculate beak, 3–4-seeded; sutures indistinct. Seeds reniform or rhomboid-reniform, compressed, 1–1.5 cm long, 8–12 mm wide, 5–5.5 mm thick, the hilum 2.5–4 mm long. Chromosome number 2n = 22.

Distr. Cultivated in the tropical and temperate regions of both hemispheres. Probably a native of tropical America.

Vern. Burma-bean, Curry bean, Haricot bean, Java bean, Lima bean, Madagascar-bean, Rangoon-bean, Sieva-bean, Tropical Lima bean, Tonga-bean (E); Bonchi (Dutch, fide Trimen); Dambala, Pothu-dambala, Dàrà-dambala (S).

Specimens Examined. KANDY DISTRICT: Gannoruwa, *Amaratunga 2057* (PDA); Peradeniya, trailing vine to 2 m long, in front of Post Office, *Maxwell 854* (NY, PDA, US).

Notes. The letter in the PDA folder cited above states, *"Phaseolus lunatus, seeds only eaten.*

1. Large podded var. Potu-Bonchi.
2. Small podded var. Potu-Bonchi.*"

There are different sized pods in the PDA folder.

Phaseolus lunatus is carried on the Seeds List (1958) as a climber with edible pods and the seeds available throughout the year.

See also "Tropical Lima Bean" in "Tropical Legumes: Resources for the Future," pp. 97–102, Nat. Acad. Sci., Washington, D.C., U.S.A., 1979. Research contacts are listed.

Description based on *Wight 727* (BM), *Wallich 5595E* (BM).

27. CAJANUS

DC., Cat. Hort. Bot. Monsp. 85, 1813, nom. cons. Type species: *C. cajan* (L.) Millsp.

Cajan Adans., Fam. Pl. 2: 326, 529. 1763.
Cytisus Schreb., Gen. Pl. ed. 8, N. 1191. 1789–1791, fide Roxb., Fl. Ind. 3: 325. 1832.

Erect shrubs; stems ridged. Leaves trifoliolate; stipulate; stipellate; leaflets gland-dotted. Inflorescences axillary, paniculate-racemose, few flowered; bracts ovate, caducous; bracteoles absent. Flowers medium; calyx somewhat 5-lobed, the upper lobes partially connate, acuminate; petals yellow with striae, about equal in length, the vexillum reflexed, orbicular; stamens diadelphous, the vexillary free, the anthers uniform; ovary about 5-ovulate, the style ascending, incurved, the upper style glabrous, the stigma capitate. Fruit dehiscent, linear-oblong, compressed, the exocarp obliquely depressed between the seeds, 3–7-seeded, rarely 1; sutures slightly swollen, with a persistent beak. Seeds ovoid, the hilum oblong, with or without arillate tissue.

A genus of possibly two or three species, wild or cultivated throughout the tropics of both hemispheres.

Cajanus cajan (L.) Millsp., Publ. Field Columbian Mus., Bot. Ser. 2: 53. 1900; Verdcourt, Fl. Trop. East Africa, Papil. 709, fig. 108. 1971; Verdcourt, Man. New Guinea Leg. 539, fig. 132. 1979; Poston, Ann. Missouri Bot. Gard. 67(3): 555, fig. 7. 1980. Types: *Hermann 2'76 & 3'30* Herm. Herb. in BM, syntypes.

Cytisus cajan L., Fl. Zeyl. No. 279. 1747; L., Sp. Pl. 2: 739. 1753; Rheede, Hort. Mal. 6, t. 13. 1686; Moon, Cat. 53. 1824; Roxb., Fl. Ind. 3: 325. 1832.
Phaseolus balicus L. in Stickman, Herb. Amboin. 23. 1754; L., Amoen. Acad. 4: 132. 1759; fide Merr., Inter. Rumph. Herb. Amboin. 282. 1917.
Cytisus pseudo-cajan Jacq., Hort. Bot. Vindob. 2: 54, t. 119. 1772.

Cajanus inodorus Medic., Vorles. Churpfälz. Phys.-Oekon. Ges. 2: 363. 1787; Alston in Trimen, Handb. Fl. Ceylon 6: 84. 1931.

Cajanus bicolor DC., Cat. Hort. Monsp. 85. n. 43. 1813; DC., Prod. 2: 406. 1825. Based on *Wallich 5577* (BM, NY).

Cajanus flavus DC., Cat. Hort. Monsp. 85. n. 43. 1813; DC., Prod. 2: 406. 1825; Burkart, Leg. Argentinas ed. 2, 397, fig. 118, c. 1952.

Cajanus indicus Spreng., Syst. Veg. 3: 248. 1826; Wight & Arn., Prod. 1: 256. 1834; Thw., Enum. Pl. Zeyl. 90. 1859; Benth. in Mart., Fl. Bras. 15(1): 199. 1859; Baker in Hook. f., Fl. Br. Ind. 2: 217. 1876; Trimen, Handb. Fl. Ceylon 2: 80. 1894 (note). Based on *Wight 764* (NY).

Cajanus cajan Merr., Philipp. J. Sci. 5: 217. 1910.

Shrubs about 1.5 m tall; stems pubescent. Leaf rachis 1–2 cm long, the petiole ½ as long or less; stipules acuminate, 2.5–5 mm long, persistent; leaflets lanceolate or narrowly elliptic, puberulent above and below, the largest to 7.5–8 × 2.8–3.5 cm, acute apically, bases similar, the venation strongly reticulate, prominent beneath. Inflorescences axillary from a single peduncle, terminating at the insertion of 1–2 pedicels or continuing for 1–3 additional nodes, rarely branching, usually slightly shorter than the leaves, mostly 2–6-flowered; bracts about 5 mm long. Flowers about 2 cm long, the pedicels to about 9 mm long; calyx tube 3–5 mm long, glandular and pubescent, the upper lobe bifid, the lowest lobe longest, about 4 mm long; vexillary petal basally inflexed biauriculate, mostly with reddish striae, bicallose in the target area, glabrous, about 14 mm long, with the claw about 4 mm long, the wings and keels adherent, the wings oblanceolate, yellow, about 12 mm long, with the claw about 4.5 mm long, the keels obliquely oblong or falcate, yellow-green, about 10 × 6 mm at the distal end, the claw about 5 mm long; vexillary filament slightly wider towards the base; ovary somewhat villous, the lower style slender, pubescent. Fruit slightly compressed, straight, to about 4 cm long, 1 cm wide, 3–4 mm thick, (1–)5-seeded; upper suture swollen, the lower indistinct, the beak downcurved. Seeds compressed, about 6 × 4 × 1.5 mm, of variable colours, the hilum linear-oblong to somewhat elliptic, about 3 mm long. Chromosome number 2n = 22.

Distr. Cultivated throughout the tropics and occasionally escaping.

Uses. Pigeon peas (seeds) are a major food source on the Indian subcontinent. The foliage is palatable to cattle.

Vern. Chick Pea (Trimen, 1894), Pigeon Pea, Dhal, Red Gram (E); Ratatora (S); Thavarai and Thovardy (T). Chick Pea is usually associated with *Cicer arietinum*.

Specimens Examined. LOCALITY UNKNOWN: *Thwaites C.P. 720* (PDA). PUTTALAM DISTRICT: Marcus Isl., *Kundu & Balakrishnan 550* (US). KURUNEGALA DISTRICT: Narammala, cult., *Amaratunga 1516* (PDA). KANDY DISTRICT: Ambagamuwa, 1846, *Gardner C.P. 720* (PDA); nr. bridge from Peradeniya, in garden, *Rudd & Balakrishnan 3323* (PDA, US); Rambukpitiya, in garden, *Maxwell & Fernando 989* (NY, PDA, US); Royal Botanic Garden (Herb. Ground), Oct. 1888, & June 1891, *s. coll. s.n.* (PDA); Exp. Sta., Peradeniya, 1916, *s. coll. s.n.* (PDA).

Notes. Trimen (1894) indicates the species is Burm. Thes. t. 37. 1737, commonly cultivated in gardens, but nowhere wild in Ceylon.

I have seen no wild collections. Wight & Arn. (1834) recognized two varieties: *C. flavus* DC., with the vexillum of uniform yellow on both sides, and *C. bicolor* DC., with the vexillum purplish and veined on the outside, yellow on the inside.

Several recent legume workers believe *C. cajan* is a cultivated form of *Atylosia*. See Lackey in Polhill & Raven, Adv. Leg. Syst. Part 1. 327. 1981, for a brief discussion of the evidence and nomenclatural problems involved.

28. ATYLOSIA

Wight & Arn., Prod. 1: 257. 1834; Hutchinson, Gen. Fl. Pl. 1: 421. 1964. Lectotype: *A. candollei* Wight & Arn., India. (=*A. trinervia* (DC.) Gamble).

Cantharospermum Wight & Arn., Prod. 1: 255. 1834.

Small erect shrubs, or twining, climbing herbs; stems somewhat angular, pubescent. Leaves trifoliolate; stipulate, with or without stipels; leaflets ovate, obovate, or lanceolate, occasionally gland-dotted beneath. Inflorescences axillary, racemose, few—many-flowered; bracts usually caducous; bracteoles absent. Flowers small or medium; calyx 4–5-lobed, frequently gland-dotted, the upper lobes partially or entirely connate, the lower usually longest; petals yellowish or purple, about the same length, the vexillum reflexed, dark striate, basally inflexed biauriculate, slightly longer than the wings; stamens diadelphous, the vexillary free, the anthers uniform; ovary 2–8-ovulate, pubescent, the style bent down, then upcurved, the upper style glabrous, the stigma terminal, small, capitate, glabrous. Fruit dehiscent, oblong, compressed, pubescent, septate, constricted transversely between the seeds, 2–6-seeded; sutures indistinct or swollen. Seeds small, suborbicular, slightly compressed, the hilum oval-elliptic, or linear-oblong, with raised white arillate tissue. Chromosome number 2n = 22.

A genus of about 35 species of the Old World tropics, probably centered in India-Asia to Australia. At least four species are found in Ceylon.

KEY TO THE SPECIES

1 Erect shrubs; leaflets elliptic, oblanceolate to linear lanceolate..............**1. A. trinervia**
1 Climbing, twining vines; leaflets obovate or orbicular
  2 Calyx lobes acute, 1–2 mm long.....................................**2. A. albicans**
  2 Calyx lobes acuminate, much longer than 2 mm
    3 Flowers about 13 mm long; leaflets sericeous beneath; pod villous; fruit usually 3–4-seeded
    ....**3. A. rugosa**
    3 Flowers about 5 mm long; leaflets cinereus beneath; pod with fulvous spreading hairs; fruit usually
    4–6-seeded..**4. A. scarabaeoides**

1. Atylosia trinervia (DC.) Gamble, Fl. Pres. Madras 1(2): 368. 1918; Alston in Trimen, Handb. Fl. Ceylon 6: 84. 1931. Type not seen.

Collaea trinervia DC., Mem. Leg. 247, t. 41. 1824; DC., Prod. 2: 240. 1825.
Odonia trinervia (DC.) Spreng., Syst. Suppl. 279. 1828.
Rhynchosia wightiana Grah. in Wall., Cat. 5500. 1831–1832. Based on *Wallich 5500* (K), nom. nud.
Atylosia candollii Wight & Arn., Prod. 1: 257. 1834; Wight, Ic. Pl. Ind. Or. t. 754. 1844; Thw., Enum. Pl. Zeyl. 91. 1859; Baker in Hook. f., Fl. Br. Ind. 2: 212. 1876; Trimen, Handb. Fl. Ceylon 2: 78. 1894. Type: *Wight 763* (K).
Atylosia major Wight & Arn., Prod. 1: 257. 1834. Based on *Wight 762,* not seen.
Cantharospermum trinervium (DC.) Taubert in Pflanzenfam. 3(3): 373. 1894.

Erect, branching shrubs, to about 2.5 m tall; stems terete, slightly ridged, fulvous-villous when young, glabrescent at maturity. Leaves digitately trifoliolate, the petioles to 1 cm long; stipules acute, about 2.5 mm long, caducous; stipels absent; leaflets oblanceolate, oblong, or narrowly elliptic, to about 4.5 × 1.5 cm, glabrous above, densely pubescent below, sparsely gland-dotted beneath, acute apically and basally, usually with 3 prominent veins inserted at the base, with 3–4 pairs of primary lateral veins. Inflorescences erect, reduced racemose (?); peduncles 10–15 mm long, slightly longer than the leaves, each 1–2-flowered, bracts at the base of the peduncles acute, about 3 mm long, caducous. Flowers yellowish, about 2 cm long, the pedicels about 9 mm long; calyx tube about 6 mm long, fulvous villous, the upper lobes connate, all lobes long, acuminate, the lower longest, 7–9 mm long; petals yellowish, the vexillum broadly oblanceolate, glabrous, about 25 × 15 mm, the claw about 5 mm long, the wings and keels adherent, slightly shorter than the vexillum, the wings narrow, about 18 × 5 mm, the claw 5 mm long, the spur acute, the keels wider, obliquely oblong, shortly beaked, with the pouch about 12 mm long; vexillary filament narrow at insertion, wider distally; ovary about 5-ovulate, villous, the lower style slender, pubescent. Fruit oblong, fulvous villous, 3–3.5 × about 1 × 0.4 cm, (2–)3–5(–6)-seeded; sutures indistinct, the beak down-curved, persistent, frequently with the stigma and style also persistent. Seeds ellipsoid, dark, about 4 × 3 × 2 mm, transverse, the hilum linear-oblong, about 2.5 mm long, with linear, white, 2-lipped arillate tissue.

Distr. In the Indian hills and montane zone in Ceylon.

Ecol. Very common, sometimes gregarious plant on the open patanas of the montane zone from 900–1800 m. The flowers are bright yellow with the back of the vexillum red; blooms year round.

Uses. Trimen (1894) characterizes the plant as an ornamental shrub (?).

Vern. Et-tora, Atta-tora (S).

Specimens Examined. LOCALITY UNKNOWN: (Kandy?), *Gardner 240*

& *s.n.* (K); 1854, *Thwaites C.P. 55* (BM, NY); received 1870, *Thwaites s.n.* (GH); *Col. Walker s.n.* (K), *176* (PDA); *Col. & Mrs. Walker s.n.*, *315* (K). KANDY DISTRICT: Pussellawa, 3000 ft, *Gardner 240* (BM); Kandy to Maturata, *Maxwell 997* (PDA, US); betw. Madugoda & Hunnasgiriya, *Rudd & Balakrishnan 3245* (K, NY, PDA, US). NUWARA ELIYA DISTRICT: Lindula, 17.11.44, *Bond s.n.* (PDA); Hakgala, *A. de Silva s.n.* (PDA), 23.8.26, *J.M. de Silva s.n.* (PDA); Hakgala, patana, *Cramer 3871* (US); A 5, nr. Hakgala, *Maxwell & Jayasuriya 885* (NY, PDA, US), *903* (PDA, US); Nuwara Eliya to Badulla, *Maxwell 1013, 1014* (NY, PDA, US); Horton Plains, open ridge top, patana, *Mueller-Dombois & Comanor 67070848* (PDA, US); Hakgala to Ambawela, *Mueller-Dombois & Comanor 67091407* (PDA, US); below Ohiya RR. Sta., *Mueller-Dombois 67091502* (PDA, US); McDonald's Valley, below Hakgala, *Rudd & Balakrishnan 3166* (NY, PDA, US); Hewaheta, *Rudd & N.H.D.W. Jayasinghe 3251* (K, NY, PDA, US); St. Coombs Est., Talawakelle, *N.D. Simpson 8886, 8887* (BM, PDA). NUWARA ELIYA or RATNAPURA DISTRICT: Galagama (Galagama Falls?), *Gardner C.P. 55* (PDA); Galagama, *Thwaites C.P. 209, C.P. 224*, (K). BADULLA DISTRICT: Boralanda, Palugama, *Amaratunga 470* (PDA); N. of Haputale, Thangamala Forest Reserve, *Maxwell & Jayasuriya 765* (PDA, US); St. Thomas' College area, Gurutalawa, *Rudd & Balakrishnan 3187* (PDA, US); nr. Boralanda, *Stone 11223* (US); Gurutalawa, *Worthington 5582* (K), *5582 B* (K).

Notes. *Atylosia lineata* Wight & Arn. & *A. sericea* Benth. are also erect shrubs found in Indian hills. Both have thinner leaflets than *A. trinervia*. Gamble (1918) indicates *A. lineata* has grey-silky pubescence beneath the leaflets; *A. sericea* has white-silky pubescence; while *A. trinervia* has brown silky pubescence. Gamble also indicates *A. sericea* is 2-seeded, *A. lineata* 2–3-seeded, and *A. trinervia* 3–4 seeded, the latter in Ceylon frequently to 5, but this character may be correlated to ecological conditions.

Bond describes and figures this species in "Wild Flowers of the Ceylon Hills," suggesting a value based on its gregarious habit: reclamation of overgrazed patana land.

Description based on *Wight 763, 771* (K).

2. Atylosia albicans (Wight & Arn.) Benth. in Miq., Pl. Jungh. 243. 1852; Thw., Enum. Pl. Zeyl. 91. 1859; Baker in Hook. f., Fl. Br. Ind. 2: 215. 1876; Trimen, Handb. Fl. Ceylon 2: 78. 1894; Gamble, Fl. Pres. Madras 1(2): 368. 1918. Type: Not seen.

Catharospermum albicans Wight & Arn., Prod. 1: 256. 1834 (excl. syn.). Based on *Wight 759* (BM, K).

Twining, climbing vines; stems frequently branching, with canescent pubescence. Leaves pinnately trifoliolate, much smaller from the branching stems than from the main, the rachis minute to absent (then the leaves digitately trifoliolate), the longer petioles about 4 cm long; stipules acute, about 1.5 mm long, persistent; lateral stipels setaceous, minute, about 1 mm long, the terminal stipels about 0.5 mm long; leaflets broadly obovate or suborbicular, to about 4.5 × 3.5 cm, dense canescent pubescence below, sparse above, rounded apically, broadly cuneate basally, minutely gland-dotted above and below, frequently with 3 prominent veins inserted at the base, with about 3 pairs of primary lateral veins, all prominent beneath. Inflorescences axillary from the main stems or from smaller side branches, some peduncles with a bract at the insertion of 1–2 pedicellate flowers, other peduncles racemose, 1.5–3(–5) cm long, 2–6-flowered; bracts ovate, 2–3 mm long, enclosing the inflorescence unit, caducous. Flowers usually in pairs, occasionally solitary, 12–17 mm long, the slender pedicels 7–10 mm long; calyx tube about 3 mm long, glabrescent or sparsely puberulent, gland-dotted, the upper lobe emarginate, obtuse, the remaining lobes acute, the lowest longest, from about 1.5 mm long; petals yellow, the vexillum obovate, glabrous, about 18 mm long, with the claw about 2.5 mm long, the wings and keels adherent, the wings narrow, about 12 × 4 mm, with a claw about 3.5 mm long, the keels slightly longer, with a short acute or obtuse beak; vexillary filament narrow at insertion, wider distally; ovary about 6-ovulate, tomentose, the lower style slender, pubescent proximally, constricted towards the middle, the upper style thick. Fruit canescent-tomentose, oblong or linear-oblong, 2.5–4 × 0.5 –1 × 0.3 cm, 4–7 seeded; sutures swollen, with a small down-curved beak. Seeds ellipsoid-oblong, vertical, dark, shining, the hilum linear-oblong, surrounded with white, 2-lipped arillate tissue. Chromosome number n = 11.

Distr. Southern India and the low country of Ceylon.

Ecol. Trimen (1894) reports it rare, found to about 2000 ft. Collectors note August through February flowering.

Vern. Wal-kollu (S).

Specimens Examined. WITHOUT EXACT LOCALITY: 1861, *Thwaites C.P. 2783* (BM); *Col. Walker s.n.* (PDA, ex. Wight Herb.); *Col. & Mrs. Walker 87* (GH). MATALE DISTRICT: Palapathwala, *Amaratunga 1565* (PDA). KANDY DISTRICT: Teldeniya, *Alston 1739* (K, PDA); Peradeniya Junction, 7.1.29, *J.W. de Sliva s.n.* (PDA); Hantane, 1858, *Gardner C.P. 2783* (BM, PDA), *Gardner 241* (BM sheet = Galle, PDA); Haragama, *Jayasuriya & Balasubramaniam 445* (PDA, US); Urugala, *Simpson 9129* (BM). NUWARA ELIYA DISTRICT: Mulhalkelle, *Worthington 5649* (BM, K). MONERAGALA DISTRICT: nr. Bibile, Jan. 1888, *s. coll. (Trimen ?) s.n.* (PDA).

Notes. This species may be difficult to distinguish from certain *Rhynchosia* species with sericeous pubescence on the underside of the leaflets; however, fruit of *Atylosia* is septate, with transverse exocarp constrictions, and usually more than 2 seeds as opposed to *Rhynchosia* fruit.

Gamble (1918) notes *A. albicans* is easily known by its bluish grey striate stems and leaflets.

Description based on *Wallich 5582, 5583* (BM, K).

3. Atylosia rugosa Wight & Arn., Prod. 1: 257. 1834; Thw., Enum. Pl. Zeyl. 91. 1859; Baker in Hook. f., Fl. Br. Ind. 2: 215. 1876; Trimen, Handb. Fl. Ceylon 2: 79. 1894; Gamble, Fl. Pres. Madras 1(2): 369. 1918. Type: *Wight 761*, not seen.

Rhynchosia velutina Grah. in Wall., Cat. 5501. 1831–1832, fide Baker (1876), nom. nud. See Wight & Arn., Prod. 1: 258. 1834 (note).

Trailing or weak climbing vines. Leaves digitately trifoliolate, of two sizes; stipules suborbicular, about 2 mm long, persistent; stipels apparently absent; leaflets ovate or obovate, rugose and pubescent above, sericeous lanate below, the larger leaflets about 3 × 3 cm, rounded apically, occasionally emarginate, cuneate basally, with 3 prominent veins inserted at the base. Inflorescences 1–3-flowered, sericeous tomentose, peduncles about equal the leaf length; bracts at the pedicel insertions of the flowers, similar to the stipules, about 1 mm long, persistent. Flowers about 13 mm long, the pedicels 4–6 mm long; calyx tube about 3 mm long, pubescent, the upper lobes connate or bifid, all lobes long-acuminate, the lower longest, about 7 mm long; petals yellow, the vexillum elliptic, very slightly emarginate, about 10 mm long, short-clawed, gland-dotted, glabrous, the wings and keel adherent, the wings narrow, with a spur, the keel obliquely oblong, with a long pouch; vexillary stamen narrow at insertion, wider distally; ovary about 4-ovulate, tomentose, the style somewhat sigmoid, the lower style sparsely bearded, slender, the upper style thickened. Fruit sericeous, 1.6–2.2 × about 0.7 × 0.2 cm, 2–4-seeded; upper suture swollen, the lower indistinct, with a short down-curved beak. Seeds ovoid, about 3.5 × 3.5 mm, dark brown, the hilum linear-oblong, about 2.5 mm long, surrounded by 2-lipped white arillate tissue.

Distr. Southern India and Ceylon.

Ecol. Trimen reports flowering in September, but additional collections show flowering is probably year round. Common in the montane zone from 900 to 1800 m.

Vern. Wal-kollu (S).

Specimens Examined. LOCALITY UNKNOWN: 1854, *Thwaites C.P. 1441* (BM), 1860, *C.P. 1441* (NY), *C.P. 1441* (K); *Col. Walker s.n. 117* (K); *Col. & Mrs. Walker s.n.* (K). DISTRICT UNKNOWN: (BADULLA ?) Bolgandawela, Uma Oya, *J.M. de Silva 233* (PDA). KANDY DISTRICT: Dunally, Galaha, (asses ears) *Worthington 2499* (K); Hantane, *Major Champion s.n.* (K), Sept. 1853, *Gardner s.n.* (PDA), *Worthington 250* (K); Elephant Plains, 6000 ft, *Gardner 239* (K); NE. of Madugoda, *Jayasuriya et al. 483* (US). NUWARA ELIYA DISTRICT: Maturata, Sept. 1853, *Gardner s.n.* (PDA); McDonald's Valley, below Hakgala, *Rudd & Balakrishnan 3174* (K, PDA, US); nr. Hewaheta,

Rudd & N.H.D.W. Jayasinghe 3250 (K, NY, PDA, US). BADULLA DISTRICT: Boralanda-Palugama, *Amaratunga 469* (PDA); Haputale to Bandarawela, *Maxwell & Jayasuriya 769, 773* (NY, PDA, US); N. of Welimada, *Maxwell & Jayasuriya 889* (PDA, US); betw. Welimada & Badulla, *Rudd & Balakrishnan 3194* (PDA, US); Welimada, patana, *Simpson 8652* (BM).

Notes. The specific epithet refers to the rugose upper leaflet surface. The sericeous undersurface also helps distinguish this species. Bond ("Wild Flowers of the Ceylon Hills") says the species is a patana herb, more common than *A. scarabaeoides* and occurs in the highest elevations.

Description based on *Wight 773* (K), *Rottler s.n.*, Mar. 1796 (K).

4. Atylosia scarabaeoides (L.) Benth. in Miq., Pl. Jungh. 243. 1852; Thw., Enum. Pl. Zeyl. 91. 1859; Baker in Hook. f., Fl. Br. Ind. 2: 215. 1876 (cum syn.); Trimen, Handb. Fl. Ceylon 2: 79. 1894; Gamble, Fl. Pres. Madras 1(2): 369. 1918; Ridley, Fl. Malay Penins. 1: 564. 1922; Verdcourt, Fl. Trop. East Africa, Papil. 707, fig. 107. 1971; Verdcourt, Man. New Guinea Leg. 540, fig. 134. 1979. Lectotype: *Hermann 2: 60* in BM.

Dolichos scarabaeoides L., Sp. Pl. 726. 1753; Moon, Cat. 53. 1824; Rcxb., Fl. Ind. 3: 315. 1832.

Phaseolus minimus Plukenet, Almagest. t. 52, fig. 3. 1691.

Catharospermum pauciflorum Wight & Arn., Prod. 1: 255. 1834. Type: *Wight 758* (BM, K).

Cajanus scarabaeoides (L.) Thouars; Grah. in Wall., Cat. 5580. 1831–1832, fide Wight & Arn. (1834).

Dolichos minutus Roxb. in E.I.C., Mus. tab. 252, f. 1. 1820.

Rhynchosia scarabaeoides (L.) DC., Prod. 387. 1825.

Rhynchosia biflora DC., Prod. 387. 1825.

Small, twining vines; stems tomentose. Leaves usually pinnately trifoliolate, the rachis 1–3 mm long; stipules acute, about 1 mm long, persistent; lateral stipels minute, the terminal absent; leaflets lanceolate, elliptic, ovate, or obovate, yellow-glandular, pubescent above and below, to about 2.8 × 1.5 cm, usually smaller, usually acute apically, the bases similar, with 3 prominent basal veins. Inflorescences 1–3-flowered, much shorter than the leaves, very short pedunculate; bracts linear, about 1.5 mm long. Flowers about 8 mm long, the pedicels about 2.5 mm long; calyx tube about 1.5 mm long, fulvous pubescent, the upper calyx lobes connate or bifid at the apex, long acuminate, the lower longest, about 5 mm long; vexillary petal obovate, about 6 mm long, yellow with reddish brown striae, short-clawed, glabrous, the wings and keels adherent, the wings yellow, narrow, lanceolate, without a spur, the keels oblanceolate, with an obtuse beak; vexillary stamen narrow at insertion, wider distally; ovary about 7-ovulate, sericeous, long-pilose, the upper style thickened. Fruit 15–19 × 6 × about 2 mm, brown, pilose with long straight hairs, about 5-seeded; upper suture swollen, the lower indistinct, with a short down-curved beak. Seeds about $3 \times 2.5 \times 1.5$ mm, reddish brown, the hilum linear-oblong, about 1.5 mm long, surrounded by white arillate tissue. Chromosome number $n = 11$.

Distr. Found throughout Asia. Introduced into Africa.

Ecol. Common in both the wet and dry zones of the low country of Ceylon. Thwaites (1859) cites hot, drier parts of the island. Flowering is probably year round, although most reports are November through March.

Vern. Wal-kollu (S). Hermann adds Woel-undu and Woel-unduwel (S), fide Trimen (1894).

Specimens Examined. LOCALITY UNKNOWN: 1854, *Thwaites C.P 1442* (BM, K); *Col. Walker s.n.*(K); *Mrs. & Col. Walker s.n.* (NY). POLON-NARUWA DISTRICT: betw. Habarane & Kantalai, *Rudd & Balakrishnan 3120* (K, PDA, US). KURUNEGALA DISTRICT: 3 mi. W. of Rambe, *Maxwell & Jayasuriya 783* (NY, PDA, US); 18 mi. W. of Rajangana Resv., *Maxwell & Jayasuriya 791* (PDA, US). MATALE DISTRICT: Mi. 48 N. of Dambulla, *Hepper & Jayasuriya 4602* (K, PDA, US); Sigiriya, *Amaratunga 256* (PDA). TRIN-COMALEE DISTRICT: *Gardner C.P. 1442* (PDA). AMPARAI DISTRICT: N. of Inginiyagala, *Maxwell & Jayasuriya 747* (NY, PDA, US); AMPARAI OR MONERAGALA DISTRICT: E. shore of Senanayake Samudra, *Mueller-Dombois & Comanor 67072610* (PDA, US); betw. Inginiyagala & Siyambalanduwa, *Rudd & Balakrishnan 3223* (K, NY, PDA, US). GALLE DISTRICT: Galle ?, Dec. 1849, *s. coll. C.P. 1442* (PDA); Hiyare, *J.M. Silva 221* (PDA), *Gardner 238* (K). HAMBANTOTA DISTRICT: from Tissa to Kataragama, *Rudd 3104* (PDA, US).

Notes. The fulvous fruit hairs, about 1.5 mm long, and smaller fruit size are good field characters in distinguishing this species from *A. rugosa*.

The Ceylon material examined has much smaller leaflets than the African, fide Verdcourt (1971, ex descr.).

29. DUNBARIA

Wight & Arn., Prod. 1: 258. 1834; Miq., Fl. Ind. Bat. 176–180. 1855. Lectotype species: *D. heynei* Wight & Arn.

Twining, climbing, woody vines; stems pubescent. Leaves pinnately trifoliolate, gland-dotted beneath; stipulate, with or without stipels. Inflorescences axillary, lax, racemose or pseudo-racemose, few–numerous-flowered; bracteate; bracteoles absent. Calyx 4–5-lobed, the upper lobes partially connate or entire, the lower longest, equalling the tube; petals usually yellow, the vexillum suborbicular, biauriculate, bicallose, longest, the wings narrow, the keels somewhat obliquely oblong, clawed; stamens diadelphous, the vexillary free, the anthers uniform; ovary 1–5-ovulate, the lower style pubescent, the upper style pubescent or glabrous, the stigma terminal, small, capitate. Fruit oblong-linear, compressed, septate or subseptate, the exocarp constricted between the seeds, 1–5-seeded. Seeds ovoid, slightly compressed, the hilum oval or oblong, usually with distinct or indistinct surrounding arillate tissue.

About 15 species from East India through Malaysia and Indonesia, reaching Australia.

KEY TO THE SPECIES

1 Calyx glabrescent or with ascending strigose hairs; leaflets with fairly dense pubescence when young; flowers 14–17 mm long; upper style pubescent . . . . **1. D. ferruginea**
1 Calyx glabrous, white-gland-dotted or with short hairs; leaflets glabrescent beneath; flowers about 20 mm long; upper style glabrous . . . . **2. D. heynei**

1. Dunbaria ferruginea Wight & Arn., Prod. 1: 258. 1834; Baker in Hook. f., Fl. Br. Ind. 2: 217. 1876; Trimen, Handb. Fl. Ceylon 2: 80. 1894; Gamble, Fl. Pres. Madras 1(2): 370. 1918. Type: *Wight 878* (p.p.), not seen.

Collaea venosa Grah. in Wall., Cat. 5573. 1831–1832, nom. nud. Based on *Wallich 5573* (BM).
Collaea gibba Grah. in Wall., Cat. 5572b, c. 1831–32, nom. nud. Based on *Wallich 5572b, c* (BM).
Dunbaria latifolia Wight & Arn., Prod. 1: 258. 1834. Type: *Wight 878* (p.p.), not seen, fide Baker (1876). Based on *Wallich 5572b, c* (BM), India, Dindygul Hills.

Stems ferruginous pubescent. Leaf rachis about 4 mm long, ½–⅓ the petiolar length or much longer; stipules caducous; stipels setaceous, about 1 mm long, semi-persistent; leaflets ovate, 3–7 × about 4 cm, or larger, pubescent above and below, acuminate apically, broadly cuneate or rounded basally, the laterals inequilateral, with 3 prominent basal veins, primary lateral veins about 4 pairs. Inflorescences paniculate or branched racemose, many-flowered, about 12 cm long, with 1–2(–3) flowers at each node, nodes oblique, not distinctly tuberculate; bracts not seen. Flowers 11–18 mm long, the pedicels 4–10 mm long; calyx tube about 5 mm long, with long ascending, fulvous-ferruginous pubescence, the upper lobes emarginate, all lobes acuminate, the lower about 7 mm long; petals yellow, the vexillum obovate-elliptic, with dark striae, reflexed, basally inflexed biauriculate, glabrous, about 11 mm long, the wings oblanceolate, about 8 mm long, auriculate, with a blunt spur, the claw about 4 mm long, the keel long-pouched, somewhat beaked, equalling the wings in length; ovary about 3-ovulate, villous, the hairs about 2 mm long, the lower style slender, sometimes bent, geniculate, the upper style thickened, pubescent. Fruit oblong, about 3 × 1 cm, with long, silky, fulvous hair, 3–4-seeded; sutures slightly swollen, with a short, downcurved beak. Seeds dark.

Distr. Ceylon and India.

Ecol. Trimen (1894) notes the species is very rare, found in the dry region and flowers in January and February.

Specimens Examined. MATALE DISTRICT: Dambulla, *Thwaites*? *C.P. 3961* (PDA). KANDY DISTRICT: Guruoya, *Jayasuriya et al. 481* (PDA, US). MONERAGALA DISTRICT: nr. Bibile, Uva, Jan. 1888, *Trimen*? *s.n.* (PDA).

Notes. *Jayasuriya et al. 481* is 16 August. There may be more than two *Dunbaria* species present in Ceylon, causing some of the difficulty in interpretation. I am using the ferruginous aspect, the ovate-lanceolate leaflets, the size of the flowers, the pubescence on the calyx and possible fruit size, seed number, upper style pubescence, and stipule retention as diagnostic characters to separate the 2 species. I follow Trimen (1894) in believing this species has smaller flowers than *D. heynei*; Gamble (1918) does not agree.

Description based on *Wallich 5573* (BM) and *Wight 770* (K), India, Nilgiris.

2. Dunbaria heynei Wight & Arn., Prod. 1: 258. 1834; Thw., Enum. Pl. Zeyl. 90, 412. 1859; Baker in Hook. f., Fl. Br. Ind. 2: 217. 1876; Trimen, Handb. Fl. Ceylon 2: 80. 1894; Gamble, Fl. Pres. Madras 1(2): 370. 1918; Alston in Trimen, Handb. Fl. Ceylon 6: 84. 1931. Type: India, hills of Mysore, *Wallich 5572a* (K).

Collaea gibba Grah. in Wall., Cat. 5572a. 1831–1832, nom. nud.
Dunbaria oblonga Arn., Nova Acta Phys.-Med. Acad. Caes. Leop.-Carol. Nat.
    Cur. 18: 333. 1836, fide Baker (1876).

Angled, strigose stems; stipules lanceolate-acute, 2.5–6 mm long, usually persistent; stipels setaceous, about 2 mm long, usually persistent; leaflets thin, about 6.5 × 3.5 cm, the terminal obovate, the laterals inequilateral, broadly lanceolate, acuminate apically, cuneate or rounded basally, both surfaces with scattered strigose hairs to nearly glabrous. Inflorescences few-flowered, nodes atuberculate; bracts not seen. Flowers about 2 cm long, the pedicels about 12 mm long, usually paired; calyx tube 6 mm long, herbaceous, glabrous, the upper lobes connate, entire, all lobes acuminate, the lower about 8 mm long; petals yellow with a purple blotch on the back of the vexillum, the vexillum with dark striations, glabrous, 15–18 × about 17 mm, the claw about 3 mm long, the wings somewhat falcate, about 16 × 5 mm, long-clawed, the keels about 13 mm long, beaked, with a very long, shallow pouch, long-clawed; ovary 3–5-ovulate, the upper style apparently glabrous. Fruit oblong-linear or oblong, straight, subseptate, about 5 × 1 × 0.4 cm, with long, villous, spreading glandular hairs, the beak 3–4 mm long, the exocarp faintly depressed between the seeds, 3–4-seeded. Seeds horizontal, ovoid-reniform, compressed, about 5 × 4 × 2 mm, mottled black and brown, the hilum oblong, about 2.5 mm long, surrounded by arillate tissue.

Distr. India and Ceylon.

Ecol. Flowering occurs from October through April. The species is common in the Central Province up to 900 m.

Specimens Examined. LOCALITY UNKNOWN: low country, up to 2000 ft, *de Silva 88* (NY); *Thwaites C.P. 1478* (BM, GH, K, NY, P); *Col. Walker s.n.* (K, P), *N207* (K), *Mrs. & Col. Walker s.n.* (P). KANDY DISTRICT: Peradeniya, *s. coll. C.P. 1478* (K); Kandy, Lady Horton's Drive, *Alston 1238* (PDA); Hantane & Peradeniya, 1854, *Gardner s.n.* (P, PDA); Hantane, 2300 ft, *Gardner 243* (BM, K); Urugala Pass, above hill, roadside patana, *N.D. Simpson 9218* (BM, PDA); Urugala, roadside jungle above the village, *Simpson 9423* (BM, PDA).

Notes. Both *Dunbaria* species bloom about the same time. *D. heynei* is perhaps more upland than *D. ferruginea*. The herbaceous, glabrous, white gland-dotted calyx of *D. heynei* easily separates it from *D. ferruginea*.

Baker (1876) describes the bracteoles as large and oblong in *Dunbaria heynei*, but I saw none in any of the specimens I examined of either species.

30. FLEMINGIA

Roxb. ex Ait. f., Hort. Kew ed. 2, 4: 349. 1812; Rudd, Taxon 19(2): 294–297. 1970, nom. cons. Lectotype species: *F. strobilifera* (L.) Ait. f.

Maughania J. St. Hil., Bull. Soc. Philom. Paris, ser. 2, 3: 216. 1813, non N.E. Br.

Moghania J. St. Hil., J. Bot. (Desvaux) 1: 16. 1813; Hui-Lin Li, Amer. J. Bot. 31: 224. 1944; Abeywickrama, Ceylon J. Sci., Biol. Sci. 2(2): 174. 1959; Nooteboom, Reinwardtia 5: 432–436. 1961; Hutchinson, Gen. Fl. Pl. 1: 422. 1964.

See Rudd (1970) for an extensive discussion of the genus synonymy and history of the conservation proposals.

Mostly erect shrubs. Leaves pinnately or digitately trifoliolate, rarely unifoliolate; stipules non-produced; stipels absent; leaflets gland-dotted, ovate, obovate, lanceolate to broadly linear. Inflorescences axillary or terminal, racemose or paniculate, few to numerous-flowered; bracts foliaceous, often persistent; bracteoles absent. Flowers medium sized; calyx 5-lobed, the lobes longer than the tube, usually gland-dotted; vexillary petal obovate, oblong or elliptic, basally auriculate, the wings obliquely obovate or oblong, the keel straight or slightly falcate; stamens diadelphous, the vexillary free, the anthers uniform; ovary 2-ovulate, the style filiform, enlarged above, the stigma small, capitate, terminal. Fruit dehiscent, oblong-ovoid, somewhat terete, the style persistent, the sutures indistinct, 1–2-seeded. Seeds globose, the hilum minute, oval, without arillate tissue.

An Asian genus of about 30 species distributed from India to southern China, throughout Malaysia and into the New Guinea region. Two species are found in Africa and species have been introduced into the West Indies.

KEY TO THE SPECIES AND VARIETIES

1 Leaves unifoliolate; bracts broadly ovate, wider than long, enclosing the inflorescence, persistent
. . . . **1. F. strobilifera**
1 Leaves trifoliolate; bracts linear or lanceolate, never wider than long, nor enclosing the inflorescence, caducous or subpersistent
  2 Inflorescences of lax panicles; leaflets 4.5–7 cm long, frequently rounded or obtuse apically
. . . . **2. F. lineata**
  2 Inflorescences racemose; leaflets 7–17 cm long, mostly long acuminate apically, except mostly obtuse in *F. wightiana*
    3 Inflorescences shorter than the petioles; petioles unwinged; leaflets glabrous above, pubescent on the veins below . . . . **3a. F. macrophylla** var. **macrophylla**
    3 Inflorescences longer than the petioles; petioles winged or unwinged; leaflets pubescent or glabrous above, densely pubescent below or only on the veins
      4 Leaflets thick, somewhat rugose above, "ferruginous" villous or tomentose below, mostly obtuse apically, petioles unwinged or slightly winged . . . .**4. F. wightiana**
      4 Leaflets thin, pubescent only on the veins below, mostly acute apically, petioles winged
. . . . **3b. F. macrophylla** var. **semialata**

1. Flemingia strobilifera (L.) Roxb. in Ait. f., Hort. Kew, ed. 2, 4: 350. 1812; Moon, Cat. 54. 1824; DC., Prod. 2: 351. 1825; Wight & Arn., Prod. 1: 243. 1834; Wight, Ic. Pl. Ind. Or. 1: t. 267. 1840; Thw., Enum. Pl. Zeyl. 92. 1859; Baker in Hook. f., Fl. Br. Ind. 2: 227. 1876; Trimen, Handb. Fl. Ceylon 2: 86. 1894; Prain, J. Asiat. Soc. Bengal 66(2): 437. 1897; Gamble, Fl. Pres. Madras 1(2): 377. 1918; Fawcett & Rendle, Fl. Jamaica 4(2): 75. 1920; Ridley, Fl. Malay Penins. 1: 565. 1922; Amshoff in Pulle, Fl. Suriname 2(2): 219. 1939; Rudd, Taxon 19(2): 297. 1970; Verdcourt, Man. New Guinea Leg. 550, fig. 136B. 1979. Lectotype: Poiret in Lam., Enc. Suppl. 5: 14, 19. Nov. 1817, No. *16*, as *Hedysarum strobilifera,* fide Rudd (1970), not seen.

Hedysarum strobilifera L., Sp. Pl. 2: 746. 1753; L., Fl. Zeyl. n. 289. 1747; Roxb., Fl. Ind. 3: 350. 1832. Lectotype: Herm. Mus. 3: 48 in BM.

Hedysarum bracteatum Roxb., Fl. Ind. 3: 351. 1832.

Flemingia fruticulosa Grah. in Wall., Cat. 5754. 1832, nomen.

Flemingia abrupta Grah. in Wall., Cat. 5755B. 1832, nomen.

Flemingia bracteata (Roxb.) Wight, Ic. Pl. Ind. Or. t. 286. 1840; Prain, J. Asiat. Soc. Bengal 66(2): 437. 1897; Gamble, Fl. Pres. Madras 1(2): 378. 1918. Based on Herb. Heyne, *Wallich 5753C* (BM).

Flemingia strobilifera (L.) R. Br. in Ait. f. var. *bracteata* (Roxb.) Baker in Hook. f., Fl. Br. Ind. 2: 227. 1876.

Moghania strobilifera (L.) St. Hil. ex Jackson, Ind. Kew 2. 1894; Li, Amer. J. Bot. 31(4): 227. 1944 (cum syn.); Nooteboom, Reinwardtia 5(4): 433. 1961; Backer & Bakh. f., Fl. Java 1: 638. 1963.

Shrubs, 1 to about 4 m tall; stems terete, the younger densely pubescent. Leaves 1-foliolate, the petiole and petiolule 5–25 mm long, with erect spreading hairs; stipules linear-lanceolate, scarious, 8–25 mm long, caducous; leaflets sub-coriaceous, ovate, elliptic to broadly lanceolate, 5–23 cm long, acuminate or acute apically, broadly cuneate, rounded, or somewhat cordate basally, the upper surface glabrous, the lower pubescent, with about 9 pairs of primary lateral veins prominent beneath. Inflorescences axillary and terminal, paniculate, 7–15 cm long, pubescent, the rachis frequently flexuose, cymules 2–4-flowered, each cyme completely enclosed by a bract; bracts large, short-petiolate, broadly ovate or cordate, wider than long, to about 2.5 cm wide, membranous, pubescent, usually with a short point apically, persistent. Flowers about 1 cm long, the pedicels about 2 mm long; calyx tube about 2 mm long, pubescent, all lobes lanceolate, 7–8 mm long; petals white or purple, the vexillum oblong-obovate, 7–8 mm long, short-clawed, the wings narrow, oblanceolate, with a large auricle, the keels equalling the wings, obliquely oblong, wings and keels long-clawed. Fruit oblong, somewhat turgid, 6–14 × about 6 mm, with a short beak, the exocarp smooth, with downy pubescence, 1–2-seeded. Seeds 2–4 × about 2 mm, mottled. Chromosome number 2n = 22.

Distr. India to the Philippines, Malaysia, and New Guinea. Introduced and naturalized in Micronesia, Polynesia and the West Indies.

Ecol. Common in the low country where it may be locally abundant. According to collectors, the flower colours are various: creamy, light yellow, and pinkish to purple. Flowering is from July through February.

Vern. Hampilla, Hampinna (S).

Specimens Examined. WITHOUT DISTRICT: *Gardner C.P. 697* (BM, PDA); *Macrae 411* (BM); *Thwaites C.P. 697* (BM); Kalagedihena, *Amaratunga 510* (PDA). POLONNARUWA DISTRICT: Polonnaruwa, near sacred area, *Dittus WD 71100701* (US); roadside, Polonnaruwa to Batticaloa, *Kostermans 24321* (BM, US); between Wewa & Elahera, *Rudd & Balakrishnan 3158* (K, NY, US); near Polonnaruwa, *Simpson 8390* (BM). MATALE DISTRICT: near Kongahawela, *Jayasuriya 351* (US). KANDY DISTRICT: Peradeniya, *J.M. de Silva 281* (NY); Kandy, *Worthington 7202* (K). KALUTARA DISTRICT: Panadura, *Amaratunga 2059* (PDA); Panadura, roadside in coconut plantation, *Comanor 995* (NY, US). RATNAPURA DISTRICT: between Panamure & Kolonna, forest border, *Balakrishnan & Jayasuriya NBK 909* (US). GALLE DISTRICT: Induruwa, coconut grove, *Rudd 3078* (K, PDA, US).

Notes. Nooteboom (1961) indicates that the combination of pubescent bracts, which are also emarginate, and longer stipules, is not consistent enough to maintain *F. bracteata* as a distinct species from *F. strobilifera*. I have seen no material from Ceylon that could be assigned to *F. bracteata* as a species or variety.

Trimen (1894) states intermediate stages between foliage leaves and bracts are commonly found at the base of inflorescence branches of *F. strobilifera*.

Description based on Herm. Mus. 3: 48 in BM.

2. Flemingia lineata (L.) Roxb. in Ait. f., Hort. Kew ed. 2, 4: 350. 1812; Moon, Cat. 54. 1824; Wight & Arn., Prod. 1: 242. 1834; Wight, Ic. Pl. Ind. Or. 2: t. 327. 1840; Thw., Enum. Pl. Zeyl. 92. 1859; Baker in Hook. f., Fl. Br. Ind. 2: 228. 1876; Trimen, Handb. Fl. Ceylon 2: 87. 1894; Gamble, Fl. Pres. Madras 1(2): 378. 1918; Verdcourt, Man. New Guinea Leg. 549. 1979. Type not seen.

Hedysarum lineatum L., Sp. Pl. 1054. 1753; Roxb., Fl. Br. Ind. 3: 341. 1832.
Flemingia parvifolia Ham. in Wall., Cat. 5752. 1831–1832, nomen.
Moghania lineata (L.) Kuntze, Rev. Gen. Pl. 1: 199. 1891; Li, Amer. J. Bot.
    31(4): 227. 1944 (cum syn.).

Small, erect shrubs; stems terete, the younger pubescent, becoming glabrescent. Leaves digitately trifoliolate, the petioles 1.5–5.0 cm long, pubescent; stipules prominent, lanceolate, about 8 × 3 mm, long-acuminate, striate, glabrous; leaflets subcoriaceous, oblanceolate, obovate or elliptic, 4–7.5 × to about 3 cm, acute or obtuse apically, cuneate basally, glabrous above, canescent pubescent below, veins prominent below, somewhat parallel, with 3 primary veins inserted at the base, conspicuously yellow gland-dotted. Inflorescences erect, axillary, short-paniculate, (2–)5–10 cm long, shorter than the leaves, several to many-flowered; bracts minute, linear or acute, about 3 mm long, caducous. Flowers about 6 mm long, the pedicels about 2 mm long; calyx tube about 2 mm long, gland-dotted, with sericeous pubescence, all lobes linear-lanceolate to acute, about 2 mm long; petals pinkish, the vexillum obovate or oblong, about 6 mm long, the wings narrow, shorter than the keels, the keels obliquely oblong. Fruit oblong, or somewhat elliptic, (5–)8–12 × 5–7.5 mm, shortly beaked, the exocarp sparsely gland-dotted, pubescent, usually 2-seeded. Seeds to about 3 × 2.5 × 2 mm. Chromosome number 2n = 20.

Distr. India, Burma, Thailand, through Southeast Asia and parts of Malaysia, also including the Philippines, and New Guinea to Australia.

Ecol. The species is found in the dry regions, but is rare. Flowering is in February and the flowers are reported pale pinkish, reddish maroon and red-veined.

Specimens Examined. WITHOUT EXACT LOCALITY: (N. Province, fide Trimen (1894)), *Gardner C.P. 1423* (K). TRINCOMALEE DISTRICT: Mahaweli Ganga, N. Forest Reserve, *Jayasuriya et al. 637* (PDA, US). BATTICALOA DISTRICT: Mar. 1858, *Thwaites C.P. 1423* (BM, K), Trimen's (1894) locality.

3. Flemingia macrophylla (Willd.) Merr., Philipp. J. Sci. 5: 130. 1910; Kuntze ex Prain, J. Asiat. Soc. Bengal 66(2): 440. 1897; Alston in Trimen, Handb. Fl. Ceylon 6: 85. 1931; Verdcourt, Fl. Trop. East Africa, Papil. 806. 1971 (note);

Verdcourt, Man. New Guinea Leg. 549, fig. 136C. 1979; non Blume ex Miq., Fl. Ind. Bat. 1(1): 163. 1855, nomen, fide Li (1944).Verdcourt (1971) maintains neither Kuntze nor Prain validly published the combination.

Crotalaria macrophylla Willd., Sp. Pl. 3: 982. 1800, non Weinm. 1828, nec. Kurtze 1876, fide Li (1944).

Flemingia congesta Roxb. in Ait. f., Hort. Kew ed. 2, 4: 349. 1812; Roxb., Fl. Ind. 3: 340. 1832; Wight, Ic. Pl. Ind. Or. t. 390. 1839. Based on *Wallich 5747* (BM).

Flemingia semialata Roxb. in Ait. f., Hort. Kew ed. 2, 4: 349. 1812; Roxb., Fl. Ind. 3: 340. 1832; Moon, Cat. 54. 1824; Wight & Arn., Prod. 1: 241. 1834; Wight, Ic. Pl. Ind. Or. 2: t. 326. 1840; Thw., Enum. Pl. Zeyl. 92. 1859; Prain, J. Asiat. Soc. Bengal 66(2): 439. 1897; Gamble, Fl. Pres. Madras 1(2): 378. 1918; Alston in Trimen, Handb. Fl. Ceylon 6: 85. 1931. Based on *Dr. Roxburgh s.n.* (K), East. Ind., in Herb. Forsyth, Purc. 1835, in the type folder (K).

Millingtonia congesta Roxb. ex Wight & Arn., Prod. 1: 241. 1834, in syn.

Millingtonia semialata Roxb. ex Wight & Arn., Prod. 1: 241. 1834, in syn.

Flemingia congesta Roxb. var. *semialata* Baker in Hook. f., Fl. Br. Ind. 2: 228. 1876; Trimen, Handb. Fl. Ceylon 2: 87. 1894.

Moghania macrophylla (Willd.) Kuntze, Rev. Gen. Pl. 1: 199. 1891; Li, Amer. J. Bot. 31(4): 226. 1944; Nooteboom, Reinwardtia 5(4): 434. 1961; Backer & Bakh. f., Fl. Java 1: 638. 1963.

Erect shrubs, 0.5–1.5 m tall; young branches 4(–3)-angled, pubescent. Leaves trifoliolate, the petioles 2.5–10 cm long, densely pubescent; stipules mostly acute-deltoid, scarious, about 12 mm long, caducous; leaflets ovate, elliptic or lanceolate-elliptic, 7–17 × 2–4 cm, long acuminate to obtuse apically, cuneate to rounded basally, the upper surfaces glabrous or with sparse, adpressed pubescence, the lower usually with sparsely adpressed pubescence mostly on the veins, 3 prominent veins inserted basally, more prominent beneath. Inflorescences axillary, clustered or single, 2.5–15 cm long, frequently in racemose, erect or lax branches, many–numerous-flowered, pubescent; bracts somewhat lanceolate, acuminate, striate, 4–8 mm long, caducous. Flowers about 8 mm long, the pedicels about 2 mm long; calyx lobes linear-lanceolate, the lower longest, much exceeding the tube, pubescent; petals pink or flesh-coloured with purple, orange or yellow striae, the vexillum oblong-obovate, 7–8 mm long, short-clawed, the wings narrow, with a large auricle, the keels about equalling the wings, the wings and keels long-clawed. Fruit somewhat elliptic, turgid, 6–17 × 6 mm, with a short beak, pubescent, usually gland-dotted and 2-seeded. Seeds 2–4 × about 2 mm. Chromosome number 2n = 22, one count of n = 10.

KEY TO THE VARIETIES

1 Inflorescences florate almost to the base, erect, usually shorter than the petioles, unbranched, nodes of the rachis condensed; petioles usually unwinged; leaflets dark green, usually long-acuminate apically, the under surface with copious and conspicuous black gland dots ....
.... var. **macrophylla**
1 Inflorescences distinctly pedunculate, erect or lax, usually longer than the petioles, frequently branched, the flowers not crowded; petioles winged; leaflets dark green or other shades, usually obtuse, acute, or occasionally acuminate apically, the under surface with a few inconspicuous yellow gland dots ....var. **semialata**

Distr. Tropical Asia, Malaysia, and into the New Guinea area.

Ecol. In Ceylon, the 3 taxa, *F. macrophylla, F. semialata* and *F. wightiana,* considered by Alston (1931) as species, are all found in the low country. *Flemingia macrophylla* is the most common, flowering January through March.

Vern. Wal-undu (S).

Specimens Examined. LOCALITY UNKNOWN: *Thwaites C.P. 2663, C.P. 3585* (p.p.) (BM); *Macrae 42* (BM). KANDY DISTRICT: "Maskeliya Ganga", Feb. 1853, *s. coll. C.P. 2663* (BM, PDA). COLOMBO DISTRICT: *Ferguson C.P. 256* (PDA). KALUTARA DISTRICT: Agalawatte, Dartonfield, 4.11.26, *s. coll. s.n.* (PDA).

Notes. Nooteboom's (1961) species concept of *F. macrophylla* includes Alston's (1931) 3 species which were segregated from Trimen's (1894) *F. congesta.* Verdcourt (1971) believes Nooteboom's concept is too broad.

4. Flemingia wightiana Grah. ex Wight & Arn., Prod. 1: 242. 1834; Baker in Hook. f., Fl. Br. Ind. 2: 228. 1876; Trimen, Handb. Fl. Ceylon 2: 88. 1894; Prain, J. Asiat. Soc. Bengal 66(2): 439. 1897; Gamble, Fl. Pres. Madras 1(2): 379. 1918; Alston in Trimen, Handb. Fl. Ceylon 6: 85. 1931. Type: *Wallich 5751* (COIMB).

Flemingia wightiana Grah. in Wall., Cat. 5751. 1832, nomen.
Flemingia ferruginea Grah. in Wall., Cat. 5750. 1832, nomen.
Flemingia ferruginea Wall. ex Benth. in Miq., Pl. Jungh. 245. 1852.
Flemingia congesta var. *wightiana* (Grah.) Baker in Hook. f., Fl. Br. Ind. 2: 229. 1876.
Moghania ferruginea (Wall. ex Benth.) Li, Amer. J. Bot. 31(4): 226. 1944.

Erect shrubs, about 0.5 m tall (fide Alston, 1931); young stems somewhat 4-angled, tomentose or somewhat puberulent. Leaves trifoliolate, petioles 3–5 cm long; stipules acuminate, scarious, about 5 mm long, caducous; leaflets ovate-oblong lanceolate, 7.5–12 × 3–7 cm, acute, obtuse apically, cuneate basally, the upper surface glabrous and somewhat rugose or with sparse adpressed hairs, the lower tomentose or densely pubescent (golden silky), black gland-dotted; venation prominent. Inflorescences axillary, racemose, to about 5 cm long, numerous-flowered; bracts ovate, acute, not exceeding the buds, caducous. Flowers crowded on the rachis, about 6 mm long, subsessile; calyx tube about 3 mm long, hirsute, (without glands, fide Wight & Arn., 1834; red glands—black when dry, fide Gamble, 1918), all lobes long-acuminate, greatly exceeding the tube; petals pinkish (?), probably similar to *F. macrophylla* but smaller. Fruit somewhat oblong, about 9 × 6 × 4 mm, light brown, the apex apiculate, without glands, 2-seeded. Seeds somewhat orbicular, about 3 × 2 mm, dark, smooth.

Distr. Ceylon and S. India.

Ecol. Alston (1931) reports the species rare in the low country, flowering in March.

Specimens Examined. WITHOUT DISTRICT: *Thwaites C.P. 3585* (p.p.) (BM). BATTICALOA DISTRICT: Mar. 1858, *Thwaites* (?) *C.P. 3584* (PDA). BADULLA DISTRICT: Mar. 1858, *Thwaites* (?) *C.P. 3584* (PDA).

Notes. Although there is ample material of the Ceylon taxa in COIMB, I was unsuccessful in finding distinctive characters to easily separate *F. macrophylla*, *F. semialata* and *F. wightiana*. I hope the key and synonymy suffice for those who favour a different classification.

31. RHYNCHOSIA

Lour., Fl. Cochinch. 425, 460. 1790; Verdcourt, Kew Bull. 25(1): 70. 1971; Verdcourt, Fl. Trop. East Africa, Papil. 711. 1971; Grear, Mem. New York Bot. Gard. 31(1): 1–168. 1978 (New World revision, cum syn.), nom. cons. Type species: China, Japan, *R. volubilis* Lour.

Cyanospermum Wight & Arn., Prod. 1: 259. 1834.
Nomismia Wight & Arn., Prod. 1: 236. 1834.
Cylista Ait., Hort. Kew ed. 1, 3: 36, 512. 1789, nom. rejic. Type: *C. villosa* Ait. = *R. hirta* (Andrews) Meikle & Verdcourt.

Climbing, trailing or prostrate perennials, occasionally annuals, occasionally small, erect shrubs; stems terete to angular. Leaves usually pinnately trifoliolate; stipulate; stipels minute or absent; leaflets usually gland-dotted beneath. Inflorescences racemose, axillary, rarely few-flowered; bracts present; bracteoles absent. Flowers small to medium; calyx 5-lobed, the upper partially connate, all lobes acuminate or oblong, occasionally accrescent, the lower usually longest; petals usually yellow, the vexillum reflexed, obovate or orbicular, lined with darker striae, basally inflexed biauriculate, the wings free, narrow, frequently with a spur, the keels similar to the wings or somewhat falcate, about equal length; stamens diadelphous, the vexillary free, the anthers uniform; ovary (1–)2-ovulate, pubescent, the lower style slender, pubescent, the upper style frequently flattened, thick, glabrous, the stigma small, capitate, glabrous. Fruit dehiscent, suborbicular, oblong or falcate, compressed, mostly non-septate, (1–)2-seeded.

Seeds suborbicular, compressed, the hilum ovoid to short-oblong, rim aril reduced, or the arillate tissue well developed.

A genus of about 200 species found throughout the tropics and subtropics, extending into temperate regions.

KEY TO THE SPECIES

1 Seeds with conspicuous, raised arillate tissue; fruit 1–2-seeded
  2 Small, erect shrubs; fruit exocarp smooth
    3 Calyx lobes acuminate; fruit 2-seeded
      4 Fruit septate; stipules linear or lanceolate..........................**1. R. suaveolens**
      4 Fruit non-septate; stipules minute, acicular...........................**2. R. cana**
    3 Calyx lobes oblong, obtuse; fruit 1-seeded..........................**3. R. rufescens**
  2 Climbing, trailing, or prostrate vines; fruit exocarp raised, reticulate
    5 Inflorescences usually with a median branch, the flowers clustered apically; fruit stipitate, 2-seeded
      ....**5. R. capitata**
    5 Inflorescences without a median branch, the flowers not conspicuously crowded; fruit sessile, usually 1-seeded, occasionally 2-seeded
      6 Leaflets about 4 × 2.5–6.5 cm; inflorescences 4–9 cm long, many-flowered; fruit suborbicular, 1-seeded ....**4. R. nummularia**
      6 Leaflets about 0.8–1.5 × 0.8–1.7 cm, whitish beneath; inflorescences 1–2 cm long, 1–6-flowered; fruit suborbicular or short-oblong, 1–2-seeded..........................**6. R. aurea**
1 Seeds lacking conspicuous arillate tissue; fruit mostly 2-seeded, occasionally 1-seeded
  7 Calyx lobes oblong, obtuse; seeds dark blue, frequently attached to the placenta after fruit dehiscence; stout, woody, climbing vines.................................**7. R. hirta**
  7 Calyx lobes acute, acuminate; seeds not dark blue, unattached; mostly smaller vines, trailers or somewhat prostrate in the cover
    8 Leaflets small, usually about 1–3.5 × 1–3 cm long
      9 Leaflets with both surfaces glabrous to glabrescent; inflorescences usually exceeding the leaves, 3–6-many-flowered; stems prostrate or weakly climbing in the cover....**10. R. minima**
      9 Leaflets sparsely pubescent above, dense grey-canescent, lanate below; inflorescences not exceeding the leaves, 2–4-flowered; stems trailing or climbing.........**11. R. velutina**
    8 Leaflets medium to large, 4.5–12 cm long
     10 Leaflets thin, surfaces glabrescent or sparsely pubescent, obtuse, acute or somewhat acuminate apically
      11 Leaflets about 4.5 cm long, surfaces glabrescent, the margins ciliate; stems densely canescent pubescent; inflorescences florate almost to the base, flowers crowded...........
      ....**8. R. densiflora**
      11 Leaflets 5–9 cm long, surfaces with sparse curly hairs; stems with short, glandular hairs; inflorescences not as above ....**9. R. viscosa**
     10 Leaflets coriaceous, surfaces fulvous pubescent, long acuminate apically, the tips 1 cm long or longer..**12. R. acutissima**

1. Rhynchosia suaveolens (L.f.) DC., Prod. 2: 387. 1825; Wight & Arn., Prod. 1: 240. 1834; Thw., Enum. Pl. Zeyl. 441. 1859; Baker in Hook. f., Fl. Br. Ind. 2: 221. 1876; Trimen, Handb. Fl. Ceylon 2: 83. 1894; Gamble, Fl. Pres. Madras 1(2): 374. 1918; Alston in Trimen, Handb. Fl. Ceylon 6: 84. 1931. Type: Madras, India, not seen.

Glycine suaveolens L. f., Suppl. 326. 1781.
Cajanus suaveolens (L.f.) Grah. in Wall., Cat. 5579. 1831–1832.

Small erect shrub; stems slender, numerous, spreading, flexuose, with fulvous, erect pubescence about 0.5 mm long. Leaves pinnately trifoliolate, the petiole 1.5–3 cm long, rachis 0.4–1.2 cm long, pubescent; stipules linear or lanceolate, 2–6 mm long, with dense fulvous, tomentose pubescence; stipels reduced to knobs, less than 1 mm long; leaflets ovate, the terminal somewhat rhomboid, to about 6 × 4 cm, acute-elongate apically, rounded or cuneate basally, the laterals similar in shape, smaller, inequilateral, both surfaces with short, white, curly hairs and dense orange gland-dots, with 3 prominent veins inserted at the base. Inflorescences few-flowered, long-pedunculate, the peduncles 1.5–4 cm long, the rachis 0.5–1 cm long; bracts at the base of the pedicels ovate, about 3 × 1.5 mm, persistent. Flowers about 8 mm long, the pedicels about 4 mm long; calyx tube about 2 mm long, with a few orange gland-dots, pubescent, all lobes acuminate, longer than the tube, the upper and laterals about 3 mm long, the lower longest, 4–5 mm long; petals yellow, about equal in length, the vexillum obovate, pubescent, about 7 mm long, the wings and keels glabrous, all claws about 2 mm long; ovary 2-ovulate, white tomentose. Fruit short-oblong, about 15 × 6–9 × 3 mm, septate, with a short beak, reddish-brown to dark brown, the exocarp with short, white pubescence, or glabrescent, somewhat depressed between the seeds, 2-seeded. Seeds flat, 2.5–3 × 2–3 × about 1.5 mm, blackish, dull, the hilum short-oblong or somewhat elliptic, with 2-lipped, white arillate tissue.

Distr. Indian peninsula and Ceylon.
Ecol. Found in the dry regions, very rare (Trimen, 1894).
Specimens Examined. WITHOUT EXACT LOCALITY: *Thwaites C.P. 3851* (BM). TRINCOMALEE DISTRICT: Trincomalee, *Glenie s.n. C.P. 3851* (K, PDA).
Notes. The septate fruit is unusual in the genus.
Description based on *Wight 760* (COIMB, K), Herb. *Wight 760, 737* (COIMB), *Wallich 5579A* (K), *5579B* (K).

2. Rhynchosia cana (Willd.) DC., Prod. 2: 386. 1825; Wight & Arn. Prod. 1: 240. 1834; Thw., Enum. Pl. Zeyl. 91. 1859; Baker in Hook. f., Fl. Br. Ind. 2: 222. 1876; Trimen, Handb. Fl. Ceylon 2: 83. 1894; Gamble, Fl. Pres. Madras 1(2): 374. 1918; Alston in Trimen, Handb. Fl. Ceylon 6: 84. 1931. Type: India, not seen.

Glycine cana Willd., Sp. Pl. 3: 1063. 1800.
Flemingia biflora Moon, Cat. 53. 1824, nomen.

Small erect shrubs; stems slender, numerous, ascending, branching, with dense spreading hairs. Leaves pinnately trifoliolate, the petiole and rachis about 2.5

cm long; stipules minute, acicular; stipels not seen; leaflets ovate, obovate, elliptic or lanceolate, of various sizes, smaller on the lateral branches than the main stems, to about 6 × 4 cm, obtuse, acute or acuminate apically, cuneate to rounded basally, pubescent and gland-dotted on both surfaces, the lower with dense, velvety pubescence, the veins prominent. Inflorescences with the peduncles shorter than the leaves, each peduncle 1–2-flowered, to about 2.5 cm long; bracts at the base of the pedicels minute, persistent. Flowers 5–8 mm long, the pedicels about 2 mm long; calyx tube 1–2 mm long, pubescent, all lobes long-acuminate, about 3 mm long; petals yellow or somewhat reddish (striae?), the vexillum obovate, about 6 mm long, the wings and keels nearly equalling the vexillum, glabrous, all claws about 2 mm long; ovary 2-ovulate, whitish tomentose. Fruit oblong or obovate, 12–16 × about 6 × 2 mm, nonseptate, with a short beak, the exocarp very finely pubescent or glabrous, somewhat depressed between the seeds, or smooth, 2-seeded. Seeds ovoid, dark, smooth, flat, about 3 × 3 × 2 mm, the hilum short-oblong, or somewhat elliptic, about 2 mm long, with 2-lipped, persistent arillate tissue.

Distr. The Indian Peninsula and Ceylon.

Ecol. Blooming December through March, the species is rather common in the low country where it frequently grows gregariously as an undershrub.

Vern. Gas-kollu (S).

Specimens Examined. LOCALITY UNKNOWN: "Ceylon" ex Herb. Beddome, *s.n.* (BM); *Macrae 358* (BM, K); *D. Kelaart s.n.* (K); "received 1870", *Thwaites C.P. 1280* (NY). DISTRICT UNKNOWN: Kumbukkan, *Worthington 5845* (K). JAFFNA DISTRICT: Chunnavil, Feb. 1890, *s. coll. s.n.* (PDA). KALUTARA DISTRICT: Nagoda, Pasdum Korale (?), March 1887, *s. coll. s.n.* (PDA), 1847, *Gardner 231* (K); "Coast Caltura" *Gardner 231* (K); "Caltura", W. Coast, *Gardner s.n.* (BM). BADULLA DISTRICT: Nilgala, Uva, Jan. 1888, *s. coll. s.n.* (PDA); Lunugala, Uva, Jan. 1888, *s. coll. s.n.* (PDA), *Gardner ? C.P. 1280* (K, PDA). MONERAGALA DISTRICT: E. of Bibile, *Fosberg & Sachet 53151* (US); Wellawaya, *Worthington 4639* (BM). HAMBANTOTA DISTRICT: 9 km post, Tissamaharama towards Kataragama, *Bernardi 14192* (PDA, US).

3. Rhynchosia rufescens (Willd.) DC., Prod. 2: 387. 1825; Wight & Arn., Prod. 1: 239. 1834; Thw., Enum. Pl. Zeyl. 91. 1859; Baker. in Hook. f., Fl. Br. Ind. 2: 220. 1876; Trimen, Handb. Fl. Ceylon 2: 82. 1894; Gamble, Fl. Pres. Madras 1(2): 373. 1918; Nooteboom, Reinwardtia 5: 439. 1961; Backer & Bakh. f., Fl. Java 1: 636. 1963. Type: not seen.

Glycine rufescens Willd., Nova Acta. Phys.-Med. Acad. Caes. Leop.-Carol. Nat. Cur. 4: 222. 1803.
Hallia trifoliata Roth., Nov. Sp. Pl. 352. 1821.
Flemingia rothiana DC., Prod. 2: 351. 1825.

Glycine pondicheriensis Spreng., Syst. 3: 196. 1826.
Lespedeza indica Spreng., Syst. 3: 202. 1826.
Cylista suaveolens Grah. in Wall., Cat. 5587. 1831–1832, nomen.
Cyanospermum javanicum Miq., Fl. Ind. Bat. 1, 1: 167. 1855.

Small, erect shrub, with trailing branches; stems slightly twining apically, densely pubescent. Leaves pinnately trifoliolate, the petioles 2.5–5 cm long, the rachis ¹/₆ or less the petiolar length; stipules minute; stipels not seen, absent or minute (Baker, 1876); leaflets ovate, elliptic or lanceolate, various sizes, the larger to 4 × 2.5 cm, usually smaller, acute or obtuse apically, rounded basally, both surfaces pubescent, with raised reticulate venation below. Inflorescences elongated, lax, (1–)4–6-flowered; bracts linear-acicular, about 2 mm long, caducous. Flowers 6–9 mm long, the pedicels about 1.5 mm long; calyx tube about 1 mm long, downy pubescent, all lobes oblong, obtuse, 4–5 mm long, incised nearly to the base; petals yellow, about equal length, vexillum obovate, pubescent, about 5 mm long, with red striae, short-clawed, the keels falcate; ovary 1-ovulate, tomentose. Fruit globose, about 8 × 5 × 2 mm, brownish, the upper margin somewhat straight, beaked, the lower margin semi-orbicular, obtuse or rounded at the base, the exocarp with fine pubescence, 1-seeded. Seeds smooth, shining, somewhat ovoid, about 3 × 3 × 2 mm, black, oblique, the hilum short-oblong, or somewhat elliptic, with 2-lipped arillate tissue.

Distr. India, Southeast Asia and scattered through Malaysia.

Ecol. The species is found in the dry regions of Ceylon and may be common locally. Flowering is February through June.

Specimens Examined. WITHOUT LOCALITY: *Major Champion s.n.* (K); *Moon s.n.* (BM); *Thwaites C.P. 1424* (BM, K); *Col. Walker s.n.* (K). ANURADHAPURA DISTRICT: Nr. Ritigala natural reserve, *Sumithraarachchi & Sumithraarachchi DBS 743* (PDA, US). POLONNARUWA DISTRICT: Dimbulagala, below Gunner's Quoin, *Jayasuriya et al. 725* (PDA, US). MATALE DISTRICT: Dambulla Rock, *Amaratunga 537* (PDA), *Rudd & Balakrishnan 3114* (K, US, NY); 8 mi. ESE. of Dambulla, *Davidse 7428* (MO, PDA, US). TRINCOMALEE DISTRICT: Foul Point, near Trincomalee, *N.D. Simpson 9663* (BM, PDA). BATTICALOA DISTRICT: Batticaloa, Mar. 1858, *Gardner C.P. 1424* (PDA), *Gardner C.P. 1424* (PDA).

Notes. Thwaites (1859) lists the species as common in Batticaloa District; Trimen (1894) indicates it rare.

Backer & Bakh. f. (1963) report, "The calyx which mostly conceal the corolla, spread towards noon at which time the flower expands. Afterwards they converge and conceal the pod."

The broad, obtuse calyx lobes are distinctive and form a primary character that Wight & Arnott (1894) used to separate species in their subgenus *Phyllomatia*. The lobes increase in size with age (accrescence).

The synonymy is from Baker (1876). Trimen (1894) cites only "*Lespedeza* (?) *indica,* Spr., Arn. Pug. 14.''

4. Rhynchosia nummularia (L.) DC., Prod. 2: 386. 1825; Baker in Hook. f., Fl. Br. Ind. 2: 221. 1876; Trimen, Handb. Fl. Ceylon 2: 82. 1894; Gamble, Fl. Pres. Madras 1(2): 374. 1918. Type: In Herb. *Burm.* fide DC. (1825), not seen.

Glycine nummularia L., Mant. 571. 1767–1771.
Nomismia nummularia (L.) Wight & Arn., Prod. 1: 237. 1834; Wight, Ic. Pl. Ind. Or. 1: t. 283. 1840; Thw., Enum. Pl. Zeyl. 91. 1859. Based on *Wight 835* (COIMB).

Trailing or prostrate annual herbs; stems to about 1 m long, terete or somewhat angular, grey pubescent. Leaves pinnately trifoliolate, the rachis about 9 mm long, about 1/6 the petiolar length, but occasionally the petioles very long, pubescent; stipules linear-lanceolate or acuminate, 3–4 mm long, striate, persistent; stipels not seen; leaflets rounded-obovate, occasionally wider than long, about 4 × 2.5–6.5 cm, truncate, slightly emarginate, or occasionally apiculate apically, broadly cuneate or rounded basally, the laterals oblique, inequilateral, about as large as the terminal, glabrescent above, finely pubescent, with minute gland-dots below. Inflorescences terminal, lax, 4–9 cm long, equalling or slightly exceeding the leaves, florate about ⅔ the length, with 1–4 distant nodes, each node 2–8-flowered; bracts ovate at the base of the pedicels, about 1.5 mm long, caducous. Flowers about 8 mm long, the pedicels about 2 mm long; calyx tube 3–4 mm long, densely pubescent, the upper lobes connate, all lobes long, lanceolate-acuminate, ciliate; petals probably yellow, long-clawed, the vexillum obovate, about 8 mm long, striate, the wings oblanceolate-falcate, the keels wider than the wings; ovary 1-ovulate, sparsely pubescent with long hairs. Fruit orbicular to somewhat squarish, 7−13 × 12−15 × 1−2 mm, usually rounded at the base, with a median, straight, sharp beak, the exocarp strongly reticulate, glabrous except for a few, long hairs, 1-seeded. Seed ovoid, about 6 × 6 mm, the hilum probably oval or somewhat elliptic, 2-lipped with arillate tissue.

Distr. India and Ceylon.
Ecol. Found rarely in the low country. The only collection date is September.
Specimens Examined. LOCALITY UNKNOWN: 23.9.30, *J.E. Senaratne s.n.* (PDA); *Thwaites C.P. 1439* (BM, COIMB, GH, K, NY). JAFFNA DISTRICT: Jaffna, *Gardner s.n. C.P. 1439* (K, PDA). COLOMBO DISTRICT: Colombo, 1865, *Ferguson C.P. 1439* (PDA).

5. Rhynchosia capitata (Roth) DC., Prod. 2: 386. 1825; Gamble, Fl. Pres. Madras 1(2): 374. 1918. Type: India Orient, not seen.

Glycine capitata Heyne in Roth, Nov. Sp. Pl. 346. 1821.

Nomismia capitata (Roth) Wight & Arn., Prod. 1: 237. 1834; Wight, Ic. Pl. Ind.
  Or. 1: t. 295. 1840. Based on *Wight 837* (NY).
Rhynchosia aurea auct. non DC., Baker in Hook. f., Fl. Br. Ind. 2: 221. 1876
  (p.p.); Trimen, Handb. Fl. Ceylon 2: 83. 1894 (note).

Small vines, climbing or trailing annuals; stems hirsute or softly villous. Leaves
pinnately trifoliolate, the rachis 6–10 mm long, the petioles about 4 cm long,
pubescent as the stems; stipules lanceolate, minute, caducous; stipels not seen;
leaflets somewhat orbicular, 3–3.5 cm long, broadly obtuse or rounded apically,
rounded or broadly cuneate basally, the laterals inequilateral, both surfaces green,
finely pubescent. Inflorescences lax, then erect, about 4.5–10 cm long, about
equalling the leaves, 6–30-flowered (Baker, 1876), the flowers clustered distally, frequently with a median branch emerging from the peduncle; bracts not seen.
Flowers exceeding 10 mm long, the pedicels about 5 mm long, flower buds with
the lobes upcurved, apiculate, calyx tube about 2 mm long, pubescent, the upper
lobes connate, slightly bifid, all lobes long-acuminate, to about 12 mm long, nearly
exceeding the petals, the lower longest; petals yellow, about equal in length, the
vexillum obovate, rounded apically, about 10 mm long, without striae (Gamble,
1918), the claw about 2 mm long, the wings somewhat oblanceolate, the keels
falcate, both with claws about 3 mm long; ovary 2-ovulate with long, pilose, hairs.
Fruit globose-oblong, flat, short-stipitate, about 10 × 8 mm, the exocarp wrinkled
with close, transverse, parallel lines, sparsely pubescent with fine, long, spreading
hairs, the apex usually with a short, hooked beak, 2-seeded. Seeds globose or
oblong-rounded, the hilum 2-lipped with arillate tissue. Chromosome number
2n = 22.

Distr. India, the plains from the foot of the Himalayas to Ceylon, but see note.
Ecol. Hills up to 450 m.
Specimens Examined. None in Ceylon.
Notes. *Rhynchosia nummularia, R. aurea,* and possibly *R. capitata* material
at PDA was mixed; a description of *R. capitata* has been added although the species
may not be present in Ceylon, or may be considered in synonymy under *R. aurea,*
following Baker (1876).
Gamble (1918) treats *R. capitata* as separate from *R. aurea* and notes "the
peduncle with a slender leafless shoot near the middle. . . "
Description based on *Wight 837, 777* (COIMB).

6. Rhynchosia aurea (Willd.) DC., Prod. 2: 386. 1825; Baker in Hook. f., Fl.
Br. Ind. 2: 221. 1876 (p.p.); Trimen, Handb. Fl. Ceylon 2: 83. 1894 (note);
Gamble, Fl. Pres. Madras 1(2): 374. 1918. Type: India Orient, not seen.

Glycine aurea Willd., Nova Acta Phys.-Med. Acad. Caes. Leop.-Carol. Nat.
  Cur. 4: 18. 1803.
Nomismia aurea (Willd.) Wight & Arn., Prod. 1: 237. 1834. Based on *Wight
  836* (NY).

Small, prostrate or trailing, annual vines; stems hirsute. Leaves pinnately trifoliolate, petioles about leaflet length, hirsute; stipules ovate-lanceolate, minute, caducous; stipels not seen; leaflets suborbicular or obtriangular-rounded, about 8 × 8 mm or somewhat larger, obtuse or rounded apically, rounded or broadly cuneate basally, both surfaces finely pubescent, whitish below. Inflorescences axillary, lax, 1–2 cm long, shorter than the leaves, 1–6-flowered; bracts reddish, somewhat scarious, about 2 mm long, persistent (?). Flowers about 10 mm long, the pedicels about 2 mm long, buds upcurved, apiculate; calyx tube about 2 mm long, densely pubescent, upper lobes connate, slightly bifid, all lobes long-acuminate, nearly exceeding the petals, the lower longest; petals bright yellow, about equal length, the vexillum obovate, broad, with dark striae, the keel falcate; ovary 1–2-ovulate. Fruit globose-orbicular (1-seeded), or short-oblong (2-seeded), about 8 × 8 mm (1-seeded), the exocarp wrinkled with close, transverse, parallel lines, sparsely pubescent, with long, white hairs, 1–2-seeded. Seeds ovoid, the hilum 2-lipped with arillate tissue. Chromosome number 2n = 22.

Distr. From the plains at the foothills of the Himalayas to Ceylon.

Ecol. Flowering is around December–January.

Specimen Examined. PUTTALAM/ANURADHAPURA DISTRICT: Wilpattu Natl. Pk., Pallugaturai, *Fosberg et al. 50904* (US).

Notes. Trimen (1894) notes the Ceylon distribution in Baker (1876) is based on a *Walker* specimen in K, labeled Ceylon by Bentham. I have not seen this specimen.

Most authors give 2-seeded fruit.

Rhynchosia aurea in the field is similar in habit to *Atylosia scarabaeoides*, except that the latter has 5–7-seeded, septate fruit.

Description based on *Gamble 21224* (COIMB) and *Wight 837* (NY).

7. Rhynchosia hirta (Andr.) Meikle & Verdcourt, Taxon 16: 462. 1967 (cum syn.); Verdcourt, Fl. Trop. East Africa, Papil. 720, fig. 109/1. 1971. Type: Specimens grown in the garden of G. Hibbert, England, from seed said to have been sent from the Cape of Good Hope, fide Verdcourt (1967), not seen.

Dolichos hirtus Andr., Bot. Rep., t. 446. 1807.

Cylista tomentosa Roxb., Pl. Corom. 3, t. 221. 1811; Moon, Cat. 53. 1824. Type: Plant grown at Calcutta with seeds sent from Mysore by Buchanan, not seen. Based on t. 221.

Cylista albiflora Sims, Bot. Mag. 12: t. 1859. 1816. Type: Plants raised by Lambert in England from seeds received from Mauritius, not seen. Based on t. 1859.

Cyanospermum tomentosum (Roxb.) Wight & Arn., Prod. 1: 260. 1834; Wight, Ic. Pl. Ind. Or. t. 84. 1838; Thw., Enum. Pl. Zeyl. 92. 1859. Based on *Wight 766* (COIMB, K, NY).

Rhynchosia cyanosperma Baker in Fl. Trop. Africa 2: 218. 1871; Baker in Hook.
f., Fl. Br. Ind. 2: 222. 1876; Trimen, Handb. Fl. Ceylon 2: 84. 1894; Gamble, Fl. Pres. Madras 1(2): 375. 1918. Type: Malawi, Manganja Hills, *Meller
s.n.* (K, lectotype), not seen.
Rhynchosia albiflora (Sims) Alston in Trimen, Handb. Fl. Ceylon 6: 85. 1931.

Coarse somewhat woody vines, twining, climbing; stems with dense fulvous-
ferruginous pubescence. Leaves pinnately trifoliolate, the rachis 2–2.5 cm long,
⅓–¼ the petiolar length, pubescence as the stems; stipules acute-lanceolate, about
6 mm long (?), caducous (?); stipels conspicuous, linear-aciculate, 4–8 mm long,
hirsute, persistent; leaflets coriaceous, ovate, the terminal frequently broadly elliptic, to about 11 × 8 cm, the laterals slightly smaller, greatly inequilateral, acute
or obtuse apically, rounded basally, densely pubescent on both surfaces, venation prominent beneath, 3 main veins inserted basally, 3–4 pairs of primary, lateral
veins. Inflorescences mostly axillary, several to many-flowered, pendant, 12–45
cm long, canescent tomentose; bracts large, ovate-acuminate, about
2.2 × 1 cm, caducous. Flowers about 1.5 cm long, the pedicels about 1 cm long;
calyx tube about 1 mm long, canescent pubescent, all lobes incised almost to the
base, acuminate, 12–15 mm long. Petals yellowish-white, or reddish, occasionally
with purple striae, long-clawed, the vexillum obovate, about 12 mm long,
emarginate, persistent, the wings somewhat oblanceolate, about 10 mm long, the
keels falcate, slightly exceeding the wings; ovary 2-ovulate, tomentose, the style
sinuate. Fruit short-oblong, about 12 × 6 × 4 mm, somewhat septate, the upper
margin straight, short-beaked, the exocarp densely pubescent, 2-seeded. Seeds
ovoid or somewhat rounded reniform, horizontal, about 5.5 × 4 × 4 mm, shining violet-blue or blue-black, usually remaining attached to the placenta by the
funiculus after dehiscence, the hilum elliptic, about 1 mm long, without arillate
tissue. Chromosome number 2n = 22.

Distr. Throughout tropical Africa, including northeastern South Africa, the
Mascarene Islands, India and Ceylon.
Ecol. Growing in the upper moist zone and intermediate low country; hills
to 4000 ft and at 4500 ft (Gamble, 1918). Verdcourt (1971) cites 0–1800 m altitude,
with plants found along forest edges, in grassland with scattered trees, hillside
bushland and cultivations. Flowering January–February in Ceylon.
Vern. Hin-garadiya (S).
Specimens Examined. LOCALITY UNKNOWN: *s. coll. C.P. 2471*
(PDA); *Thwaites s.n.* (BM); 1854, *Thwaites s.n.* (BM), *Thwaites C.P. 2471* (GH);
Col. Walker s.n. (K, PDA). DISTRICT UNKNOWN: Sangilikanadarawa, *Simp-
son 9240* (BM). NUWARA ELIYA & KEGALLE DISTRICTS: Hanguranketa,
July 1839 & Allagalla, July 1855, *s. coll. C.P. 2471* (PDA, on 1 sheet).
BADULLA DISTRICT: N. of Welimada, *Maxwell & Jayasuriya 894* (US, NY).
Notes. This species falls within section *Cyanospermum* (Wight & Arn.) Benth.,
characterized by blue seeds lacking arillate tissue and remaining attached to the

pod, the long inflorescences and the conspicuous calyx about equalling the petals in length. Trimen (1894) reports the species as rare and occurring at Ambagamuwa, Etampitiya and Four Korales besides the localities noted above.

Description based on *Wight 766, 786* (NY), *Wallich 5585* (K).

8. Rhynchosia densiflora (Roth) DC., Prod. 2: 386. 1825; Wight & Arn., Prod. 1: 239. 1834; Baker in Hook. f., Fl. Br. Ind. 2: 226. 1876; Trimen, J. Bot. 23: 144. 1885; Trimen, Handb. Fl. Ceylon 2: 85. 1894; Gamble, Fl. Pres. Madras 1(2): 376. 1918; Verdcourt, Kew Bull. 25: 72. 1971a; Verdcourt in Fl. Trop. East Africa, Papil. 723, fig. 109/4. 1971b. Type: *Heyne s.n.* (K), not seen.

Glycine densiflora Roth, Nov. Pl. Sp. 348.1821.

Climbing, twining vines, ascending or somewhat trailing; stems slender, angled, densely canescent pubescent. Leaves pinnately trifoliolate, the rachis about 4 mm long, the petioles mostly 3–3.5 cm long; stipules membranous, ovate to lanceolate, about 4 mm long, pubescent, persistent; stipels setaceous, about 1 mm long, persistent; leaflets thin, ovate, obovate, acute or obtuse apically, cuneate or rounded basally, the surfaces somewhat glabrescent, densely orange or black gland-dotted below, the margins ciliate, the terminal leaflet frequently rhomboid, to about 4.5 × 4 cm, the laterals smaller, greatly inequilateral. Inflorescences about 4 cm long, the peduncles about 0.5–3 cm long, the rachis crowded with 10–15 flowers; bracts lanceolate, 5–8 mm long, gland-dotted and with long, whitish, ciliate hairs, caducous. Flowers about 12 mm long, the pedicels about 2 mm long; calyx tube about 2 mm long, pubescent, ciliate, sparsely gland-dotted, the upper lobe deeply bifid, all lobes long-acuminate, 8–13 mm long, exceeding the petals; petals yellow, with red or purple striae, the vexillum oblong to somewhat elliptic, glabrous, about 10 × 6 mm, short-clawed, persistent, the wings and keels oblanceolate, about 6 mm long, the claw 1−2 mm long, ovary 1-2-ovulate, villous. Fruit short-oblong, or somewhat elliptic, 12–14 mm long, about 6 mm wide, 2–3 mm thick, turning dark, non-septate, the upper margin somewhat straight, short-beaked, the exocarp gland-dotted, usually covered with long, spreading, white hairs, (1–)2-seeded. Seeds oblong-rounded or suborbicular, 2.5–4 × 2–2.5 × 1–1.5 mm, reddish-brown with black mottling, or blackish, horizontal, the hilum oblong, about 1 mm long, without arillate tissue, or with a scale-like piece of rim aril, or some rudimentary arillate tissue. Chromosome number 2n = 22.

Distr. Tropical and South Africa, southern India, especially Madras, and upper Burma and Ceylon.

Ecol. Trimen (1894) found the species in the dry regions and reports it as very rare. Flowering is December through February. Verdcourt (1971a) reports the subsp. *densiflora* in India (Madras), Ceylon and upper Burma; mostly in wet areas.

Specimens Examined. HAMBANTOTA DISTRICT: Near Tissamaha-rama, 19 Dec. 1882, *s. coll. s.n.* (PDA), *Trimen s.n.* (K); Tissamaharama, *Alston 1624* (K, PDA).

Note. Description based on *Wight 833* (COIMB).

9. Rhynchosia viscosa (Roth) DC., Prod. 2: 387. 1825; Baker in Hook. f., Fl. Br. Ind. 2: 225. 1876; Trimen, Handb. Fl. Ceylon 2: 85. 1894; Gamble, Fl. Pres. Madras 1(2): 375. 1918; Nooteboom, Reinwardtia 5: 441. 1961; Backer & Bakh. f., Fl. Java 1: 637. 1963; Verdcourt, Kew Bull. 25(1): 82–87. 1971a; Verdcourt, Fl. Trop. East Africa, Papil. 733. 1971b. Type: *Heyne,* fide Verd-court, (p. 733, 1971b), not seen.

Glycine viscosa Roth, Nov. Sp. Pl. 349. 1821.
Flemingia viscosa Moon, Cat. 54. 1824, nomen.
Dolichos glutinosus Roxb., Fl. Ind. 3: 312. 1832; Wight & Arn., Prod. 1: 248. 1834.
Dolichos tomentosus Roth, Nov. Sp. Pl. 345. 1821; Wight & Arn., Prod. 1: 248. 1834. Based on *Wight 1152* (COIMB).
Rhynchosia villosula Thw., Enum. Pl. Zeyl. 412. 1864. Type: Trincomalee, Mar. 1863, *Glenie s.n. C.P. 3790* (PDA).

Climbing, twining or trailing vines, viscous, herbaceous perennials to about 2.5 m long; stems pubescent, with short glandular hairs. Leaves pinnately trifoliolate, the rachis 1–2 cm long, ⅓–¼ the petiolar length, viscous-hairy; stipules narrowly triangular to lanceolate, 3.5–10 mm long, with 4 striae, persis-tent; stipels minute, setaceous tufts, 1–2 mm long, persistent (?); leaflets broadly ovate, 5–9 × 5–8.5 cm, frequently slightly lobed, acute or somewhat acuminate apically, rounded, very broadly cuneate, or truncate basally, the surfaces with sparse, short, erect, curly hairs above and below, orange gland-dotted below, the terminal rhomboid, the laterals greatly inequilateral. Inflorescences 4–35 cm long, many–numerous-flowered; bracts lanceolate, 2–3 mm long, caducous. Flowers 12–15 mm long, the pedicels about 1 mm long; calyx tube 2–3 mm long, pubescent with short, glandular hairs, the upper lobe deeply bifid, all lobes acute to acuminate, the lower longest, about 4 mm long; petals yellow, the vexillum oblong to somewhat elliptic, densely pubescent, about 12 × 8 mm, short-clawed, the wings and keels oblanceolate, about 12 mm long; ovary 2-ovulate, densely pubescent. Fruit oblong-falcate, 2.5–3 × about 0.8 × 0.3 cm, the upper margin occasionally long-beaked, the beak 7–10 mm long, the exocarp puberulent and also covered with fulvous or whitish, gland-based hairs, about 2 mm long, 2-seeded. Seeds globose or orbicular, horizontal, about 5 × 4 × 2 mm, smooth, black, the hilum oval, about 1 mm long, usually with a scale-like piece of persis-tent rim-aril. Chromosome number 2n = 22.

Distr. Tropical Africa, including West Africa, Cameroun, Central African

Republic, Congo, Sudan, Angola; Madagascar, India, Java, Thailand and introduced into the Mascarene Islands.

Ecol. The plant is found in the low country, flowers January through March, and is apparently still rare.

Specimens Examined. TRINCOMALEE DISTRICT: Trincomalee, *Glenie s.n. C.P. 3790* (BM, K, P, PDA). AMPARAI DISTRICT: near Komari teak plantation, *Jayasuriya et al. 727* (PDA, US). COLOMBO DISTRICT: Colombo, Jan. 1866, *Ferguson s.n. C.P. 3895* (PDA), without date, *C.P. 3895* (PDA), *Ferguson s.n., Thwaites C.P. 3895* (K); *Thwaites C.P. 3895* (P).

Notes. Verdcourt (1971b) notes that the presence of *R. viscosa* subsp. *viscosa* var. *breviracemosa* (Hauman) Verdcourt in Ceylon may have come about from introductions as the var. is primarily African. The var. *breviracemosa* in Ceylon is based on *Ferguson C.P. 3895* collections near Colombo. These collections have mostly smaller flowers, 8–10 mm long, congested at the inflorescence apex. The *Glenie* collections and those studied at Coimbatore fall into the subsp. *viscosa* var. *viscosa* category with larger flowers, 10–14 mm long, scattered more or less evenly along the elongate inflorescences.

Description based on *Wight 834, 1152* (COIMB).

10. Rhynchosia minima (L.) DC., Prod. 2: 385. 1825; Benth. in Mart., Fl. Bras. 15(1): 204, t. 54, fig. 2. 1859; Thw., Enum. Pl. Zeyl. 91. 1859; Baker in Hook. f., Fl. Br. Ind. 2: 223. 1876 (cum syn.); Trimen, Handb. Fl. Ceylon 2: 84. 1894; Cooke, Fl. Bombay 1: 389. 1902; Gamble, Fl. Pres. Madras 1(2): 375. 1918; Urban, Symb. Ant. 4: 307. 1903–1911 & 8: 310. 1920–1921; Burkart, Leg. Argentinas ed. 2, 394. 1952; Nooteboom, Reinwardtia 5: 439. 1961; Backer & Bakh. f., Fl. Java 1: 637. 1963; Verdcourt, Kew Bull. 25: 101. 1971a; Verdcourt, Fl. Trop. East Africa, Papil. 756, fig. 109/42. 1971b; Verdcourt, Man. New Guinea Leg. 553. 1979; Grear, Mem. New York Bot. Gard. 31(1): 76. 1978. Type: Jamaica, West Indies, *Sloane* 3: 79 (BM lectotype), not seen.

Phaseolus exiguus Burm., Thes. Zeyl. 188, t. 88. 1737.
Dolichos minimus L., Sp. Pl. 726. 1753.
Dolicholus minimus (L.) Medik. in Vorles. Churpfälz. Phys.-Oekon. Ges. 2: 354. 1787.
Glycine rhombifolia Willd., Sp. Pl. 3: 1065. 1802.
Dolichos medicagineus Lam., Enc. Meth. Bot. 2: 297. 1786; Moon, Cat. 53. 1824.
Rhynchosia nuda DC., Prod. 2: 385. 1825; Wight & Arn., Prod. 1: 238. 1834.
    Type: India, *Rottler* (G-DC, holotype), not seen. Based on *Wight 831* (p.p.).
Rhynchosia ervoidea DC., Prod. 2: 385. 1825.
Rhynchosia rhombifolia (Willd.) DC. var. *timoriensis* DC., Prod. 2: 386. 1825.
Rhynchosia rhombifolia (Willd.) DC., Prod. 2: 386. 1825.
Rhynchosia microphylla Heyne in Wall., Cat. 5497. 1831–1832, nomen.

Rhynchosia ternicaulis Grah. in Wall., Cat. 5495. 1831–1832, nomen.

Rhynchosia prostrata Grah. in Wall., Cat. 5496. 1831–1832, nomen.

Rhynchosia medicaginea (Lam.) DC., Prod. 2: 386. 1825; Wight & Arn., Prod.
1: 238. 1834. Type: Ceylon, *Burm*. Thes. Zeyl. 188. t. 88. 1737, not seen.

Dolichos scarabaeoides auct., Roxb., Fl. Ind. 3: 315. 1832, desc. only.

Small vines, prostrate or trailing, occasionally twining, perennials; stems to
several m long, slender, numerous from a woody root-stock, glabrescent or finely puberulent. Leaves pinnately trifoliolate, the rachis 2–12 mm long, ⅓–¼ the
petiolar length; stipules linear-lanceolate, about 3.5 mm long; stipels not seen;
leaflets ovate, broadly oval or rhomboid-oval, 1–2.5(–6) × 1–2(–5) cm, rounded, acute to acuminate apically, rounded or truncate basally, both surfaces glabrous
or glabrescent, gland-dotted below. Inflorescences long, lax, usually exceeding
the leaves, 3–22 cm long, 3–6–many-flowered; bracts acute, about 2 mm long,
deciduous. Flowers 5–10 mm long, the pedicels about 1 mm long; calyx tube
about 4 mm long, pubescent, the upper lobes connate, mostly entire, or to about
½ incised, bifid, all lobes acute or triangular, 3–5 mm long; petals yellowish,
the vexillum obovate, with red striae, 5–6.5 × 3–6 mm, pubescent, glandular,
short-clawed. Fruit oblong-falcate, 1–2 cm long (2-seeded), 3–5 mm wide, non-
septate, the upper margin slightly upcurved, or straight, the exocarp finely
puberulent or often with long hairs, glandular, (1–)2-seeded. Seeds oblong-
reniform, 2.5–3.2 × 2–2.5 × 1–2 mm, grey or black, the hilum oblong, lacking
arillate tissue. Chromosome number 2n = 22.

Distr. Pantropical, America, the Cape and tropical Africa, East Asia,
Malaysia, the New Guinea region, and Australia. The most extensive distribution of any species in the genus.

Ecol. In Ceylon the species is found in the low country; common in the dry
region and occasionally found in moist regions. Flowering all the year.

Vern. Maha-wal-kollu (S).

Specimens Examined. LOCALITY UNKNOWN: *Thwaites* (?) *C.P. 1440*
(K). JAFFNA DISTRICT: Jaffna, 1848, *Gardner C.P. 1440* (PDA); Point Pedro,
Simpson 9272 (BM, K). VAVUNIYA DISTRICT: between Medawachchiya &
Mannar, *Jayasuriya et al. 582* (US). ANURADHAPURA DISTRICT: Wilpattu
Natl. Pk., near gate, *Cooray 70020230R* (PDA, US); Isurumuni Vihare, *Hepper
& Jayasuriya 4650* (K, US); near Anuradhapura, *Fosberg & Balakrishnan 53437*
(US, NY); near Rajangana Resv., *Maxwell & Jayasuriya 788* (PDA, US).
POLONNARUWA DISTRICT: "Mineri", Sept. 1885, *s. coll. s.n.* (PDA).
MATALE DISTRICT: "Damboul" (Dambulla ?), March 1868, *Beckett s.n.*
(PDA). TRINCOMALEE DISTRICT: Kanniya, nr. hot wells, *Sumithraarachchi
& Sumithraarachchi DBS 849* (US). COLOMBO DISTRICT: Colombo, *Ferguson
C.P. 1440* (PDA). HAMBANTOTA DISTRICT: Kirinde, Hambantota, on shore,
Dec. 1882, *s. coll. s.n.* (PDA, on same sheet as *Beckett s.n.* above), *Alston 1235*

(PDA); Ruhuna Natl. Pk., Palatupana Rd., *Cooray 70032514R* (PDA, US); Ruhuna Natl. Pk., Komawa Wewa, *Cooray 69121018R* (PDA, US), *Mueller-Dombois et al. 67093030* (PDA, US), *Mueller-Dombois 69030901* (NY, PDA, US); beyond Tissa to Kataragama, *Rudd 3103* (PDA, US).

Notes. Most of the Ceylon and southern Indian specimens would fall within var. *nuda* (DC.) Kuntze (Rev. Gen. Pl. 1: 204. 1891).

The synonymy is selected for the Flora area. See the references cited above for more extensive synonymy and discussion. See Verdcourt for a discussion of the variation within the taxon which makes this species one of the most taxonomically difficult in the genus.

11. Rhynchosia velutina Wight & Arn., Prod. 1: 238. 1834; Baker in Hook. f., Fl. Br. Ind. 2: 224. 1876; Gamble, Fl. Pres. Madras 1(2): 375. 1918; Verdcourt, Kew Bull. 25(1): 92. 1971; Verdcourt, Fl. Trop. East Africa, Papil. 738. 1977. Type: India, Madras, Negapatam, *Wight 832* (BM, K holotype).

Trailing or climbing herb, to about 60 cm long; stems ridged, with erect, somewhat dense, canescent pubescence. Leaves pinnately trifoliolate, the petioles 2–3 cm long, the rachis about 6 mm long; stipules lanceolate or linear, about 2.5 mm long, pubescent; stipels not seen; leaflets sparsely pubescent above, grey-canescent, lanate below when young, sparsely gland-dotted below, 3 veins inserted at the base, with about 2 pairs of primary lateral veins, venation prominent beneath, the terminal leaflets broadly rhomboid, about 3.5 × 3 cm, the laterals suborbicular, oblique, about equalling the terminal in size. Inflorescences 2–4-flowered, the peduncle usually shorter than the rachis, both canescent tomentose, about equalling the leaves in length; bracts acute, about 1.5 mm long, pubescent, apparently caducous. Flowers about 12 mm long, the pedicels about 4 mm long; calyx tube about 3 mm long, gland-dotted, densely pubescent, the upper lobes deeply divided, about 2.5 mm long, the lower longest, about 5 mm long, all lobes acute or acuminate; petals yellow, the vexillum obovate, densely pubescent, about 9 mm long, the wings and keels about equalling the vexillum in length, glabrous, all claws about 3 mm long; ovary 2-ovulate, white tomentose, the upper style about 4–5 mm long, the upper portion thickened for about 3 mm. Fruit straight or falcate, about 3 × 0.7 cm, with short, white hairs, non-septate, narrowed at the base, with a small, downcurved beak, 2-seeded. Seeds usually somewhat oblong-reniform, 5–6.3 × 3.5–4.5 × 2.5–3 mm, (fide Verdcourt, 1976), the hilum oblong, without a distinct aril.

Distr. The species is found in India, Africa, Malagasy Republic, and the Comoro Islands.

Specimens Examined. HAMBANTOTA DISTRICT: Yala, next to Uraniya Lagoon, *Nowicke & Jayasuriya 394* (P, PDA); Bundala, scrub jungle, in shore dunes near the coast, *Simpson 9953* (BM).

Notes. The Simpson collection has flowers and immature fruit and was

collected 13.8.1932. The seeds are immature, but do not show an aril development. The calyx lobes are rather long for Verdcourt's description, but fit Gamble's key and specimens from Madras examined at COIMB.

The species has the habit, leaflets, indumentum and general aspect of *Atylosia
albicans,* but the 2-seeded fruit without partitions is typical of *Rhynchosia.*

12. Rhynchosia acutissima Thw., Enum. Pl. Zeyl. 413. 1864; Baker in Hook.
f., Fl. Br. Ind. 2: 226. 1876; Trimen, Handb. Fl. Ceylon 2: 85. 1894; Prain,
J. Asiat. Soc. Bengal 66(2): 436. 1897; Gamble, Fl. Pres. Madras 1(2): 375.
1918. Type: Central Province, Hantane, 3000 ft, *Thwaites* ? *C.P. 3445* (PDA,
holotype).

Stout, climbing, twining vines; stems short fulvous tomentose. Leaves pinnately trifoliolate, the rachis 1–2.2 cm long, petioles 5–9 cm long, fulvous pubescent; stipules lanceolate, deciduous; stipels setaceous, minute, persistent; leaflets
somewhat coriaceous, ovate, trapezoid-oblong or rhombic, 8.5–12 × 5.5–7.5
cm, long acuminate apically, the tips 1 cm or longer, cuneate basally, both surfaces fulvous pubescent, with 3 prominent veins inserted at the base, the laterals
smaller, inequilateral. Inflorescences equalling or shorter than the leaves, 2–11
cm long, many-flowered, flowers paired or single on the rachis nodes; bracts
lanceolate, about 5.5 mm long, caducous. Flowers about 12 mm long, the pedicels
about 7 mm long; calyx tube about 2 mm long, pubescent and sparsely white
gland-dotted, the upper lobes slightly bifid, about 3.5 mm long, the lateral lobes
acute, about 2.5 mm long, the lower acuminate, about 4 mm long; petals pale
yellow, the vexillum ovate-oblong, red striate, pubescent, short-clawed, the wings
oblanceolate or somewhat falcate, shorter than the vexillum and keels, the keels
obliquely oblong, upcurved into a somewhat obtuse beak or erostrate, long-clawed;
ovary 2-ovulate, canescent villous, the style straight then upcurved. Fruit oblong
or somewhat elliptic, about 2.5 × 0.6 cm, narrowing at the base, the upper margin
down-curved into a beak, the lower margin constricted between the seeds, the
exocarp covered with canescent pubescence, 2-seeded. Seeds dark, ovoid, horizontal, about 3.5 × 3 mm (immature), the hilum oval, about 2 mm long, without
fleshy arillate tissue persisting.

Distr. India: Travancore, Sikkim, and Assam, as well as Ceylon.
Ecol. Flowering in December.
Specimen Examined. KANDY DISTRICT: Hantane, 3000 ft., *Thwaites*
? *C.P. 3445* (PDA).
Notes. Known only in Ceylon from the type collection, made Dec. 1859.
Much of the above description is based on *M. Rama Rao 737* (COIMB) from
Kurmalai, in the evergreen forests of the low country of Travancore, India, and
cited by Gamble (1918).

32. ERIOSEMA

(DC.) G. Don, Gen. Hist. 2: 347. Oct. 1832; Verdcourt, Fl. Trop. East Africa, Papil. 761. 1971; Grear, Mem. New York Bot. Gard. 20(3): 1–98. 1970 (New World Revision); Grear & Dengler, Brittonia 28: 281–288. 1976, nom. cons. Type Species: *Eriosema rufum* (H.B.K.) G. Don.

Rhynchosia DC. sect. Eriosema DC., Prod. 2: 388. 1825.
Euriosma Desv., Ann. Sci. Nat. Bot. Ser. 1. 9(1): 421. 1826, nom. rejic. Type:
   E. sessiliflora Desv.
Pyrrhotrichia Wight & Arn., Prod. 1: 238. 1834 (note), nomen ?

Small, perennial herbs or subshrubs, erect or prostrate, rarely twining. Leaves 1–3(–6)-foliolate, mostly pinnately trifoliolate, short-petiolate or subsessile; stipulate; stipels present or often absent; leaflets often gland-dotted beneath. Inflorescences axillary or terminal, racemose, rarely paniculate, short and condensed or elongate; bracts ciliate, persistent or caducous; bracteoles absent. Flowers small; calyx 5-lobed, often glandular, the upper lobes free or partially connate; petals yellow, the vexillum reflexed, bicallose, often auriculate, with or without lobed appendages, the wings free, narrowly oblong, without auricles, the keels falcate, about equalling the wings; stamens diadelphous, the vexillary free, the anthers uniform; ovary 2-ovulate, mostly villous, the style ascending or geniculate, the upper style glabrous, the stigma minute, glabrous. Fruit dehiscent, broadly oblong to ovoid, compressed, variously pubescent to glabrescent, 2-seeded. Seeds reniform to oblong or oval, the hilum linear, frequently with whitish rim arillate tissue. Chromosome number $2n = 22$.

A genus of about 130 species occurring throughout the tropics.

Eriosema chinense Vogel, Nova Acta Phys.-Med. Acad. Caes. Leop.-Carol. Nat. Cur. 19. Suppl. 1: 31. 1843; Baker in Hook. f., Fl. Br. Ind. 2: 219. 1876; Trimen, Handb. Fl. Ceylon 2: 81. 1894; Ridley, Fl. Malay Penins. 1: 565. 1922; Verdcourt, Man. New Guinea Leg. 554, fig. 138. 1979. Type: China, *Meyen* (types lost at B, probably at KIEL), not seen.

Crotalaria tuberosa DC., Prod. 241. 1825.
Rhynchosia virgata Grah. in Wall., Cat. 5503. 1831–1832, nom. nud.
Rynchosia grahami Wall, Cat. 5504. 1831–1832, fide Baker (1876), nomen.
Pyrrotrichia tuberosa Wight & Arn., Prod. 1: 238. 1834. (Herb. Arn.), not seen,
   fide Baker (1876).

Erect perennial herbs; stems terete, occasionally branched, 15–20 cm tall, with erect hirsute, ferruginous hairs 1.0–2.5 mm long. Leaves unifoliolate, the petiolule about 2 mm long; stipules linear, about 7 mm long; leaflets linear, lanceolate to oblong, 2.5–4.8 cm long, 6.0–7.5 mm wide, the upper surface dull, pubescent, the lower sericeous, tomentose, with a prominent midrib, acute apically, cuneate basally. Inflorescences axillary, to about 1.5 cm long, 1–4-flowered, much shorter than the leaflets; bracts not seen. Flowers about 6.5 mm long, the pedicels about 3 mm long; calyx tube about 2.5 mm long, ferruginous-hirsute, all lobes about equal, acute to acuminate, 1.5–2.0 mm long; petals yellow, the vexillum obovate, emarginate, about 6 × 4 mm, pubescent, biauriculate, the auricles with membranous ovoid appendages, short-clawed, the wings oblong, about 5 × 1.5 mm, the keels somewhat falcate, about 4 × 2 mm, the wing and keel claws about 2 mm long. Fruit about 1 cm long, villous to hirsute, with light-coloured hairs about 1 mm long, turning ferruginous, 2-seeded. Seeds reniform to oblong, the hilum linear with arillate tissue.

Distr. *Eriosema chinense* is a wide ranging species, occurring from India through Malaysia, Indonesia and the Philippines to northern Australia.

Ecol. Trimen (1894) states that Walker's specimens are very dwarf and look as if they were grown in dry patana land; he suggests they may have come from E. Bengal. Verdcourt (1979) cites the species as "mostly strictly erect herbs 8–90 cm tall" and found in grasslands, savannas. Baker (1876) notes the species is found up to 6000 ft.

Specimen Examined. LOCALITY UNKNOWN: *Col. Walker s.n.* (K).

Notes. I have found no specimens other than Walker's that could be attributed to Ceylon. My description is based on *Wallich 5503* (K), *Dr. Buchanan s.n.,* "Nepaul" (BM), and also on Chinese specimens which are similar to the specimens in the type folder at Kew.

FLAGELLARIACEAE

(by Neil A. Harriman*)

Dumort., Anal. Fam. Pl. 59, 60. 1829, nom. cons. Type genus: *Flagellaria* L., Sp. Pl. ed. 1, 333. 1753.

Lofty perennial climbers, the stems herbaceous. Leaves sheathing at base, the sheaths closed or open. Blades strongly divergent, with many closely parallel veins, the leaf-tip modified into a tendril which is coiled in the same plane as the blade. Inflorescences terminal, the flowers in amply developed panicles, perfect, hypogynous, virtually sessile, each from the axil of a tiny bract; tepals 6, biseriate, the members of each series somewhat imbricate. Stamens 6, exsert, anthers linear, dehiscent lengthwise. Ovary trigonous, style short (or none), the 3 stigmas much longer. Fruit a globose drupe, 1–3 seeded, red at maturity.

The family as here treated includes only the genus *Flagellaria*, with but one species in Ceylon. For comments on the circumscription of the family, and the exclusion *of Hanguana* Blume (i.e., *Susum* Blume in Schult. f. as treated by Hooker in Trimen, Handbook to the Flora of Ceylon, part 4: 317) see Tomlinson, P.B. and A.C. Smith. 1970. Joinvilleaceae, a new family of monocotyledons. Taxon 19 (6): 887–889, and the many references cited therein.

FLAGELLARIA

L., Sp. Pl., ed. 1, 333. 1753. Type species: *Flagellaria indica* L., op. cit.

Characters of the family. There are three species in the Old World tropics, only the following one in Ceylon.

Flagellaria indica L., Sp. Pl., ed. 1 333. 1753; Trimen, Handb. Fl. Ceylon 4: 317. 1898.

Herbaceous perennial, glabrous throughout, high-climbing by its tendrillate leaf tips or a straggling shrub when support is lacking. Leaves alternate, sheathing at base, the sheaths longer than the internodes, the many-nerved blades 10–16 cm long, 1.4–1.6 cm wide. Inflorescences of terminal shortly peduncled panicles 6–10 cm high, panicle branches alternate, well separated. Flowers virtually sessile

*Biology Department, University of Wisconsin-Oshkosh, Oshkosh, Wisconsin 54901, USA.

in the axils of scarious bracts; tepa?s 6, membranous, pale yellow or white, barely 2.5 mm long, the 3 outer with a basal gibbosity, otherwise differing from the 3 inner only their point of insertion. Stamens 6, the filaments eventually 2.5 mm long, the anthers as long, linear, longitudinally dehiscent. Ovary trigonous, about 1.5 mm long at anthesis; style very short or none, the stigmas long and spreading. Fruit globose, 7 mm high and wide, red, 1–3-seeded, the perianth and filaments becoming chaffy and persisting beneath the fruit.

Distr. Throughout tropical Africa, Asia, Australia, and western Oceania.

Ecol. Low country. Flowers December.

Vern. Goyi-wel.

Note. The illustration in Hutchinson (Families of flowering plants, ed. 2, 2: 567, fig. 360. 1959; ed. 3, 2: 702, fig. 360. 1973) shows the habit extremely well; however, the flower is shown as having stamens with tiny oval anthers; in fact, the anthers are linear and as long as the tepals.

Specimens Examined. PUTTALAM DISTRICT: c. 6 miles from Kalpitiya, *Sumithraarachchi 701* (US); Mellawa Forest Reserve, *Sumithraarachchi 685* (OSH, US). GALLÉ DISTRICT: Bona Vista, *Cooray 70029097R* (US) and *Balakrishnan 945* (US).

HYDROPHYLLACEAE

(by Neil A. Harriman*)

R. Br., Bot. Reg. 3: sub t. 242. 1817, nom. cons. Type genus: *Hydrophyllum* L.

Herbs, sometimes woody. Leaves alternate (or opposite), simple and entire (or variously divided or compound), estipulate. Flowers hypogynous, actinomorphic, perfect, complete, 5-merous; calyx tubular or divided to the base; corolla gamopetalous, rotate to campanulate; stamens 5, epipetalous, alternate with the lobes of the corolla; ovary (1- or) 2-celled; styles 2, separate to the base (or connate part way up), stigmas capitate. Ovules many (or few), placentation axile. Fruit capsular, septicidally (or loculicidally) dehiscent.

In Ceylon, the family includes only two species of *Hydrolea*. Worldwide, there are 18 genera with c. 250 species (Airy Shaw, H.K. 1973. A dictionary of the flowering plants and ferns, 8th ed.). In the now-classic monograph of the family (Brand, H. 1913. Hydrophyllaceae in Pflanzenreich 59: 1–210), there are recognized 230 species in 18 genera.

HYDROLEA

L., Sp. Pl. ed. 2, 328. 1762, nom. cons. Type species: *Hydrolea spinosa* L.

Characters of the family.

KEY TO THE SPECIES

1 Spiny; stems hirsute-glandular; plants stipitate-glandular throughout . . . . **1. H. spinosa**
1 Unarmed; stems glabrous; plants stipitate-glandular only in the inflorescence . . . .
. . . . **2. H. zeylanica**

1. Hydrolea spinosa L., Sp. Pl. ed. 2, 328. 1762.

Stout, erect, very spiny perennial, the spines representing modified branch tips. Stems hirsute glandular, 0.6–1.4 m tall. Leaves alternate, simple, entire, lanceolate to ovate, very veiny, 3–5 cm long, nearly sessile or on a petiole to 5 mm long, glandular hairy both above and below. Flowers blue, showy, pedicelled in terminal and axillary leafy clusters. Calyx short tubular at base, bristly

*Biology Department, University of Wisconsin-Oshkosh, Oshkosh, Wisconsin 54901, USA.

glandular-hirsute like the stems, 6 mm high; corolla open campanulate, 9–10 mm high, glabrous. Stamens 5, epipetalous, slightly exsert, filaments glabrous, anthers deeply sagittate at base and thereby appearing to be almost versatile on the filament. Ovary 2-locular, shortly stipitate-glandular above, glabrous below, 2.5 mm high; styles 2, separate to base or joined for less than half a millimetre, exceeding the calyx, stigmas slightly expanded. Capsule 6 mm high, as long as the persistent calyx, septicidally dehiscent; seeds numerous on axile placentae.

Distr. Widespread and weedy in tropical and subtropical America, in moist situations, and in Java; only recently detected as an adventive in Ceylon—see Jayasuriya, A.H.M., 1976. New records of some native and naturalized plant species in Ceylon. Ceylon J. Sci., Biol. Sci. 12(1): 15–22.

Ecol. "in gregarious populations, very locally naturalised at the semi-swampy base of Eriyativimalai hill"—Jayasuriya l.c.

Specimens Examined. TRINCOMALEE DISTRICT: Swamp, Eriyativumalai, 5 February 1972, *Jayasuriya 672* (US); marshy base of Eriyativumalai, 7 April 1974, *Jayasuriya 1618* (US).

2. Hydrolea zeylanica (L.) Vahl, Symb. Bot. 2: 46. 1791; Trimen, Handb. Fl. Ceylon 3: 191. 1895.

Nama zeylanica L., Sp. Pl. ed. 1, 226. 1753. (For complete synonymy, see Backer, C.A. 1951. Hydrophyllaceae in Fl. Mal., series 1. 4(3): 207–209.

Soft, unarmed, herbaceous perennial, but flowering the first year from seed, therefore seemingly annual, 7–100 cm tall or more, often repent and repeatedly branched, rooting from the lower nodes. Stems glabrous below the flowering branches. Leaves glabrous or very thinly puberulent on the petiole, veiny below, lanceolate to elliptic, 2–12 cm long, 6–12 mm wide, on winged petioles up to 5 mm long. Flowers in irregularly bracted racemes or panicles, the inflorescence branches, pedicels, and calyces stipitate glandular (or merely puberulent, or even entirely glabrous, this eglandular phase apparently rare in Ceylon). Calyx 5–6.5 mm long, divided nearly to base; corolla blue, glabrous, 5 mm long, lobes spreading. Stamens not or barely exsert, filaments glabrous, anthers deeply sagittate at base. Ovary ellipsoid, hairy above, 2.5 mm high in flower. Styles separate to base, stigmas capitate. Fruit capsular, 4–4.5 mm high, septicidally dehiscent, seated in the persistent calyx; seeds numerous, less than 0.5 mm long.

Distr. In Ceylon, widespread in moist situations around ponds and tanks, in irrigation ditches, and in wet depressions. Elsewhere, from India through Malaysia and the Philippines to (allegedly) northern Australia. In Trimen, H. 1895. A handbook to the flora of Ceylon, part 3: 91, the species is said to occur in the tropics of both hemispheres. There is a record of its occurrence in Peru: *Mathews 3046,* no date, no collection data (GH), cited by Davenport, L.J.

Monograph of *Hydrolea,* unpublished Ph.D. dissertation, University of Alabama, USA; one suspects a labelling error.

Ecol. Wet places, margins of tanks, etc., in the low country, especially the dry region. Flowers March, September.

Vern. Diya kirilla.

Specimens Examined. ANURADHAPURA DISTRICT: Kotagala, *Balakrishnan 623* (US); Wilpattu National Park, *Cooray 70020226R* (US). POLONNARUWA DISTRICT: Ellewewa, *Balakrishnan 352* (US); Polonnaruwa Sacred Area, *Ripley 465* (US). KURUNEGALA DISTRICT: Kirallabokke, *Cramer 3198* (US). AMPARAI DISTRICT: Andunoruwa Wewa, *Fosberg 51132* (US). BADULLA DISTRICT: Mahiyangana, *Hepper 4711* (US). HAMBAN-TOTA DISTRICT: Hambantota, Tissamaharama Road, *Cramer 2819* and *2833* (US); near Meynet Wewa, *Cooray 70032302R* (US). VAVUNIYA DISTRICT: Kandy to Jaffna, *Townsend 73/73* (OSH, US).

JUNCACEAE

(by Neil A. Harriman*)

Juss., Gen. Pl. 43. 1789, nom. cons. Type genus: *Juncus* L., Sp. Pl. ed. 1, 325. 1753.

Annual or perennial glabrous herbs, stems tufted, sedge-like in aspect. Leaves sheathing at base, blade flat or convolute or terete (and then ± nodose-septate), or the leaves represented only by basal sheaths, the blades then reduced to a terminal awn or apiculus on the summit of the sheath. Flowers numerous, hypogynous, perfect, 3-merous; inflorescences always terminal but appearing to arise laterally in species with a single erect involucral leaf (*J. effusus* in Ceylon), pedicelled or nearly sessile but always from the axil of a ± scarious bract; flowers with two small bracteoles (commonly called prophylls) immediately beneath the perianth, or these lacking (always absent when the leaves are nodose-septate). Perianth chaffy, never with green sepals and coloured petals, green or brown, the three outer tepals (conventionally called sepals) differing scarcely at all from the three inner tepals (conventionally called petals), persistent in fruit. Stamens 3 or 6, one stamen always opposite each sepal, and one opposite each petal when the flower is hexandrous. Ovary trilocular, the style with three stigmatic branches, but the septa variously developed, the ovary sometimes seemingly unilocular. Fruit a capsule, dehiscent from above into 3 valves; seeds numerous, apiculate (not tailed in the Ceylon species), reddish-brown.

The family in Ceylon is represented only by three species of *Juncus*; worldwide, there are nine genera and perhaps 400 species, best developed in north-temperate areas.

JUNCUS

L., Sp. Pl. ed. 1, 325. 1753. Type species: *Juncus acutus* L., op. cit.

Characters of the family.

*Biology Department, University of Wisconsin-Oshkosh, Oshkosh, Wisconsin 54901, USA.

KEY TO THE SPECIES

1 Tufted annuals, with slender fibrous roots only, the very narrow leaves flat or somewhat involute and almost setaceous; inflorescences half to two-thirds or more the height of the plant; flowers bibracteolate, the bracteoles ¼ to ⅓ the length of the sepals; flowers well separated or somewhat congested but never headlike, nor ever appearing to arise laterally from the stem..........
....**1. J. bufonius**
1 Tufted perennials from definite (though often short) rhizomes, leafless or leafy (and then the blades septate), the inflorescences much smaller relative to the height of the plant, the flowers either in definite headlike clusters or seeming to arise laterally from the stem, bibracteolate or not
  2 The bibracteolate flowers in apparently lateral inflorescences, the involucral leaf appearing as an erect continuation of the stem......................................**2. J. effusus**
  2 The ebracteolate flowers in definite headlike clusters; stems leafy, the blades septate (easily detected in dried material by running one's finger up the blade lengthwise)....**3. J. prismatocarpus**

1. Juncus bufonius L., Sp. Pl. ed. 1, 328. 1753.

Tufted annuals 4–30 cm tall, leaves very narrow, flat or involute, barely 0.5 mm wide. Inflorescences half to two-thirds or more the height of the plant, usually elongated and the flowers well separated and the flowers secund on the branches, or somewhat congested and the flowers then in small clusters of 2–6, but the individual flowers readily apparent nonetheless; flowers sessile or nearly so, always bibracteolate immediately at the base of the flower (in addition to the more distant bract on the inflorescence branch proper). Sepals acuminate-aristate, greenish or brown in age, 4–5 mm long, the petals similar but only 2.5–3.5 mm long, always obviously shorter than the sepals. Stamens 3 or 6 (variable even on the same plant). Capsule always shorter than the perianth, dehiscent to the base by 3 valves; seeds numerous, reddish-brown, minutely apiculate at either end, less than 0.5 mm long.

Distr. In Ceylon, the few records are all from mountainous areas, along roadsides, in gardens, etc. The species is not given for Ceylon by Hooker in Trimen, H. 1898. A handbook to the flora of Ceylon, part 4: 318–319, nor in Abeywickrama, B.A. 1959. A provisional check list of the flowering plants of Ceylon. Ceylon J. Sci., Biol. Sci. 2(2): 119–240. It seems likely that *J. bufonius* is a rather recent introduction and may be expected to spread over the moister, cooler parts of the island. While the species is very weedy, it does not become a pest in cultivated fields, and poses no threat to agriculture. It is not at all shade tolerant, nor does it withstand much competition, and therefore it should not displace native species.

Specimens Examined. KANDY DISTRICT: Adam's Peak, *Fosberg 58108A* (OSH), *Fosberg 58109* (OSH, US), *Davidse 8656* (US). NUWARA ELIYA DISTRICT: Horton Plains, *Clayton 5499* (US), *Balakrishnan 408* (OSH, US), *Wheeler 12894* (US).

2. Juncus effusus L., Sp. Pl. ed. 1, 326. 1753; Trimen, Handb. Fl. Ceylon 4: 318. 1898.

Densely tufted hard-stemmed perennial from tough, short, scaly rhizomes (these often lacking in herbarium specimens), commonly up to a metre tall. Stems terete, indistinctly striate, clothed at base with several red-brown bladeless sheaths, the blades reduced to a fragile terminal bristle. Inflorescences many flowered, appearing to arise from the side of the stem, the erect involucral leaf simulating a continuation of the stem and overtopping the inflorescence by 14–25 cm. Flowers on pedicels up to 2 mm long, from sheathing bracts on the inflorescence branches, with a pair of bracteoles at the summit of the pedicel and embracing the perianth. Sepals 3, sharply keeled over the angles of the trigonous capsule, 3 mm long; petals as long as the sepals or 0.25 mm shorter, appressed to the flat sides of the capsule. Stamens 3, opposite the sepals only, and barely half as long. Capsule trigonous, dehiscent to the base; seeds numerous, elliptical, apiculate, not tailed, c. 0.5 mm long.

Distr. Continually moist or wet situations in the highlands of the island. Worldwide in temperate zones, in mountainous districts in the tropics, not in Australia (teste Backer, C.A. 1951. Juncaceae in Flora Malesiana, Series 1. 4(3): 210–215, where this and the following species are nicely illustrated).

Notes. Thwaites (Enumeratio Plantarum Zeylaniae, p. 340. 1864) mistakenly called this species *Juncus glaucus* Ehrh., which is now reduced to the synonymy of *Juncus inflexus* L. *Juncus inflexus* differs from *J. effusus* by its much more prominently ribbed stems, its perianth 3.5–5 mm long, and its flowers with 6 stamens, not 3 as in *J. effusus*. *Juncus inflexus* is credited to Ceylon by Backer (op. cit.) apparently on the strength of the Thwaitesian misnaming, even though Hooker (in Trimen, op. cit.) had set the record straight in 1898. Abeywickrama (op. cit.) picked up Backer's report of its occurrence in Ceylon, and properly so. But there are to my knowledge no authentic herbarium specimens to vouch for its occurrence in Ceylon. The record may on present evidence be deleted.

Specimens Examined. NUWARA ELIYA DISTRICT: Hakgala, *Sohmer 10090* (US); 4 miles N. of Nuwara Eliya, *Davidse 8900* (US); Horton Plains, *Read 2137* (OSH, US), *Koyama 13524*—distributed as *J. glaucus*—(US); Kandepola FR, *Sohmer 8343* (US); Gregory Lake, *Sohmer 8463* (US); along Nuwara Eliya-Hakgala Road, *Sohmer 10143* (US); Hakgala Botanical Garden, *Davidse 7982* (US); Kandy-Nuwara Eliya Road, *Sohmer 8327* (US); Westward Ho Estate, *Fosberg & Sachet 53230* (US); roadside swamp on route A-5, *Read 2280* (US). BADULLA DISTRICT: marshy place, railway station, Ambawela, *Fosberg & Sachet 53241* (US); one mile from Ambawela toward Nuwara Eliya, *Comanor 979* (US).

3. Juncus prismatocarpus R. Br., Prod. Fl. Nov. Holland 259. 1810, incl. *J. leschenaultii* J. Gay in LaHarpe, Monog. Jonc. 137. 1827, and var. *leschenaultii*

(J. Gay in LaHarpe) Buchenau, Bot. Jahrb. Syst. 6: 205. 1885; Trimen, Handb. Fl. Ceylon 4: 319. 1898.

Tufted rhizomatous perennial, readily falling over and rooting from the nodes and also emitting leafy outgrowths from the flower clusters which may then bend over and give rise to new plants (as neither of the other two species in Ceylon does), 15–70 cm tall. Leaves basal and cauline, long-sheathing, the sheaths auriculate at the summit, compressed, the edge of the leaf toward the stem, blade septate-nodulose, 5–15(–20) cm long. Inflorescence lax and widespreading, of 3–10 or more hemispheric to globose heads, each head with 3–25 or more nearly sessile ebracteolate flowers. Sepals and petals homomorphous, 2.5–3.5 mm long. Stamens 3, opposite the sepals only, and barely half their length. Capsule trigonous, strongly exsert, 4.5 mm long, with a prominent sterile beak, dehiscent throughout by 3 valves, the valves not cohering at the summit. Seeds. numerous, elliptical, brown, barely over 0.5 mm long.

Distr. In continually wet places, often in a few inches of standing water, at higher elevations in Ceylon. New Zealand and Australia, west to Ceylon and north into the Philippines, southeastern and eastern Asia.

Specimens Examined. KANDY DISTRICT: Adam's Peak, *Davidse 8642* and *8674* (US); roadside ditch, 2 miles E. of Madugoda, *Wheeler 12060* (US); Knuckles Range, *Lazarides 7203* (US); NW. of Knuckles Peak, *Davidse 8323* (US); near Pussellawa, *Sumithraarachchi 543* (US). NUWARA ELIYA DISTRICT: slopes of Mt. Pidurutalagala, *Davidse 7994* (US); Horton Plains behind Farr Inn, *Davidse 7605* (OSH, US); ascent to Horton Plains, *Jayasuriya 184* (US); NW. of Nuwara Eliya, *Davidse 7947* (US); Horton Plains, *Read 2138* (OSH, US); road from Diyagama Tea Estate to Horton Plains, *Sohmer 10028* (OSH, US); Gregory Lake, *Sohmer 8471* (US); road to Farr Inn, *Sohmer 8539* (US); Westward Ho Estate, *Fosberg 53229* (OSH, US). RATNAPURA DISTRICT: east of Deniyaya, *Davidse 7879* (OSH, US); 2 miles from Rasagalla, *Sohmer 10495* (US) and *10497* (OSH, US). AMPARAI DISTRICT: Potuvil-Moneragala Road, *Sumithraarachchi 893* (US). BADULLA DISTRICT: Wet patana about 1 mile from Farr Inn, *Sohmer 8587* (US); Horton Plains, *Clayton 5518* (US). MATARA DISTRICT: Enselwatte, *Waas 1487* (US).

LEEACEAE

(by Neil A. Harriman*)

Dumort., Anal. Fam. Pl. 21, 27. 1829, nom. cons. Type genus: *Leea* Royen ex L.

Shrubs or treelets, but never vines. Stems warty-roughened or smooth, sometimes puberulous, etendrillate. Stipules sheathing and fused with the petiole, deciduous, leaving a scar encircling the stem. Leaves imparipinnately once or twice (or thrice) compound, alternate. Inflorescences terminal or leaf-opposed compound cymes. Flowers hypogynous, perfect, complete, actinomorphic, 5-merous. Calyx obconical, open in bud; corolla sympetalous, the lobes valvate in bud, reflexed at anthesis. Stamens 5, epipetalous, opposite the corolla lobes, filaments appearing to be fused somewhat to the staminodial tube, the anthers held inside the staminodial tube. Staminodial tube within the androecium, attached to the corolla tube such that the staminodial tube has an upper, longer free portion and a lower, much shorter free portion. Staminodial tube with 5 free lobes, each lobe notched at the summit, the lobes alternate with the stamens, at anthesis the filaments of the functional stamens elongating, carrying the anthers to a position over the top of the staminodial tube, such that the anthers all lie horizontally with their tips to the center of the flower. Ovary one, superior, central in the calyx tube, 4–6 locular, with one basal ovule per locule; style one, jointed to a short beak on the summit of the ovary, stigma capitate or punctiform. Fruits (4–)6-lobed, depressed-globose, initially dry, eventually fleshy and black. Seeds with ruminate endosperm (i.e., the inner layer of the seed coat intruded into the endosperm).

The family includes only the genus *Leea,* with but one species in Ceylon. The genus was included in the Vitaceae in Trimen, H. 1893. A handbook to the flora of Ceylon, part 1: 297–298. Modern opinion regards Leeaceae as a distinct, monogeneric family; see, for example: Cronquist, A. 1981. An integrated system of classification of flowering plants, pp. 746–748; Airy Shaw, H.K. 1973. A dictionary of the flowering plants and ferns, ed. 8. While it is probably allied with the Vitaceae, its erect, etendrillate habit, its one-seeded ovary locules, and its curiously developed staminodial tube make the family distinct. For an excellent

*Biology Department, University of Wisconsin-Oshkosh, Oshkosh, Wisconsin 54901, USA.

overview of the family, with 34 species in its sole genus, see Ridsdale, C.E. 1974. A revision of the family Leeaceae. Blumea 22: 57–100; and also Ridsdale, C.E. 1976. Leeaceae in Flora Malesiana, series 1, 7(4): 755–782.

LEEA

Royen ex L., Syst. Nat. ed. 12. 2: 627. 1767; L., Mant. 17, 124. 1767. Type species: *Leea aequata* L., typus cons. A complete generic synonymy is given in Ridsdale, 1974, supra.

One species in Ceylon.

Leea indica (Burm. f.) Merr., Philipp. J. Sci. 14: 245. 1919; Alston in Trimen, Handb. Fl. Ceylon 6: 54. 1931.

Staphylea indica Burm. f., Fl. Ind. 75, t. 24, f. 2. 1768.
Leea sambucina Willd., Sp. Pl. 1: 1177. 1797; Trimen, Handb. Fl. Ceylon 1: 297. 1893.

For an extensive listing of further synonymy, see Ridsdale, 1974, supra. Most of the synonymous names have never been applied in the literature to Ceylon plants.

Shrubs or small trees, 1–5 m tall, stems warted or smooth, glabrous throughout or nearly so. Stipules fused with the petiole and encircling the stem, promptly deciduous, leaving a scar around the stem at each node. Leaves alternate, once or twice (or thrice) imparipinnately compound, 19–50 cm long (or much more; Trimen, op. cit., gives 1.5–3 feet—that is, 45–90 cm). Petioles 5–17 cm long. Leaflets opposite, petiolulate, toothed, ovate-lanceolate to elliptic, rounded to cuneate at base, acute to caudate-acuminate at the tip, 3–18(–21) cm long, 4–13 cm wide, pinnately veined, the major venation raised both above and below. Flowers in leaf-opposed (or terminal) peduncled compound ± flat-topped cymes. Flowers actinomorphic, hypogynous, perfect, complete, green to white in colour, on pedicels 2 mm long; calyx 2 mm long, tubular below, the 5 deltoid lobes each 1 mm long; corolla sympetalous, 3 mm long, the 5 lobes each 1.5 mm long, cucullate at tip within, reflexed. Stamens 5, 2.5 mm long, opposite the corolla lobes. Staminodial tube about 2 mm high, about 1.5 mm high above its point of attachment in the corolla tube, 5-lobed, each lobe retuse at the summit, the lobes alternate with the stamens. Ovary superior, style one, stigma punctiform. Fruit 6-locular, seated atop the persistent calyx, at first dry, with age fleshy and black, 4.5–5 mm high, 9–10 mm wide, flattened from the ends, 6-lobed, each locule one-seeded, placentation basal. Seeds with stony testa, endosperm ruminate.

Distr. At lower elevations throughout much of Ceylon; elsewhere, from India to China, throughout Malesia, Palau, Australia, and Fiji.

Vern. Burulla, Gurulla, Otta-nali, Nyckki.

Note. To judge from the specimens at hand, flowering occur· from October through May. The flowers of this species are illustrated in detail i·. both the papers by Ridsdale cited above; a more ample illustration is in Wight, R. 1838. Ic. Pl. Ind. Or. 1: 78.

Specimens Examined. ANURADHAPURA DISTRICT: Ritigala Strict Nature Reserve, *Jayasuriya 1138* (US). POLONNARUWA DISTRICT: Alutoya, *Jayasuriya 678* (US); 4 miles from Somawathiya, *Sumithraarachchi 362* (OSH, US). MATALE DISTRICT: Sigiriya, *Wambeek & Wanntorp 2693* (US). KANDY DISTRICT: Hantana, *Nooteboom 3043* (US); Gannoruwa Hill, *Wirawan 619* (US); trail to Nitre Cave, *Sohmer 10671* (US); banks of Hingula Oya, *Cramer 4210* (US). BATTICALOA DISTRICT: 6 km ante Mutur, de Batticaloa, *Bernardi 15677* (US). KALUTARA DISTRICT: ½ mile N. of Pahala Hewessa, *Bremer 811* (US). RATNAPURA DISTRICT: Adam's Peak, *Cramer 4683* (US); 3 miles SE. of Rakwana, *Fosberg 57283* (OSH, US). BADULLA DISTRICT: Uma Oya, *Balakrishnan 836* (US). MONERAGALA DISTRICT: Moneragala Estate, *Waas 687* (US); banks of Diyaluma Oya, *Cramer 4750* (US); c. 3 miles W. of Wellawaya, *Davidse 7725* (US); Koslanda-Wellawaya Road, *Faden 76/594* (US); de Kataragama 10 km ante Buttala, *Bernardi 15531* (OSH, US). GALLE DISTRICT: Hiniduma, *Bernardi 15471* (US). KEGALLE DISTRICT: Lower slopes of Alagalla Mt., *Fosberg 57844A* (OSH, US). LOCALITY UNKNOWN: 29 Oct. 1970, *Kundu & Balakrishnan 452A* (US).

MALPIGHIACEAE

(by Neil A. Harriman*)

Juss., Gen. Pl. 252. 1789, nom. cons. Type genus: *Malpighia* L.

Woody vines, scandent shrubs, or erect shrubs or small trees when no support is available, ± clothed with dolabriform (pick-shaped) hairs throughout. Leaves opposite, entire, estipulate, petiolate, simple, biglandular on blade margin at summit of petiole. Flowers in axillary and terminal racemes, these sometimes congested and collected into panicles. Pedicels jointed, readily disarticulating at the joint when fruits fail to form. Flowers hypogynous, perfect, complete, slightly irregular, 5-merous. Calyx lobes imbricate, with one large brown gland decurrent onto the pedicel; petals imbricate, fimbriate or erose, the adaxial (upper) the largest. Stamens 10, one much larger and strongly exsert. Ovary tricarpellate, each locule one-ovuled; style one (the other two abortive), long-exserted, circinate, stigma punctiform. Fruit a 3-winged samara, potentially three such samaras per flower, but commonly only one developing; a smaller dorsal crest may develop in addition. (Description limited to Ceylon species.)

A family of c. 60 genera and 1200 or more species in the tropics and subtropics of both hemispheres, with relatively few species in the Old World; in Ceylon, represented only by two species of *Hiptage*.

HIPTAGE

Gaertn., Fruct. 2: 169. 1791.

Characters of the family.

KEY TO THE SPECIES

1 Leaves closely silvery pubescent below, the blades mostly well over 8.5 cm long and over 4 cm broad; fruit with a dorsal crest in addition to the 3 wings, the longest wing more than 3 cm long; common . . . .**1. H. benghalensis**
1 Leaves glabrous beneath, almost always less than 8 cm long, and always less than 4 cm broad; fruit 3-winged only, dorsal crest lacking, longest wing less than 3 cm long; rare . . . .**2. H. parvifolia**

*Biology Department, University of Wisconsin-Oshkosh, Oshkosh, Wisconsin 54901, USA.

1. Hiptage benghalensis (L.) Kurz, J. Asiat. Soc. Bengal 43: 136. 1874. Based on *Banisteria benghalensis* L., Sp. Pl. ed. 1, 427. 1753. (Kurz gives "Sp. Pl. 356"—the page number is a lapsus calami, surely.); Alston in Trimen, Handb. Fl. Ceylon 6: 34. 1931.

Hiptage madablota Gaertn., Fruct. 2: 169. 1791; Trimen, Handb. Fl. Ceylon 1: 193. 1893.

Gaertnera racemosa (Cav.) Roxb., Pl. Corom. 1: 19 & t. 18. 1795, non *Gaertnera* Lam., nom. cons., Rubiaceae; based on *Molina racemosa* Cav., Diss. 9: 435 & t. 362. 1791.

Hiptage obtusifolia (Roxb.) DC., Prod. 1: 583. 1824; based on *Gaertnera obtusifolia* Roxb., Hort. Beng. 32. 1814.

Vine, scandent shrub, or shrub or small tree when support is lacking. Fully developed leaves on thick petioles c. 1 cm long, blades lanceolate to ovate to almost rotund, rounded to cuneate at base, acute to acuminate at tip, 9–15 cm long, 4–8(–9) cm broad, reticulate-veiny both above and below, with fine, appressed dolabriform hairs across the entire lower surface. Flowers in axillary and terminal racemes, these sometimes collected into panicles. Pedicels jointed near the middle or above, bibracteolate at the joint. Sepals c. 5 mm long, oblong, densely strigose with dolabriform hairs, tawny, the single linear sepalar gland decurrent onto the pedicel. Petals strigose-hairy without, glabrous within, fimbriate, rotund, clawed, c. 1 cm long. Stamens exsert, filaments somewhat persistent, the 9 short ones c. 7 mm long, the longest one c. 12 mm long. Ovary trilocular, hairy like the sepals, style c. 13 mm long, circinate at tip. Samaras potentially three per flower, but usually only one developing, each 3-winged, the wings papery, thinly hairy, the largest obovate, to 5 cm long, the 2 smaller wings c. 2.5 cm long, the dorsal crest (simulating a fourth wing) 3.5–11 mm high or rarely undeveloped, one or more of the wings sometimes lobed, the entire range of variation in fruits sometimes occurring on a single individual.

Distr. Very widespread, from Ceylon, throughout India, to Taiwan, the Philippines, and Sumatra.

Ecol. Low country, moist and dry region. Flowers in March.

Vern. Puwak-gediya-wel.

Note. *Hiptage obtusifolia,* listed in Abeywickrama, B.A. 1959. A provisional check list of the flowering plants of Ceylon. Ceylon J. Sci., Biol. Sci. 2(2): 119–240, is here reduced to the synonymy of *H. benghalensis,* as was earlier done in Jacobs, M. 1955. Malpighiaceae in Flora Malesiana, series 1. 5(2): 125–145, where two photographs depict the species very well. In the work by Niedenzu (Malpighiaceae in Pflanzenreich 141 (not 91): 79. 1928) there are no reliable characters by which *H. obtusifolia* can be separated.

Specimens Examined. ANURADHAPURA DISTRICT: Ritigala Strict Natural Reserve, *Jayasuriya 1308* (US). KURUNEGALA DISTRICT: Gonagama,

Cramer 3586 (US). MATALE DISTRICT: Niyandagala, *Cramer 5132* (US); Dambulla, *Cramer 4854* (US). KANDY DISTRICT: Hunasgiriya, *Tirvengadum et al. 77* (US); Kandy-Mahiyangane Road, *Sohmer et al. 8243, 8245, 8252* (OSH, US) and *Sohmer et al. 8325* (US); Corbet's Gap, *Cramer et al. 3935* (US); between Hasalaka and Madugoda, *Huber 461* (US); along Lady Blake Road, along the Mahaweli ganga, *Wirawan 1226* (OSH, US); Mahaweli ganga, *Kostermans 26867*—two sheets—(US); Madugoda, *Balakrishnan 374* (US); Guru Oya, *Balakrishnan & Jayasuriya 1217* (US); Kandy-Mahiyangane Road, *Tirvengadum & Waas 437* (US); Medamahanuwara, *Waas & Tirvengadum 785* (US). NUWARA ELIYA DISTRICT: near Ramboda Pass above Kandy, *Stone 11303* (US). BADULLA DISTRICT: by Rawanaella Falls, *Waas 79* (OSH, US); Boralande-Tangamalai Road, *Cramer 4344* (US). MONERAGALA DISTRICT: between Wellawaya and Monaragala, *Balakrishnan 579* (US).

2. Hiptage parvifolia Wight & Arn., Prod. 107. 1834; Trimen, Handb. Fl. Ceylon 1: 193. 1893.

Vine, climbing high into canopy trees, generally smaller in all its parts than *H. benghalensis*. Fully developed leaves on slender petioles to 1 cm long, blades narrowly to broadly lanceolate, rounded to cuneate at base, acute to shortly acuminate at tip, 4–7.5(–8) cm long, 2–3.5 cm wide, reticulate-veiny on both surfaces, quite glabrous both above and below. Flowers in panicles of racemes, on hairy pedicels jointed above the middle, the joint not marked by a pair of opposite bracteoles, but these scattered and variously developed on the proximal portion of the pedicel. Sepals c. 3 mm long, oblong, densely strigose with tawny dolabriform hairs, the single sepalar gland decurrent onto the pedicel. Petals strigose-hairy without, glabrous within, erose, rotund, clawed, c. 8 mm long. Stamens exsert, filaments ± persistent, the 9 short ones c. 6 mm long, the longest one c. 9 mm long. Ovary trilocular, hairy like the sepals, style c. 11 mm long, circinate at tip. Samaras potentially three per flower (and 2 or 3 commonly developing in some flowers), each 3-winged, the wings papery, thinly hairy, the largest wing linear to elliptic, to 2.1 cm long, the 2 smaller wings c. 1 cm long, the dorsal crest never formed, the wings often shallowly lobed.

Distr. Ceylon and India.
Ecol. Dry region. Flowers in March.
Specimens Examined. KANDY DISTRICT: west of Madugoda, *Jayasuriya 2127* (US). AMPARAI DISTRICT: between Panama and Okanda, *Jayasuriya 2004* (OSH, US). MONERAGALA OR HAMBANTOTA DISTRICT: Ruhuna National Park, *Cooray 70032713R* (US) and *Cooray 68053102R* (US).

SALVADORACEAE

(by Neil A. Harriman*)

Lindl., Nat. Syst. Bot. ed. 2. 269. 1836, nom. cons. Type genus: *Salvadora* L.

Trees or shrubs. Leaves opposite, simple, entire, petiolate, stipulate. Flowers hypogynous, actinomorphic, perfect or unisexual (and the plants then dioecious), polypetalous (or the petals slightly connate at base), 4-merous. Ovary one- or two-celled, stigma sessile. Fruit a drupe, with one or two seeds.

The family includes three small genera, with two genera (each with one species) in Ceylon. The Ceylon representatives of the family are so different in aspect as to suggest their segregation into separate families, but no modern author has taken such a step.

KEY TO THE GENERA

1 Rigidly spiny shrub, with 0–2 spines from the leaf axils, hence 0–4 spines at each node; stems quadrangular and hairy; flowers unisexual, sessile, the plants dioecious, petals separate to base; ovary 2-celled . . . . **1. Azima**
1 Unarmed tree with drooping branches; stems terete, glabrous; flowers pedicelled, perfect, in axillary and terminal racemes, petals connate at base; ovary one-celled . . . . **2. Salvadora**

1. AZIMA

Lam., Enc. 1: 343. 1783.

Very spiny dioecious shrubs, the paired spines arising from the axils of opposite leaves, not stipular in position, each node potentially with four spines. Flowers axillary, sessile, the pistillate single, the staminate clustered and passing into interrupted spikes on vigorous branches. Fruits globose, white.

Four species, from South Africa to Hainan Island, the Philippines, and the Lesser Sunda Islands, a single species in Ceylon.

Azima tetracantha Lam., Enc. 1: 343. 1783; Trimen, Handb. Fl. Ceylon 3: 121. 1895.

*Biology Department, University of Wisconsin-Oshkosh, Oshkosh, Wisconsin 54901, USA.

Monetia barlerioides L'Her., Stirp. Nov. 1, t. 1. 1785.

Spiny shrubs with hairy, quadrangular stems, potentially 4-spinous at each node (whence the epithet) but one or more suppressed at many nodes, spines up to 3 cm long but usually much shorter. Leaves opposite, leathery, entire, very shortly petioled, venation pinnate, major lateral veins arising from the midrib in the basal third of the blade, reticulum evident; blade oval, acute at both ends, tipped with a sharp, brown mucro, 3–4(–7) cm long, 1.5–3 cm broad. Stipules very tiny or lacking, the stem not encircled by a stipular ridge. Flowers unisexual, sessile; in the staminate plants: flowers 2–6 at the nodes, calyx hairy, 1.5–2 mm long, lobes 4, reflexed; corolla of 4 distinct ciliate-hairy petals, exsert c. 0.5 mm beyond the calyx; stamens 4, hypogynous, alternate with the petals, pistillode minute. In the pistillate plants: flowers always solitary at a leaf axil, hence paired at a node; calyx and corolla as in the staminate flowers, staminodes with very short filament but antheroid as large as the functional anthers in the staminate flowers; stigma sessile, vaguely 2-lobed; fruit globose, white, fleshy, to 10 mm in diameter, seeds one or two.

Distr. Southern Africa to India and Ceylon.

Ecol. Dry region. Flowering all the year.

Vern. Iyanku, Ichanku.

Specimens Examined. ANURADHAPURA DISTRICT: Medawachchiya, *Meijer 733* (US) and *Dassanayake 429* (US). PUTTALAM DISTRICT: Wilpattu National Park, Kolankanatta Beach, *Wirawan et al. 945* (US); junction to Periya Naga Villu, *Wirawan et al. 926* (US); Karapitiya, *Kundu & Balakrishnan 364* (US); S. end of Puttalam Lagoon, *Fosberg & Jayasuriya 52801* (US); Karaitivu Island, *Grupe 115* (US). MATALE DISTRICT: Dambulla Road, *Meijer 331* (US). TRINCOMALEE DISTRICT: Foul Point, S. side of Koddiyar Bay, *Fosberg 56413* (US). BATTICALOA DISTRICT: Keeli-Kudah, *Waas 2147* (US). MATARA DISTRICT: Medawatta, *Sumithraarachchi 1042* (OSH, US). HAMBANTOTA DISTRICT: Ruhuna National Park, *Fosberg & Mueller-Dombois 50145* (US); Udamalella, *Meijer 245* (US); Ruhuna National Park, Menik Ganga, *Comanor 394* (US); Kirinda, *Fosberg 50382* (US); Bundala Sanctuary, *Bernardi 14171* (US); beach about one mile E. of Bundala, *Davidse 7765* (OSH, US).

2. SALVADORA

L., Sp. Pl. ed. 1, 122. 1753.

Unarmed trees; flowers pedicelled, perfect, in axillary and terminal panicles of racemes. Corolla shortly tubular, stamens alternate with lobes, epipetalous. Fruit a one-celled, one-seeded drupe, red to black at maturity.

The genus includes five species, distributed from the warmer parts of Africa through southern Asia, with only the following in Ceylon.

Salvadora persica L., Sp. Pl. ed. 1, 122. 1753; Trimen, Handb. Fl. Ceylon 3: 120. 1895.

Salvadora wightiana Planch. ex Thw., Enum. Pl. Zeyl. 190. 1860.

Glabrous tree with drooping branches; leaves with a peppery flavour, with tiny deltoid stipules, the twigs encircled by a stipular ridge. Flowers perfect, complete, synsepalous, sympetalous, pedicelled, opposite in racemes, the racemes mostly panicled. Calyx 1 mm high, the lobes very shallow; corolla 2 mm high, its lobes strongly reflexed; stamens alternate with the corolla lobes, exserted by virtue of the corolla lobe reflexion. Ovary globose, stigma sessile. Fruit a one-seeded drupe, maturing red or black.

Distr. Ceylon and India to Africa.

Ecol. Dry region, especially by the coast. Flowering all the year.

Vern. Uvay, Viyay.

Specimens Examined. JAFFNA DISTRICT: Vicinioribus Pooneryn, *Bernardi 14276* (US); Karativu Ferry, *Bernardi 15309* (US); Delft, *Cramer 3361* (US). MANNAR DISTRICT: Madhu Road, *Dassanayake 421* (US); Giant Tank, *Meijer 808* (US); coastal sand dunes, *Kostermans 24900* (US). VAVUNIYA DISTRICT: Alampil, *Meijer 750* (US). PUTTALAM DISTRICT: Puttalam Lagoon, *Fosberg & Jayasuriya 52800* (US); Karativu Island, *Grupe 116* (US); Madura Odai, *Wirawan et al. 936* (US); near Erige Ara, *Wirawan 68091407* (US); at Manikepola Uttu Park Bungalow, *Mueller-Dombois 67051802* (US); Palavi, *Jayasuriya & Lazarides 4* (US). BATTICALOA DISTRICT: Keeli-Kudah, *Waas 2150* (US). HAMBANTOTA DISTRICT: Patangala, *Cooray 69101902R*—two sheets—(US); Bundala Sanctuary, *Bernardi 14172* (US); Yala, *Waas 66* (US); Hambantota-Tissamaharama Road, *Sohmer et al. 8847* (US); Yala National Park, *Hepper & de Silva 4743* (US); near Patanagala, *Cooray 69091302R* (US); Palaputana, *Cooray 68060704R* (US), *Fosberg 50378* (US); Komawa Wewa, *Comanor 406* (US); main Yala Road, *Comanor 1164* (US); Yala National Park, *Meijer 212* (US) and *Bernardi 15542* (OSH, US); along the road from Welligate to Bundala, *Davidse 7762* (OSH, US).

TAMARICACEAE

(by Neil A. Harriman*)

Link, Enum. Hort. Berol. Alt. 1: 291. 1821. Type genus: *Tamarix* L.

Shrub or small tree, leaves alternate, simple, estipulate, scale-like. Flowers perfect, pentamerous, hypogynous, regular, polysepalous, polypetalous, in panicles of slender racemes. Sepals imbricate; petals imbricate; stamens from between the lobes of a hypogynous disc. Ovary compound, unilocular, styles 3; fruit capsular, dehiscing by 3 valves; placentation basal; seeds several, rostrate, the beak beset with long hairs, wind-disseminated.

The family includes four–five genera, with about 100 species; in Ceylon, there is only *Tamarix,* with one native species.

TAMARIX

L., Sp. Pl. ed. 1, 270. 1753. Lectotype species: *Tamarix gallica* L.

Characters of the family; with but one species in Ceylon.

Tamarix indica Willd., Ges. Naturf. Freunde Berlin Neue Schriften 4: 214. 1803.

Tamarix gallica sensu Trimen, Handb. Fl. Ceylon 1: 91. 1893, non L.
Tamarix gallica L. var. *indica* (Willd.) Ehrenb., Linnaea 2: 268. 1827.

Small tree or shrub, 2–5 m tall. Leaves scale-like, 1–1.5 mm long, auriculate-clasping at base, deltoid, acute to acuminate, glandular-punctate, imbricated on the young growth, well separated on the older stems and then up to 2 mm long. Flowers pink to white, pedicellate, in racemes 6–10 cm long, 3.5–4 mm wide, each pedicel from the axil of an acuminate bract (as long as or) longer than the pedicel, the racemes collected in drooping panicles. Sepals 0.5–0.75 mm long, ovate or rotund, membranous, erose-denticulate. Petals 1.5–2 mm long, elliptic to elliptic-obovate, very caducous. Stamens 5, exsert, filaments inserted in shallow sinuses between the lobes of a minute, horizontal hypogynous disc. Ovary at anthesis globose basally, beaked; styles 3, separate to base, stigmas irregularly

*Biology Department, University of Wisconsin-Oshkosh, Oshkosh, Wisconsin 54901, USA.

toothed. Fruit capsular, lanceolate, 3 mm long, polished, chestnut-brown in colour; seeds numerous, beaked, the rostrum densely beset with long brown hairs.

Distr. Ceylon, India, Pakistan, Bangladesh, Afghanistan.

Ecol. On the sea-coast, confined to the dry districts of the Northwest and the North.

Vern. Kiri.

Note. Critical features of the species are illustrated in Baum, B.R. 1966. Monographic revision of the genus *Tamarix* (*T. indica*, pp. 59–60) and in Baum, B.R. 1978. The genus *Tamarix* (*T. indica,* pp. 62–65). Baum's keys and diagnoses make very clear the distinctness of *T. indica* from *T. gallica*; indeed, the 2 species fall into different series in sectio *Tamarix*.

Specimens Examined. JAFFNA OR MANNAR DISTRICT: sea coast, along the Jaffna-Mannar Road, *Kostermans 25128* (US). PUTTALAM DISTRICT: Wilpattu National Park, *Davidse & Sumithraarachchi 8235* (OSH, US); 5.5 miles S. of Kalpitiya, *Fosberg & Jayasuriya 52775* (US); 1.5 km S. of Palavi, *Wheeler 12981* (US); Puttalam, *Kundu & Balakrishnan 375* (US); Puttalam-Palavi Road, *Cooray 69100605R* (US); salt flats near Mampuri, *Sumithraarachchi 704* (US).

TILIACEAE

(by Andre Robyns* and Willem Meijer**)

Juss., Gen. 289. 1789.

Trees or shrubs, sometimes herbaceous. Leaves in general in spirals, blades lobed or not, base sometimes oblique, often cordate and with strongly developed basal lateral nerves, margin often dentate, often with stellate tomentum or scales, stalks often swollen at apex. Stipules often present, in most cases soon deciduous. Bark in general with strong sclerenchyma fibres and with slime cavities. Flowers in general in inflorescences basically cymose, sometimes leaf-opposed, axillary or terminal, cymes arranged into racemiform, paniculiform or umbelliform systems or solitary, provided with bracts and bracteoles. Flowers in general bisexual, sometimes unisexual, (plants in general monoecious, but in some genera dioecious), regular (actinomorphic), most commonly 5-merous but sometimes 4-merous, rarely trimerous, though sepals can vary from 2–8, whorls of androecium in general multiples of the number of sepals or petals. Some genera or subgenera with epicalyx, formed by bracts of condensed cymes. Calyx in bud always valvate, free sepals or forming a tube at base, often persistent, sometimes accrescent in the fruiting stage. Petals sometimes missing, in general prominent and coloured whitish, yellow, pink, purplish or greenish. Petals free or rarely connected, often with glands at their base. Aestivation variable, contort, imbricate or valvate. Stamens inserted above the base of the petals or on a gonophore, often above a hairy disklike ring. In general many stamens but in some cases only 10 (or 5), in many cases partly staminodial or reduced in functional female flowers, free or united at base or in phalanges (5 or 10). Anthers in general dorsifixed, rarely basifixed, with two thecae, opening with lateral slits or apical, sometimes united at apex of the stamens. Connectivum with or without appendages. Pollen in general 3-colporate, with reticulate or tegillate sexine. Ovary in general superior, syncarpous, with 2–5(10) carpels or in some cases apocarpous with 5–6 carpels, sometimes with false septa, each cavity with one to many seeds. Placenta in general axile. Ovule anatropous, usually ascending. Style only one, stigma punctiform, capitate or divided into as many rays as the carpels in the ovary, rarely sessile. Fruits in general 2–many loculate, with one to many seeds per locule, capsules

*National Botanic Garden of Belgium, Meise, Belgium.
**University of Kentucky and Missouri Botanic Garden, U.S.A.

dehiscent loculidally or septicidally or a dry or somewhat fleshy drupe with one or several pyrenes or a schizocarp with dehiscent or indehiscent mericarps. Fruits sometimes winged, sometimes provided with spines, spinules, hairy or smooth. Seeds sometimes provided with hairs or wings or with arils. Endosperm copious to scanty or absent, embryo usually straight. Cotyledons in general leaflike, often lobed or dissected.

A family of worldwide distribution with about 42 genera and around 400 species represented in Ceylon by six genera and 22 species. Sometimes the Elaeocarpaceae and the tribe Prockieae (now in Flacourtiaceae) are included in it.

KEY TO THE GENERA

1 Fruits unarmed
  2 Calyx with the sepals free to the base
    3 Trees or shrubs; fruits drupaceous
      4 Inflorescences of umbellate cymes or flowers solitary.................... **1. Grewia**
      4 Inflorescences paniculate, the ultimate ramifications cymose, the cymes usually 3-flowered
      ....**2. Microcos**
    3 Herbs; fruits capsular..**3. Corchorus**
  2 Calyx gamosepalous, campanulate or urceolate-campanulate, irregularly splitting at the apex or 5-lobate
    5 Flowers without staminodes; capsules largely 6(–8)-alate.................... **4. Berrya**
    5 Flowers with 5 staminodes; capsules obovoid, conspicuously 5-angulate-lobate, verrucose, densely lepidote .... **5. Diplodiscus**
1 Fruits densely covered with spines topped by hyaline spinules............. **6. Triumfetta**

1. GREWIA

L., Sp. Pl. 964. 1753. Type species: *G. occidentalis* L.

Trees or shrubs or sometimes scandent shrubs, the indumentum of stellate hairs. Leaves alternate, distichous, petiolate, stipules entire or sometimes divided, more or less caducous; blade usually serrate-dentate, 3–6-nerved from the base. Inflorescences axillary, super-axillary or oppositifolious, sometimes (sub)-terminal, cymose-umbellate, solitary or few-fasciculate, rarely flowers solitary. Flowers hermaphrodite or rarely polygamous, actinomorphic, pedicellate; bracteate; sepals 5, distinct, pubescent without, glabrous within; petals 5, shorter than the sepals, usually provided with a gland at the base within, the gland usually barbellate marginally, sometimes glandless; gonophore present, sometimes very short, usually glabrous at base, densely pubescent at the apex; stamens ∞, the filaments distinct, usually glabrous, the anthers dorsifixed, 2-thecate, longitudinally dehiscent; ovary superior, sessile, 2–4-celled, with to 2–8, or rarely more, ovules in each cell; style slender, the stigma usually lobulate. Fruit drupaceous, more or less carnose or fibrous, entire or more or less lobed, with 1–4 pyrenes, 1-several-seeded; seeds exalate, the endosperm copious or not; cotyledons foliaceous.

A palaeotropic genus of about 280–300 species; 10 species occur within Ceylon.

KEY TO THE SPECIES

1 Flowers solitary. Leaf blades only about 0.5–2.5 cm long, circular............ **8. G. tenax**
1 Flowers in pedunculate umbels. Leaf blades larger
  2 Leaf blades more or less rotund
    3 Petioles 0.5–1.5 cm long. Flowers up to 10 mm long. Petals in general with obovate glands at their base. Gonophore up to 2 mm long. Cultivated fruit tree........... **7. G. asiatica**
    3 Petioles 1–3.5 cm long. Flowers up to 7.5 mm long. Petals in general without glands at base. Gonophore up to 0.5 mm long. Wild, common in intermediate and dry zones............
    .... **2. G. damine** (=*G. tiliifolia*)
  2 Leaf blades ovate to elliptic or oblong
    4 Leaf blades whitish hairy on the lower face, very unequal at base, acuminate and serrulate towards the apex. Flowers 6–8 mm long, fruits 5 mm in diameter...................**1. G. rothii**
    4 Leaf blades not whitish on lower face, of different shape and fruits larger
      5 Mature flower buds up to 6 mm long. Leaf blades with more or less oblique base and acute apex, narrowly oblong-ovate, softly hairy at touch on upper face......... **3. G. hirsuta**
      5 Mature flower buds 15–22 mm long. Leaf blades on upper face rough at feel with fingertips or glabrous
        6 Leaf blades on upper face rough (scabridulous) at feel, ovate-elliptic. Flowers up to 18 mm long. Fruits often unlobed, depressed at apex or lobed, up to 1.5 cm in diameter.......
        ....**4. G. orientalis**
        6 Leaf blades glabrous on upper face, form more ovate
          7 Flowers up to 15 mm long. Fruits conspicuously 4-lobed, black, each lobe about 5–7 mm
          ....**5. G. bracteata**
          7 Flowers up to 22 mm long. Fruits brown, about 1.5–2 cm diameter, shallowly lobed or unlobed
          ....**6. G. carpinifolia**

1. Grewia rothii DC., Prod. 1: 509. 1824.

Grewia bicolor Roth, Nov. Pl. Sp. 240. 1821, non Juss., Ann. Mus. Natl. Hist. Nat. 4: 90, Tab 50, fig. 2. 1804.
Grewia polygama Roxb., Hort. Beng. 1814 ?; Roxb., Fl. Ind. 2: 588. 1832; Masters in Hook. f., Fl. Br. Ind. 1: 391. 1874; Trimen, Handb. Fl. Ceylon 1: 177. 1893; Worthington, Ceylon Trees 90. 1959.
Grewia helicterifolia sensu Thw., Enum. Pl. Zeyl. 31. 1858.

Tree to 10 m tall, the branchlets minutely stellate-tomentellous to stellate-puberulous. Leaves with slender petioles to 7 mm long, minutely stellate-tomentellous, the stipules subulate, to 6 mm long, fugacious; blade usually markedly inequilateral, narrowly oblong-ovate, narrowly ovate, or ovate, rounded on one side and obtuse on the other side at the base, acuminate to long-acuminate apically, the margins serrulate, to 11 cm long and 3.7 cm wide, membranous, slightly discolorous, minutely stellate-puberulous above, glaucous-stellate-tomentellous beneath, 3- or 4-nerved from the base, the costa and lateral veins prominulous beneath. Inflorescences axillary, 1–few in each axil, the umbels

3–5-flowered, the peduncles slender, 7–12 mm long, the pedicels slender, 5–8 mm long, both minutely tomentellous, the bracts to 2.5 mm long. Flowers 7–8 mm long, hermaphrodite in all Ceylon specimens examined (sometimes polygamous?); sepals narrowly oblong-ovate, acute apically, 7–8 mm long and c. 1.5 mm wide, thin-carnose, stellate-tomentellous outside, minutely puberulous inside; petals oblong, acute apically, 3.25–4 mm long and 1–1.3 mm wide, membranous, tomentellous below to pilose towards the apex without, glandular-thickened at the base within, the glands c. ⅓ as long as the petals, marginally barbellate; gonophore c. 1.5 mm long, 5-subangulate at the apex, minutely tomentellous apically; stamens indefinite, the filaments unequal, 1.75–5 mm long, glabrous, the anthers small; ovary c. 1.5 mm long, densely appressed-villosulous, the style to 4 mm long, glabrous, the stigma 2-lobulate, the lobes densely papillate. Fruits globose, c. 5 mm in diam., dark purplish when mature, tawny when dry, tomentellous.

Distr. India, Ceylon and Australia.

Ecol. A plant of thickets and forest in the dry zone below 300 m altitude; rather common.

Uses. The fruits are eaten by monkeys. The whitish bark is used for making string.

Vern. Boru-daminiya (S); Taviddai (T).

Specimens Examined. ANURADHAPURA DISTRICT: Anuradhapura, Oct. 1883, *s. coll. s.n.* (PDA); Mihintale, on hill, Sept. 1937, *Simpson 8534* (PDA). POLONNARUWA DISTRICT: Polonnaruwa, close to Gal Vihare Statuary in sacred area, Feb. 1969, *Robyns 6922* (BR, CAMB, E, PDA, US); Mannampitiya, Sept. 1953, *Worthington 6408* (PDA). PUTTALAM DISTRICT: Karativu Island, Aug. 1883, *s. coll. s.n.* (PDA). MATALE DISTRICT: Dambulla, Dec. 1951, *Worthington 5556* (PDA), Nov. 1961, *Amaratunga 29* (PDA); Nalanda, in jungle, Oct. 1947, *Worthington 3071* (PDA). DISTRICT UNKNOWN: *Thwaites C.P. 1113* (annotated Dambulla, Matale Distr., and Hanguranketa, Nuwara Eliya Distr.) (PDA); *s. loc., Alston A37* (PDA).

2. Grewia damine Gaertn., Fruct. 12: 113. t. 106. f. 7. 1790; non auct. Florae Indiae.

Grewia tiliifolia Vahl, Symb. 1: 35. 1791; Moon, Cat. 41. 1824; Thw., Enum. Pl. Zeyl. 32. 1858; Masters in Hook. f., Fl. Br. Ind. 1: 386. 1874; Trimen, Handb. Fl. Ceylon 1: 186. 1893; Worthington, Ceylon trees 91. 1959.

Large shrub to small tree attaining 10 m, the branchlets minutely stellate-puberulous, soon glabrous. Leaves alternate, distichous, the petioles 1–3.5 cm long, stellate-puberulous, the stipules inequilateral, 7.5–11 mm long and to 3.8 mm wide, the lower half auriculate and glabrous or not, the upper half linear to narrowly triangular, acute apically, minutely stellate-puberulous on both sides;

blade subinequilateral to markedly inequilateral, rotund, ovate or obovate, the base more or less cordate to truncate to rounded, the apex rounded to emarginate, sometimes obtuse or even acute, the margins crenate-serrate, up to 18 cm long and 12 cm wide, chartaceous, sparsely and minutely stellate-puberulous to glabrous above, minutely stellate-puberulous especially along the veins beneath, sometimes sparsely so, 5–6-nerved from the base, the costa and lateral veins prominent to prominulous beneath. Inflorescences axillary, up to 6 pedunculate umbels in each axil, the umbels usually 3-flowered, the peduncles slender, 7–11 mm long, stellate-puberulous, the pedicels slender, shorter than the peduncles, 5–8 mm long, stellate-puberulous, the bracts 2.5–5 mm long, stellate-puberulous without. Flowers yellow; sepals narrowly oblong-obovate or narrowly oblong-elliptic, acute apically, 6–7.5 mm long and c. 2 mm wide, tomentellous outside, glabrous or almost so inside; petals obovate to oblong-obovate, emarginate to undulate at the apex, c. 3 mm long and 1–1.3 mm wide, glabrous on both sides, eglandular; gonophore very short, c. 0.5 mm long, apically 5-lobulate and hirtellous to almost glabrous; stamens unequal, 1.5–5 mm long, glabrous; ovary c. 1 mm long, densely appressed-hirsute, the style c. 3.5–4.5 mm long, glabrous, the stigma discoid-lobulate. Fruits depressed-globose, conspicuously 2-lobed (when dry), 6–10 mm wide, black (when dry) and sparsely hirsute.

Distr. India, Ceylon and Burma.

Ecol. In lowland dry zone woodlands, patanas, and hill slopes.

Uses. According to Worthington (l.c.) the wood is tough and is used for making oars, shafts and gunstocks, while the bark is good for making string.

Vern. Daminiya (S); Cadachi (T).

Specimens Examined. ANURADHAPURA DISTRICT: Near Kekirawa, July 1887, *s. coll. s.n.* (PDA); 200 m alt., spreading tree, 6 m tall, flowers greenish yellow, with orange-yellow stamens, 25 Sept. 1969, *van Beusekom 1592* (PDA, US). MATALE DISTRICT: Dambulla, June 1932, *Simpson 9825* (PDA), Dec. 1951, *Worthington 5555* (K); below Dambulla Hill, West side, May 1931, *Simpson 8113* (PDA); near the Rock, 280 m alt., 9 Dec. 1971, fr, *Cramer 3536* (PDA, US); 8 Dec. 1970, fr and fl (yellow, stamens turning reddish), *Fosberg 53384* (PDA, US, MO); 29 Sept. 1969, fl yellow, *Cooray 69092401R* (PDA, US); around the rock base, common, tree 8 m, diam. 15 cm, bark brown smooth, living bark 2 mm, outside striped green and orange-red, flowers yellow, stamens brown-orange, 15 May 1969, *Kostermans 23549* (PDA, US, identified by Robyns as G. sp. aff. *G. tiliifolia* Vahl, note conspicuous glands at the inner base of the petals. Leaf shape and flowers further like *G. tiliifolia* and thus treated by me only as an aberrant variety—W.M.); Naula-Karawilahena, 350 m alt., 10 July 1971, fl, (flowers yellow, vern. name "Daminna" in Sinhala), *Jayasuriya 356* (PDA, US). BATTICALOA DISTRICT: Batticaloa area near Unichchai, in woodland area with burned undergrowth, on sandy soil, 14 Aug. 1967, fl, *Mueller-Dombois 67081408* (US); Palakadu, in thicket beside Vakerai-Trincomalee Road, at sea-level, 5 Dec. 1976, fr, *Cramer 4774* (PDA, US, MO). BADULLA DISTRICT: Kuruminiya Kandawura, Uma-Oya, Dec. 1927, *J.M. Silva 262* (PDA); Uma-Oya, Talpitigala, side of stream, 1400 m alt., 15 Sept. 1971, fl (yellow), *Balakrishnan and Jayasuriya 837* (PDA, US, MO); Passara, talawa, June 1939, *Worthington 464* (PDA); Ella, hillside, alt. 2600 ft, Aug. 1947, *Worthington 2946* (PDA); Passara-Moneragala Rd., 5/20 marker, 11 July 1974, fl yellow, *Waas 696* (PDA, US, KY); Mahiyangana, 150 m, 15 Apr. 1973, fl, (yellow, petals darker), *Stone 11157* (PDA, US). MONERAGALA DISTRICT: Bibile, July 1924, *J.M. Silva s.n.* (PDA); Bibile to Nilgala, deeply fissured bark, fr red-brown, 29 June 1973, *Kostermans 25161* (PDA, US); Mahiyangana-Bibile, dry forest, 27 June 1973, fl and fr, *Kostermans 25159* (PDA, US); Bibile, roadside, 29 July 1974, fl (yellow-orange. Stamens orange-yellow), *Kostermans 25299* (PDA, US). POLONNARUWA DISTRICT: Along Ambanganga River, near Wasgomuwa Nature Reserve, 10 Sept. 1977, fl, *Meijer and Jayasuriya 1309* (PDA, US, KY, MO). GALLE DISTRICT: Galle, 19 Nov. 1973, fl, *Sohmer, Waas and Eliezer 8900* (PDA, US, MO). NUWARA ELIYA DISTRICT: Lamasuriya-Ella, 18 Oct. 1973, fr, *Waas 208* (PDA, US, KY); Mulhalkelle, 1000 m alt., tree 30 m, outer bark cracked into strips. Petals deep yellow, reflexed, 18 July 1975, *Cramer s.n.* (PDA, US). KANDY DISTRICT: 2 miles East of Urugala on Kandy-Mahiyangana Rd, steep watercourse along road, 10 Oct. 1973, *Sohmer, Jayasuriya & Eliezer 8250* (PDA, US). DISTRICT UNKNOWN: *Thwaites C.P. 621*, pro parte, annotated Ramboda, Apr. 1846, *Gardner* and Teldeniya, Ragawella, July 1853 (PDA); s. loc, in 1849, *Fraser 117* (US).

Note. *G. damine* was described by Gaertner in volume 2 of his book De fructibus in 1790 on page 113 and illustrated on plate 106. The fruits came from the Bot. Garden in Leiden. Gaertner wrote: *G. salvifolia* Linn.? That species occurs in West Pakistan and India and is in fact the same as *G. bicolor* of Jussieu. It is very unlikely that the seeds came from India or Pakistan. Gaertner wrote also: Zeylonens.

Conclusion: one of the Ceylon species must be identical with *G. damine*. The most likely candidate is the common *G. tiliifolia* of Vahl, Symb. Bot. 1: 35. Aug.–Oct. 1790. After intensive restudy of all the Ceylon collections by one of us (Meijer) we are convinced that *G. tiliifolia* fruits are the only ones really fitting the illustrations given by Gaertner of *G. damine*. Added to this argument can be the common name which fits this species also exclusively.

3. Grewia hirsuta Vahl, Symb. Bot. 1: 34. 1790; Thw., Enum. Pl. Zeyl. 31. 1858; Masters in Hook. f., Fl. Br. Ind. 1: 391. 1874; Trimen, Handb. Fl. Ceylon 1: 176. 1893.

Shrub, the branchlets densely and softly stellate-hirsute. Leaves short-petiolate, the petiole to 5 mm long, densely and softly hirsute, the stipules subulate, longer than the petioles, to 8 mm long, long persistent, densely hirsute; blade more or less inequilateral, ovate to narrowly ovate to narrowly oblong-ovate, the base more or less oblique and rounded to acute, the apex acute, the margins serrulate, to 10 cm long and 2.5–3.8 cm wide, chartaceous, stellate-velvety-hirsute above, very densely so below, 3-nerved from the base, the nervation slightly impressed above, prominent beneath. Inflorescences axillary, the umbels (1–)3-flowered, the peduncles to 10 mm long, the pedicels to 6 mm long, the peduncles and pedicels densely hirsute. Flowers to 6 mm long; sepals narrowly oblong, acute apically, to 6 mm long and c. 1.5 mm wide, densely hirsute without, glabrous within; petals ovate, rounded apically, c. 1.5 mm long, hirsute without on the lower half, the glands c. 1/2 as long as the petals, densely hirsute marginally; gonophore less than 1 mm long, 5-lobulate, hirsute apically; stamens to 3 mm long, the filaments glabrous; ovary densely appressed-hirsute. Fruits depressed-globose, shallowly 4-lobed, to 10 mm in diam., dark brown when dry, shining, sparsely hirsute; pyrenes 4.

Distr. Nepal, India, Burma and Ceylon. Common and variable.

Ecol. A plant of arid regions.

Specimens Examined. LOCALITY UNKNOWN: s. coll. C.P. 3554 (PDA, US) agrees well with coll. seen and photogr. by W.M. at the BM. Apparently a rare species in Ceylon!

Note. Following the species concepts of the Flora of West Pakistan, Fam 75, Tiliaceae by Abdul Ghafoor, 1974, this might be considered identical with: G. helicterifolia Wall. ex G. Don., Gen. Syst. 1: 548. 1831; Burret in Notizbl. Bot. Gart. Berlin-Dahlem 9: 713. 1926. This is G. polygama auct. non Roxb.: Masters in Hook. f., Fl. Br. Ind. 1: 391. 1874, pro parte. Syn. G. hirsuta Vahl var. helicterifolia (Wall. ex G. Don) Haines, For. Fl. Chota Nagpur 196. 1910; Parker, For. Fl. Punjab 52. 1918; ed. 3: 51. 1956.

4. Grewia orientalis L., Sp. Pl. 964. 1753. Lectotype: *Hermann 324,* Ceylon (BM).

Grewia columnaris Smith in Rees, Cyclop. 17. 1811; Thw., Enum. Pl. Zeyl. 31. 1858; Masters in Hook. f., Fl. Br. Ind. 1: 383. 1874; Trimen, Handb. Fl. Ceylon 1: 176. 1893.

Shrub to 3–4 m high, the branchlets scabridulous, stellate-tomentellous to stellate-puberulous. Leaves with the petiole to 8 mm long, with the same indumentum as the branchlets, the stipules subulate, fugacious; blade ovate to elliptic, subcordate to obtuse at the base, acute to subobtuse at the apex, sometimes acuminate, the margins bluntly and irregularly crenulate, to 8 cm long and 4 cm wide, membranous to chartaceous, scabridulous and stellate-puberulous on both sides, glabrescent above, 3-nerved from the base, the costa and lateral veins slightly impressed above and prominent beneath. Inflorescences axillary, supra-axillary or oppositifolious, the umbels unusually 3-florous, the peduncles 6–15 mm long,

the pedicels 10–15 mm long, both stellate-tomentellous, the bracts ovate to narrowly so, to 5 mm long. Flowers to 18 mm long; buds ovate-oblong, tawny when
dry; sepals recurved, narrowly ovate, acute apically, to 7 mm long and 3.5 mm
wide, membranous, glabrous, densely tomentellous; gonophore to 2.5–3 mm long,
5-angulate-lobulate, the upper half minutely tomentellous; stamens indefinite, the
filaments free, unequal, 7–12 mm long, glabrous; ovary ovoid, c. 2.5 mm long,
densely appressed-villous, the style 10–12 mm long, glabrous, the stigma dilated,
lobulate. Fruits on the slightly accrescent gonophore, depressed-globose, more
or less 4-lobulate, to 13 mm in diam., black with brown dots when dry, almost
glabrous when mature.

Distr. India, Ceylon, and East tropical Africa to Java.

Ecol. In Ceylon rather common in the dry zone, in forest, scrub, or grassland.

Uses. According to Hermann, the wood is particularly adapted for bows
(Trimen l.c.).

Vern. Wel-keliya, Wel-mediya (S); Taviddai (T).

Specimens Examined. MATALE DISTRICT: 9 miles S.E. of Kandalama,
road to Dambulla, 17 Sept. 1974, fr, *Sumithraarachchi 468* (PDA, US, MO);
near Sigiriya Rock, small tree, fl green-white, calyx strongly recurved, fruits
orange-brown, shiny, 5 July 1973, *Nowicke, Fosberg and Jayasuriya 373* (PDA,
US, MO); Inamaluwa, turn off to Sigiriya on road to Trincomalee, edges of
thicket, 22 Oct. 1976, *Fosberg 56380* (PDA, US, MO); Dambulla-Trinco Road,
mile marker 99–101, fl, *Comanor 744* (PDA, US, MO); Dambulla, foot of Rock,
common shrub, 2–3 m, flowers white, 17 Jan. 1974, *Wambeek 2705* (US).
ANURADHAPURA DISTRICT: c. 3 miles N. of Mihintale, along road
Mihintale-Rambewa, in forest, March 1969, *Robyns 6971* (BR, CANB, E, PDA,
US, MO); Dambulla-Kekirawa Road, mile 51, common in cattle grazing area,
scrambling shrub, 8 ft, calyx outside green, inside white, revolute, corolla green,
a 5-lobed disc, ripe fruit yellowish, 6 July 1971, *Meijer 715* (PDA, US, MO);
near Kekirawa, July 1887, *s.coll. s.n.* (PDA); Anuradhapura-Trincomalee Road
near marker 65/4, 5 Oct. 1973, fl, *Sohmer 8143* (PDA, US, MO), near marker
73/3, *Sohmer 8148* (PDA, US, MO); Ritigala, 11 Jan. 1974, fr, *Waas 326* (PDA,
US). POLONNARUWA DISTRICT: along road Habarana-Polonnaruwa, just east
of Habarana, in forest, Feb. 1969, *Robyns 6933* (BR, CANB, E, PDA, US, MO);
about one mile NE. of Elahera, along the Amban Ganga, gallery forest on sandy
river bank, 400 ft alt., *Davidse 7370* (PDA, US, MO); sacred area, 60 m alt.,
8 Oct. 1971, fl, *Dittus 61* (PDA, US); 6 Sept. 1971, *Dittus 71090601* (PDA,
US); NW. shore of Minneriya Tank, near Kadawatte-Bendimulla, c. 80°52′ E.
and 8°4′ N., 300 ft alt., 27 Feb. 1977, fl and fr, *B. and K. Bremer 917* (PDA,
US, S); Gallale, near the railway station, 13 Oct. 1970, sterile, *Kundu and
Balakrishnan 237* (PDA, US). KANDY DISTRICT: Hantane, *Gardner 79* (Herb.
R. Wight, PDA); Syambalagastenne, Urugala, Sept 1926, *Alston 481* (PDA);
Hakkinda, Nov. 1927, *Alston 1688* (PDA); Kandy-Mahiyangane, 25/11 mile post,

primary montane forest, 18 Oct. 1973, fr, *Waas 211* (PDA, US). TRIN-COMALEE DISTRICT: c. 17 miles N. of Trinco, road to Kuchchaveli, along road in forest, Feb. 1969, *Robyns 6938* (BR, PDA, US, MO); Trincomalee, Nov. 1941, *Worthington 1141* (PDA); Kantalai, Aug. 1885, *s. coll. s.n.* (PDA); Foul Point, south side of Koddiyar Bay, across from Trincomalee, common on low berm of coral sand and gravel back of low beach ridge, covered by unevenly closed scrub forest, 23 Oct. 1976, *Fosberg 56430* (PDA, US, MO); north of Trin-comalee, last ferry before Pulmoddai, shrub forest, 7 m alt., 15 Jan. 1968, *Com-anor 792* (PDA, US, MO); Morawewa Tank, 1 Dec. 1977, fr, *Fosberg and Jaya-singhe 57137* (PDA, US, MO); Trincomalee-Habarana Road, woods near marker 152/4, 6 Oct. 1973, *Sohmer 8175* (PDA, US, MO); near Trincomalee, in NW. direction to Peryakulam, straggling shrub, yellow fl, 10 Oct. 1975, *Bernardi 15295* (PDA, US, MO). BATTICALOA DISTRICT: S. of Batticaloa, June 1931, *Simp-son 8297* (PDA); NW. of Unichchai Tank in Batticaloa area, thorn scrub-grassland, Aug. 1967, *Mueller-Dombois 67081418* (US); Batticaloa-Kalmunai Road, near bridge, dry zone, 19 May 1973, fl, *Kostermans 24835A* (PDA, US); Palakadu, in thicket beside Vakerai-Trincomalee Road, at sea level, shrub about 1.5 m, corolla white, 5 Dec. 1976, *Cramer 4773* (PDA, US, MO); Kathiraveli, 30 m alt., flowers white, fruits yellow-orange, *Bernardi 15670* (PDA, US). VAVUNIYA DISTRICT: Medawachchiya-Mannar Road, near 118 mile post, June 1975, fl, *Sumithraarachchi 756* (PDA, US). PUTTALAM DISTRICT: South end of Puttalam Lagoon, rare in thicket on partially vegetated saline flats, 15 Nov. 1970, *Fosberg and Jayasuriya 52809* (PDA, US, MO); Wilpattu National Park, Maduru Odai, plot W 20, straggling shrub, 2 m tall, 2 July 1969, *Wirawan, Cooray and Balakrishnan 935* (PDA, US). KURUNEGALA DISTRICT: Bad-degama, in thicket close to Pitadeniyagala by the Tittawela-Gonagama Road, 140 m alt., 20 Jan. 1972, *Cramer 3588* (PDA, US, MO). RATNAPURA DISTRICT: about 10 miles SE. of Godakewela on the Pelmadulla-Hambantota Road, on a sheer bare rock outcrop with small sparsely wooded watercourse, 270 m alt., 24 Nov. 1974, fl, *Davidse and Sumithraarachchi 8793* (PDA, US, MO). MONERAGALA DISTRICT: Bibile, June 1924, *J.M. Silva s.n.* (PDA); near Kuda Oya, roadside, *Grierson 1147* (BR, E, PDA, US). AMPARAI DISTRICT: near Lahugala near temple Magul Maha Vihara, straggling shrub, 10 m alt., 30 Oct. 1975, *Bernardi 15558* (PDA, US, MO); near Pottuvil, 2–5 m alt., 14 Dec. 1975, fl, *Bernardi 16021* (PDA, US, MO); Lahugala Reserve, 20 m alt., 15 Dec. 1975, *Bernardi 16048* (PDA, US, MO). HAMBANTOTA DISTRICT: Tissamaharama, Jan. 1927, *Alston 1192* (PDA), Jan. 1928, *Alston 1640* (PDA); Yala-Ruhuna National Park, Block 1, Bambowa, 17 Nov. 1969, fl, *Cooray 6911173R* (PDA, US, MO); Patanagala, near Smithsonian camp, on deep sandy soil between the rocks, straggling shrub, sterile coll. [leaves in all respects not different from *G. orientalis*: W.M.], Oct. 1968, *Wirawan 686* (US, MO). BADULLA DISTRICT: waterfall of Dunhinda, 200 m alt., 4 Aug. 1974, *Koster-mans 25328* (PDA, US). JAFFNA, PUTTALAM & KANDY DISTRICTS:

Thwaites C.P. 1111 (annotated Jaffna 1846, Puttalam 1848, *Gardner*, Haragama, July 1854, PDA), LOCALITY UNKNOWN: *Amaratunga 1004* (PDA, US); in 1850, *Fraser 59* (US).

Note. The best test between *G. rhamnifolia* and *G. orientalis* is the roughness of the upper face of the leaves in *G. orientalis*, which can be tested by rubbing them lightly with the fingertips.

Conclusion: Both the Ratnapura, Davidse collections belong to *G. rhamnifolia* and both species can be shrubs or straggling woody climbers.

5. Grewia bracteata Heyne ex Roth, Nov. Pl. Sp. 243. 1821; Hook. f., Fl. Br. Ind. 1: 389. 1874; Burret, Notizbl. Bot. Gart. Berlin-Dahlem 9: 697. 1926; Matthew, Fl. Tamilnadu Carnatic III(1): 164. 1983. Type: Heyne coll. India. (B, †).

Grewia diplocarpa Thw., Enum. Pl. Zeyl. 31. 1858; Masters in Hook. f., Fl. Br. Ind. 1: 390. 1874: Trimen, Handb. Fl. Ceylon 1: 175. 1893. Type: Ceylon, *Thwaites C.P. 2568.*

Shrub to small tree, the branchlets sparsely pilose, soon glabrescent. Leaves rather long-petiolate, the petiole slender, to 1–2 cm long, appressed-pilose, the stipules subulate, shorter than the petioles, to 8 mm long, appressed-pilose, blade somewhat inequilateral, ovate, oblique and more or less rounded at the base, long-acuminate at the apex, the margins crenulate, to 14 cm long and 6.5 cm wide, membranous, very sparsely appressed-stellate-puberulous to almost glabrous above, sparsely appressed-pilose mainly along the veins beneath. Inflorescences axillary, solitary, the umbels 3-flowered, the peduncle slender, 7–30 mm long, the pedicels slender, 10–15 mm long, both sparsely appressed-pilose to almost glabrous, the bracts to 4 mm long. Flowers to 15 mm long, hermaphrodite; buds oblong, sepals linear-oblong, acute apically, to 15 mm long and 2 mm wide, thin-carnose, tomentellous outside, glabrous or almost so inside; petals triangular, acute apically, c. 4.5 mm long and 2.5 mm wide, dorsally in the middle and marginally stellate-pilosulous, the glands c. ½ as long as the petals or even somewhat longer, thick, marginally densely barbellate; gonophore c. 2 mm long, tomentellous at the apex; stamens indefinite, the filaments free, unequal, c. 6–9 mm long, glabrous; ovary globose, c. 1.5 mm in diam., densely appressed-villosulous, the style to 9 mm long, stellate-puberulous at the base to glabrous at the apex, the stigma 4-lobulate. Fruits conspicuously 4-lobed, each lobe to 7 mm long, black when dry and sparsely appressed-pilose; seeds 4.

Distr. Abyssinia to Malay Peninsula.

Ecol. Lower montane forest in intermediate zone of Ceylon.

Specimens Examined. KANDY DISTRICT: near Madugoda, streamside, Jan. 1932, *Simpson 9138* (BM, PDA); about 8 miles NE. of Hunnasgiriya, near mile post 29/10 along the Kandy-Mahiyangana Road, low montane woody vegetation, slender tree 3 m tall, petals and filaments white, anthers pale yellow, 810 m alt., 14 Nov. 1974, *Davidse and Jayasuriya 8406* (PDA, US, MO); Kandy-Mahiyangana Road, 24 mile marker, culvert 6, steep mountain side, sepals white within, green on the outside, petals small and white, stamens white, 10 Oct. 1973, *Sohmer, Jayasuriya and Eliezer 8254* (PDA, US, MO). MATALE DISTRICT: Karakolagastenna, Illukkumbura, 80°47′ N. and 7°33′ E., sparse forest, 530 m alt., treelet 7 m tall, fl white, 1 Dec. 1971, *Jayasuriya, Dassanayake and Balasubramaniam 423* (PDA, US); Dikpatana, 80°45′ N. and 7°33′ E., patch of submontane jungle by small stream, semi-sun, 900 m alt., 1 Dec. 1971, *Jayasuriya, Dassanayake and Balasubramaniam 431* (PDA, US); Kandy-Dambulla Road, 36 mile marker, culvert 4, east side of road in dry brush land, 4 Oct. 1973, fl, *Sohmer 8065* (PDA, US, MO). NUWARA ELIYA DISTRICT: Maturata, *Thwaites C.P. 2568* (type of *G. diplocarpa* Thwaites, K, PDA); Rikiligaskada, Nov. 1961, *Amaratunga 92* (PDA). BADULLA DISTRICT: Uma Oya, June 1883, *s. coll. s.n.* (PDA); road to Dunhinda Falls, in open forest on steep slope, about 650 m alt., 16 Oct. 1971, *Balakrishnan and Jayasuriya 865* (PDA, US, MO).

6. Grewia carpinifolia Juss., Ann. Mus. Natl. Hist. Nat. 4: 91. n. 20, Tab. 51, fig. 1, 1804.

Grewia orientalis L. as misapplied by Moon, Cat. 41. 1824; Thw., Enum. Pl. Zeyl. 31. 1858; Masters in Hook. f., Fl. Br. Ind. 1: 384. 1875; Trimen, Handb. Fl. Ceylon 1: 176. 1893.
Grewia rhamnifolia Heyne ex Roth, Nov. Pl. Sp. 244. 1821. Type: Heyne coll. India?

Shrub, often scandent, 1.5–3 m tall, the branches divaricate, the branchlets stellate-puberulous, becoming glabrous. Leaves alternate, distichous, the petioles slender, 5–8 mm long, stellate-tomentellous to stellate-puberulous, the stipules fugacious; blade equilateral to slightly inequilateral, from narrowly ovate-elliptic to elliptic, sometimes oblong-elliptic or subobovate, more or less rounded at the base, acute to acuminate at the apex, crenulate-serrulate along the margins, to 12.5 cm long and 5.5 cm wide, chartaceous (when dry), green and shining above, somewhat paler and dull beneath, sparsely stellate-puberulous to almost glabrous on both sides, the indument mainly along the veins, 3-nerved from the base, the costa and lateral veins slightly prominent beneath. Inflorescences axillary, oppositifolious (cf. A. Robyns 6934 & 6937) or axillary, the umbels usually 3-flowered, sometimes 5-flowered, the peduncles 10–30 mm long, usually shorter than the pedicels, the pedicels 20–30 mm long, the peduncles and pedicels stellate-puberulous, often densely so, the bracts to 5 mm long, appressed-stellate-pilosulous. Flowers to 22 mm long; sepals reflexed, narrowly ovate, acute apically, to 22 mm long and 4.5 mm wide, subcarnose, fulvous to ferruginous when dry, densely stellate-puberulous outside, glabrous inside; petals ovate, acute apically, to 7 mm long and 3–3.5 mm wide, membranous, glabrous, the glands to 3 mm long, densely barbellate marginally; gonophore c. 3 mm long, 5-lobulate near the middle, the lower part (below the lobes) glabrous, the upper part appressed-tomentellous; stamens unequal, to 12 mm long, glabrous, ovary ovoid, c. 2 mm long, densely appressed-sericeous, the style to 11 mm long, glabrous, the stigma lobulate. Fruits depressed-globose, shallowly 2-lobed or unlobed, 15–20 mm diam., yellowish green when fresh, tawny and wrinkled when dry, tomentellous.

Distr. India and Ceylon.

Ecol. From sea level up to about 1500 m. From dry to wet montane zone in all kinds of shrubby and secondary forest vegetation types.

Specimens Examined. PUTTALAM DISTRICT: 1.5 miles beyond Kalpitiya, occasional in open scrub with *Borassus*, 13 Nov. 1970, fr, *Fosberg and Jayasuriya 52743* (PDA, US, KY). ANURADHAPURA DISTRICT: c. 23 miles N. of Mihintale, along road Mihintale-Rambewa, in forest, March 1969, *Robyns 6967* (BR, US, PDA); Ritigala Kande: Weweltenna, among boulders, 10 Oct. 1977, fl, *Huber 420* (PDA, US); Kekirawa-Anuradhapura Road, mile 62/3, climber with hanging branches, fruits subglobose, with 4 seeds, 12 Jan. 1973, *Tirvengadum and Jayasuriya 250* (PDA, US); Ritigala, 100–200 m alt., dry mixed evergreen forest, 6 March 1971, fr, *Robyns 7267* (PDA, US). POLON-NARUWA DISTRICT: along road Habarana to Kantalai at Gal Oya junction, 22 Feb. 1969, fr, *Robyns 6934* (BR, CANB, E, PDA, US, MO). TRIN-COMALEE DISTRICT: along road Trinco-Kuchchaveli, in forest, 23 Feb. 1969, fr, *Robyns 6937* (BR, CANB, E, PDA, US, MO); Mutur, Dec. 1944, *Worthington 1141* (PDA); west of Trincomalee, near Kantalai in small chena, 12 Jan. 1968, fr, *Comanor 755* (PDA, US, KY). MATALE DISTRICT: Dambulla Rock, Nov. 1961, *Amaratunga 28* (PDA), 7 Oct. 1975, *Bernardi 15245* (PDA, US, MO); Wiltshire Forest, Kandagedara, 2 July 1974, fr, *Sumithraarachchi 398* (PDA, US MO); Dambulla, in jungle along rocky slope on western side of Rock, 280 m alt., 9 Dec. 1971, fr, *Cramer 3535* (PDA, US); near Rock, 150–300 m alt., straggling shrub, 3–4 m high, fruits yellowish, to 1.6 cm diam., 23 Feb. 1971, *Robyns 7209* (PDA, US); mile post 37 between Lagala and Ilukkumbura, 1000 m alt., 10 Apr. 1971, fr, *Jayasuriya 272* (PDA, US); Kandy-Dambulla Road, mile marker 36, culvert 4, 4 Oct. 1973, fl, *Sohmer 8065* (PDA, US). KANDY DISTRICT: Murutenne, near Laxapana Falls, margin of forest near tea estate, 2800 ft alt., March 1969, *Grierson 1053* (BR, E, PDA, US); Laxapana-Maskeliya, 850 m alt., liana, fruit reddish green, 13 Sept. 1977, *Nooteboom and Huber 3098* (PDA, US); Hantane, May 1924, *J.M. Silva s.n.* (PDA); East slope, edge of forest 960 m alt., 15 Feb. 1971, *Robyns 7186* (PDA, US); Hunnasgiriya, June 1926, *J.M. Silva 53* (PDA); same loc, 1000 m alt., 28 July 1974, fr, *Kostermans 25275* (PDA, US); Rondara group, Ginigathena Road, Nov. 1931, *Simpson 8549* (PDA); Dehigama, Murutalawa, March 1953, *s. coll. s.n.* (com. J.M. Appuhamy) (PDA); Peradeniya, Gannoruwa hill, sec. forest, 9 July 1978, *Meijer 1725*

(PDA, US, MO); Deltota-Hewaheta, along streambed, 24 Dec.1968, fr, *Wirawan 771* (PDA, US, MO, KY); Loolecondera Tea Estate, Hewawisse Korale Adm. Div, 4600 ft alt., 30 Oct. 1978, fr, *Fosberg 58174* (PDA, US, MO); Dara Oya stream, wooded ravine, 2 miles east of Loolecondera, 30 Oct. 1978, *Fosberg 58165* (PDA, US, MO); Doluwa, 500 m alt., 29 July 1974, fr, *Albert 1779*, coll. nr. *Jayasuriya* (PDA, US, MO); Kandy-Mahiyangana Road, 28 mile marker, submontane forest on steep hillside, 10 Oct. 1973, *Sohmer, Jayasuriya and Eliezer 8289* (PDA, US, MO); about 2 miles N. of Hunnasgiriya, near mile post 234, Cardamom plantation, 1260 m alt., 17 Nov. 1974, fr, *Davidse 8441* (PDA, US, MO); Kadugannawa, on top of rock near tunnel on Kandy road, fairly common, 550 m alt., 31 May 1974, *Cramer 4219* (PDA, US); Kadugannawa, along rocky slopes in secondary forest close to Pahala-Kadugannawa road, fairly common, 600 m alt., 28 Aug. 1972, fr up to 2 cm in diam., *Cramer 3842* (PDA, US); Madugoda, 1000 ft alt., 13 Oct. 1977, fr, climber, *Nooteboom 3365* (PDA, US); Rangala to Corbets Gap, 100 m alt., 13 May 1969, fl, *Kostermans 23482* (PDA, US); Madugoda-Mahiyangana Road, rather dry low forest, 600 m alt., 6 Aug. 1978, *Kostermans 26770* (PDA, US). NUWARA ELIYA DISTRICT: Hakgala, *J.M. Silva s.n.* (PDA). RATNAPURA DISTRICT: Rasagala, along borders of secondary forest near culvert 2/11, Rasagala-Ratnapura Road, 600 m alt., 1 Nov. 1970, fr, *Cramer 3266* (PDA, US, MO); Adam's Peak trail, NE. of Carney, montane forest, 900–1350 m alt., 23 Nov. 1974, leaves larger than in the lowland form, small tree, fruit orange, partly lobed, some stamens still attached to base of young fruits, *Davidse and Sumithraarachchi 8775* (PDA, US, MO); about 13 miles NE. of Deniyaya along Highway A 17 to Ratnapura, at mile post 64, in forest on cliff, vine, fruit greenish-brown, 1050 m alt., *Davidse 7915* (PDA, US MO); north of Pinnawala, Adam's Peak Sanctuary, 1600 m, 21 June 1972, *Maxwell and Jayasuriya 914* (PDA, US). KEGALLE DISTRICT: Dolosbage, forest edge, 1800 ft alt., 14 May 1970, *Balakrishnan 338* (PDA, US). HAMBANTOTA DISTRICT: Tissamaharama, Jan. 1928, *Alston 1641* (PDA); Ruhuna National Park, Yala, Block 1, near Situlpahuwa ruins, on roadside to Katagamuwa Tank, Nov. 1968, *Wirawan 686A* (US); Block I, mile 11 at Yala Road, in scrub vegetation on sand mixed with clay, straggling vine, flower creamy white, 18 Oct. 1968, *Mueller-Dombois 68101841* (PDA, US); same loc, 11 Jan. 1969, straggling shrub, fruits with persistent stamens, 11 Jan. 1969, *Cooray 69011104R* (PDA, US); Buttawa bungalow, in low scrub on granite rocks, flowers white, 24 Nov. 1970, *Fosberg and Sachet 52877* (PDA, US). BATTICALOA DISTRICT: Mankerni, secondary sandy shore vegetation, treelet 2.5 m, flower yellow, 2 May 1977, *Waas 2099* (PDA, US, leaves almost smooth, shape as *G. carpinifolia*, an intermediate hybrid form ? W.M.); Kalkudah, along boundary of block 5, Kennedy Estate beside railway track, sea level, 5 Nov. 1969, fl white, petals reflexed with age, *Cramer 2743* (PDA, US). AMPARAI DISTRICT: Pottuvil, 2–5 m alt., 14 Dec. 1975, fr, *Bernardi 16021* (PDA, US). BADULLA DISTRICT: Mahiyangana, in forest, 1500 ft, 8 May 1971, fl, *Balakrishnan 336* (PDA, US).

WITHOUT EXACT LOCALITY: *s. coll. C.P. 1110* (US).

7. Grewia asiatica L., Mant. Pl. 1: 122. 1767; Moon, Cat. 41. 1824; Masters in Hook. f., Fl. Br. Ind. 1: 385. 1875; Trimen, Handb. Fl. Ceylon 1: 174. 1893. Type: Suratte, India, Braad Coll. (LINN ?).

Grewia subinaequalis DC., Prod. 1: 511. 1824.

Grewia rotundifolia Juss. as misapplied by Thw., Enum. Pl. Zeyl. 402. 1864.

Grewia orbiculata Rottl. as misapplied by Trimen, J. Roy. Asiat. Soc., Ceylon Branch 9: 12. 1855. (cf. Narayanaswami & Rao, J. Indian Bot. Soc. 29 (4): 177–190. 1950)

Small tree, the branchlets stellate-tomentellous, becoming glabrous. Leaves alternate, distichous, the petioles rather stout, 6–15 mm long, densely and minutely stellate-puberulous to stellate-tomentellous, the stipules slightly falcate, inequilateral, narrowly ovate, acute apically, to 12 mm long and 2.5 mm wide, stellate-puberulous without, sparsely so within; blade slightly inequilateral to almost equilateral, more or less rotund, subcordate or truncate at the base, acute at the apex, coarsely and irregularly serrate along the margins, to 13.5 cm long and 10 cm wide, chartaceous, minutely stellate-puberulous above, dull and stellate-tomentellous beneath, 5-nerved from the base, the costa and lateral veins prominent beneath, the transverse ones parallel and prominulous. Inflorescences axillary, up to 4, pedunculate, umbels 3–6-flowered, the peduncles slender, to 2.5 cm long, stellate-tomentellous, the bracts linear, acute apically, to 4 mm long, early caducous. Flowers to 10 mm long; buds oblong-obovoid; sepals narrowly oblong, acute at the apex, 9.5–10 mm long and c. 2 mm wide, tomentellous without, glabrous within; petals obovate, subemarginate at the apex; stamens unequal, to 5 mm long, glabrous; ovary c. 1 mm long, densely appressed-hirsute, the style c. 4 mm long, glabrous, the stigma 4-lobulate, the lobes spreading. Fruits more or less globose, 7–8 mm in diam., black (when dry) and sparsely appressed-stellate-pilose.

Distr. and Ecol. No collections, other than the historical Koenig coll. (BM) and Thwaites, *C.P. 3785* (PDA) known from Ceylon. Localities unknown, probably formerly cultivated in gardens but no longer seen in the Island. Also sometimes recorded from India and Nepal.

Note. The type is from a cultivated plant collected at Surat in India by Christian Henrik Braad, who was in the service of the Royal Swedish East India Company and a correspondent of Linnaeus. Narayanaswami and Rao (1950) repeated the story of Drummond (1911) and others that *G. asiatica,* cultivated at Pondicherry and described by Linnaeus from a cultivated tree from a garden at Surat, collected by the Swede Christian Henrik Braad, would only be known in cultivation and possibly may originate from outside India. He went against Roxburgh's assertion that it is a native of India and often cultivated in gardens (in Flora Indica ed. 2, 431. 1832).

Abdul Ghafoor in his revision of Tiliaceae for the Flora of West Pakistan takes the more commonsense point of view that it is not necessary to substitute the name *G. subinaequalis* for *G. asiatica* L. Vahl (1790) pointed out the rather short petioles and the fact that the flowers of this species are about twice as large as those of the similar looking *G. damine*. Trimen (1893) reported *G. asiatica* from Ceylon, but called it an extremely rare species, probably always cultivated. In the British Museum, Nat. History, herbarium I (WM) studied and photographed Feb. 1981 a specimen collected by Koenig in Ceylon that agrees well with the description of *G. asiatica*. On the label is a note: "habitat in sylvis, cult. in hospis zeylana frequentia". Dr. Robyns cited *Thwaites C.P. 3785* from PDA in his draft revision. He describes petioles as being 6–15 mm long, stipules slightly falcate, up to 12 mm long, flowers up to 10 mm long, petals with obovate glands, gonophore only 2 mm long, 5-lobulate at apex, and the fruits globose, 7–8 mm in diameter. This contrasts with the 10–35 mm long petioles of *G. damine* and its flowers 6–7.5 mm long, petals without glands, gonophore even shorter, only 0.5 mm, and fruits 6–10 mm wide. There is no way to deny that the two species are very closely related and easily confused.

The fruit of *G. damine* is also said to be eaten. Possibly *G. asiatica* is a cultivar of *G. damine* selected by man ? There is an excellent collection of this species made by the late William T. Gillis, from P.I. *116841*, M8830 from coll. *Koelz 785* from Lahore, Punjab. Flowers yellow, turning deep orange with age. Sprouts from stump vigorously. Berry reportedly used in flavouring sherbet and beverages. "Phalsi". This is the Indian name for *G. asiatica*, Coll. *Gillis 10412*, 23 March 1971, Fairchild Trop. Garden Herbarium.

8. Grewia tenax (Forssk.) Fiori, Agric. Colon. 5 (suppl.): 23. 1912; Burret, Notizbl. Bot. Gart. Berlin-Dahlem 9: 689. 1926; Alston in Trimen, Handb. Fl. Ceylon 6: 33. 1931; Matthew, Flora Tamilnadu Carnatic III (1): 172. 1983.

Chadara tenax Forssk., Fl. Aegypt.-Arab. 114. 1775. Type: "in sched. herb. Firsk." acc. Christensen, Dansk. Bot. Ark. 4(1): 23. 1922.
Grewia populifolia Vahl, Symb. Bot. 1: 33. 1790; Trimen, J. Bot. 143. 1885; Masters in Hook. f., Fl. Br. Ind. 1: 385. 1874; Trimen, Handb. Fl. Ceylon 1: 178. 1893.

Shrub, the branches divaricate, the branchlets sparsely stellate-puberulous, soon glabrous. Leaves alternate, distichous, or fascicled on abbreviated branches, the petioles slender, 0.5–2.5 cm long, stellate-puberulous to sparsely stellate-puberulous, the stipules small, linear, soon caducous; blade more or less circular, sometimes sub-3-lobed, rounded to slightly cordate at the base, obtuse to sometimes acute apically, irregularly and coarsely crenate-serrate along the margins, the 2 or 4 lowermost teeth usually glandular above, to 4.6 cm long and 4.2 cm wide, thin-chartaceous, glabrous above, sparsely stellate-puberulous to almost glabrous below, 3-nerved from the base, the costa and lateral nerves prominulous below. Flowers solitary (or sometimes geminate ?), extra-axillary or often (sub) terminal on abbreviated branches, white; pedicels slender, to 1.5 cm long in flower, to 3 cm long in fruit, very sparsely stellate-puberulous, the bracts narrowly oblong, acute, c. 3 mm long and 1 mm wide; sepals linear, acute apically, 10–14 mm long and 2 mm wide, tomentellous on both sides; petals linear, shortly bilobed at the apex, c. 7.5 mm long and 1 mm wide, glabrous except for the minutely papillate outer base, c. 2.5 mm long and 1.8 mm wide (indumentum included), marginally densely barbellate-hirtellous; gonophore c. 2 mm long, lobulate and hirtellous at the apex, somewhat accrescent; stamens unequal, to 7 mm long, glabrous; ovary globose, c. 1 mm diam., glabrous except for the appressed-hirtellous base, the style to c. 8 mm long, glabrous, the stigma discoid-peltate. Fruits depressed-globose, markedly 4-lobed, to 10–15 mm wide, the pericarp brownish (when dry), shining, thin, brittle, and glabrous.

Distr. From Africa (Egypt) to Arabia-Afghanistan to southern India and to the dry zone of Ceylon.

Vern. Katu-peratti, Achu (T).

Specimens Examined. VAVUNIYA DISTRICT: Matalan, Feb. 1889, *Nevill s.n.* (PDA). PUTTALAM DISTRICT: Puttalam near mangrove swamp, Nov. 1926, *Alston 1193* (PDA); Chilaw, Feb. 1882, *Nevill s.n.* (PDA, K); Wilpattu National Park, west end of Mail Villu, 1.5–2 m bush, white flowers with yellow stamens, 30 Apr. 1969, *Wirawan 69043026* (PDA, US, MO); South end of Puttalam Lagoon, common in thickets of partially vegetated saline flats, shrub 2 metres, flowers white, fr. green, 15 Nov. 1970, *Fosberg and Jayasuriya 52808* (PDA, US, MO). MANNAR DISTRICT: about 2 miles SW. of Mantai along road to Mannar, near mile marker 4/4, sandy tidal flats with low woody vegetation on slightly elevated places, shrub 3 m, petals and filaments white, anthers yellow, fruit green, *Davidse and Sumithraarachchi 9154* (PDA, US, MO). HAMBANTOTA DISTRICT: Kirinda, Dec. 1882, *s. coll. s.n.* (PDA); Bundala, Dec. 1882, *s. coll. s.n.* (PDA); Ruhuna National Park, just outside gate, 28 Oct. 1968, fl, *Cooray 68102801R* (PDA, US, MO).

2. MICROCOS

L., Sp. Pl. 1: 514. 1753.

Grewia L. subgenus *Omphacarpus* (Korth.) Kuntze, Rev. Gen. Pl. 1: 83. 1891.

Trees or shrubs, sometimes scandent shrubs, resembling the genus *Grewia* in habit. Leaves alternate, distichous, petiolate, stipulate, the stipules more or less caducous; blade simple, entire-margined or almost so, 3-nerved from base. Inflorescences terminal or axillary, fasciculate, the ultimate ramifications cymose, the cymes 3-flowered; bracts involucrate, each cyme surrounded by 3 3-fid bracts and 3 narrow and mostly entire bracts. Flowers hermaphrodite, actinomorphous, pedicellate; sepals 5, distinct, tomentellous without and within; petals 5, rarely absent, smaller than the sepals, provided with a gland at the base within, the gland barbellate marginally; gonophore present, villosulous at the apex; stamens indefinite, rarely 5, the filaments distinct, pilosulous below, the anthers 2-thecate, longitudinally dehiscent; ovary superior, sessile, usually 3-celled, each cell with (2)4–8 ovules; style narrowed towards the apex, the stigma punctiform or inconspicuously lobulate. Fruits drupaceous, unlobed, the mesocarp fibrous, the endocarp coriaceous, seeds exalate, with endosperm.

Fifty-three species in Indo-Malesia, Fiji.

Microcos paniculata L., Sp. Pl. 1: 514. 1753; Gaertn., Fruct. 1: 273, t. 57, fig. 4. 1788; Moon, Cat. 41. 1824; Nair and Henry, Fl. Tamilnadu, Ser. 1, 1: 45. 1983. Lectotype: *Hermann* (BM).

Grewia microcos L., Syst. ed. 12. 602. 1767, nom. illeg. superfl. (see Panigrahi, Taxon 34: 702. 1985); Thw., Enum. Pl. Zeyl. 32. 1858; Masters in Hook. f., Fl. Br. Ind. 1: 392. 1874; Trimen, Handb. Fl. Ceylon 1: 177. 1893; Ramamoorthy in Saldanha & Nicolson, Fl. Hassan Distr., 136. 1976. Type: same as for *Microcos paniculata* L.
Grewia nervosa (Lour.) Panigrahi, Taxon 34: 702–704. 1984.
Fallopia nervosa Lour., Fl. Cochinch. 336. 1790.

Shrub to 1.8 m tall, the branchlets nearly glabrous. Leaves alternate, distichous, the petioles to 12 mm long, slightly pulvinate at the apex, minutely stellate-puberulous, the stipules narrowly triangular, 5–6 mm long, minutely stellate-puberulous, caducous; blade slightly inequilateral, narrowly ovate to narrowly elliptic, slightly oblique and more or less obtuse at the base, usually long-acuminate at the apex, the margins somewhat repand and subentire (remotely and inconspicuously serrulate), to 25 cm long and 7 cm wide, subcoriaceous, shining and glabrous or almost so above, dull and stellate-puberulous especially along the veins beneath, 3-nerved from the base, the costa and lateral veins prominent beneath, the transverse veins almost parallel and prominulous beneath. Inflorescences terminal and axillary towards the apex of the branchlets, paniculate, to 10 cm long, the ultimate ramifications cymose, the cymes 3-flowered, each cyme surrounded by 3 deeply (almost to the base) trifid bracts up to 4–4.5 mm long and 3 narrowly lanceolate (one of them sometimes 2-lobed) to subulate bracts to 3 mm long, the inflorescence axes minutely stellate-puberulous and often also scattered-pilose (2 distinct tiers of hairs), the pedicels very short. Flowers greenish yellow; sepals obovate, erect, convex-cucullate, acute apically, the margins sinuate, c. 6 mm long and almost 3 mm wide, tomentellous without and within; petals narrowly ovate-oblong, emarginate to sinuate apically, c. 2.8 mm long and 1 mm wide, tomentellous on the lower half to gradually glabrous towards the apex without, glandular at the base within, the glands obovate, c. 1–1.5 mm long, marginally villosulous; gonophore obconical, 5-angular, almost 1 mm long,

minutely villosulous at the apex; stamens ∞, the filaments unequal, 2.5–4.5 mm
long, the lower part pilosulous; ovary subglobose, c. 1 mm in diameter, glabrous,
3-celled, the style to c. 4 mm long, gradually tapering towards the apex, glabrous,
the stigma indistinct. Fruits ellipsoid to obovoid, to 8 mm long, the pericarp black
(when dry) and glabrous.

Distr. China, Burma, Indo-Malesia.

Ecol. A tree of secondary forests and hedges.

Vern. Kohu-kirilla, Keliya (S); Kapila (T).

Specimens Examined. PUTTALAM DISTRICT: Puttalam, *Gardner s.n.*
(Thwaites C.P. 1109) (PDA). KURUNEGALA DISTRICT: Doluwa Kande, June
1886, *s. coll. s.n.* (PDA); Ibbagamuwa, 8 km NE. of Kurunegala, 600 ft, July
1948, *Worthington 4035* (K); Wewa-gala, rocky area, sec. forest, 11 Aug 1974,
fl, *Waas 757* (PDA, US); Palliyadde, in sec. forest reserve, 13 Sep. 1974, fl
yellow, fr, *Waas and Tirvengadum 823* (PDA, US); Galewela, mile marker 78,
border of dense intermediate dry zone forest, in shade, 13 Sep. 1974, fl,
Tirvengadum et al. 501 (PDA, US); Kankaniyanmulla forest, sec. intermediate
forest, 11 Aug. 1974, fr, *Waas 746* (PDA, US). KANDY DISTRICT: Peradeniya,
Royal Botanic Garden, June 1891, *s. coll. s.n.* (PDA); Kadugannawa, 1600 ft,
jungle edge, May 1939, *Worthington 360* (K). COLOMBO DISTRICT: Mirigama,
June 1927, *Alston 679* (PDA). RATNAPURA DISTRICT: Udakarawita, in forest,
1500 ft, 1 May 1970, fl, *Balakrishnan 327* (PDA, US); Rakwana, 600 m alt.,
26 May 1969, fl, fr, *Kostermans 23637* (PDA, US). MATARA DISTRICT: Don-
dra Head, 20 Nov. 1973, fl, *Sohmer et al. 8874* (PDA, US). GALLE DISTRICT:
Galle, 15 Aug. 1975, fl, *Sumithraarachchi and Austin 980* (PDA, US). DISTRICT
UNKNOWN: s. loc., *Col. Walker s.n.* (PDA); *Frazer s.n.* (US).

3. CORCHORUS

L., Sp. Pl. 529. 1753; L., Gen. Pl. ed. 5, 234. 1754.

Herbs, suffrutices or shrubs, the pubescence of simple or stellate hairs. Leaves
alternate, petiolate, 2-stipulate, the blade usually simple, rarely 3–5-lobed, the
margins serrate, the 2 lowermost serrations sometimes produced into caudae,
3–5-nerved from the base. Flowers small, oppositifolious or axillary, solitary
to few-fasciculate, hermaphrodite, actinomorphic, pedicellate, bracteolate; sepals
4–5, free, the apex apiculate or not, more or less caducous; petals 4–5, distinct,
generally imbricate, sometimes unguiculate, glandless; gonophore present or not,
in Ceylon annulate; stamens ∞, rarely 5 or twice the number of the sepals, all
fertile, the filaments distinct or (not in Ceylon) more or less long-connate; anthers 2-thecate, longitudinally dehiscent; ovary superior, sessile, 2–6-celled, each
cell many-ovulate; style simple, the stigma lobulate. Capsules globose or elongated
and silique-like, loculicidally 2–6-valvate, the valves sometimes internally
transversely septate, the cells many-seeded; seeds small, pendulous or horizontal,

exalate, albuminous; embryo generally curved; cotyledons cordate, foliaceous.

A pantropical genus of about 60 species; only six species occur in Ceylon. *Corchorus olitorius* L. and *C. capsularis* L. are cultivated in many tropical countries for the important fibre, jute.

KEY TO THE SPECIES

1 Capsules depressed-globose, conspicuously longitudinally sulcate, coarsely verrucose, 8–13 mm long, 5- or 10-valvate...**1. C. capsularis**
1 Capsules linear-cylindrical or narrowly oblong-cylindrical
  2 Valves of the capsule transversely septate
    3 Capsules 3-valvate, deflected, more or less 3-angular to terete, 1.5–2.5 cm long; style about as long as the ovary; lowermost serrations of leaf blades not produced into caudae.......
    ....**2. C. urticifolius**
    3 Capsules 5-valvate, erect, longitudinally costellate, to 6.5 cm long; style less than 1/4 as long as the ovary; lowermost serrulations of leaf blades prolonged into deflexed caudae to 10 mm long
    ....**3. C. olitorius**
  2 Valves of the capsule without marked transversal septa; capsules 3-valvate, rarely 4-valvate
    4 Capsules longitudinally broadly 6(–8)-alate, each pair of wings prolonged into a conspicuous patent bifid appendage...**4. C. aestuans**
    4 Capsules without longitudinal wings
      5 Capsules c. 1.2 cm long, rather long-beaked, the beak to 2–2.5 mm long, densely downy; lowermost serrations of leaf blades not produced into caudae; sepals c. 2.5 mm long; stamens 5–10
      ....**5. C. fascicularis**
      5 Capsules 2–3.5 cm long, short-beaked, the beak with 3 spreading teeth, glabrous; lower-most serrations of leaf-blades prolonged into deflexed caudae to 4 mm long; sepals to 4 mm long; stamens 14–20...**6. C. trilocularis**

1. Corchorus capsularis L., Sp. Pl. 529. 1753; Moon, Cat. 42. 1824; Thw., Enum. Pl. Zeyl. 31. 1858; Masters in Hook. f., Fl. Br. Ind. 1: 397. 1874; Trimen, Handb. Fl. Ceylon 1: 181. 1893. Lectotype selected here: Icones, Plate 261 (BM-HERM).

Herb, annual, robust, erect, attaining 1–2 m, the stems glabrous. Leaves with the petióles to 2 cm long, puberulous adaxially, the stipules very narrowly ovate-filiform, to 8 mm long, caducous; blade narrowly ovate, to 12 cm long and 3.5 cm wide, truncate to rounded at the base, acuminate at the apex, serrulate-crenulate along the margins, the lowermost serrulations produced into deflexed caudae to 8 mm long, thin-chartaceous, glabrous, 3–5-nerved from the base, the venation prominulous beneath. Flowers oppositifolious, 2–3-fasciculate or sometimes flowers solitary, short-pedicellate, the pedicels glabrous; sepals narrowly oblong-obovate, cucullate at the apex and shortly apiculate, c. 3–3.5 mm long and 1 mm wide, minutely ciliolate along the margin near the base, the apiculus c. 0.5 mm long; petals narrowly obovate, shortly unguiculate, slightly longer than the sepals, c. 4 mm long and 1.5 mm wide, the claw minutely ciliolate; disc annulate, very short; stamens indefinite, c. 2.5–3 mm long, glabrous; ovary obovoid, truncate-emarginate at the apex, c. 1.2 mm long, glabrous, 5- or 10-locular, the style about

½ as long as the ovary, the stigma 2-lobulate. Capsules depressed-globose, emarginate at the apex, conspicuously longitudinally sulcate, coarsely verrucose, 8–13 mm long, glabrous, 5- to 10-valved, the valves without marked transversal septa; seeds indefinite, more or less rhomboid or cuneate, 2.5–3 mm long, dark brown.

Distr. Wild and cultivated in India and Pakistan and escaped as a weed in most tropical areas of the world.

Ecol. A weed of moist soils.

Specimens Examined. COLOMBO DISTRICT: Agric. School Colombo, May 1894, *Drieberg s.n.* (PDA). BATTICALOA DISTRICT: Thannamulai near a paddy field, at sea level, 8 Nov. 1969, fl, *Cramer 2749* (PDA, US); road from Maha Oya to Elwakumbura, close to paddy fields, 9 Sept. 1970, *Balakrishnan 375* (PDA, US). TRINCOMALEE DISTRICT: Southern edge of Kantale Tank, sunny weedy flood plain, 14 March 1973, fr, *Jayasuriya and Townsend⁻1187* (PDA, US). HAMBANTOTA DISTRICT: Ruhuna National Park, Block 2, at Walaskema at edge of rock pool, 31 July 1969, fr, *Cooray 69073118R* (PDA, US); near Meynet Wewa, 23 March 1970, fr, *Cooray 70032307R* (PDA, US). KALUTARA AND BADULLA DISTRICTS: *Thwaites C.P. 1105* (PDA, annotated Kalutara, *Moon* & Bintenna, *Gardner*).

2. Corchorus urticifolius Wight & Arn., Prod. 73. 1834; Thw., Enum. Pl. Zeyl. 31. 1858; Masters in Hook. f., Fl. Br. Ind. 1: 397. 1874; Trimen, Handb. Fl. Ceylon 1: 182. 1893; Matthew, Mat. Fl. Tamilnadu Carnatic 157. 1981; Matthew, Fl. Tamilnadu Carnatic III(1): 161, plate 14 f. 1983.

Herb, annual, 30–90 cm tall, branched or not, the stems sparsely and finely hirsute. Leaves slender-petiolate, the petiole 0.5–4 cm long, sparsely hirsute, the stipules narrowly triangular-filiform, c. 7 mm long, hirsute, long-persistent; blade ovate, truncate to rounded at the base, acute at the apex, serrate-dentate along the margins, the lowermost serrations not produced into caudae, to 7.5 cm long and 4.2 cm wide, thin-chartaceous, very sparsely appressed-puberulous to almost glabrous above, softly appressed-puberulous, especially along the veins beneath, usually 5-nerved from base, the costa and lateral veins prominulous beneath. Flowers oppositifolious, 2–3-fasciculate, yellow, shortly pedicellate, the pedicel to 1.5 mm long, puberulous; sepals narrowly oblong, apiculate, 5–5.5 mm long and 1–1.2 mm wide, hirtellous without, glabrous within, the apiculus c. 0.5–0.8 mm long; petals obovate, shortly unguiculate, rounded at the apex, c. 3.5 mm long and 1.2–1.4 mm wide, the claw ciliolate; disc annulate, very short; stamens many, to 3.5 mm long, glabrous; ovary narrowly oblong, c. 1.8–2 mm long, densely appressed-pilose, 3-locular, the style about as long as the ovary. Capsules deflected, linear-cylindric, more or less 3-angular to terete, beaked, 1.5–2.5 cm long and c. 3 mm wide, sparsely and finely hirsute, the beak erect, entire, to 4 mm long, 3-valvate, the valves with transverse septa; seeds many, angular, wrinkled, 1–1.3 mm long, black.

Distr. India-Ceylon.

Ecol. Lowland species from the dry zone.

Specimens Examined. PUTTALAM DISTRICT: Puttalam, Nov. 1881, *Nevill s.n.* (PDA). TRINCOMALEE DISTRICT: Trincomalee, *Thwaites C.P. 2758* (PDA). HAMBANTOTA DISTRICT: Ruhuna National Park, Patanagala, 14 Nov. 1969, *Cooray 69111411R* (US).

3. Corchorus olitorius L., Sp. Pl. 529. 1753; Moon, Cat. 42. 1824; Thw., Enum. Pl. Zeyl. 31. 1858; Masters in Hook. f., Fl. Br. Ind. 1: 397. 1874; Trimen, Handb. Fl. Ceylon 1: 182. 1893; Bennet, Fl. Howrah District 234.1979; Matthew, Mat. Fl. Tamilnadu Carnatic 156. 1981; Matthew, Ill. Fl. Tamilnadu Carnatic t. 85. 1982. Matthew, Fl. Tamilnadu Carnatic III(1): 160. 1983. Lectotype selected here: vol IV, Folius 41, left hand No 213 (BM-HERM).

Herb, often suffrutescent, erect or prostrate, 0.6–1.2 m tall, branched, the stems glabrous. Leaves with the petioles to 5 cm long, with 2 longitudinal rows of short hairs, the stipules very narrowly ovate-filiform, to 10 mm long, long-persistent; blade thin chartaceous, glabrous, rarely with a few appressed hairs along the veins beneath, narrowly ovate, truncate to more or less acute at the base, acuminate at the apex, serrulate along the margins, the 2 basal serrulations prolonged into deflexed caudae up to 10 mm long, to 12 cm long and 3.5 cm wide, 3–5-nerved from the base, the veins prominulous beneath. Flowers op-positifolious, solitary or 2–3-fasciculate, orange-yellow, the pedicels very short, c. 1 mm long, glabrous; sepals 5, narrowly obovate-oblong, the apex somewhat cucullate and apiculate (the apiculi to 1 mm long), c. 6 mm long and 1.5 mm wide, glabrous; petals 5, obovate, shortly unguiculate, c. 5 mm long and 2 mm wide, the claw ciliolate; disc annulate, shallowly 5-undulate, c. 0.5 mm long; stamens many, the filaments glabrous; ovary narrowly cylindrical, obscurely longitudinally 10-costellate, 2–2.5 mm long, densely puberulous, 5-locular, the style less than ¼ as long as the ovary, glabrous, the stigma 2-lobulate. Capsules erect, linear-cylindrical, longitudinally 10-costellate, tapering towards the apex and minutely 5-dentate at the apex, to 6.5 cm long and 0.5–0.7 cm in diam., almost glabrous, 5-valvate, the valves transversely septate; seeds many, angulate, to 2 mm long, black.

Distr. Originally wild probably in the Indo-Malesian region. One of the two species grown for jute and the mainstay of the export economy of Bangladesh. Now naturalized in many tropical parts of the world. In Ceylon found as a weed in the dry zone but apparently rather rare or overlooked by collectors. Thwaites (l.c.) considered it common in the hotter parts of the island. Trimen (l.c.) though calling it a common roadside weed, did actually not see much of it and we never encountered this species in the field though one of us (Robyns) was hunting for Tiliaceae.

Specimens Examined. POLONNARUWA DISTRICT: Minneriya, July 1848 (?), 5 Feb. 1858, *Gardner s.n.* (*C.P. 1107,* PDA). ANURADHAPURA DISTRICT: on a levee in a rice field, 5–6 miles northeast of Anuradhapura, 9 Dec. 1970, fl, *Fosberg and Balakrishnan 53449* (PDA, US). TRINCOMALEE DISTRICT: three miles West of Mahaweli Ganga Ferry Road, 23 Nov. 1975, *Sohmer and Sumithraarachchi 10792* (PDA, US). VAVUNIYA DISTRICT: near marker 120, Parayanalankulam, 80°11′ N and 8°45′ E on exposed grassy wayside, 24 Jan. 1972, fr, *Jayasuriya et al. 620* (PDA, US).

4. Corchorus aestuans L., Syst. Nat. ed. 10, 1079. 1759; Alston in Trimen, Handb. Fl. Ceylon 6: 33. 1931; Bennet, Fl. Howrah District 234. 1979; Matthew, Mat. Fl. Tamilnadu Carnatic 156. 1981; Matthew, Fl. Tamilnadu Carnatic III(1): 157. 1983. Lectotype selected here: 691.4 (LINN).

Corchorus acutangulus Lam., Enc. 2: 104. 1786; Moon, Cat. 42. 1824; Thw., Enum. Pl. Zeyl. 31. 1858; Masters in Hook. f., Fl. Br. Ind. 1: 398. 1874; Trimen, Handb. Fl. Ceylon 1: 183. 1893.

Herb, annual, sometimes suffrutescent, erect or prostrate, 30–60 cm tall, often much branched (from the base), the stems sparsely puberulous to glabrous. Leaves slender-petiolate, the petioles to 3 cm long, shorter towards the apex of the stems, puberulous, especially adaxially, the stipules narrowly ovate-filiform, 6–10 mm long, persistent; blade ovate, sometimes narrowly so, rounded at the base, acute at the apex, serrate along the margins, the 2 lowermost serrations prolonged or not into caudae to 3 mm long (in Ceylon collections, lowermost serrations seldom prolonged into caudae), to 6 cm long and 3.5 cm wide, often much smaller, thin-chartaceous, glabrous to infrequently sparsely appressed-puberulous above, sparsely appressed-puberulous especially along the veins beneath, 3–5-nerved from the base, the costa and lateral veins subprominulous beneath. Flowers oppositifolious, solitary or 2–3-fasciculate, yellow, short-pedicellate, the pedicel 1–1.5 mm long, elongated in fruit and up to 6 mm long; sepals narrowly oblong-obovate, apiculate, c. 4 mm long and 0.8 mm wide, glabrous, the apiculus c. 0.5 mm long; petals narrowly obovate, shortly unguiculate, longer than the sepals, to 5 mm long and c. 2 mm wide, the claw ciliolate, disc annular, very short; stamens indefinite, the filaments c. 4 mm long, glabrous; ovary narrowly oblong, c. 1.5 mm long, sparsely appressed-puberulous, usually 3-locular, infrequently 4-locular; style about as long as the ovary, glabrous. Capsules erect, narrowly oblong-cylindrical, longitudinally broadly 6(–8)-alate, each pair of wings prolonged into a conspicuous patent bifid appendage, 1.5–3.5 cm long, 3–5 mm wide, glabrous, 3(–4)-locular, the valves without marked transversal septa; seeds many, more or less rhomboid, c. 1.5 mm long, blackish.

Distr. Original area in Latin America and transported by man to other regions of the tropics.

Ecol. Weed; found in Ceylon in the dry zone.

Specimens Examined. JAFFNA DISTRICT: Jaffna, Feb. 1890, *s. coll. s.n.* (PDA); Karainagar, paddy field boundary, *Kundu and Balakrishnan 655* (PDA, US). ANURADHAPURA DISTRICT: Anuradhapura, Dec. 1881, *s. coll. s.n.* (PDA); Tissa wewa, along the shores of the tank, no date given, *Sohmer 8960* (PDA, US). POLONNARUWA DISTRICT: Rankoth Vehera sacred area, exposed open sandy soil with annual burning, 61 m alt., 26 Dec. 1970, fl, *Ripley 398* (PDA, US). TRINCOMALEE DISTRICT: Fishing village 14 miles NW. of Trincomalee, 4 miles SE. of Kuchchaveli Ferry, on sand flat of beach, 1 Dec. 1977, fl, *Fosberg 57090* (PDA, US). KURUNEGALA DISTRICT: Near Melsiripura, Kurunegala-Dambulla Road, 16 Nov. 1961, *Amaratunga 37* (PDA). PUTTALAM DISTRICT: Wilpattu National Park, near Eerige Ara confluence with Moderagama Ara, in open short grass with patches of spiny scrub (MD plot W. 14), 30 m alt., 29 Dec. 1968, *Fosberg et al. 50774* (PDA, US). MONERAGALA DISTRICT: Ekiriankumbura, Jan. 1888, *s. coll. s.n.* (PDA). HAMBANTOTA DISTRICT: Ruhuna National Park, Block 1, near Yala camp in forest scrub, vegetation on dune in plot 26, Jan. 1967, *Mueller-Dombois & Cooray 67013027* (PDA, US), 13 Dec. 1969, *Cooray 69121306R* (PDA, US); near Buttawa Bungalow in plot R38, 9 Jan. 1969, fr, *Mueller-Dombois et al. 69100907R* (PDA, US); Uraniya in short-grass cover over sand over claypit in plot R36, 5 Oct. 1969, fr, *Cooray et al. 69100531* (PDA, US); Patanagala, near camp, 27 Nov. 1969, fr, *Cooray 69112705R* (PDA, US); Block I between Andunoruwa and Komawa Wewa in plot R13, in deciduous scrub on reddish brown earth, 7 Jan. 1969, *Mueller-Dombois 69010722 and 69010726* (PDA, US); Rakina Wewa, near Gonalabbe Lewaya, in open tall scrub, 2 m alt., 5 Apr. 1968, fr, *Fosberg 50261* (PDA, US). DISTRICT UNKNOWN: *Thwaites C.P. 1106* (PDA, US) (annotated Jaffna 1846, Colombo, Batticaloa); s. loc., *Alston A25* (PDA).

5. Corchorus fascicularis Lam., Enc. 2: 104. 1786; Thw., Enum. Pl. Zeyl. 40. 1864; Masters in Hook. f., Fl. Br. Ind. 1: 398. 1874; Trimen, Handb. Fl. Ceylon 1: 183. 1893; Matthew, Mat. Fl. Tamilnadu Carnatic 156. 1981; Matthew, Fl. Tamilnadu Carnatic III(1): 158, Plate 14 b. 1983. Type at Paris ?

Herb, suffrutescent, annual, erect or prostrate, to 50 cm tall, branched, the stems glabrous or almost so. Leaves with the petioles slender, to 10 mm long, puberulous adaxially, the stipules narrowly ovate, acuminate, c. 3 mm long, long-persistent; blade ovate to usually narrowly ovate, rounded to more or less acute at the base, acute at the apex, serrulate-crenulate along the margins, the lowermost serrulations not prolonged into caudae, to 5 cm long and 1–2 cm wide, thin-chartaceous, glabrous, usually 3-nerved from the base, the costa prominulous beneath. Flowers oppositifolious, 2–5-fasciculate, almost sessile, the pedicel less than 1 mm long, slightly longer in fruit, glabrous; sepals narrowly ovate, c. 2.5 mm long; petals narrowly obovate, c. 2–2.5 mm long; disc very short; stamens 5–10, shorter than the sepals; ovary narrowly oblong-cylindrical, longitudinally 3–6-costellate, densely puberulous, 3-locular, the style shorter than the ovary, glabrous. Capsules erect, narrowly oblong, rather long-beaked, c. 1.2 cm long and 2 mm wide, the beak to 2–2.5 mm long, densely downy, 3-valved, the valves without conspicuous transverse septa; seeds many, rhomboid, to 1.5 mm long, blackish.

Distr. Tropical Africa, India, Ceylon, Australia.

Specimens Examined. JAFFNA DISTRICT: Near Jaffna Fort, 14 Jan. 1974, *Wambeek s.n.* (US). COLOMBO DISTRICT: Near Colombo, Sep. 1862, *Ferguson s.n.* (*C.P. 3773*) (PDA). HAMBANTOTA DISTRICT: Ruhuna National Park, Andunoruwa Vela (Wewa) just west of Yala Bungalow, Mueller-Dombois Plot R3, in openings between scrub patches, on stiff dark clay soil, 7 Jan. 1969, fr, *Fosberg et al. 51154* (US, PDA).

6. Corchorus trilocularis L., Syst. Nat. ed. 12, II: 369. 1767. Type: 6124 HU (Hortus Upsalensis) (LINN).

Corchorus tridens sensu auct. non L., Mant. 2, Appendix 566. 1771; Thw., Enum. Pl. Zeyl. 401. 1864; Masters in Hook. f., Fl. Br. Ind. 1: 398. 1874; Trimen, Handb. Fl. Ceylon 1: 183. 1893.

Since no type of *C. tridens* L. could be found in the Linnaean herbarium in London, we have to refer to the citation in Mantissa. Linnaeus described this species there clearly as having bifid divaricate style horns. This character has been illustrated by Manuel Martinez in his unpublished Masters Thesis, School of Biol. Sc., University of Kentucky (1981), for collections identified under this species from the Caribbean and tropical South America. The citation of Linnaeus of Plates in Plukenet's Phytographica and Burman's Flora Indica has caused the mistaken idea that *C. tridens* occurs in India and Ceylon. If we base the interpretation of this species solely on the description by Linnaeus and not on his citation of two plates then the concept of the other authors cited above refers to the African-Asian species *Corchorus trilocularis* L., cited in the original description only from a Forsskal collection from Arabia and grown by Linnaeus in Hortus Upsalensis according to his citation and as written on the type, *6124* in the Linnaean herbarium, London. This old record from Ceylon fits also with the ecology and distribution of this species, one of the desert plants which are all the way distributed from Africa through Arabia via India to the dry zone of Ceylon like *Salvadora* and *Grewia tenax*.

Herb, annual, erect, 20–60 cm tall, slightly branched, the stems glabrous. Leaves short-petiolate, the petiole to 7–8 mm long, puberulous adaxially, the stipules narrowly ovate-setuliform, to 6 mm long, glabrous, long-persistent; blade narrowly ovate, more or less truncate at the base, acute at the apex, serrate-crenate along the margins, the lowermost serrations prolonged into deflexed caudae to 4 mm long, to 6 cm long and 1.2 cm wide, thin-chartaceous, glabrous, 3–5-nerved from the base, the costa and lateral veins prominulous beneath. Flowers oppositifolious, solitary or 2–4-fasciculate, yellow, subsessile, the pedicels c. 0.5 mm long, glabrous; sepals narrowly elliptic to narrowly ovate, apiculate, to 4 mm long, glabrous; petals narrowly obovate, shortly unguiculate, to 4 mm long; disc annulate, very short; stamens indefinite, the filaments glabrous; ovary narrowly cylindrical, obsoletely longitudinally 3–6-costellate, puberulous, 3-locular, rarely 4-locular, the style shorter than the ovary, glabrous. Capsules erect, linear-cylindrical, longitudinally striate, short-beaked apically, the beak with 3 spreading teeth, 2–3.5 cm long and 0.2 cm in diam., glabrous, 3-valvate, rarely 4-valvate, the valves without transversal septa; seeds many, narrowly oblong, more or less truncate at both ends, to 1.8 mm long and 0.5–0.7 mm in diam., blackish.

Distr. Tropical Africa, Arabia, tropical Asia and Australia; found in Ceylon in the dry zone; apparently very rare or overlooked.

Specimen Examined and identified as *C. tridens* L. by Robyns: TRINCOMALEE DISTRICT: Trincomalee, Nov. 1863, *Glenie (C.P. 3791)* (PDA).

4. BERRYA

Roxb., Pl. Corom. 3: 60, f. 264. 1820, correction A.P. de Candolle, Prod. 1: 517. 1824 (nom. cons. against *Berria* Roxb.) non *Berrya* Knowlton, Prof. Papers U.S. Geol. Survey 155: 133. 1930; also Roxb., Hort. Beng. 42. 1842; Maza, Anales Soc. Esp. Hist. Nat. 19: 215. 1890; Benth. and Hook. f., Gen. Pl. 1: 232. 1862; Baill., Hist. Pl. 4: 184. 1873; Schum. in Pflanzenfam. III. 6: 16. 1895; Hutch., Gen. Fl. Pl. 2: 492. 1967; Kostermans, Reinwardtia 7: 423–424. 1969; Backer and Bakh. f., Fl. Java 1: 389. 1963. Type: *Berrya cordifolia* (Willd.) Burret (syn. *B. ammonilla* Roxb.).

Carpodiptera Griseb., Mem. Amer. Acad. Arts Series 2, 8: 163. 1861; Benth. and Hook. f., Gen. Pl. 1: 232. 1862; Baill., Hist. Pl. 4: 185. 1873; Masters, Fl. Trop. Africa 1: 241. 1868; Schum. in Pflanzenfam. III, 6: 15. 1895; Wild, Fl. Zambesiaca 2: 33, f. 1. 1963; Capuron, Adansonia NS III(1): 97–98. 1963; Hutch., Gen. Fl. Pl. 2: 493. 1967. Type: *Carpodiptera cubensis* Griseb. in Mem. Amer. Acad. Arts Series 2, 8: 164. 1861.
Espera Willd., Ges. Naturf. Freunde Berlin Neue Scriften 3: 449. 1801.
Hexagonotheca Turcz., Bull. Soc. Imp. Naturalistes Moscou 19(2): 505. 1846.
Pterocoellion Turcz., Bull. Soc. Imp. Naturalistes Moscou 36: 572. 1863.

Trees. Leaves alternate, petiolate, stipulate, the stipules small and caducous, the blade symmetrical to sometimes slightly asymmetrical, entire-margined, cordate to subcordate at the base, 5–7-nerved from the base. Inflorescences terminal and axillary, paniculate (sometimes axillary cymose ?). Flowers hermaphrodite, actinomorphous, pedicellate; calyx campanulate to cupulate, irregularly 3–5-lobed, persistent; petals 5, infrequently 3, 4 or 6, free, glandless, persistent; gonophore present or absent; stamens indefinite, free to hardly connate at the base, filaments filiform and persistent, the anthers dorsifixed, 2-thecate, longitudinally dehiscent; ovary superior, sessile, 3–5-celled, with 1–6 ovules in each cell, style subulate, the stigma inconspicuously lobed. Capsules surrounded at the base by persistent calyx, corolla and filaments, with 3–5 pairs of longitudinal erect to widely patent large wings, loculicidally 3–5-valvate, the cells 1–4-seeded; seeds densely bristly; endosperm copious; cotyledons foliaceous.

Five species in the Indo-Malesian and Pacific area. One species in Ceylon.

Berrya cordifolia (Willd.) Burret, Notizbl. Bot. Gart. Berlin-Dahlem 9: 606. 1926; Alston in Trimen, Handb. Fl. Ceylon 6: 32. 1931; Worthington, Ceylon Trees 89. 1959.

Espera cordifolia Willd., Ges. Naturf. Freunde Berlin, Neue Schriften 3: 449. 1801.
Berrya ammonilla Roxb. [Hort. Beng. 42. 1814, nom. nud.] Pl. Corom. 3: 60. t. 264. 1819; Moon, Cat. Add. 1824; Thw., Enum. Pl. Zeyl. 32. 1858; Masters in Hook. f., Fl. Br. Ind. 1: 383. 1874; Trimen, Handb. Fl. Ceylon 1: 173. 1893.

Tall tree to 25 m in height, the trunk straight, terete, to 40 cm in diam., the bark smooth, greyish brown, the branchlets glabrescent. Leaves long-petiolate, the petiole slender, 1.5–7 cm long, sparsely stellate-puberulous near the apex, glabrescent, the stipules subulate, 4–7 mm long, stellate-puberulous, caducous; blade symmetrical or sometimes somewhat asymmetrical, broadly ovate to broadly elliptic, cordate at the base, shortly and acutely or obtusely acuminate at the apex, entire-margined, 7–23 cm long and 5–16 cm wide, thin-chartaceous, pale green, somewhat shining and glabrous above, with tufts of hairs in the primary nerve-axils beneath, 5–7-nerved from the base, the costa and lateral veins prominent beneath. Panicles terminal and axillary, usually large, multi- and laxiflorous, the rachis glabrescent, the lateral axes from stellate-puberulous to stellate-tomentellous, the bracts narrowly ovate, to 8 mm long at the base of the inflorescence, gradually smaller towards the apex. Flowers with the pedicels slender, 10–12 mm long, stellate-tomentellous; calyx campanulate, irregularly 3–4-lobate, 4–5.5 mm long, minutely stellate-tomentellous without, glabrous within, the lobes acute to obtuse, to c. 2 mm long; petals 5, white, obovate, sometimes narrowly so, emarginate apically, to 8 mm long and 3.5 mm wide, glabrous; gonophore none; stamens indefinite, the filaments hardly united at the base, of unequal length, to 5 mm long, glabrous; ovary longitudinally 3-alate, sometimes 4-alate, densely tomentellous, 3–(4–)-celled, each cell 4-ovulate, the style slender, to 4 mm long, the stigma capitellate. Capsules surrounded at the base by the persistent calyx, corolla and stamens, largely 6-alate, infrequently 8-alate, the body stellate-tomentellous, the wings chartaceous, asymmetrical, rather acute apically, to 2.5 cm long and 1.3 cm wide, green, becoming red, stellate-puberulous, 3(–4)-valved; seeds 1–2 in each cell, angular, to 4.5 mm long, provided with numerous, long, tawny, caducous bristles.

Distr. Also Southern India, Andamans, Java, Burma.

Ecol. Forest, often alluvial, of the dry zone, rarely extending into the intermediate and moist low country.

Uses. A tree excellently suited for reforestation in the dry and intermediate zones, but being of local origin so far not much valued by local foresters or foreign "experts". The wood, known as Trincomalee wood, is pale red-yellow, pliable and easily worked; it is universally useful, especially for boats, bent work, beams, etc.

Vern. Halmilla (S); Chavandalai (T).

Specimens Examined. JAFFNA DISTRICT: Vavuniya-Jaffna, 21 June 1973, fl, *Kostermans 25114* (PDA, US). PUTTALAM DISTRICT: Wilpattu Natl. Park, Malimaduwa, between Magul Illaimá and Malimaduwa, 13 July 1969, fl, *Wirawan et al. 1132* (PDA, US); North of Puttalam, dry zone, 24 July 1974, fr, *Kostermans 25248* (PDA, US). RATNAPURA DISTRICT: Illukwatte, along road to Carney, about 1.5 miles from Gilimale, 90 m, 15 Feb. 1969, fr, *Robyns 6919* (PDA, US). MATALE DISTRICT: 8 miles east of Naula, 19 Aug. 1978, fr, *Huber 780* (PDA, US); Sigiriya, 3 Oct. 1969, *Reitz 30023* (PDA, US). TRINCOMALEE DISTRICT: About 7 miles SW. of Trinco along road from Kantalai, Ceylon Forest College, Monkey Bridge, in remnants of forest, 0–30 m, 23 Feb. 1969, fr, *Robyns 6936* (PDA, US). POLONNARUWA DISTRICT: Angamedilla, N. of Wasgamuwa Strict Nature Reserve, 10 Sep. 1977, fr, *Meijer and Jayasuriya 1308* (PDA, US); sacred area, section 2B, east, sandy soil, 61 m, 29 Nov. 1970, *Ripley 226* (PDA, US), 30 May 1971, fl, *Kostermans 24311* (PDA, US). ANURADHAPURA DISTRICT: Ritigale Hill, primary forest, 12 Jan. 1974, fr, *Waas 353* (PDA, US), 9, Oct. 1975, fr, *Bernardi 15273* (PDA, US). HAMBANTOTA DISTRICT: Ruhuna Natl. Park, Kumbukkan Oya, main ford, at rivers edge in high forest, 1–2 m, 19 June 1967, *Comanor 379* (PDA, US); Menik Ganga, one mile north of Waran corner, 20 m, 22 June 1967, *Mueller-Dombois 67062205* (PDA, US); Block 1, alluvial forest along Menik Ganga, 4 Apr. 1969, fr, *Robyns 6994* (PDA, US), 29 June 1970, fl, *Meijer 228* (PDA, US); Tissamaharama, road from resthouse to Kiri Vihare, 24 May 1968, fl. and fr, *Cooray 68052401R* (PDA, US). JAFFNA, KURUNEGALA & TRINCOMALEE DISTRICTS: annotated Jaffna, Kurunegala, *Gardner,* Trincomalee, *Glenie, Thwaites, C.P. 1108* (PDA).

5. DIPLODISCUS

Turcz., Bull. Soc. Imp. Naturlistes Moscou 31(1): 235. 1858; Kosterm., Reinwardtia 5: 256. 1960 & 5: 372. 1960. Type species: *D. paniculatus* Turcz. (Philippines).

Pityranthe Thw., Enum. Pl. Zeyl., 29. 1858; Masters in Hook. f., Fl. Br. Ind. 1: 382, 1874; Trimen, Handb. Fl. Ceylon 1: 173. 1893.

Trees, the branchlets lepidote. Leaves alternate, petiolate, the blade symmetrical or not, entire-margined, sometimes with small protruding glands along the margins, penninerved. Inflorescences terminal or axillary, paniculate. Flowers hermaphrodite, actinomorphic, pedicellate; calyx campanulate or urceolate-campanulate, 3–5-lobed, persistent; petals 5, free, spathulate to obovate, longer than the calyx; gonophore absent; stamens ∞, free to hardly connate at base, the anthers versatile, 2-thecate, the thecae divergent at base; staminodes 5, forming an inner row inside the androecium, foliaceous; ovary superior, sessile, 5-celled, each cell 2-ovulate; style simple or 3–5-branched at the apex; stigma inconspicuous. Capsules surrounded at the base by the persistent calyx and filaments, longitudinally 5-costate to conspicuously 5-angulate-lobate, loculicidally valvate, the cells 1–5, each with one pendulous seed, the separating septa thin-chartaceous; seed (in Ceylon sp.) stellate-lepidote.

A tropical Asian genus of seven species distributed in Ceylon, the Malay Peninsula, the Philippines and in Borneo; only one endemic species occurs in Ceylon.

Diplodiscus verrucosus (Thw.) Kosterm., Reinwardtia 5: 372. 1960.

Pityranthe verrucosa Thw., Enum. Pl. Zeyl. 29. 1859; Masters in Hook. f., Fl. Br. Ind. 1: 382. 1874; Trimen, Handb. Fl. Ceylon 1: 172. 1893; Worthington, Ceylon Trees 88. 1959. Type: Ceylon, *Thwaites s.n.* (*C.P. 1144*).

Tree to 10 m tall, the trunk straight, to 60–75 cm in girth, the bark smooth, greyish, thin, the branchlets scabrous, lepidote, the scales small and fimbrillate. Leaves with the petiole 1–2.2 cm long, densely lepidote, the scales fimbriate (sometimes almost also stellate-lepidote); blade ovate to elliptic, sometimes narrowly so, obtuse to rounded or sometimes subcordate at base, rounded to obtuse to sometimes acute at the apex, the margins more or less repand, remotely and minutely glandular, to 17 cm long and 9 cm wide, rigid-chartaceous, somewhat shining and sparsely stellate-tomentellous beneath, the costa prominent and the lateral veins prominulous beneath. Panicles terminal and axillary, erect, large, much-branched, multiflorous, the axes densely lepidote, the scales fimbrillate, the bracts caducous. Flowers with the pedicels to 5 mm long, densely lepidote, the scales fimbrillate; calyx urceolate-campanulate, 5-lobed, c. 3–3.5 mm long, densely and minutely lepidote outside, the upper minutely tomentellous inside, persistent, the lobes acute, c. 1 mm long; petals 5, pink-white (acc. to Worthington l.c.), narrowly obovate, tapering toward base, about twice as long as the calyx (acc. to Trimen l.c., and Kostermans l.c.), glabrous; stamens 15, the filaments rather thick, almost free, to 3 mm long, glabrous; staminodes 5, appressed against the ovary, linear-spatulate, shorter than the stamens, 1.5–2 mm long; ovary subglabrous, more or less 5-angulate, c. 1 mm in diam., densely and minutely lepidote, the scales fimbrillate, 5-celled, each cell 2-ovulate; style to 1.5 mm long, glabrous, the stigma inconspicuous. Capsules surrounded at the base by the persistent calyx and filaments, obovoid, conspicuously 5-angulate-lobate, to 2.3 cm long and 1.6 cm wide, yellowish brown when fresh, ochraceous when dry, subligneous, verrucose, densely lepidote, loculicidally 5-valvate; seeds usually 1 per locule, pendulous, subglobose, c. 5 mm long, the testa blackish, covered with numerous stellate scales.

Distr. Endemic.

Ecol. Dry zone, in monsoon forest, locally very common.

Uses. According to Worthington (l.c.) the timber is tough but of small size; it sinks instantly in water and is used for anchors and axles.

Vern. Dik-wenna, Dik-andhe (S); Vid-pani, Yakada-maran (T).

Specimens Examined. ANURADHAPURA DISTRICT: Korapotana (Horowupotana ?), Aug. 1885, *s. coll. s.n.* (PDA); Madawachchiya, Feb. 1890, *s. coll. s.n.* (Sayaneris coll.) (PDA). POLONNARUWA DISTRICT: along road Habarane to Kantalai at Gal Oya junction, in forest, very common, Feb. 1969, *Robyns 6935* (BR, CANB, E, PDA, US); Mineriya, Sept. 1885, *s. coll. s.n.* (PDA). BATTICALOA DISTRICT: s. loc., *A.M. Walker s.n.* (PDA). BADULLA DISTRICT: Ekiriyankumbura, along road Padiyatalawa to Bibile, in forest, April 1969, *Robyns 6984* (BR, CANB, E, PDA, US). MONERAGALA DISTRICT: near Bibile, Jan. 1888, *s. coll. s.n.* (PDA). DISTRICT UNKNOWN: *Thwaites C.P. 1144* (2 sheets, one sheet annotated 'Batticaloa Distr., Feb. 1858, the other sheet annotated 'Jaffna, Trincomalee, *Gardner*', PDA).

6. TRIUMFETTA

L., Sp. Pl. 444. 1753; L., Gen. Pl. ed. 5, 203. 1754. Type species: *T. lappula* L.

Herbs, shrubs, or (not in Ceylon) trees, the pubescence usually of stellate hairs. Leaves alternate, petiolate, stipulate, the blade entire or 3–5-lobed or rarely (not in Ceylon) 5–7-partite, the margins usually serrate, 3–7-nerved from the base. Inflorescences axillary or oppositifolious, rarely terminal, the cymules usually 3-flowered, clustered, often collected into terminal, interrupted spiciform inflorescences. Flowers actinomorphous, pedicellate, generally 5-merous; sepals valvate, free, with apical appendages; petals sometimes absent, free, imbricate, shorter than the sepals, usually ciliate basally, glandless; gonophore short, sometimes obsolete, provided with 5 glands opposite the petals, sometimes glandless, usually crowned by a ciliate membranous urceolus surrounding the stamens; stamens 5–many, the filaments free, the anthers 2-thecate, introrse, more or less medifixed, longitudinally dehiscent; ovary superior, usually sessile, setose, sometimes tuberculate, 2–5-celled, the cells 2-ovulate; style filiform; stigma entire or shortly 2–5-parted. Fruit capsular, globose or ellipsoid, tuberculate, or bristly, or (in Ceylon) covered with numerous spines topped by hyaline, straight to uncinate spinules, the spines glabrous or not, indehiscent or loculicidally dehiscent, 2–5-celled, the cells usually 2-seeded, sometimes (not in Ceylon) falsely 6–10-celled, the cells 1-seeded, exceptionally 1-celled and 1-seeded by abortion; seeds pendulous, glabrous, albuminous; embryo straight; cotyledons flat, foliaceous.

A large pantropical genus of about 150 species, some of them being pantropical weeds; four species are reported from Ceylon.

KEY TO THE SPECIES

1 Fruits more than 1 cm across, dehiscent when completely ripe, hyaline spines patent hairy; leaves with rather long stalks, never lobed; plants pilose-tomentose hairy all over; flowers not cucullate at apex, stamens 8–10..**1. T. pilosa**
1 Fruits less than 1 cm across, not dehiscent, spines hairy or not; leaf blades, at least the lower ones, often 3-lobed; plants less dense pilose-tomentose hairy; flowers in general cucullate at apex
  2 Lower face of leaves with a dense indumentum of tiny stellate hairs, greyish-whitish to pale ochre, fruits more or less ovoid, flowers with sepals about 7–7.5 mm, stamens 20....**3. T. glabra**
  2 Lower face of leaves with much larger pilose appressed hairs; stamens 5–15
    3 Sepals 5.5–6 mm long, stamens 15; fruits more or less globose, hairs only on the fruit body
    ....**2. T. rhomboidea**
    3 Sepals 2–3.5 mm long, stamens 5–10; fruits more or less ovoid, with a pointed apex, hyaline spines with hairs more or less arranged like combs..................**4. T. pentandra**

1. Triumfetta pilosa Roth, Nov. Pl. Sp. 223. 1821; Thw., Enum. Pl. Zeyl. 31. 1858; Masters in Hook. f., Fl. Br. Ind. 1: 394. 1874; Trimen, Handb. Fl. Ceylon 1: 179. 1893; Matthew, Mat. Fl. Tamilnadu Carnatic t. 87. 1982; Matthew, Fl. Tamilnadu Carnatic III(1): 176. 1983.

Triumfetta tomentosa Bojer, Ann. Sci. Nat. Bot. ser. 2, 20: 103. 1843; Thw., Enum. Fl. Zeyl. 31. 1858; Masters in Hook. f., Fl. Br. Ind. 1: 394. 1874; Trimen, Handb. Fl. Ceylon 1: 179. 1893.

Herb, perennial, erect, branched, the branches more or less densely and softly hirsute. Leaves with the petioles to 5 cm long, gradually shorter towards the apex of the branches, more or less densely and softly hirsute, the stipules subulate, 5–6 mm long, hirsute; blade ovate to narrowly so, cordate to truncate to rounded basally, acute to acuminate at the apex, serrate-margined, to 12 cm long and 6–7 cm wide, gradually smaller towards the apex of the branches, thin-chartaceous, more or less densely stellate-pubescent above, softly stellate-tomentose beneath, 3–5(–7)-nerved from the base, the costa and lateral veins prominent especially near the base beneath. Flowers yellow or orange, in oppositifolious and axillary clustered cymules, the cymules usually 3-flowered, the peduncles to 10 mm long, the pedicels 3–4 mm long, the bracts c. 2.5 mm long, the peduncles, pedicels and bracts shortly hirsute; sepals 5, linear-lanceolate, subcucullate and acute apically, to 10–11 mm long and c. 1 mm wide, the appendages less than 1 mm long, sparsely short-stellate-hirsute without; petals narrowly oblong-obovate, c. 6–7 mm long and 1.3–1.8 mm wide, the claw ciliate; gonophore short, c. 0.5 mm long, the urceolus c. 0.2–0.3 mm long, ciliate; stamens 10, the filaments to 8 mm long, glabrous, the anthers oblong-elliptic, c. 0.8 mm long; ovary subglobose, c.1 mm in diam., densely spinulose. Fruit depressed-globose, dehiscent, the body c. 4 mm long, glabrous or hirsute, the spines 7–8 mm long, hirsute throughout or only on the lower part, the apex of the spines and the spinules uncinate, or rarely straight, 4-locular, each locule 2-seeded; seeds c. 2.5 mm long.

Distr. Throughout India, Ceylon, Malay Peninsula, and tropical Africa.
Ecol. Roadside weed in mid- and up-country.
Note. After study of many specimens from Africa and Indo-Malesia we are convinced that *T. tomentosa* is nothing else than a freak form of *T. pilosa*, with fruit spines not or hardly curved. Though this character plays a role in sectional differences in the genus its taxonomic value breaks down in this species. The "*tomentosa*" form was apparently common around Peradeniya around the middle of last century.

Specimens Examined. MATALE DISTRICT: Elkaduwa Road, Dec. 1931, *Simpson 8963* (PDA). KANDY DISTRICT: Gampola, Dec. 1963, *Amaratunga 760* (PDA); Peradeniya, river bank, Jan. 1928, *Alston 1992* (PDA); Royal Bot. Garden, *Thwaites C.P. 2902* (PDA as cited by Thwaites l.c.), Jan. 1858, *Macrae* (PDA), Nov. 1925, *J.M. Silva s.n.* (PDA); Hewaheta-Talatuoya Road, in jungle, 19 Feb. 1974, *Sumithraarachchi 121* (PDA, US); Madulkele, in secondary forest, 1200 m, 17 Jan. 1975, fr, *Waas 942* (PDA, US); 3 miles NE. of Madugoda, 80°54′ N and 7°20′ E, 825 m, 1 June 1972, fl, *Jayasuriya 482* (PDA, US). MONERAGALA DISTRICT: Gamawella, c. 9 miles from Passara along road to Moneragala, on steep banks of road, 300–6000 m, 3 Apr. 1969, fr, *Robyns 6986* (PDA, US). KALUTARA DISTRICT: Welipenna, scrubland, 28 Jan. 1972, fr, *Balakrishnan 1172* (PDA, US). BADULLA DISTRICT: Bandarawela, along a forest path, 1200 m, 14 Feb. 1981, *Pittoni s.n.* (US); Welimada, scrubland, 12 March 1971, 1800 m, *Balakrishnan 444* (PDA, US); between Ohiya and Boralanda, roadside weed, 1500 m, 13 March 1969, fr, *Hoogland 11560* (PDA, BR, E, US); Keppetipola-Boralanda Road, in damp embankment, 1200 m, 2 March 1975, fl, *Cramer 4433* (PDA, US). NUWARA ELIYA DISTRICT: Open roadside of Hakgala Rd., close to RC church, 20 June 1974, fl, *Cramer 4235* (PDA, US); Boragas, 3 Nov. 1971, *Cramer 3513* (PDA, US); 63rd mile post on Nuwara Eliya to Welimada Road, open scrubland, 30 March 1970, fr, *s. coll. s.n.* (PDA, US); Horton Plains to Ohiya, 2070 m, 6 Feb. 1971, fl, *Robyns 7146* (PDA, US); Fort Macdonald valley, patana, March 1906, *A.M. Silva s.n.* (PDA); Hakgala, Apr. 1906, *A.M. Silva s.n.* (PDA), *Willis s.n.* (PDA). KEGALLE

DISTRICT: Mawanella, Feb. 1881, *s. coll. s.n.* (PDA). RATNAPURA DISTRICT: Between Udakarawita and Kalawana, culvert 19/4, 8 Feb. 1970, *Cooray 70020805R* (PDA, US). DISTRICT UNKNOWN: *G. Thomson s.n.* (US); *Thwaites C.P. 730* (PDA).

2. Triumfetta rhomboidea Jacq., Enum. Pl. Carib. 22. 1760; Jacq., Select. 47. 1763; Masters in Hook. f., Fl. Br. Ind. 1: 395. 1874; Trimen, Handb. Fl. Ceylon 1: 179. 1893; Bennet, Fl. Howrah District 232. 1979; Matthew, Mat. Fl. Tamilnadu Carnatic 159. 1981; Matthew, Fl. Tamilnadu Carnatic III(1): 176. 1983.

Triumfetta bartramia L. nom illegit.; Alston in Trimen, Handb. Fl. Ceylon 6: 33. 1931.
Triumfetta angulata Lam. sensu Thw., Enum. Pl. Zeyl. 31. 1858.

Herb, often woody at the base, or shrub to 1.2 m tall, the branchlets sparsely pubescent, becoming glabrous. Leaves slender-petiolate, the petioles of the basal leaves up to 5.5 cm long, gradually shorter towards the apex of the branchlets, pubescent, the stipules c. 4 mm long, early caducous; blade of the lower leaves deeply 3-lobed, that of the higher ones less deeply lobed to unlobed, from more or less rounded to narrowly ovate, the apex (of the blade or lobes) acuminate to acute, the base subcordate to rounded-cuneate to obtuse, the margins irregularly and often more or less bluntly serrate, to 7 cm long and 6 cm wide, gradually smaller towards the apex of the branchlets, thin-chartaceous, the upper surface sparsely short-pubescent, the lower surface softly pubescent, 3–5-nerved from the base, the costa and lateral veins prominulous beneath. Flowers yellow or orange, in axillary 3–5-clustered cymules, the peduncles 1–2 mm long, the pedicels c. 1–1.5 mm long, articulate in the middle, the bracts narrowly ovate, to 2.5 mm long, the peduncles, pedicels and bracts sparsely puberulous; sepals 5, linear, cucullate and acute apically, c. 5.5–6 mm long and c. 0.8–1 mm wide, the appendages c. 0.5 mm long or less, stellate-puberulous dorsally, glabrous within; petals 5, narrowly oblong-obovate, rounded apically, c. 4.5–5 mm long and 1–1.2 mm wide, the claw ciliate; gonophore short, hardly 0.2 mm long, with 5 distinct rotund glands, the urceolus undulate, c. 0.2 mm long, ciliate; stamens 10, the filaments 4–5.5 mm long, the anthers c. 0.8 mm long; ovary subglobose, less than 1 mm in diameter, densely setose, 2-locular, each locule 2-ovulate; style filiform, c. 4 mm long, glabrous, the stigma 2-lobulate. Fruit more or less globose, indehiscent, the body c. 3 mm in diameter, cinereous-tomentellous, the spines c. 1.5 mm long, glabrous, the spinules uncinate.

Distr. A pantropical weed distributed around harbours and weedy places all around the world.

Ecol. A weed of roadsides and other ruderal areas.

Specimens Examined. KANDY DISTRICT: Gelioya-Galaha, roadside,

24 Dec. 1968, fr, *Wirawan 757* (PDA, US); Peradeniya below University Circuit Bungalow, on footpath to creek, 890 m, 9 Nov. 1967, fl, *Mueller-Dombois 67110928* (PDA, US), 5 July 1975, *Jayasuriya and Pemadasa 2216* (PDA, US); Peradeniya, Nov. 1854, *Thwaites C.P. 2759* (PDA); in waste ground, 14 Dec. 1925, *Alston 1109* (PDA). MONERAGALA DISTRICT: Gamawella c. 9 miles from Passara along road to Moneragala, 300–600 m, on steep banks along road, growing in rock crevices, Apr. 1969, *Robyns 6989* (BR, CANB, E, PDA, US). GALLE DISTRICT: Akmimana, in waste place, March 1927, *Alston 110* (PDA).

3. Triumfetta glabra Rottler in Spreng. in L., Syst. Veg. ed. 2. 450. 1825; Masters in Hook. f., Fl. Br. Ind. 1: 395. 1874; Trimen, Handb. Fl. Ceylon 1: 180. 1893.

Triumfetta conspicua Trimen, J. Bot. 23: 206. 1885.

Suffrutex to 1.5 m tall, much-branched, the branchlets stellate-tomentose to stellate-pubescent. Leaves slender-petiolate, the petioles 1.5–4.5 cm long, from stellate-tomentellous to stellate-puberulous, the stipules narrowly ovate, acuminate, to 6 mm long, stellate-puberulous outside; blade subcircular to broadly ovate, slightly 3-lobed to unlobed, subcordate to truncate or rounded at the base, acute at the apex, the margins irregularly crenate-dentate, to 6.5 cm long and wide, thin-chartaceous, slightly discolorous (at least when dry), the upper surface sparsely to very sparsely stellate-puberulous, the lower surface softly stellate-tomentose, usually 5-nerved from the base, the costa and lateral veins prominent beneath. Flowers yellow, the cymules usually 3-flowered and clustered among the leaflets of the erect upper part of the branchlets, the peduncles and articulate pedicels short, shortly stellate-tomentose, the bracts longer than the pedicels, to 2.5 mm long, stellate-tomentellous; sepals 5, linear to linear-obovate, acute and cucullate apically, c. 7.5 mm long and 1 mm wide, stellate-tomentellous to stellate-puberulous outside, glabrous inside, the appendages c. 1 mm long; petals 5, broadly spatulate, rounded apically, unguiculate basally, c. as long or slightly shorter than the sepals, the claw ciliolate; gonophore c. 0.5 mm long, with 5 distinct subrotund glands, the urceolus c. 0.3–0.4 mm long, finely ciliolate; stamens 20, the filaments c. 6–6.5 mm long, glabrous, the anthers c. 0.6 mm long; ovary broadly ovoid, c. 0.8 mm long, densely setose, 4-locular, each locule 1-ovulate; style slender, to 5 mm long, glabrous. Fruit broadly ovoid, the body 4–5 mm long and 3–5 mm wide, fulvous-tomentellous, usually 2-celled and 2-seeded, the spines 2–4 mm long (the upper ones are the longest), shortly and finely hirsute on the lower part, the spinules uncinate; seeds broadly pyriform, c. 2–3 mm long.

Distr. Ceylon endemic.

Ecol. A plant from the dry zone.

Specimens Examined. PUTTALAM DISTRICT: Wilpattu Natl. Park, Occapu Kallu, in plot 32, small woody perennial, up to 75 cm tall, 11 July 1969, fr, *Wirawan et al. 1086* (US); in plot W34, open scrub, 28 March 1969, fl, *Mueller-Dombois et al. 69032819* (PDA, US); near Sadpuda Kallu, 2 miles east of Kattankandal Kulam, semi-open spaces in shrub forest, 29 Dec. 1968, fl, *Fosberg et al. 50803* (PDA, US). HAMBANTOTA DISTRICT: Ruhuna National Park, between Yala and Kosgasmankada, in wallow, 21 March 1970, fr, *Cooray 70032102R* (PDA, US); Butawa Modera, rocky point, 5 April 1968, fr, *Fosberg 50306* (PDA, US); Tissamaharama, Dec. 1882, extremely abundant, *s. coll. s.n.* (PDA), Jan. 1928, *Alston 22* (PDA); Bata-ata, March 1927, *Alston 1108* (PDA). POLONNARUWA DISTRICT: In shade of trees in jungle close to the bank of Parakrama Samudra, 60 m alt., 26 Feb. 1972, *Cramer 3678* (PDA).

4. Triumfetta pentandra A. Rich. in Guill., Perr. & A. Rich., Fl. Seneg. Tent. 1: 93, t. 19. 1831; Sprague & Hutch., J. Linn. Soc. Bot. 39: 267. 1909; Alston in Trimen, Handb. Fl. Ceylon 6: 33. 1931; Matthew, Mat. Fl. Tamilnadu Carnatic 158. 1981; Matthew, Fl. Tamilnadu Carnatic III(1): 175. 1983. Type: herb. P.

Triumfetta neglecta Wight & Arn., Prod. 75. 1834; Masters in Hook. f., Fl. Br. Ind. 1: 396. 1874; Trimen, Handb. Fl. Ceylon 1: 181. 1893.
Triumfetta pilosula Thw., Enum. Pl. Zeyl. 401. 1864. Type: Ceylon, *Glenie s.n.* (*C.P. 2901*) (PDA).

Herb, erect, to 0.8–1.2 m tall, sometimes suffrutescent, unbranched or sparsely branched, the stem and branchlets stellate-pubescent, the younger parts sometimes stellate-pubescent and long hirsute with simple hairs, glabrescent. Leaves slender-petiolate, the petiole of the lower leaves up to 5 cm long, becoming shorter towards the apex of the stem and branchlets, puberulous, the stipules narrowly ovate, the apex long-acuminate to subulate, 2.5–5 mm long and 0.6–0.8 mm wide, sparsely pilose outside; blade from depressed-circular to subrotund to broadly ovate to ovate-elliptic, that of the lower ones usually 3-lobed, narrowly cordate to obtuse at the base, acute at the apex, the margins rather coarsely and irregularly serrate-dentate, to 6 cm long and 5.5 cm wide, thin-chartaceous, sparsely appressed-pilose on both sides, the hairs simple and stellate, the indumentum usually somewhat denser beneath, 3–5 nerved from the base, the costa and lateral veins slightly prominulous beneath. Inflorescences axillary, oppositifolious, or seemingly terminal, the cymules 3-flowered, frequently collected into terminal interrupted spiciform inflorescences. Flowers small, yellow to yellow-orange, the peduncles and pedicels short, to 1.5 mm long, the latter articulate near the apex, both puberulous, the bracts and bracteoles narrowly ovate, acuminate, 1.5–2.5 mm long; sepals 5, oblong, somewhat constricted near the middle, cucullate and acute apically, c. 2.7–3.5 mm long and 0.5–0.7 mm wide, stellate-puberulous outside, the appendage c. 0.5 mm long; petals 5, broadly spatulate to spatulate, rounded apically, unguiculate at the base, c. 1.6–2.5 mm long and 0.7–0.8 mm wide, the claw sparsely pilosulose; gonophore very short, with 5 rotund glands;

stamens 5–10, the filaments c. 1.2–2 mm long, glabrous, the anthers c. 0.5 mm long; ovary ovoid, c. 1 mm long, densely setose, the style slender, 0.8–2 mm long, glabrous, the stigma 2-lobulate. Fruit broadly ovoid, the body c. 3 mm long and 2.5 mm wide, hirtellous, indehiscent, the spines 1–2.5 mm long (the longest spines at the apex of the fruit), densely hirtellous, the hyaline spinules uncinate; seeds c. 2.5 mm long.

Distr. Tropical Africa, India and Ceylon.

Ecol. A plant of dry and arid country, grasslands and roadsides.

Vern. Kapu-kinissa (acc. *Simpson 9797*), Epala (acc. *Alston 2403*) (S).

Specimens Examined. PUTTALAM DISTRICT: Wilpattu Natl. Park, Occapu Kallu, in plot W. 34, in open forest-scrub on shallow reddish brown earth, 28 Apr. 1969, fr, *Mueller-Dombois et al. 69042820* (PDA, US). KANDY DISTRICT: Boyagoda Kanda near Galagedera, 7°23′ N., 80°30′ E., 1000 ft, 13 Feb. 1975, fr, *Sumithraarachchi 596* (PDA, US). MATALE DISTRICT: Dambulla Rock, steep slope, in shade of forest, 150–300 m, 23 Feb. 1971, *Robyns 7202* (PDA, US), March 1927, *Alston 1003* (PDA); east of Dambulla, roadside, June 1932, *Simpson 9797* (PDA); Nalanda, May 1928, *Alston 2403* (PDA); Dambulla, *s. coll. C.P. 2901 pp* (PDA). POLONNARUWA DISTRICT: Habarane, May 1927, *Alston 516* (PDA); Polonnaruwa, near town under shade of trees in jungle close to banks of Parakrama Samudra, 60 m, 26 Feb. 1972, fr, *Cramer 3678* (PDA, US); along road Polonnaruwa to Valachchenai, c. 7 miles E. of bridge over Mahaweli Ganga, common weed along roadside in forest, 30–150 m, 21 Feb. 1969, fr, *Robyns 6926* (BR, E, PDA, US); sacred area close to Lotus bath, grassland, 30–150 m, grazed by cattle, 20 Feb. 1969, fr, *Robyns 6924* (BR, CANB, E, PDA, US); Archaeological Quarters yard, sec. growth in garden, 61 m, 18 Dec. 1970, fr, *Ripley 313* (PDA, US). TRINCOMALEE DISTRICT: Exact locality unknown, in 1860, *Glenie C.P. 2901*, type of *T. pilosula* Thwaites (PDA); Eriyativumalai Hill, c. 2.5 miles south of Muthur, in rock crevices, 0–80 m, 7 March 1971, *Robyns 7272* (PDA, US); southern edge of Kantalai tank, sunny weedy flood plain, 14 March 1973, fr, *Jayasuriya and Townsend 1185* (PDA, US); 0.5–1.0 mile west of Kantalai tank (west of Trinco), in small clearing, 12 Jan. 1968, fr, *Comanor 758* (PDA, US). HAMBANTOTA DISTRICT: Ruhuna Natl. Park, Block I, near Yala Camp site, in forest-scrub vegetation on loose sand in plot R 26, 30 Jan. 1968, *Mueller-Dombois and Cooray 68013030* (PDA, US), plot R 28, *Mueller-Dombois 68012802* (US); plot R25, at Karaugaswala, Feb. 1968, *Mueller-Dombois & Cooray 68022808* (US); between Andunoruwa and Komawa Wewa, in plot R13, deciduous scrub on reddish brown earth, 7 Jan. 1969, *Mueller-Dombois 69010723* (PDA, US); Patanagala, on deep sandy soil in rocky area, 28 Oct. 1968, *Wirawan 696* (PDA, US), 3 Apr. 1968, fr, *Fosberg and Mueller-Dombois 50128* (PDA, US), 6 Apr. 1968, fr, *Fosberg 50360* (PDA, US); Main Yala Road, Plot R28, near mile marker 7, in sand, open scrub, 6–10 m, 22 Nov. 1967, fr, *Comanor 824* (PDA, US), 16 Jan. 1969, fl, *Cooray*

69011601R (PDA, US); Butawa Modera, in sheltered rock crevices, 5–8 m, 5 Apr. 1968, *Fosberg 50298* (PDA, US); Bambowa on roadside, 14 Dec. 1969, fl, *Cooray 69121404R* (PDA, US). BADULLA DISTRICT: Ekiriyankumbura, along road from Padiyatalawe to Bibile, in forest, 150–300 m, 2 Apr. 1969, fr, *Robyns 6985* (BR, CANB, E, PDA, US).

VAHLIACEAE

(by Neil A. Harriman*)

Dandy in Hutchinson, Fam. Fl. Pl. ed. 2, 1: 461, 462. 1959. Type genus: *Vahlia* Thunb., nom. cons.

Erect annual herbs from a slender taproot. Leaves opposite, simple, entire, estipulate. Flowers epigynous, actinomorphic, perfect, complete, 5-merous, on simple axillary pedicels or the pedicels once bifurcate and then the geminate flowers pedicellate on a short peduncle. Ovary hemispherical, unilocular; calyx lobes 5, valvate; petals 5, imbricate in bud; stamens 5, alternate with the petals, filaments inserted on the margin of an epigynous disc. Styles 2, stigmas capitellate. Fruit capsular, with two large placentae hanging from the summit of the ovary, with many minute seeds on each placenta.

The family includes only the genus *Vahlia*, with but one species in Ceylon. There is an obvious near relationship with the Saxifragaceae (where *Vahlia* was treated by Trimen, H. 1894. A handbook to the flora of Ceylon, part 2: 143). Modern opinion varies: the family is held to be distinct by, for example, Abeywickrama, B.A. 1959. A provisional check list of the flowering plants of Ceylon. Ceylon J. Sci., Biol. Sci. 2(2): 119–240; Airy Shaw, H.K. 1973. A dictionary of the flowering plants and ferns, ed. 8; and of course Hutchinson, op. cit. By contrast, this and a number of other segregate families are reduced to a broadly construed Saxifragaceae in Cronquist, A. 1981. An integrated system of classification of flowering plants.

VAHLIA

Thunb., Nov. Gen. Pl. 2: 36. 1782, nom. cons. Typus cons.: *Vahlia capensis* (L.f.) Thunb.

Bistella Adans., Fam. Pl. 2: 226. 1763, nom. rej.
(That the priorable *Bistella* is a nom. rej. is overlooked in Airy Shaw, op. cit., resulting in a somewhat confused application of generic names.)

Characters of the family. There are perhaps three species in the genus, from tropical Africa to Pakistan and India, only the following species in Ceylon.

*Biology Department, University of Wisconsin-Oshkosh, Oshkosh, Wisconsin 54901, USA.

Vahlia dichotoma (Murr.) Kuntze. Rev. Gen. 227. 1891. *Heuchera dichotoma* Murr., Comm. Gotting. 64, t. 1. 1772.

Vahlia oldenlandioides Roxb., Fl. Ind. 2: 89. 1832; Trimen, Handb. Fl. Ceylon 2: 143. 1894.

For a complete synonymy, see Bullock, A.A. 1966. *Bistella* Adans. versus *Vahlia* Thunb. (Vahliaceae). Acta Botanica Neerlandica 15: 84–85. Only the names given here have been applied in the literature to the Ceylon plant. As pointed out by Trimen (op. cit.), *Vahlia digyna* (Retz.) Kuntze (i.e., the synonymous *V. viscosa* Roxb.) does not occur in Ceylon.

Taprooted annual herb, 3–10 cm tall, viscid glandular hairy throughout. Leaves opposite, estipulate, sessile, linear to narrowly oblanceolate, entire, acute, 8–12 mm long, 1–2.5 mm wide. Flowers c. 3.5 mm high, paired or solitary from the leaf axils, sometimes half to two-thirds of the plant floriferous. Ovary hemispherical, 1.5 mm high. Calyx lobes deltoid, 1 mm high. Petals yellow-white, scarcely or not at all exceeding the calyx lobes. Stamens as long as the calyx lobes or a little shorter. Styles 2, 0.5 mm long. Capsule 3 mm high, 2-valved at summit, calyx lobes persistent; seeds numerous, 0.25 mm long.

Distr. In Ceylon, in sandy areas in the dry northern parts of the island; elsewhere, from India to tropical Africa.

Ecol. Dry region, flowers February, March.

Note. The similarity of this species to certain Rubiaceae has often been mentioned in the literature. The lack of stipules and the separate petals will serve to exclude this species from that family. Our species is very amply and accurately depicted in Hutchinson, op. cit.

Specimens Examined. JAFFNA DISTRICT: Nagercoil, *Jayasuriya 1350* (US). VAVUNIYA DISTRICT: just south of Mullaittivu, *Fosberg 53499* (US).

9 789061 915515